# Methods in Enzymology

Volume 372
LIPOSOMES
Part B

# METHODS IN ENZYMOLOGY

EDITORS-IN-CHIEF

## John N. Abelson    Melvin I. Simon

DIVISION OF BIOLOGY
CALIFORNIA INSTITUTE OF TECHNOLOGY
PASADENA, CALIFORNIA

FOUNDING EDITORS

## Sidney P. Colowick and Nathan O. Kaplan

*Methods in Enzymology*

*Volume 372*

# Liposomes

*Part B*

EDITED BY

*Nejat Düzgünes*

DEPARTMENT OF MICROBIOLOGY
UNIVERSITY OF THE PACIFIC SCHOOL OF DENISTRY
SAN FRANCISCO, CALIFORNIA

ELSEVIER
ACADEMIC
PRESS

AMSTERDAM • BOSTON • HEIDELBERG • LONDON
NEW YORK • OXFORD • PARIS • SAN DIEGO
SAN FRANCISCO • SINGAPORE • SYDNEY • TOKYO

Academic Press is an imprint of Elsevier

Academic Press
*An Elsevier Imprint.*
525 B Street, Suite 1900, San Diego, California 92101-4495, USA
http://www.academicpress.com

Academic Press
84 Theobald's Road, London WC1X 8RR, UK
http://www.academicpress.com

International Standard Book Number: 0-12-182275-3

PRINTED IN THE UNITED STATES OF AMERICA
03   04   05   06   07   08     9   8   7   6   5   4   3   2   1

# Table of Contents

## Section I. Liposomes in Biochemistry

## Section II. Liposomes in Molecular Cell Biology

# Contributors to Volume 372

Article numbers are in parentheses and following the names of contributors.
Affiliations listed are current.

ALICIA ALONSO (3), *Unidad de Biofísica and Departamento de Bioquímica, Universidad Del País Vasco, Aptdo. 644, 48080 Bilbao, Spain*

BRUNO ANTONNY (151), *CNRS-Institut de Pharmacologie Moleculaire et Cellulaire, 660 Route des Lucioles, 06560 Sophia Antipolis-Valbonne, France*

JOHN D. BELL (19), *Department of Physiology and Developmental Biology, Brigham Young University, Provo, Utah 84602*

ROBERT BITTMAN (374), *Department of Medical Microbiology, Molecular Virology Section, University of Groningen, Ant. Deusinglaan 1, 9713 AV Groningen, The Netherlands*

PIERRE BONNAFOUS (408), *Crucell Holland BV, Archimedesweg 4, P.O. Box 2048, Leiden, The Netherlands*

MAURO DALLA SERRA (99), *CMR-ITC Institute of Biophysics, Section at Trento, Via Sommarive 18, Povo, Trento 38050, Italy*

DAVID W. DEAMER (133), *Department of Chemistry and Biochemistry, University of Californi-Santa Cruz, Santa Cruz, California 95064*

PIETRO DE CAMILLI (248), *Department of Cell Biology, Howard Hughes Medical Institute, Yale University School of Medicine, 295 Congress Avenue, New Haven, Connecticut 06510*

JEANINE DE KEYZER (86), *University of Groningen, Department of Microbiology, P. O. Box 14, Haren 9750AA, The Netherlands*

SUE E. DELOS (428), *Department of Cell Biology, UVA Health System, School of Medicine, P.O. Box 800732, Charlottesville, Virginia 22908*

ARNOLD J. M. DRIESSEN (86), *University of Groningen, Department of Microbiology, P. O. Box 14, Haren 9750AA, The Netherlands*

NEJAT DÜZGÜNEŞ (260), *Department of Microbiology, School of Dentistry, University of the Pacific, 2155 Webster Street, San Francisco, California 94115*

LAURIE J. EARP (428), *Department of Cell Biology, UVA Health System, School of Medicine, P.O. Box 800732, Charlottesville, Virginia 22908*

RAQUEL F. EPAND (124), *Department of Biochemistry, McMaster Health Sciences Center, Hamilton, Ontario L8N 3Z5, Canada*

RICHARD M. EPAND (124), *Department of Biochemistry, McMaster Health Sciences Center, Hamilton, Ontario L8N 3Z5, Canada*

SHIROH FUTAKI (349), *Faculty of Pharmaceutical Sciences, The University of Tokushima, Shomachi 1-78-1, 770–8505 Tokushima, Japan*

YVES GAUDIN (392), *Laboratoire de Genetiquie des Virus du CNRS, Gif sur Yvette Cedex 91198, France*

RÉMY GIBRAT (166), *Plant Biochemistry and Molecular Biology, Agro-M/CNRS/ONRA/UMII, ENSA-INRA, Montpellier, 34060 Cedex 1, France*

FÉLIX M. GOÑI (3), *Unidad de Biofísica and Departamento de Bioquímica, Universidad Del País Vasco, Aptdo. 644, 48080 Bilbao, Spain*

CKAYDE GRIGNON (166), *Plant Biochemistry and Molecular Biology, Agro-M/CNRS/ONRA/UMII, ENSA-INRA, Montpellier, 34060 Cedex 1, France*

HIDEYOSHI HARASHIMA (349), *Faculty of Pharmaceutical Sciences, The University of Tokushima, Shomachi 1-78-1, 770–8505 Tokushima, Japan*

THEODORE L. HAZLETT (19), *Laboratory for Fluorescence Dynamics, University of Illinois at Urbana-Champaign, Urbana, Illinois 61801*

LORRAINE D. HERNANDEZ (428), *Department of Cell Biology, UVA Health System, School of Medicine, P.O. Box 800732, Charlottesville, Virginia 22908*

ANDREAS HOFFMAN (186), *Macromolecular Crystallography Laboratory, NCI at Frederick, 539 Boyles Street, Frederick, Maryland 21702*

ROBERT HUBER (186), *Institute of Cell and Molecular Biology, University of Edinburgh, Michael Swann Building, The King's Building Mayfield Road, EH9 3JR Edinburgh, Scotland*

HIROSHI KIWADA (349), *Faculty of Pharmaceutical Sciences, The University of Tokushima, Shomachi 1-78-1, 770–8505 Tokushima, Japan*

KYUNG-DALL LEE (319), *Department of Pharmaceutical Sciences, College of Pharmacy, University of Michigan, 428 Church Street, Ann Arbor, Michigan 48109*

TATIANA S. LEVCHENKO (339), *Department of Pharmaceutical Sciences, Northeastern University, 360 Huntington Avenue, Boston, Massachusetts 02115*

DANIEL LÉVY (65), *Institut Curie, UMR-CNRS 168 and LRC-CEA 34V, 11 Rue Pierre et Marie Curie, 75231 Paris Cedex 05, France*

SONG LIU (274), *Department of Biochemistry and Cell Biology, Rice University, Houston, Texas 77005*

MANAS MANDAL (319), *Department of Pharmaceutical Sciences, College of Pharmacy, University of Michigan, 428 Church Street, Ann Arbor, Michigan 48109*

ELIZABETH MATHEW (319), *Department of Pharmaceutical Sciences, College of Pharmacy, University of Michigan, 428 Church Street, Ann Arbor, Michigan 48109*

JAMES A. MCNEW (274), *Department of Biochemistry and Cell Biology, Rice University, Houston, Texas 77005*

THOMAS J. MELIA (274), *Cellular Biochemistry and Biophysics Program, Memorial Sloan-Kettering Cancer Center, New York, New York 10021*

GIANFRANCO MENESTRINA (99), *CMR-ITC Institute of Biophysics, Section at Trento, Via Sommarive 18, Povo, Trento 38050, Italy*

PIERRE-ALAIN MONNARD (133), *Department of Chemistry and Biochemistry, University of Californi-Santa Cruz, Santa Cruz, California 95064*

JOSÉ L. NIEVA (3, 235), *Unidad de Biofísica and Departamento de Bioquímica, Universidad Del País Vasco, Aptdo, 644, 48080 Bilbao, Spain*

SHLOMO NIR (235), *Seagram Center for Soil and Water Sciences, Faculty of Agricultural, Food and Environmental Quality Sciences, Rehovot 76100, Israel*

OLIVIER NOSJEAN (216), *Pharmacology Moleculaire et Cellulaire, Institut de Recherches Servier, Crossy-sur-Seine, France*

CHRISTIAN OKER-BLOM (418), *University of Jvaskyla, Department of Biological and Environmental Sciences, P.O. Box 35, FIN 40351 Jyvaskyla, Finland*

FRANK OPITZ (48), *University of Leipzig, Institute for Medical Physics and Biophysics, Liebigstrasse 27, Leipzig D-04103, Germany*

SERGIO GERARDO PEISAJOVICH (361), *Department of Biological Chemistry, Weigmann Institute of Science, Rehovot 76100, Israel*

JENS PITTLER (48), *University of Leipzig, Institute for Medical Physics and Biophysics, Liebigstrasse 27, Leipzig D-04103, Germany*

CHESTER PROVODA (319), *Department of Pharmaceutical Sciences, College of Pharmacy, University of Michigan, 428 Church Street, Ann Arbor, Michigan 48109*

RAM RAMMOHAN (339), *Department of Pharmaceutical Sciences, Northeastern University, 360 Huntington Avenue, Boston, Massachusetts 02115*

JEAN-LOUIS RIGAUD (65), *Institut Curie, UMR-CNRS 168 and LRC-CEA 34V, 11 Rue Pierre et Marie Curie, 75231 Paris Cedex 05, France*

KARINE ROBBE (151), *CNRS-Institut de Pharmacologie Moleculaire et Cellulaire, 660 Route des Lucioles, 06560 Sophia Antipolis-Valbonne, France*

STÉPHANE ROCHE (392), *Laboratoire de Genetiquie des Virus du CNRS, Gif sur Yvette Cedex 91198, France*

BERNARD ROUX (216), *Physico-Chemie Biologique, Universite C Bernard-Lyon 1, Villeurbanne, France*

SUSANA A. SANCHEZ (19), *Laboratory for Fluorescence Dynamics, University of Illinois at Urbana-Champaign, Urbana, Illinois 61801*

BRENTON L. SCOTT (274), *Department of Biochemistry and Cell Biology, Rice University, Houston, Texas 77005*

YECHIEL SHAI (361), *Department of Biological Chemistry, Weigmann Institute of Science, Rehovot 76100, Israel*

JOLANDA M. SMIT (374), *Department of Medical Microbiology, Molecular Virology Section, University of Groningen, Ant. Deusinglaan 1, 9713 AV Groningen, The Netherlands*

JAMES E. SMOLEN (300), *Department of Pediatrics, Baylor College of Medicine, 1100 Bates, Room 6014, Houston, Texas 77030*

TOON STEGMANN (408), *Crucell Holland BV, Archimedesweg 4, P.O. Box 2048, Leiden, The Netherlands*

REIKO TACHIBANI (349), *Faculty of Pharmaceutical Sciences, The University of Tokushima, Shomachi 1-78-1, 770-8505 Tokushima, Japan*

VLADIMIR P. TORCHILIN (339), *Department of Pharmaceutical Sciences, Northeastern University, 360 Huntington Avenue, Boston, Massachusetts 02115*

CHRIS VAN DER DOES (86), *University of Groningen, Department of Microbiology, P. O. Box 14, Haren 9750AA, The Netherlands*

MARTIN VAN DER LAAN (86), *University of Groningen, Department of Microbiology, P.O. Box 14, Haren 9750AA, The Netherlands*

JEFFREY S. VAN KOMEN (274), *Department of Biochemistry and Cell Biology, Rice University, Houston, Texas 77005*

ANA V. VILLAR (3), *Unidad de Biofisica and Departamento de Bioquímica, Universidad Del País Vasco, Aptdo. 644, 48080 Bilbao, Spain*

NATALIA VOLODINA (339), *Department of Pharmaceutical Sciences, Northeastern University, 360 Huntington Avenue, Boston, Massachusetts 02115*

MATTI VUENTO (418), *University of Jvasky-la, Department of Biological and Environmental Sciences, P.O. Box 35, FIN 40351 Jyvaskyla, Finland*

BARRY-LEE WAARTS (374), *Department of Medical Microbiology, Molecular Virology Section, University of Groningen, Ant. Deusinglaan 1, 9713 AV Groningen, The Netherlands*

THOMAS WEBER (274), *Department of Molecular, Cell, and Developmental Biology and Carl C. Icahn Institute for Gene Therapy and Molecular Medicine, Mount Sinai School of Medicine, New York, New York 10029*

MARKUS R. WENK (248), *Department of Cell Biology, Howard Hughes Medical Institute, Yale University School of Medicine, 295 Congress Avenue, New Haven, Connecticut 06510*

JUDITH M. WHITE (428), *Department of Cell Biology, UVA Health System, School of Medicine, P.O. Box 800732, Charlottesville, Virginia 22908*

JAN WILSCHUT (374), *Department of Medical Microbiology, Molecular Virology Section, University of Groningen, Ant. Deusinglaan 1, 9713 AV Groningen, The Netherlands*

OLAF ZSCHÖRNIG (48), *University of Leipzig, Institute for Medical Physics and Biophysics, Liebigstrasse 27, Leipzig D-04103, Germany*

# Preface

The origins of liposome research can be traced to the contributions by Alec Bangham and colleagues in the mid 1960s. The description of lecithin dispersions as containing "spherulites composed of concentric lamellae" (A. D. Bangham and R. W. Horne, J. Mol. Biol. 8, 660, 1964) was followed by the observation that "the diffusion of univalent cations and anions out of spontaneously formed liquid crystals of lecithin is remarkably similar to the diffusion of such ions across biological membranes (A. D. Bangham, M. M. Standish and J. C. Watkins, J. Mol. Biol. 13, 238, 1965). Following early studies on the biophysical characterization of multilamellar and unilamellar liposomes, investigators began to utilize liposomes as a well-defined model to understand the structure and function of biological membranes. It was also recognized by pioneers including Gregory Gregoriadis and Demetrios Papahadjopoulos that liposomes could be used as drug delivery vehicles. It is gratifying that their efforts and the work of those inspired by them have lead to the development of liposomal formulations of doxorubicin, daunorubicin and amphotericin B now utilized in the clinic. Other medical applications of liposomes include their use as vaccine adjuvants and gene delivery vehicles, which are being explored in the laboratory as well as in clinical trials. The field has progressed enormously in the 38 years since 1965.

This volume includes applications of liposomes in biochemistry, molecular cell biology and molecular virology. I hope that these chapters will facilitate the work of graduate students, post-doctoral fellows, and established scientists entering liposome research. Subsequent volumes in this series will cover additional subdisciplines in liposomology.

The areas represented in this volume are by no means exhaustive. I have tried to identify the experts in each area of liposome research, particularly those who have contributed to the field over some time. It is unfortunate that I was unable to convince some prominent investigators to contribute to the volume. Some invited contributors were not able to prepare their chapters, despite generous extensions of time. In some cases I may have inadvertently overlooked some experts in a particular area, and to these individuals I extend my apologies. Their primary contributions to the field will, nevertheless, not go unnoticed, in the citations in these volumes and in the hearts and minds of the many investigators in liposome research.

I would like to express my gratitude to all the colleagues who graciously contributed to these volumes. I would like to thank Shirley Light of Academic Press for her encouragement for this project, and Noelle Gracy of Elsevier Inc. for her help at the later stages of the project.

I am especially thankful to my wife Diana Flasher for her understanding, support and love during the endless editing process, and my children Avery and Maxine for their unique curiosity, creativity, cheer, and love. I wish to dedicate this volume to Diana, Avery and Maxine.

NEJAT DÜZGÜNES

# METHODS IN ENZYMOLOGY

VOLUME 90. Carbohydrate Metabolism (Part E)
*Edited by* WILLIS A. WOOD

VOLUME 91. Enzyme Structure (Part I)
*Edited by* C. H. W. HIRS AND SERGE N. TIMASHEFF

VOLUME 92. Immunochemical Techniques (Part E: Monoclonal Antibodies and General Immunoassay Methods)
*Edited by* JOHN J. LANGONE AND HELEN VAN VUNAKIS

VOLUME 93. Immunochemical Techniques (Part F: Conventional Antibodies, Fc Receptors, and Cytotoxicity)
*Edited by* JOHN J. LANGONE AND HELEN VAN VUNAKIS

VOLUME 94. Polyamines
*Edited by* HERBERT TABOR AND CELIA WHITE TABOR

VOLUME 95. Cumulative Subject Index Volumes 61–74, 76–80
*Edited by* EDWARD A. DENNIS AND MARTHA G. DENNIS

VOLUME 96. Biomembranes [Part J: Membrane Biogenesis: Assembly and Targeting (General Methods; Eukaryotes)]
*Edited by* SIDNEY FLEISCHER AND BECCA FLEISCHER

VOLUME 97. Biomembranes [Part K: Membrane Biogenesis: Assembly and Targeting (Prokaryotes, Mitochondria, and Chloroplasts)]
*Edited by* SIDNEY FLEISCHER AND BECCA FLEISCHER

VOLUME 98. Biomembranes (Part L: Membrane Biogenesis: Processing and Recycling)
*Edited by* SIDNEY FLEISCHER AND BECCA FLEISCHER

VOLUME 99. Hormone Action (Part F: Protein Kinases)
*Edited by* JACKIE D. CORBIN AND JOEL G. HARDMAN

VOLUME 100. Recombinant DNA (Part B)
*Edited by* RAY WU, LAWRENCE GROSSMAN, AND KIVIE MOLDAVE

VOLUME 101. Recombinant DNA (Part C)
*Edited by* RAY WU, LAWRENCE GROSSMAN, AND KIVIE MOLDAVE

VOLUME 102. Hormone Action (Part G: Calmodulin and Calcium-Binding Proteins)
*Edited by* ANTHONY R. MEANS AND BERT W. O'MALLEY

VOLUME 103. Hormone Action (Part H: Neuroendocrine Peptides)
*Edited by* P. MICHAEL CONN

VOLUME 104. Enzyme Purification and Related Techniques (Part C)
*Edited by* WILLIAM B. JAKOBY

VOLUME 105. Oxygen Radicals in Biological Systems
*Edited by* LESTER PACKER

VOLUME 106. Posttranslational Modifications (Part A)
*Edited by* FINN WOLD AND KIVIE MOLDAVE

VOLUME 140. Cumulative Subject Index Volumes 102–119, 121–134

VOLUME 141. Cellular Regulators (Part B: Calcium and Lipids)
*Edited by* P. MICHAEL CONN AND ANTHONY R. MEANS

VOLUME 142. Metabolism of Aromatic Amino Acids and Amines
*Edited by* SEYMOUR KAUFMAN

VOLUME 143. Sulfur and Sulfur Amino Acids
*Edited by* WILLIAM B. JAKOBY AND OWEN GRIFFITH

VOLUME 144. Structural and Contractile Proteins (Part D: Extracellular Matrix)
*Edited by* LEON W. CUNNINGHAM

VOLUME 145. Structural and Contractile Proteins (Part E: Extracellular Matrix)
*Edited by* LEON W. CUNNINGHAM

VOLUME 146. Peptide Growth Factors (Part A)
*Edited by* DAVID BARNES AND DAVID A. SIRBASKU

VOLUME 147. Peptide Growth Factors (Part B)
*Edited by* DAVID BARNES AND DAVID A. SIRBASKU

VOLUME 148. Plant Cell Membranes
*Edited by* LESTER PACKER AND ROLAND DOUCE

VOLUME 149. Drug and Enzyme Targeting (Part B)
*Edited by* RALPH GREEN AND KENNETH J. WIDDER

VOLUME 150. Immunochemical Techniques (Part K: *In Vitro* Models of B and T Cell Functions and Lymphoid Cell Receptors)
*Edited by* GIOVANNI DI SABATO

VOLUME 151. Molecular Genetics of Mammalian Cells
*Edited by* MICHAEL M. GOTTESMAN

VOLUME 152. Guide to Molecular Cloning Techniques
*Edited by* SHELBY L. BERGER AND ALAN R. KIMMEL

VOLUME 153. Recombinant DNA (Part D)
*Edited by* RAY WU AND LAWRENCE GROSSMAN

VOLUME 154. Recombinant DNA (Part E)
*Edited by* RAY WU AND LAWRENCE GROSSMAN

VOLUME 155. Recombinant DNA (Part F)
*Edited by* RAY WU

VOLUME 156. Biomembranes (Part P: ATP-Driven Pumps and Related Transport: The Na, K-Pump)
*Edited by* SIDNEY FLEISCHER AND BECCA FLEISCHER

VOLUME 157. Biomembranes (Part Q: ATP-Driven Pumps and Related Transport: Calcium, Proton, and Potassium Pumps)
*Edited by* SIDNEY FLEISCHER AND BECCA FLEISCHER

VOLUME 352. Redox Cell Biology and Genetics (Part A)
*Edited by* CHANDAN K. SEN AND LESTER PACKER

VOLUME 353. Redox Cell Biology and Genetics (Part B)
*Edited by* CHANDAN K. SEN AND LESTER PACKER

VOLUME 354. Enzyme Kinetics and Mechanisms (Part F: Detection and Characterization of Enzyme Reaction Intermediates)
*Edited by* DANIEL L. PURICH

VOLUME 355. Cumulative Subject Index Volumes 321–354

VOLUME 356. Laser Capture Microscopy and Microdissection
*Edited by* P. MICHAEL CONN

VOLUME 357. Cytochrome P450, Part C
*Edited by* ERIC F. JOHNSON AND MICHAEL R. WATERMAN

VOLUME 358. Bacterial Pathogenesis (Part C: Identification, Regulation, and Function of Virulence Factors)
*Edited by* VIRGINIA L. CLARK AND PATRIK M. BAVOIL

VOLUME 359. Nitric Oxide (Part D)
*Edited by* ENRIQUE CADENAS AND LESTER PACKER

VOLUME 360. Biophotonics (Part A)
*Edited by* GERARD MARRIOTT AND IAN PARKER

VOLUME 361. Biophotonics (Part B)
*Edited by* GERARD MARRIOTT AND IAN PARKER

VOLUME 362. Recognition of Carbohydrates in Biological Systems (Part A)
*Edited by* YUAN C. LEE AND REIKO T. LEE

VOLUME 363. Recognition of Carbohydrates in Biological Systems (Part B)
*Edited by* YUAN C. LEE AND REIKO T. LEE

VOLUME 364. Nuclear Receptors
*Edited by* DAVID W. RUSSELL AND DAVID J. MANGELSDORF

VOLUME 365. Differentiation of Embryonic Stem Cells
*Edited by* PAUL M. WASSAUMAN AND GORDON M. KELLER (in preparation)

VOLUME 366. Protein Phosphatases
*Edited by* SUSANNE KLUMPP AND JOSEF KRIEGLSTEIN (in preparation)

VOLUME 367. Liposomes (Part A)
*Edited by* NEJAT DÜZGÜNEŞ

VOLUME 368. Macromolecular Crystallography (Part C)
*Edited by* CHARLES W. CARTER, JR., AND ROBERT M. SWEET (in preparation)

VOLUME 369. Combinational Chemistry (Part B)
*Edited by* GUILLERMO A. MORALES AND BARRY A. BUNIN (in preparation)

VOLUME 370. RNA Polymerases and Associated Factors (Part C)
*Edited by* SANKAR L. ADHYA AND SUSAN GARGES (in preparation)

# Section I

# Liposomes in Biochemistry

# [1]  Interaction of Phospholipases C and Sphingomyelinase with Liposomes

By Félix M. Goñi, Ana V. Villar, José L. Nieva, and Alicia Alonso

## Introduction

The conventional classification of membrane proteins as intrinsic or integral, and as extrinsic or peripheral, has on the whole been superseded by a more complex pattern in which a continuum of possibilities is considered, from the integral protein firmly embedded in the bilayer to the soluble protein that contacts the membrane only transiently for a specific function. Phospholipases stand in a class of their own as membrane proteins because, irrespective of their more or less "peripheral" location, they perturb the physical properties of the membrane through chemical modification of its lipid components. Thus it is not their mere binding and/or insertion into the bilayer, but the chemical reactions they catalyze, that determines ultimately the nature of their interaction with the membrane.

In this laboratory we have examined the membrane interactions of phosphatidylcholine (PC)-preferring phospholipase C (PC-PLC), and of the sphingomyelin-specific phospholipase C usually known as sphingomyelinase. More recently, we have explored the effects of a phosphatidylinositol (PI)-specific phospholipase C (PI-PLC). The effects of these enzymes occur essentially through their lipid end-products, diacylglycerol or ceramide. Depending on the enzyme, and on the bilayer lipid compositions, a variety of effects can be observed. Enzyme activity is commonly followed by vesicle–vesicle aggregation, and, under certain conditions, by intervesicular lipid mixing, and by mixing of vesicular aqueous contents. Observation of intervesicular contents mixing is always accompanied by detection of mixing of lipid inner monolayers, indicative of vesicle–vesicle fusion. Moreover, efflux of vesicle contents, whether or not accompanied by other effects, is observed often as a result of phospholipase C treatment. All of the above-described phenomena can be monitored conveniently through the use of fluorescence spectroscopy techniques, as detailed below. A summary of the results obtained by these methods in our laboratory is presented in a review.[1]

---

[1] F. M. Goñi and A. Alonso, *Biosci. Rep.* **20,** 443 (2000).

Materials

*Enzymes*

Phospholipase C (EC 3.1.4.3) from *Bacillus cereus* (MW, ~23,000) is usually obtained from Roche Molecular Biochemicals (Indianapolis, IN) and used without further purification. Routine sodium dodecyl sulfate-polyacrylamide gel electrophoresis (SDS–PAGE) controls reveal that the enzyme preparations supplied by this company are ≥90% pure. The enzyme shows broad specificity (see below) and is active on glycero-phospholipids in a variety of aggregational states, for example, monomeric in solution, dispersed in detergent-mixed micelles, and in model bilayers. Roche Molecular Biochemicals has discontinued the sale of this enzyme. Other suppliers provide equivalent enzymes, but they have not been tested thoroughly in our laboratory.

Phosphatidylinositol-specific phospholipase C (EC 4.6.1.13) from *B. cereus* is supplied by Molecular Probes (Eugene, OR) and used without further purification. Sphingomyelinase (EC 3.1.4.12) from *B. cereus* is purchased from Sigma (St. Louis, MO). As indicated by the manufacturer, preparations of this enzyme often contain significant phospholipase C contamination, in amounts that vary from batch to batch. We have been unable to separate the PC-PLC impurity from sphingomyelinase, using a variety of chromatographic methods. In our case, and with the exception of those experiments in which the simultaneous activities of PC-PLC and sphingomyelinase are required, the PC-PLC inhibitor *o*-phenanthroline is used routinely in sphingomyelinase assays (see below). In the absence of PLC activity, sphingomyelinase is found to cleave specifically sphingomy-elin, and not any glycerophospholipid. Activity on sphingophospholipids other than sphingomyelin, for example, ceramide phosphorylethanolamine, has not been tested.

*Substrates*

Egg phosphatidylcholine (PC), egg phosphatidylethanolamine (PE), and wheat germ phosphatidylinositol (PI) are grade I from Lipid Products (South Nutfield, Surrey, UK). Egg sphingomyelin (SM) is from Avanti Polar Lipids (Alabaster, AL). The purity of the above-described lipids is checked by running 0.1 mg of lipid on a thin-layer chromatography plate that is later revealed by charring in an oven under conditions that allow detection of 1 μg of lipid. Dihexanoylphosphatidylcholine (DHPC) and cholesterol are supplied by Sigma. All these lipids are used without further purification. Glycosylphosphatidylinositol (GPI) is purified from rat liver according to Varela-Nieto *et al.*[2] GPI is stored at −20° and used within the following 2 weeks. Oxidation or other forms of degradation are detected after long-term storage. DHPC is used below its critical micellar concentration (i.e., below 10 m$M$) to obtain dispersions of monomeric phospholipid. However, enzyme assays on defined substrates are usually carried out with phospholipid vesicles (liposomes).

For liposome production, phospholipid dispersions are prepared by rehydrating lipid films dried from organic solvents. Solvents are evaporated thoroughly under a current of N$_2$, and then left for at least 2 h under high vacuum to remove solvent traces.

Small unilamellar vesicles are prepared by sonication[3] from aqueous phospholipid dispersions, consisting mainly of multilamellar vesicles (MLVs). Samples on ice are treated in a Soniprep 150 probe sonicator (MSE, Crawley, Surrey, UK) with 10- to 12-$\mu$m pulses for 30 min, alternating on and off periods every 10 s. Probe debris and MLV remains are pelleted by centrifugation at 6000 $g$ and 4° for 10 min.

Large unilamellar vesicles (LUVs) are prepared by the extrusion method.[4] To obtain these vesicles aqueous lipid suspensions (MLVs) are extruded 10 times through two stacked Nuclepore (Pleasanton, CA) polycarbonate filters (pore diameter, 0.1 $\mu$m). The extruder is supplied by Northern Lipids (Vancouver, BC, Canada). Extrusion takes place at room temperature, except for LUVs consisting of pure SM, in which case the extruder is equilibrated at 42° with a temperature regulation accessory. Average vesicle diameters are measured by quasi-elastic light scattering (QELS), using a Zetasizer instrument (Malvern Instruments, Malvern, Worcestershire, UK). LUV mean diameters are ~ 100–115 and ~160–190 nm for PC-based liposomes and SM-based liposomes, respectively.

To ascertain that the extrusion procedure does not alter the lipid composition of the systems under study, the lipid mixtures are quantitated occasionally after the extrusion treatment. For that purpose, the resulting LUV suspensions are extracted with chloroform–methanol (2:1, v/v). The organic phase is concentrated and separated on thin-layer chromatography (TLC) Silica Gel 60 plates, using successively in the same direction the solvents chloroform–methanol–water (60:30:5, v/v/v) for the first 10 cm and petroleum ether-ethyl ether-acetic acid (60:40:1, v/v/v) for the whole plate. After charring with a sulfuric acid reagent, the spot intensities are quantified with a dual-wavelength TLC scanner (CS-930; Shimadzu, Tokyo, Japan). The results of these studies have shown that, under our

[2] I. Varela-Nieto, L. Alvarez, and J. M. Mato, "Handbook of Endocrine Research Techniques," p. 391. Academic Press, San Diego, CA, 1993.
[3] A. Alonso, R. Sáez, A. Villena, and F. M. Goñi, *J. Membr. Biol.* **67,** 55 (1982).
[4] L. D. Mayer, M. H. Hope, and P. R. Cullis, *Biochim. Biophys. Acta* **858,** 161 (1986).

conditions, the extrusion procedure does not significantly modify the lipid composition of the LUVs with respect to the original mixture.

Human erythrocyte ghosts are also used occasionally as lipase substrates. Ghost membranes are obtained from erythrocyte concentrate, as supplied by a blood bank, using the procedure of Steck and Kant.[5]

*Fluorescent probes*

The following fluorescent probes are purchased from the suppliers indicated in each case, and used without further purification. Octadecylrhodamine B (R18), N-(7-nitrobenz-2-oxa-1,3-diazol-4-yl) phosphatidylethanolamine (NBD-PE), N-(Lissamine rhodamine B sulfonyl) phosphatidylethanolamine (Rh-PE), 1-aminonaphthalene-3,6,8-trisulfonic acid (ANTS), and N,N'-p-xylene-bis(pyridinium bromide) (DPX) are provided by Molecular Probes. 6-Carboxyfluorescein (6-CF) is supplied by Eastman Kodak (Burnaby, BC, Canada). Fluorescein isothiocyanate-derivatized dextrans (FITC-dextrans) are purchased from Sigma.

*Buffers*

For phospholipase C assays, liposomes are usually hydrated and assayed in 10 m$M$ HEPES, 200 m$M$ NaCl, 10 m$M$ CaCl$_2$, pH 7.0. In experiments involving PI-PLC, the buffer is 10 m$M$ HEPES, 150 m$M$ NaCl, pH 7.5. For sphingomyelinase studies, the hydration and assay buffer is 10 m$M$ HEPES, 200 m$M$ NaCl, 10 m$M$ CaCl$_2$, 2 m$M$ MgCl$_2$, pH 7.0. Experiments from this and other laboratories have shown that PLC requires >5 m$M$ Ca$^{2+}$, and that sphingomyelinase requires 10 m$M$ Ca$^{2+}$ and 2 m$M$ Mg$^{2+}$, for optimal catalytic activity under our conditions, whereas no divalent cations are required for PI-PLC.

Methods

*Asymmetric Incorporation of Glycosylphosphatidylinositol Lipids into Large Unilamellar Vesicle Outer Lipid Monolayers*

Glycosylphosphatidylinositol asymmetric model membranes are created by incorporation of GPI into the external monolayer of preformed liposomes.[6] GPI in methanol is mixed with liposomes (LUVs) in buffer (10 m$M$ HEPES, 300 m$M$ NaCl, pH 7.5). For successful incorporation,

[5] T. L. Steck and J. A. Kant, *Methods Enzymol.* **31,** 172 (1974).
[6] A. V. Villar, A. Alonso, C. Pañeda, I. Varela-Nieto, U. Brodbeck, and F. M. Goñi, *FEBS Lett.* **457,** 71 (1999).

the methanolic glycolipid solution must be kept at 5% of the total reaction volume. This solvent allows GPI to bind the membrane and become stabilized there. The incorporation does not disrupt membrane stability.[6] Once the symmetric LUV liposomes are prepared as described above, they are diluted to a final lipid concentration of 0.3 m$M$. GPI methanolic solution is then made to reach a concentration of 0.03 m$M$ (10% of total lipid concentration), in a volume that is 5% of the total reaction volume (250 $\mu$l). The GPI methanolic sample is injected with a microsyringe into the buffered liposomal suspension. Liposomes are vortexed (2 min) after the GPI injection. The sample is then kept at room temperature for 15 min before the assays.

The asymmetric nature of the resulting vesicles may be shown in an experiment in which liposomes are treated with $\alpha$-galactosidase.[6] This enzyme degrades the glycosylated part of the lipid, yielding free monosaccharides from GPI. Enzyme action reaches equilibrium when 63% of GPI sugars are hydrolyzed. GPI hydrolysis occurs in the external membrane monolayer. After 60 min of enzyme treatment, when equilibrium has been established, Triton X-100 addition (0.1%, w/v) permeabilizes the membrane, allowing the enzyme to act inside the liposome. An additional 30 min of reaction over all the membrane lipids does not lead to further GPI hydrolysis. Therefore, most, if not all, GPI molecules are inserted in the outer monolayer of the vesicles, all of them being accessible to $\alpha$-galactosidase from the outside.

*Enzyme Assays*

Aqueous suspensions of liposomes (or phospholipid–detergent mixed micelles, or phospholipid monomers) and enzyme are incubated under the desired conditions. Typically 3 ml of a suspension (0.3 m$M$ lipid) at ~37° is incubated with enzyme (PC-PLC or sphingomyelinase, 1.6 U/ml; or PI-PLC, 0.16 U/ml), with continuous stirring. At defined times, aliquots (0.6 ml) are collected and the reaction is quenched by low-temperature rapid extraction with an ice-cold chloroform–methanol–concentrated HCl (66:33:1, v/v/v) mixture (3 ml). A reduced-volume assay may be performed with a 0.25-ml total volume and removal of 50-$\mu$l aliquots. After gentle vortexing to ensure partitioning, the extraction mixtures containing the reaction aliquots are subjected to centrifugation to optimize phase separation. We have found that in phase-separated samples enzyme activity is completely abolished; nevertheless, these samples are either processed immediately or, when required, stored at −20° before phosphorus determination.

Phosphorus content is determined in aliquots obtained from the aqueous phase, or from the organic phase, or both. Phosphorus is assayed by the ammonium heptamolybdate method according to Bartlett[7] (see protocol in [15] in this volume[7a]) or by the modified version described by Böttcher et al.[8] The latter is used with the assays in small volume. We have found that, when the modified assay is performed on a single phase, phosphorus quantification in the water phase is advantageous for several reasons. First, there is lower variability in the experimental data. Second, phosphorus determination from the organic phase requires the previous complete evaporation of the solvents in the samples in order to avoid interference with the phosphomolybdate colorimetric assay. Finally, in stored samples the volume in the aqueous phases does not change appreciably, whereas the organic phases might eventually evaporate in part.

On occasion, simultaneous measurements of phosphate release and diacylglycerols present in PLC-treated liposomes are carried out, always with good correlation. In these cases, the enzyme activity is stopped at the appropriate times by increasing the pH to about pH 10; enzyme-treated vesicles are then collected by centrifugation and diacylglycerols are quantitated, using the radioenzymatic assay for diacylglycerol kinase (Amersham Biosciences, Piscataway, NJ), essentially following the method of Preiss et al.[9] The assay is based on the conversion of detergent-solubilized diacylglycerol to [$^{32}$P]phosphatidic acid, employing Escherichia coli diacylglycerol kinase and [$\gamma$-$^{32}$P]ATP. After enzymatic phosphorylation of diacylglycerol, [$^{32}$P]phosphatidic acid is separated chromatographically from unreacted [$\gamma$-$^{32}$P]ATP, and determined by liquid scintillation counting.

All three enzymes described here show latency periods (lag times) in their activities, which are more evident when the substrate is in the form of LUVs. Lag times are sensitive to bilayer lipid composition, bilayer curvature, and temperature, among other factors. No latency periods are observed when the substrate is in the form of short-chain phospholipid monomers, or phospholipid–detergent mixed micelles. A detailed investigation of the lag times of PC-PLC indicates that, during the latency period, diacylglycerol is produced at slow rates. When the diacylglycerol concentration in the bilayer reaches 10 mol% of total lipid, a rapid burst of activity is seen that correlates with the start of vesicle aggregation.[1]

[7] G. R. Bartlett, J. Biol. Chem. 234, 466 (1959).
[7a] N. Düzgüneş, Methods Enzymol. 372, [15], 2003 (this volume).
[8] C. S. F. Böttcher, C. M. Van Gent, and C. Fries, Anal. Chim. Acta 1061, 297 (1961).
[9] J. Preiss, C. R. Loomis, W. R. Bishop, R. Stein, J. E. Niedel, and R. M. Bell, J. Biol. Chem. 261, 8597 (1986).

*Vesicle Aggregation*

Diacylglycerol (or ceramide) accumulation in liposomes through the action of PLC (or sphingomyelinase) induces their aggregation.[10–12] Enzyme-induced vesicle aggregation can be monitored continuously in a spectrophotometer as an increase in sample turbidity (absorbance at 400–500 nm). Vesicle suspensions, typically 0.3 m$M$ lipid, at 37°, are placed in a spectrophotometer cuvette with continuous stirring, and turbidity (absorbance) is recorded continuously. Turbidity-versus-time curves typically display a sigmoidal shape, the maximal slope of which may be used to estimate enzyme activity. The slope ($\Delta$absorbance min$^{-1}$) may be measured conveniently from the first derivative maximum of the curves. The initial low increase in turbidity corresponds to the latency period detected when enzyme activity is assayed through chemical analysis of water-soluble phosphates, as described above.

The time required to accomplish maximal aggregation rate (lag phase) appears to reflect the necessity of attaining certain levels of diacylglycerol generated in the membrane to induce the observed effect. In PC LUVs a lag phase is observed as a zero-level effect on turbidity, while the chemical activity still operates at a slow rate. When the level of accumulated diacylglycerol reaches ~10% a sudden burst in vesicle aggregation is observed. Maximal rates of chemical activity and vesicle aggregation then run in parallel. The level of diacylglycerol required in the membrane to induce aggregation as determined chemically depends on the average size of the vesicles (lower for smaller sizes) and on the lipid concentration in the reaction mixture (lower for higher concentrations), but not on enzyme concentration, temperature, or lipid composition. Although the adherence properties of the vesicles appear to be modulated by the accumulation of diacylglycerol, it is ultimately the enzyme-generated product that induces the process, because inclusion of even 20% diacylglycerol in the lipid composition generates stable, dispersed vesicles.

Assay of phospholipase C or sphingomyelinase activity by turbidity measurements is a rapid and easy method, and a good relationship between aggregation and chemical activity has been shown for many types of vesicles. In the case of PC-PLC and PC small unilamellar vesicles (SUVs), both activities are shown to be linearly correlated in Fig. 1. At low and high lipid concentrations, however, the correlation is lost. It must be kept in mind that aggregation is a second-order process affected

[10] J. L. Nieva, F. M. Goñi, and A. Alonso, *Biochemistry* **28,** 7364 (1989).
[11] M. B. Ruiz-Argüello, G. Basañez, F. M. Goñi, and A. Alonso, *J. Biol. Chem.* **271,** 26616 (1996).
[12] A. V. Villar, A. Alonso, and F. M. Goñi, *Biochemistry* **39,** 14012 (2000).

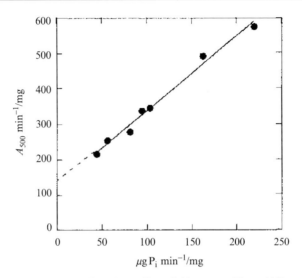

FIG. 1. Correlation between PC-PLC-specific activities assayed by turbidity measurements ($A_{500}$ min$^{-1}$/mg) and by determination of water-soluble phosphorus ($\mu$g P$_i$ min$^{-1}$/mg). PC SUVs (2 m$M$) were used as the substrate (J. L. Nieva, unpublished data, 2002).

specifically by parameters such as the lipid concentration or adherence properties of vesicles.[13]

Vesicle aggregation can also be estimated as an increase in scattered light at 90° in a spectrofluorometer, by fixing the excitation and emission wavelengths at 520 nm, with results that parallel those of turbidity measurements. Light scattering is preferred when vesicle concentrations are low (e.g., below 0.1 m$M$ lipid).

It is sometimes observed, with light scattering more often than with turbidity, that the aggregation activity-versus-time curve reaches a maximum, and then decreases. It may even be observed that higher enzyme doses lead to lower rates of increase in light scattering. This paradoxical response is due to the fact that scattering increases with particle size (i.e., aggregation) as long as the incident light wavelength is larger than the scattering particles (the so-called Rayleigh condition). With incident light of wavelength 400–500 nm and particles originally of diameter 100 nm, it is not difficult to surpass the Rayleigh limit, and reach a condition in which an increased particle size leads actually to a decrease in light scattering.[14] This problem

[13] A. V. Villar, F. M. Goñi, and A. Alonso, *FEBS Lett.* **494,** 117 (2001).
[14] A. R. Viguera, A. Alonso, and F. M. Goñi, *Colloids Surf. Biointerfaces* **3,** 263 (1995).

can be overcome, but only partially, by increasing the incident light wavelength. Alternatively, vesicle concentration or enzyme activity must be decreased.

*Intervesicular Lipid Mixing*

Lipase-induced vesicle aggregation leads usually to intervesicle lipid mixing. Enzyme-induced lipid mixing is measured by two procedures, one based on fluorescence dequenching and the other on fluorescence resonance energy transfer (FRET).

Octadecylrhodamine B (R18) is a probe whose fluorescence is self-quenched to an extent that depends on its concentration in the bilayer.[15] Mixing of lipids from a vesicle containing a high concentration of R18 with lipids from a probe-free vesicle leads to R18 dilution, and subsequent dequenching, that is detected as an increase in fluorescence. When the probe is incorporated into vesicles as a lipid component of the bilayer (i.e., mixed with phospholipids in organic phase before evaporation), at concentrations ranging from 1 to 9 mol% with respect to total lipid, the efficiency of self-quenching is proportional to its surface density (% quenching $\approx 9 \times$ mol% R18). Dilution of the probe on fusion of labeled and unlabeled vesicles results in a proportional increase in fluorescence intensity that can be monitored continuously in a spectrofluorometer (excitation and emission wavelengths of 560 and 590 nm, respectively).[10]

In our assays, the 0% fluorescence level (or 0% mixing) is determined from a 1:4 mixture of 8 mol% R18-containing liposomes and R18-free liposomes. The fluorescence of the same amount of liposomes with the diluted probe uniformly distributed, that is, 1.6 mol% R18-containing liposomes, is taken as the 100% fluorescence level, or 100% lipid mixing or 0% quenching. Alternatively, 0% quenching value (or infinite dilution) can be inferred from Triton X-100-solubilized samples. The 100% lipid mixing fluorescence value in a particular experiment can then be estimated from the percent quenching-versus-mole percent R18 curve.

In FRET-based lipid-mixing assays two fluorescent phospholipid derivatives are used: *N*-(7-nitrobenz-2-oxa-1,3-diazol-4-yl) phosphatidylethanolamine (NBD-PE, energy donor) and *N*-(Lissamine rhodamine B sulfonyl) phosphatidylethanolamine (Rh-PE, energy acceptor). Dilution due to membrane mixing results in an increase in donor NBD-PE fluorescence.[16]

[15] D. Hoekstra, T. de Boer, K. Klappe, and J. Wilschut, *Biochemistry* **23**, 5675 (1984).
[16] D. K. Struck, D. Hoekstra, and R. E. Pagano, *Biochemistry* **20**, 4093 (1981).

In our case, vesicles containing 0.6 mol% NBD-PE and 0.6 mol% Rh-PE are mixed with probe-free liposomes at a 1:4 ratio. NBD-PE emission is monitored at 530 nm (excitation wavelength at 465 nm) with a cut-off filter at 515 nm; 0% mixing is set as the fluorescence emission in the absence of enzyme; 100% mixing is set after addition of Triton X-100 to a final concentration of 1 m$M$.[12]

The R18 assay has been criticized because the spontaneous tendency of the probe to exchange between vesicles may give rise to false high values of lipid mixing. Although this is certainly a problem to be kept in mind, in our hands R18 and FRET assays measure similar rates of lipid dilution when compared in similar systems. This may be because our enzyme-induced lipid-mixing rates are fast in comparison with the spontaneous rates of R18 exchange. Moreover, PI-PLC and sphingomyelinase can be assayed with either probe, but PC-PLC recognizes as substrates the PE-based probes of FRET, and cleaves them rapidly, so that with PC-PLC only the R18 method is accessible.

*Intervesicular Mixing of Inner Monolayer Lipids*

Vesicle aggregation usually leads to some extent of intervesicular lipid mixing, but this can either be limited to lipids from the outer monolayer (in the case of "hemifusion" or "close apposition"[17,18]) or else involve lipids located in the vesicle inner monolayers. The latter phenomenon occurs only when a fusion pore opens between two apposed vesicles, and the aqueous contents intermix, that is, when true fusion occurs.

We have developed a novel and simple one-step method for the assay of inner monolayer lipid mixing, based on FRET between NBD-PE and Rh-PE.[12] Vesicles composed of PI–PE–PC–cholesterol (Ch) (40:30:15:15, mole ratio), containing 0.6 mol% of each probe, are prepared as described above. Fluorescence probes are thus located in both membrane layers. Fluorescence from the outer monolayer is quenched by addition of 0.2% (w/v) bovine serum albumin (BSA) and 10 m$M$ dithiothreitol (DTT). Addition of BSA and DTT chenches NBD-PE fluorescence without any membrane structural perturbation. Thus, the inner monolayer fluorescence remains unaffected. This method is based on the ability of BSA molecules to extract NBD-PE from vesicle membranes[19–21] and the modulation of

[17] L. V. Chernomordik, A. Chanturiya, J. Green, and J. Zimmerberg, *Biophys. J.* **69,** 922 (1995).
[18] A. R. Viguera, M. Mencia, and F. M. Goñi, *Biochemistry* **32,** 3708 (1993).
[19] J. Connor and A. J. Schroit, *Biochemistry* **27,** 848 (1988).
[20] G. Morrot, P. Hervé, A. Zachowski, P. Fellmann, and P. F. Devaux, *Biochemistry* **28,** 3456 (1989).
[21] H. N. T. Dao, J. C. McIntyre, and R. G. Sleight, *Anal. Biochem.* **196,** 46 (1991).

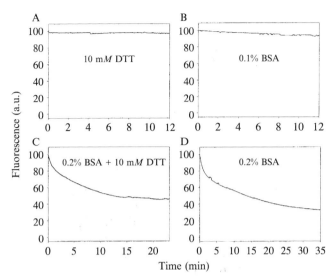

FIG. 2. Effects of bovine serum albumin (BSA) and dithiothreitol (DTT) on the fluorescence emission of the lipid probe NBD-PE in large unilamellar vesicles composed of PI–PE–PC–Ch (40:30:15:15, mole ratio). The total lipid concentration was 0.3 m$M$. NBD-PE was present at 0.6 mol% in the bilayer, and its fluorescence emission intensity was considered 100%. (A) Effect of 10 m$M$ DTT. (B) Effect of 0.1% (w/v) BSA. (C) Combined effect of 0.2% (w/v) BSA and 10 m$M$ DTT. (D) Effect of 0.2% (w/v) BSA (A. V. Villar, unpublished data, 2002).

activity by DTT, which reduces BSA disulfide bonds, decreasing BSA lipid extraction capacity. The BSA:DTT mole ratio is critical to achieve external but not internal quenching. When the BSA concentration is ≥0.2% (w/v) membrane structure desestabilization occurs, whereas lower concentrations do not promote NBD-PE extraction (Fig. 2). If the DTT concentration is increased, extensive BSA reduction leads to inefficient quenching. The 0.2% (w/v) BSA–10 m$M$ DTT system produces the right action over membrane fluorescence and at the same time preserves membrane structural integrity. The lack of perturbation of membrane integrity by BSA–DTT under our conditions has been shown by the fact that the treatment does not induce intervesicular mixing of aqueous contents, or vesicle leakage, or spontaneous mixing of inner monolayer lipids (Villar *et al.*[12] and A. V. Villar, unpublished data, 2002). BSA–DTT addition is followed by a 15-min incubation at 39°, with continuous stirring. At this point vesicles are labelled only internally. In Fig. 2C, a 50% reduction in fluorescence signal is shown. This would correspond to about one-half the incorporated probes, the ones in the outer layer.

After the incubation period, fusion is started by addition of PI-PLC. Inner monolayer lipid mixing is measured in a spectrofluorometer, with excitation and emission wavelengths of 465 and 530 nm, respectively, and a cutoff filter at 515 nm. The 100% fluorescence value ($F_{100}$) is fixed with a labeled liposome population containing 0.12% of each probe[12] and treated with BSA–DTT as described above. The stabilized initial signal ($F_0$) and the fluorescence data ($F_f$) after addition of enzyme at time zero are recorded. The equation for the final analyzed data is the following:

$$\text{Percent lipid mixing} = (F_f - F_0)/(F_{100} - F_0) \times 100$$

Other published methods describe removal of the fluorescence from outer monolayers, either with BSA[19–21] or by the use of dithionite as a reducing agent.[22] However, in these methods excess reagent must be removed either by centrifugation or by gel filtration, whereas this step is not required in our case. Moreover, we have found that for certain vesicle compositions (e.g., those containing PI, PE, and Ch) dithionite permeates rapidly across the bilayer, thus quenching the fluorescence of probes in both the inner and outer monolayers.[12] Fusion of vesicles [PC–PE–Ch, 2:1:1 (mole ratio)] induced by PC-PLC is a system in which both the two-step method of McIntyre and Sleight[22] and our own single-step procedure can be applied. As shown in Fig. 3, both procedures allow the observation of mixing of inner monolayer lipids, with virtually identical results.

*Intervesicular Mixing of Aqueous Contents*

Under certain conditions, PC-PLC,[10] PI-PLC,[12] and a mixture of sphingomyelinase and PC-PLC,[23] but not sphingomyelinase alone, can induce liposome fusion. This is indicated by the simultaneous mixing of lipids (particularly inner monolayer lipids) and vesicle contents.

Mixing of vesicular aqueous contents induced by phospholipases C is measured by the ANTS–DPX assay.[10] This assay is based on the quenching of 1-aminonaphthalene-3,6,8-trisulfonic acid (ANTS) by $N,N'$-$p$-xylene-bis(pyridinium bromide) (DPX).[24] ANTS and DPX are encapsulated in two different vesicle populations. Coalescence of internal aqueous contents (true fusion) results in quenching of ANTS fluorescence. A certain amount of concomitant release of the probes to the medium does not interfere with the fusion signal because dilution of DPX in the medium prevents quenching of ANTS fluorescence outside the liposomes. Assays based on the use

[22] J. C. McIntyre and R. G. Sleight, *Biochemistry* **30,** 11819 (1991).
[23] M. B. Ruiz-Argüello, F. M. Goñi, and A. Alonso, *J. Biol. Chem.* **273,** 22977 (1998).
[24] H. Ellens, J. Bentz, and F. C. Szoka, *Biochemistry* **24,** 3099 (1985).

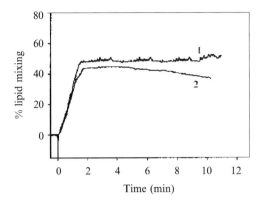

FIG. 3. Observation of intervesicular mixing of lipids located in the inner monolayers. LUVs (0.3 m*M*) were composed of PC–PE–Ch (2:1:1, mole ratio). Vesicle fusion was induced by PC-PLC.[10] Lipid mixing was monitored by the fluorescence resonance energy transfer method.[16] Fluorescence arising from probes in the outer monolayers was eliminated either by our BSA–DTT method (curve 1) or by the dithionite method[22] (curve 2) (A. V. Villar, unpublished data, 2002).

of these probes has turned out to be of great applicability in phospholipase C studies. None of these compounds interfere with enzyme activity and our assay conditions do not appreciably affect their fluorescence. In addition, ANTS does not bind to the external side of the vesicle membranes and does not permeate across them even in the presence of 20 mol% diacylglycerol, most likely because this compound is hydrophilic, a characteristic conferred by its three sulfonic acid groups with p$K_a$ values between 0 and 1.

Three liposome preparations are prepared and loaded with (1) 50 m*M* ANTS, 90 m*M* NaCl, 10 m*M* HEPES, pH 7.0; (2) 180 m*M* DPX, 10 m*M* HEPES, pH 7.0; or (3) 25 m*M* ANTS, 90 m*M* DPX, 45 m*M* NaCl, 10 m*M* HEPES, pH 7.0 (in PI-PLC studies the pH is 7.5). Divalent cations (10 m*M* CaCl$_2$ and/or 2 m*M* MgCl$_2$) are added according to the requirements of the enzyme to be used. Nonencapsulated material is removed from the vesicles, using a Sephadex G-75 column, with an equiosmotic elution buffer. The same buffer is also used in all the fusion and enzyme assays. The osmolalities of all solutions are measured in a cryoscopic osmometer (Osmomat 030; Gonotec, Berlin, Germany) and adjusted to 0.4 osmol/kg by adding NaCl. The lipid concentration in the assays is usually 0.3 m*M* (0.15 m*M* ANTS liposomes plus 0.15 m*M* DPX liposomes). The process is started by adding enzyme.

Fluorescence scales are calibrated for content-mixing assays as follows. The 100% fluorescence level (or 0 fusion) is set by using a 1:1 mixture of

ANTS (a) and DPX (b) liposomes. The fluorescence level corresponding to 100% mixing of contents is determined from 0.3 m$M$ liposomes (c) containing coencapsulated ANTS and DPX. Corrections for differences in the amount of entrapped solutes in the various vesicle preparations are routinely carried out after measuring the ratio of ANTS fluorescence before and after the addition of excess detergent (5 m$M$ Triton X-100). The fluorescence change of a preparation containing 0.15 m$M$ ANTS liposomes plus 0.15 m$M$ "empty" liposomes (i.e., buffer loaded) is subtracted routinely from the ANTS–DPX fluorescence signal, in order to account for scattering and other possible artifacts. Because under our measuring conditions the aggregates may involve a large number of vesicles, fusion rates and maximal fusion values are directly estimated from the degree of ANTS quenching at the required time point. No further corrections are made over those values. The excitation monochromator is adjusted to 355 nm, and the emission monochromator is adjusted to 520 nm. An interference filter (450 nm) is used to avoid scattered excitation light.

Mixing of aqueous contents could also, in principle, be assayed with a different pair of water-soluble reagents, namely, terbium ions and dipicolinic acid.[25] Both reagents interact to give a highly fluorescent compound. However, PC-PLC and sphingomyelinase activities are strongly inhibited by dipicolinic acid, which appears to complex divalent cations ($Ca^{2+}$, $Mg^{2+}$) essential for enzyme activity. This fact precludes the use of this otherwise useful system with the above-described enzymes. PI-PLC does not require divalent cations for optimal activity, and thus it could be used with terbium–dipicolinic acid, but to our knowledge this possibility has not been tested experimentally.

*Phospholipase-Induced Vesicle Leakage*

PC-PLC induces fusion, but not leakage, from PC-containing liposomes. Sphingomyelinase alone causes efflux of vesicular contents, but no fusion. PI-PLC, in turn, induces both fusion and release of aqueous contents. The latter phenomenon can, in principle, be assayed with the same systems used in the content-mixing assays (see previous section) but, for the reasons detailed above, only the ANTS–DPX system has been applied in our case.

For leakage assays ANTS and DPX are coencapsulated in a single liposome population, so that DPX quenches most of the ANTS fluorescence. Release of the probes to the medium may then be followed by an increase in fluorescence due to the relief of DPX quenching on dilution.[24]

[25] J. Wilschut, N. Düzgüneş, R. Fraley, and D. Papahadjopoulos, *Biochemistry* **19**, 6011 (1980).

Liposomes (LUVs) are prepared in 25 m$M$ ANTS, 90 m$M$ DPX, 45 m$M$ NaCl, 10 m$M$ CaCl$_2$, 2 m$M$ MgCl$_2$, 10 m$M$ HEPES, pH 7.0 (when PI-PLC is the enzyme, CaCl$_2$ and MgCl$_2$ are substituted by NaCl and the pH is 7.5). Nonencapsulated material is removed with a Sephadex G-75 column, and osmolalities are adjusted as detailed in the previous section. The LUVs are diluted as required (usually to 0.3 m$M$) with assay buffer, and the fluorescence is recorded continuously (excitation, 355 nm; emission, 520 nm; 450-nm interference filter). The basal signal obtained under these conditions is considered as 0% leakage. The 100% fluorescence level for leakage is obtained by detergent lysis of the liposomes (5 m$M$ Triton X-100).

Vesicle leakage can also be assayed on the basis of carboxyfluorescein dequenching. In this method, 6-carboxyfluorescein (6-CF) is entrapped at self-quenching concentrations in the vesicles, according to the method described by Weinstein et al.[26] Liposomes are prepared in 50 m$M$ 6-CF, 100 m$M$ NaCl, and 50 m$M$ HEPES, pH 7.0, plus divalent cations as required, according to the enzyme to be assayed. Nonencapsulated probe is removed from the vesicles with a Sephadex G-50 column, with 50 m$M$ HEPES, 300 m$M$ NaCl, pH 7.0 (plus divalent cations), as the elution buffer. Dilution of the probe after being released to the medium results in an increase in quantum yield. The maximum dilution (or 100% leakage) value is obtained by solubilizing the liposomes with Triton X-100. 6-CF fluorescence is continuously registered (excitation and emission wavelengths of 492 and 520 nm, respectively) and increases after addition of the enzyme, indicating that the phospholipase activity destabilizes the overall organization of the bilayer, thereby allowing the release of encapsulated solutes. However, the use of 6-CF is not free from problems. Its relatively nonpolar character gives it a certain affinity for the membrane matrix, and this may in turn perturb in several ways the release measurements.

Release of vesicle aqueous contents by sphingomyelinase has also been measured with fluorescein isothiocyanate-derivatized dextrans (FITC-dextrans; molecular mass, 4.4–20 kDa).[27] FITC-dextrans possess fluorescence self-quenching properties, and thus their fluorescence intensity increases when they are released to the external medium. In this case LUVs are prepared in a medium containing 4.36 m$M$ FITC-dextran in the appropriate buffer, and excess dextran is removed by passage through

[26] J. N. Weinstein, S. Yoshikami, P. Henkant, R. Blumenthal, and W. A. Hagins, *Science* **195**, 489 (1977).

[27] H. Ostolaza, B. Bartolomé, I. O. Zárate, F. de la Cruz, and F. M. Goñi, *Biochim. Biophys. Acta* **1147**, 81 (1993).

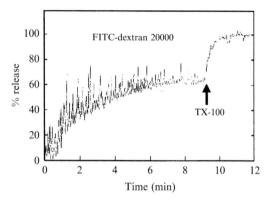

Fig. 4. Release of vesicle-entrapped FITC-dextran 20000 induced by sphingomyelinase action on SM–PE-Ch (2:1:1, mole ratio) LUVs (modified from Montes et al.[28]).

a column (30 × 2.5 cm) of Sephacryl S-300. To compensate for colloid osmotic effects of FITC-dextrans inside the vesicles, the assay medium, in which the LUVs are diluted to 0.3 m$M$ or other convenient concentration, contains the same concentration of nonderivatized dextran as the FITC-dextran concentration inside (4.36 m$M$). The fluorescence excitation and emission wavelengths are 465 and 520 nm, respectively, with a 495-nm interference filter. The use of FITC-dextrans of increasing molecular masses allows an estimation of the size of the channel, pore, or other bilayer discontinuity created by the enzyme. Figure 4[28] shows the efflux of a high molecular mass FITC-dextran (20 kDa) from LUVs containing 50 mol% sphingomyelin, in the presence of sphingomyelinase. The enzyme end-product ceramide causes membrane restructuring and release of vesicle aqueous contents.

Finally, leakage can also be measured as the release of entrapped glucose.[29] Glucose release is determined by the glucose oxidase plus peroxidase method, with phenol and amino-4-antipyrine as the color reagent. Note that these enzymes are added externally to the liposomes and do not have access to the entrapped glucose. This method is less sensitive than the fluorescence methods and more expensive because it requires the use of enzymes, but may find application in specific cases.

[28] R. Montes, M. B. Ruiz-Argüello, F. M. Goñi, and A. Alonso, J. Biol Chem. 277, 11788 (2002).
[29] F. M. Goñi, M. A. Urbaneja, and A. Alonso, in "Liposome Technology" (G. Gregoriadis, ed.), vol. II, p. 261. CRC Press, Boca Raton, FL, 1992.

Concluding Remarks

Phospholipases are enzymes whose substrates are, under physiological conditions, in the liquid crystalline state, in contrast to the vast majority of enzymes that work on substrates in solution. Phospholipases themselves exist most often in aqueous suspensions and must transiently dock the membranes to exert their catalytic actions. A second peculiar property of the phospholipases is that, through their activity, they can modify profoundly the physical properties of the substrate, and of the substrate environment. The latter effect is largely due to the extensive mesomorphism exhibited by lipids. A third, unique property of phospholipases is their small size as compared with the aggregates of which their substrates almost invariably make a part. These three properties are at the heart of most of the kinetic and mechanistic peculiarities of that group of enzymes. In phospholipases C and D one of the end-products is usually water soluble, making those phospholipases amenable, to a certain extent, to assay by more or less conventional procedures. All other aspects of phospholipase activity require a specific technology for their study.

In this chapter we have reviewed a number of methods that can be used to assay, and, most importantly, to evaluate the structural changes brought about by the phospholipases C, including sphingomyelinase. Apart from direct and indirect methods to assay enzyme activity, we have described applications of several fluorescence-based techniques to the study of phospholipase-induced vesicle aggregation, intervesicular lipid mixing, and intervesicular content mixing. Moreover, we have described here methods developed in our laboratory for the preparation of vesicles with asymmetric lipid distribution, and for the detection of intervesicular mixing of inner leaflet lipids. The latter permits the detection of vesicle–vesicle fusion even under the most leaky conditions. This collection of methods should allow the detailed characterization of the interaction of any phospholipase C with liposomes.

# [2] Liposomes in the Study of Phospholipase A$_2$ Activity

*By* John D. Bell, Susana A. Sanchez, and Theodore L. Hazlett

Introduction

Phospholipase A$_2$ (PLA$_2$) catalyzes hydrolysis of the sn-2 acyl chain of phospholipids. Physiologically, it appears to be involved in diverse functions such as digestion, membrane homeostasis, production of precursors for synthesis of several lipid mediators, defense against bacteria, clearing of dead or damaged cells, and as ligands for receptors.[1,2] Three basic types have been identified: secretory, cytosolic, and intracellular $PLA_2$ ($sPLA_2$, $cPLA_2$, and $iPLA_2$, respectively).[3,4] These three types differ in several ways including the compartment in which they act, their structure, and their dependence on calcium. Both $cPLA_2$ and $iPLA_2$ are large (40–85 kDa) intracellular enzymes; $sPLA_2$ is a small (14 kDa) secretory enzyme that acts extracellularly. Secretory $PLA_2$ requires calcium as a cofactor at micromolar to millimolar concentrations depending on the experimental conditions.[5,6] In contrast, activation of $iPLA_2$ does not require calcium, and $cPLA_2$ is regulated by calcium at the low concentrations applicable to the cytosol.[3,4] In addition, several isozymes of $sPLA_2$ have been identified and are classified into various groups based on characteristic structural differences.[2] This chapter focuses on methods that have been employed primarily with $sPLA_2$. Nevertheless, many of them can be applied to studies of $cPLA_2$.[7,8]

The action of these enzymes toward liposomes has been studied for a variety of reasons. First, liposomes have provided a convenient *in vitro* reconstitution system for efforts to identify the basic enzymology of $sPLA_2$. Second, liposomes have been useful in studies of potential inhibitors of the enzyme. More broadly, the $sPLA_2$–liposome system has been used widely as a model for studying the fundamentals of lipid–protein interactions. This is true both from the perspective of basic studies to investigate the principles of catalysis at an aqueous–lipid interface as well as efforts to elucidate mechanisms by which physical properties of lipid aggregates influence the behavior of enzymes that bind reversibly to those aggregates.

*Mechanism of Action of Secretory PLA₂ at Membrane Interface*

In general, two steps are involved in the action of $sPLA_2$ at interfaces (Fig. 1). In the first step, the enzyme adsorbs to the bilayer surface. The second step appears to be an activation step resulting in a productive complex between the bound enzyme and a phospholipid monomer from the membrane. Evidence from X-ray diffraction experiments and studies using

[1] I. Kudo, M. Murakami, S. Hara, and K. Inoue, *Biochim. Biophys. Acta* **1170,** 217 (1993).
[2] G. Lambeau and M. Lazdunski, *Trends Pharmacol. Sci.* **20,** 162 (1999).
[3] E. A. Dennis, *J. Biol. Chem.* **269,** 13057 (1994).
[4] M. F. Roberts, *FASEB J.* **10,** 1159 (1996).
[5] B. K. Lathrop and R. L. Biltonen, *J. Biol. Chem.* **267,** 21425 (1992).
[6] E. D. Bent and J. D. Bell, *Biochim. Biophys. Acta* **1254,** 349 (1995).
[7] T. Bayburt and M. H. Gelb, *Biochemistry* **36,** 3216 (1997).
[8] A. M. Hanel, S. Schuttel, and M. H. Gelb, *Biochemistry* **32,** 5949 (1993).

$$PLA_2 + Liposome \rightleftharpoons Bound\ PLA_2 + PL_m \overset{Ca^{2+}}{\rightleftharpoons} Bound\ PLA_2{\cdot}PL_m$$

FIG. 1. Proposed mechanism for interaction between sPLA$_2$ and substrate in liposomes. PL$_m$, Membrane phospholipid.

polymerized phospholipids have suggested that for sPLA$_2$ this second step involves physical migration of phospholipids upward from the plane of the bilayer into the enzyme active site.[9–11] For sPLA$_2$, the second step appears to be the one that requires calcium.[12–14] Because the two steps are linked thermodynamically, the presence of calcium promotes the adsorption of the enzyme to the membrane surface.[14]

### Choice of Experimental System

Both steps shown in Fig. 1 appear to be highly dependent on the physical properties of the membrane. This observation is the basis for much of the interest in studying sPLA$_2$, but it has also led to considerable confusion in the interpretation of apparently conflicting results. The choice of the liposome system and the details of the reaction conditions, then, are especially important when conducting investigations with this enzyme. In fact, the physical properties of the membrane contribute much more toward determining the level of activity measured than the identity of the phospholipid substrate per se. Some of the properties that appear to be critical are the charge, curvature, phospholipid phase, and the presence of compositional heterogeneity in the membrane. These properties are considered here individually, but they do not operate independently in determining the degree to which a given liposome will be vulnerable to hydrolysis by sPLA$_2$. Therefore, investigators must consider them collectively when making decisions regarding experimental conditions and interpretation of results.

[9] D. L. Scott, S. P. White, Z. Otwinowski, W. Yuan, M. H. Gelb, and P. B. Sigler, *Science* **250,** 1541 (1990).

[10] S. K. Wu and W. Cho, *Biochemistry* **32,** 13902 (1993).

[11] C. E. Soltys, J. Bian, and M. F. Roberts, *Biochemistry* **32,** 9545 (1993).

[12] B. Z. Yu, O. G. Berg, and M. K. Jain, *Biochemistry* **32,** 6485 (1993).

[13] J. D. Bell and R. L. Biltonen, *J. Biol. Chem.* **264,** 225 (1989).

[14] J. B. Henshaw, C. A. Olsen, A. R. Farnbach, K. H. Nielson, and J. D. Bell, *Biochemistry* **37,** 10709 (1998).

*Charge*

In general, sPLA$_2$ adsorbs with much higher affinity to membranes that carry a net negative charge than to those composed of zwitterionic or neutral lipids.[14,15] This observation is true regardless of whether the source of the charge is substrate or nonsubstrate lipids in the membrane. Therefore, charge affects the first step in Fig. 1 directly. The basis for this effect appears to involve both the conformation and the electrostatics of the enzyme.[16,17] Importantly, negative charge can either promote or inhibit sPLA$_2$ activity depending on the experimental conditions. For example, liposomes composed entirely of anionic phospholipids tend to be susceptible to the enzyme and are rapidly and completely hydrolyzed when the enzyme concentration is significantly higher than the bulk liposome concentration (i.e., multiple enzyme molecules bound to each liposome). Alternatively, liposomes that contain a mixture of anionic and zwitterionic phospholipids are only partially hydrolyzed. A similar result occurs if the concentration of sPLA$_2$ is less than the concentration of liposomes. In these cases, the enzyme becomes trapped on the first liposome to which it binds and/or on regions of the membrane dominated by high concentrations of anionic lipids, preventing it from gaining access to the remaining available substrate.[18]

One specific advantage of using liposomes composed of anionic lipids is that it simplifies the kinetics of phospholipid hydrolysis.[19] This is true because step 1 in Fig. 1 can be eliminated from consideration, because enzyme molecules appear to remain adsorbed to the same bilayer surface throughout the reaction and hydrolyze lipids only along that surface. Accordingly, estimation of traditional kinetic constants, studies of substrate specificity, and evaluation of potential inhibitors of the enzyme become less ambiguous compared with similar investigations with membranes to which the enzyme does not bind tightly.[20,21] Interpretation of results from such experiments does, of course, assume that lipids do not exchange

[15] M. K. Jain, B. Z. Yu, and A. Kozubek, *Biochim. Biophys. Acta* **980,** 23 (1989).

[16] D. L. Scott, A. M. Mandel, P. B. Sigler, and B. Honig, *Biophys. J.* **67,** 493 (1994).

[17] F. Ghomashchi, Y. Lin, M. S. Hixon, B. Z. Yu, R. Annand, M. K. Jain, and M. H. Gelb, *Biochemistry* **37,** 6697 (1998).

[18] W. R. Burack and R. L. Biltonen, *Chem. Phys. Lipids* **73,** 209 (1994).

[19] M. K. Jain, J. Rogers, D. V. Jahagirdar, J. F. Marecek, and F. Ramirez, *Biochim. Biophys. Acta* **860,** 435 (1986).

[20] O. G. Berg, B. Z. Yu, J. Rogers, and M. K. Jain, *Biochemistry* **30,** 7283 (1991).

[21] M. K. Jain, F. Ghomashchi, B. Z. Yu, T. Bayburt, D. Murphy, D. Houck, J. Brownell, J. C. Reid, J. E. Solowiej, S. M. Wong, V. Mocek, R. Jarrell, M. Sasser, and M. H. Gelb, *J. Med. Chem.* **35,** 3584 (1992).

among liposomes, an issue that can be resolved by appropriate assessment of partition coefficients and exchange rates.[6,22]

## Curvature

Membrane curvature has profound effects on the susceptibility of phospholipid aggregates to hydrolysis by sPLA$_2$. For example, micelles and small unilamellar vesicles (SUVs; diameter, $\sim$200 Å or less) composed of phosphatidylcholine (PC) are hydrolyzed immediately on addition of sPLA$_2$.[23] In contrast, the membranes of large unilamellar vesicles (LUVs) or multilamellar vesicles (diameter, $\sim$700 Å or more) resist catalysis unless contaminated with other molecules such as the products of phospholipid hydrolysis, fatty acid and/or lysophospholipid.[14,24] In the absence of calcium, sPLA$_2$ binds to membranes with high curvature, suggesting that curvature promotes step 1 in Fig. 1.[14] Binding to membranes with low curvature generally requires calcium and/or negative charge in the membrane.[14,15,18]

## Phospholipid Phase

The membrane state relative to the main thermotropic transition between gel (or solid ordered) and liquid crystalline (or liquid disordered) lamellar phases has profound influence on the susceptibility of the bilayer to sPLA$_2$. The relationship between membrane phases and the action of sPLA$_2$ are complicated by effects on both steps in Fig. 1. The adsorption of sPLA$_2$ to the surface of PC SUVs in the absence of calcium requires the membrane to be in the gel phase.[13] Alternatively, for LUVs, the rate of turnover of substrate by bound enzyme is higher when the membrane is near the phase transition or in the liquid crystalline state.[25] Likewise, the ability of contaminating molecules such as fatty acid and lysophospholipid to promote step 2 in Fig. 1 appears to be optimal when the LUV membrane is at or above the transition temperature.[25] Accordingly, the temperature dependence of liposome hydrolysis can be complex. For example, hydrolysis of LUVs of saturated PC is biphasic with a so-called lag or latency period of slow hydrolysis followed by a period of rapid hydrolysis.[24,26–28] The length of this lag period is longest for temperatures

[22] S. D. Brown, B. L. Baker, and J. D. Bell, *Biochim. Biophys. Acta* **1168**, 13 (1993).

[23] M. Menashe, G. Romero, R. L. Biltonen, and D. Lichtenberg, *J. Biol. Chem.* **261**, 5328 (1986).

[24] D. Lichtenberg, G. Romero, M. Menashe, and R. L. Biltonen, *J. Biol. Chem.* **261**, 5334 (1986).

[25] J. D. Bell, M. L. Baker, E. D. Bent, R. W. Ashton, D. J. Hemming, and L. D. Hansen, *Biochemistry* **34**, 11551 (1995).

below the main lipid-phase transition. As the temperature is raised through the phase transition, the latency period reaches a minimum and then becomes longer again when the liquid crystalline phase is dominant.[24,27,29]

*Membrane Heterogeneity*

The tendency of the thermotropic phase transition of the membrane to promote hydrolysis by sPLA$_2$ probably relates to the dynamic heterogeneity present under that condition.[24,29] The coexistence of domains with distinct physical properties is likely to enhance hydrolysis by promoting step 2 in Fig. 1. Evidence in support of this interpretation has come from other studies in which the ability of contaminating neutral lipids such as lysophospholipid, protonated fatty acid, diacylglycerol, and triacylglycerol to promote hydrolysis of phospholipid bilayers has been examined.[14,25,30,31] In each case, the ability of these molecules to increase membrane susceptibility to sPLA$_2$ appears linked to the existence of lipid domains in the membrane.

*Other Conditions*

*Calcium Concentration.* It is clear that calcium is an obligatory cofactor for hydrolysis catalyzed by sPLA$_2$; in the complete absence of calcium (i.e., in the presence of a chelator), no activity is observed. The binding site for calcium is well defined in the crystal structure of many species of sPLA$_2$.[9,16,32–34] It is commonly stated that sPLA$_2$ requires millimolar concentrations of calcium. Nevertheless, the calcium requirement is influenced by the choice of substrate, and some of the influences of calcium on the reaction reflect interactions of the ion with the membrane. In fact, studies examining the calcium dependence of the enzyme suggest that the $K_m$ of calcium is actually about 20 $\mu M$ as opposed to the millimolar requirement

---

[26] R. Apitz-Castro, M. K. Jain, and G. H. de Haas, *Biochim. Biophys. Acta* **688,** 349 (1982).

[27] M. Menashe, D. Lichtenberg, C. Gutierrez-Merino, and R. L. Biltonen, *J. Biol. Chem.* **256,** 4541 (1981).

[28] J. D. Bell and R. L. Biltonen, *J. Biol. Chem.* **267,** 11046 (1992).

[29] T. Honger, K. Jorgensen, R. L. Biltonen, and O. G. Mouritsen, *Biochemistry* **35,** 9003 (1996).

[30] W. R. Burack, M. E. Gadd, and R. L. Biltonen, *Biochemistry* **34,** 14819 (1995).

[31] J. D. Bell, M. Burnside, J. A. Owen, M. L. Royall, and M. L. Baker, *Biochemistry* **35,** 4945 (1996).

[32] S. P. White, D. L. Scott, Z. Otwinowski, M. H. Gelb, and P. B. Sigler, *Science* **250,** 1560 (1990).

[33] D. L. Scott, Z. Otwinowski, M. H. Gelb, and P. B. Sigler, *Science* **250,** 1563 (1990).

[34] D. L. Scott, S. P. White, J. L. Browning, J. J. Rosa, M. H. Gelb, and P. B. Sigler, *Science* **254,** 1007 (1991).

commonly stated.[5] Nevertheless, the requirement for calcium can appear to be higher depending on the experimental system. For example, hydrolysis of dipalmitoylphosphatidylcholine (DPPC) at calcium concentrations less than millimolar underestimates the amount of hydrolysis by some methods because of the ability of calcium to bind the fatty acid released during hydrolysis.[5] Special caution must be exerted when using substrates composed of anionic lipids. The presence of calcium can cause changes in the phase state of the membrane as well as aggregation of liposomes. In these cases, calcium concentrations at or below 100 $\mu M$ may be advisable.

*pH.* The pH optimum for sPLA$_2$ is broad and tends to be at maximum around neutral pH. The enzyme performs well at high pH, and values in the range of pH 8.0 to 9.0 are commonly used. One reason for selecting high pH values is the choice of the pH-stat assay for measuring hydrolysis (see below). Use of pH at or above pH 8.0 ensures that the fatty acid will remain ionized during hydrolysis, allowing a stoichiometric release of protons reflecting the number of hydrolysis events. This decision is especially important for substrates with long-chain fatty acids because the $pK_a$ for fatty acids partitioned in the membrane is much higher than that for free fatty acids.[35–37]

*Temperature.* Choice of experimental temperature is a critical issue for experiments with sPLA$_2$ for the reasons given above in the discussion of phase transitions. Importantly, the selection must be made relative to an understanding of the phase behavior of the system rather than on the basis of some external reference such as physiological temperature. For example, the kinetics of hydrolysis of DPPC at 37° will be dramatically different compared with dimyristoylphosphatidylcholine (DMPC) at the same temperature. Superficially, the investigator might be tempted to draw conclusions about substrate specificity. However, the differences will be due almost entirely to the phase state of the two substrates. At 37°, DPPC is in the gel state, about 5° below the main phase transition, whereas DMPC is in the liquid crystalline state, about 15° above its phase transition temperature.

*Summary*

The interactions of the various factors discussed in this section must be emphasized. For example, the importance of membrane heterogeneity in membrane hydrolysis is minimal or absent for neutral phospholipid

[35] M. S. Fernandez, M. T. Gonzalez-Martinez, and E. Calderon, *Biochim. Biophys. Acta* **863**, 156 (1986).
[36] M. Ptak, M. Egret-Charlier, A. Sanson, and O. Bouloussa, *Biochim. Biophys. Acta* **600**, 387 (1980).
[37] G. Cevc, J. M. Seddon, R. Hartung, and W. Eggert, *Biochim. Biophys. Acta* **940**, 219 (1988).

micelles, greater for SUVs, and critical for LUVs. Evidence suggests that the situation may be different still for the membranes of giant unilamellar vesicles (GUVs) that possess negligible curvature relative to the size of the enzyme. In this case, enzyme binding and substrate hydrolysis appear entirely confined to liquid regions of the membrane (see below). Conversely, when membranes are composed of anionic lipids, curvature and membrane heterogeneity are commonly of little importance. The impact of membrane heterogeneity contributed by contaminating molecules will also depend on the identity of the major phospholipid in the bilayer. This is true both in terms of specific interactions that affect the miscibility of components, as well as general properties such as the partition coefficient of the contaminants. The latter is especially relevant to the effects of fatty acid and lysophospholipid as contaminants resulting from the hydrolysis reaction.[6,22]

These examples are not exhaustive in terms of the variety of phenomena that have been observed. Nevertheless, they serve to illustrate several key issues with respect to choices of experimental conditions, interpretation of results, and comparison with results obtained by other investigators. Importantly, conclusions reached with one experimental system do not necessarily apply to other systems. Instead, a careful comparison of the various physical and chemical properties of the two systems under comparison must be made. A simple change such as the length of the phospholipid acyl chains makes a large difference in the resulting behavior of sPLA$_2$, as in the example of DPPC and DMPC suggested above. In addition to contrasts in the phase transition temperature, differences in the ability of hydrolysis products to partition in the bilayer also affect greatly the hydrolysis kinetics.[6,22] In summary, it is advisable to use a holistic approach to interpretation of data that considers all the contributions of liposome morphology, state, and composition.

## Sources of Materials

### Liposomes

Procedures for producing the liposomes useful in studies of sPLA$_2$ are generally well established and are described here only briefly. A few suggestions regarding issues of importance when working with an enzyme as sensitive to bilayer properties as is sPLA$_2$ are included.

*Small Unilamellar Vesicles.* Depending on the lipid mixture chosen, SUVs generally have a diameter of 100 to 200 Å. Lipids and fluorescent probes, if any (see below), are codispersed in a glass tube in organic solvent, usually chloroform. The solvent is removed by evaporation either under vacuum or under a stream of an inert gas to avoid oxidation of lipids. The tube should be rotated during removal of organic solvent so that the dried lipid is spread in a fine film on the bottom of the tube. Multilamellar vesicles are then formed by hydrating the lipid in aqueous diluent. We commonly use 35–50 m$M$ KCl, 3 m$M$ NaN$_3$, and 0–10 m$M$ CaCl$_2$ with 10 m$M$ buffer at an appropriate pH depending on the desired experimental conditions. NaN$_3$ is included in all solutions to prevent bacterial growth. When experiments will involve the pH-stat assay (see below), the buffer is not included. Samples are incubated for 1 h in the aqueous diluent at a temperature well above the phase transition temperature of the lipids (e.g., 45–55° for samples of pure DPPC) with periodic rigorous agitation (generally for 5 to 10 s every 5 to 10 min) on a vortex mixer. SUVs are then formed by sonicating the hydrated multilamellar vesicles three times for 3 min with a titanium probe. It is critical that the sonication be done at a temperature above the phase transition temperature of the lipids. Titanium granules can be removed by centrifuging the mixture for 30 s at 13,000 rpm in a microcentrifuge. The concentration of vesicles is usually expressed in terms of the bulk phospholipid concentration, which can be determined by assay of the phosphate content.[38] In general, it is best to use SUVs composed of saturated lipids within 1 day of preparation because of aggregation and fusion of vesicles. Stability can be improved by storing the sample above the phase transition temperature. SUVs composed of unsaturated phospholipids are stable for longer periods of time as long oxidation of lipids is prevented, because their phase transition temperature is usually below the freezing point of water.

*Large Unilamellar Vesicles.* LUVs are formed by extrusion.[39] Lipids are dispersed in organic solvent, dried, and hydrated as multilamellar vesicles as described for SUVs. The suspension is then extruded through polycarbonate filters 10 times at a temperature well above the lipid phase transition temperature. We routinely use filters with a 0.1-$\mu$m pore size, which generally produces vesicles with a diameter of 800–1000 Å.[39] Different pore sizes can be used to adjust vesicle size.[40]

Most lipids and/or fluorescent probes should be included in the original organic mixture before formation of multilamellar vesicles. However, some molecules can be added directly to aqueous suspensions of LUVs, allowing flexibility in experimental conditions without having to produce multiple liposome preparations. For example, lysophospholipid and/or fatty acid (dispersed in aqueous methanol) can be added directly to the vesicle

[38] G. R. Bartlett, *J. Biol. Chem.* **234,** 466 (1959).
[39] M. J. Hope, M. B. Bally, G. Webb, and P. R. Cullis, *Biochim. Biophys. Acta* **812,** 55 (1985).
[40] L. D. Mayer, M. J. Hope, and P. R. Cullis, *Biochim. Biophys. Acta* **858,** 161 (1986).

sample. This procedure has been used extensively in attempts to mimic at steady state the conditions that exist during vesicle hydrolysis.[6,22,28,41] Control experiments have demonstrated that lysophospholipid and fatty acid incorporate rapidly into the bilayer and produce a steady state that remains stable over the course of the experiment.[15,25,28] Furthermore, as assessed by prodan and laurdan fluorescence, the properties of the bilayer produced by this method resemble those observed during vesicle hydrolysis both quantitatively and qualitatively.[42] In such experiments, the final methanol concentration should not exceed 0.5% (v/v), an amount that does not interfere with the results.[42] Interpretation of results must take into consideration the partition coefficients for the lipids. Methods for assessing those partition coefficients have been described.[6,22,25]

*Secretory PLA$_2$*

*Snake Venom and Pancreatic.* Secretory PLA$_2$ from a variety of snake venoms or mammalian pancreas is available commercially. The exact choice of sPLA$_2$ source should be based on the properties of the enzyme, obtainable from the extensive literature on the subject. In general, the purity of commercial enzyme preparations is relatively low. Depending on the purpose of the experiments, these commercial enzymes may be used directly; however, it is usually preferable to purify them further. Nevertheless, purification of enzymes directly from venom or pancreatic extracts is usually simple and is clearly the economical preference if large amounts (i.e., more than milligrams) of enzyme are needed. Methods have been published for a wide variety of sources including mammalian pancreas and *Crotalus atrox, Agkistrodon piscivorus piscivorus,* and *Naja naja* venoms.[43–46] Most of these enzymes can be stored long term as a lyophilized powder at −20°. The enzyme can be dissolved in stock solutions in 50 m$M$ KCl and 3 m$M$ NaN$_3$ and stored at 4° before use. The concentration of sPLA$_2$ is assessed by the absorbance at 280 nm (e.g., the absorbance coefficient for *A. p. piscivorus* enzyme is 2.2 mg ml$^{-1}$ cm$^{-1}$). The useful final sPLA$_2$ concentration in experiments with the methods described here should be in the range of about 0.1 to 10 $\mu$g/ml.

[41] M. K. Jain and G. H. de Haas, *Biochim. Biophys. Acta* **736,** 157 (1983).
[42] M. J. Sheffield, B. L. Baker, D. Li, N. L. Owen, M. L. Baker, and J. D. Bell, *Biochemistry* **34,** 7796 (1995).
[43] W. Nieuwenhuizen, H. Kunze, and G. H. de Haas, *Methods Enzymol.* **32,** 147 (1974).
[44] Y. Hachimori, M. A. Wells, and D. J. Hanahan, *Biochemistry* **10,** 4084 (1971).
[45] J. M. Maraganore, G. Merutka, W. Cho, W. Welches, F. J. Kezdy, and R. L. Heinrikson, *J. Biol. Chem.* **259,** 13839 (1984).
[46] T. L. Hazlett and E. A. Dennis, *Toxicon* **23,** 457 (1985).

*Human.* Several human sPLA$_2$ enzymes have been cloned and can be synthesized with bacterial transfection systems followed by folding and purification of the recombinant proteins. Group IIA,[47] group V,[48] and group X isozymes can be obtained in this manner.[49]

## Assays of Hydrolysis

It is advisable to measure phospholipase activity over time rather than at a single point because the kinetics are rarely linear and are commonly complex. If a single point assay is desirable for screening a large number of compounds as potential inhibitors or for assaying column fractions during purification, it is recommended that either a charged or a micellar substrate be used. Several methods exist for assaying hydrolysis as a function of time. Here, we discuss indirect methods using pH titration and fluorescence spectroscopy and direct methods including chromogenic substrates and thin-layer chromatography.

### pH-Stat

The use of an autotitrator to assay hydrolysis by the pH-stat technique has been detailed in a previous volume of this series.[43] This technique offers certain advantages and disadvantages. It has the advantage of providing quantitative real-time data. Figure 2 illustrates the level of detail obtainable with the pH-stat assay. A second advantage is that the technique can be coupled with fluorescence spectroscopy to display simultaneously the time course of hydrolysis and enzyme fluorescence (Fig. 2) and/or fluorescence of a membrane probe such as laurdan or dansyl-labeled phospholipids.[31,42] Methods for combining fluorescence with the pH-stat assay have been reported previously in this series.[50] The critical issues in mating the two techniques are to avoid optical interference from the autotitrator dispenser tip and electrode while maintaining adequate stirring to ensure rapid mixing of components as the protons released during the reaction are titrated.

The primary disadvantage of the pH-stat method is that reaction conditions are confined to those that ensure complete ionization of the fatty acid released during hydrolysis. This means that experiments must be

[47] Y. Snitko, R. S. Koduri, S. K. Han, R. Othman, S. F. Baker, B. J. Molini, D. C. Wilton, M. H. Gelb, and W. Cho, *Biochemistry* **36,** 14325 (1997).
[48] S. K. Han, E. T. Yoon, and W. Cho, *Biochem. J.* **331,** 353 (1998).
[49] S. Bezzine, R. S. Koduri, E. Valentin, M. Murakami, I. Kudo, F. Ghomashchi, M. Sadilek, G. Lambeau, and M. H. Gelb, *J. Biol. Chem.* **275,** 3179 (2000).
[50] J. D. Bell and R. L. Biltonen, *Methods Enzymol.* **197,** 249 (1991).

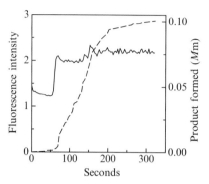

FIG. 2. Concurrent observation of sPLA$_2$ tryptophan fluorescence (solid curve) during hydrolysis of DPPC LUVs (dashed curve). Fluorescence intensity was observed at an excitation of 280 nm and emission at 340 nm. Reaction conditions were 400 $\mu M$ DPPC, 360 n$M$ sPLA$_2$ (from *A. p. piscivorus* venom), 40°, and 1 m$M$ calcium. Hydrolysis was assayed simultaneously by the pH-stat method.

conducted at basic pH. Even then, incomplete ionization (especially at low calcium concentrations) may be an issue because the $pK_a$ of fatty acid in the bilayer may be higher than pH 8.0. Sensitivity may also be a concern with the pH-stat assay. Substrate concentrations must usually be greater than 0.1 m$M$. Low hydrolysis rates are difficult to discern from a baseline caused by atmospheric carbon dioxide. The baseline can be reduced by maintaining a nitrogen stream over the reaction vessel, but evaporation then becomes a challenge when reactions are slow and time courses are therefore prolonged.

*Fluorescent Fatty Acid-Binding Proteins*

Hydrolysis can also be monitored in real time by measuring the release of free fatty acids using a fluorescent fatty acid-binding protein[51] or by displacement of a fluorescent fatty acid from a binding protein.[52] Acrylodan-labeled intestinal fatty acid-binding protein (ADIFAB) is available commercially. It is added directly to the reaction mixture containing liposomes, buffer, and calcium at the desired concentration. It is best if the calcium concentration is at or below about 2 m$M$ depending on the phospholipid (less if anionic). Furthermore, the liposome concentration should be at 100 $\mu M$ or less to avoid interference from scattered light. Fluorescence is monitored at two wavelengths (0.2 $\mu M$ ADIFAB; excitation,

[51] G. V. Richieri and A. M. Kleinfeld, *Anal. Biochem.* **229,** 256 (1995).
[52] D. C. Wilton, *Biochem. J.* **266,** 435 (1990).

390 nm; emission, 432 and 505 nm[51,53]) for a period of time to establish the baseline. Enzyme is then added (0.5–5 $\mu$g/ml) to initiate hydrolysis. Simultaneous monitoring of two emission wavelengths can be accomplished by using a spectrofluorometer, with dual-emission detectors using bandpass filters or monochromators in the "T" format or by automated rapid switching of monochromator mirrors between wavelengths.[14,31,42] Relative amounts of hydrolysis can be quantified by calculation of the generalized polarization.[53,54] Estimation of the absolute amounts of hydrolysis requires knowledge of the partition coefficient of the specific fatty acid involved.[51]

An advantage to using these fatty acid-binding proteins to monitor hydrolysis is that, like the pH-stat assay, liposome hydrolysis can be monitored in real time without requiring the presence of contaminating probes in the membrane. These assays are more sensitive than the pH-stat assay and can be conducted at neutral pH. The primary disadvantages are the cost of the assay and challenges associated with quantifying the results. If more than one type of fatty acid is involved, only relative hydrolysis kinetics can be measured. Also, calcium concentration can be an issue because calcium competes with the binding protein for interaction with fatty acids.

## Chromogenic Substrates

Several substrate analogs with absorbance or fluorescence sensitive to hydrolysis have been reported for use in assaying sPLA$_2$.[55–58] In general, they provide the same advantage as the fatty acid-binding proteins, that is, they provide sensitive real-time data. However, use of nonnatural substrates may be a concern for some investigations. Also, the results report only hydrolysis of the chromogenic substrate and do not provide information concerning other substrates present in the liposome.

## Thin-Layer Chromatography

Thin-layer chromatography (TLC) is an effective means for separating substrate from product and is therefore an effective means for quantitative assessment of hydrolysis. Phospholipids and lysophospholipids can be

[53] H. A. Wilson, W. Huang, J. B. Waldrip, A. M. Judd, L. P. Vernon, and J. D. Bell, *Biochim. Biophys. Acta* **1349,** 142 (1997).
[54] T. Parasassi, G. De Stasio, G. Ravagnan, R. M. Rusch, and E. Gratton, *Biophys. J.* **60,** 179 (1991).
[55] C. Balet, K. A. Clingman, and J. Hajdu, *Biochem. Biophys. Res. Commun.* **150,** 561 (1988).
[56] F. Radvanyi, L. Jordan, F. Russo-Marie, and C. Bon, *Anal. Biochem.* **177,** 103 (1989).
[57] T. Bayburt, B. Z. Yu, I. Street, F. Ghomashchi, F. Laliberte, H. Perrier, Z. Wang, R. Homan, M. K. Jain, and M. H. Gelb, *Anal. Biochem.* **232,** 7 (1995).
[58] W. Cho, M. A. Markowitz, and F. J. Kézdy, *J. Am. Chem. Soc.* **110,** 5166 (1988).

visualized on TLC plates by reaction with iodine vapor and quantified by densitometry or phosphate assay.[59] Improved precision and sensitivity are obtained by using radiolabeled substrates. It is generally easy to resolve phospholipids and lysophospholipids, allowing one to track both degradation of substrate and formation of product. Thus, it is usually preferable to use substrates labeled on the head groups rather than the acyl chains. There are no limitations in reaction conditions in terms of pH, calcium concentration, or membrane composition. If the kinetics of hydrolysis of more than one phospholipid species is desired, multiple radiolabeled substrates can be used in the same reaction because phospholipids and lysophospholipids are separated on the basis of both head group and acyl chain composition.

Label substrates are codispersed with other lipids during liposome formation. Reactions are initiated by addition of sPLA$_2$ (0.05–1 $\mu$g/ml) to the liposome suspension (100–200 $\mu M$, at 1000–2000 cpm/nmol is appropriate) in a buffer and calcium solution appropriate for the particular experiment. At the desired time points, 10-$\mu$l aliquots are removed from the mixture and added to 10 $\mu$l of methanol containing 10 m$M$ unlabeled substrate. Fifteen microliters of this solution (2000–3000 cpm) is applied to a TLC plate (K6 silica) and dried. For substrate composed of a single head group, the plate can be developed in CHCl$_3$–CH$_3$CO$_2$H (95:5, v/v). Phosphatidylcholine, phosphatidylethanolamine, and phosphatidylserine can be separated effectively with chloroform–methanol–acetic acid (6.5:2.5:1, v/v/v). If substrate mixtures more complicated than these are used, separation may require two-dimensional TLC.[60] Spots are identified with standards. Lipid content in each spot is quantified by liquid scintillation counting of scraped or excised spots.

### Assays of Binding

It is important for binding studies that the enzyme be inhibited. If hydrolysis were allowed to proceed, the high activity of the enzyme in cleaving bilayer phospholipids would quickly alter the bilayer composition and change the nature of the interface. There are several methods to inactivate the sPLA$_2$: site-directed mutagenesis of active site residues, chemical modification of the active site histidine residue (e.g., 100 $\mu M$ $p$-bromophenacyl bromide), the use of nonhydrolyzable substrates (e.g., ether phospholipids),

[59] H. A. Wilson, J. B. Waldrip, K. H. Nielson, A. M. Judd, S. K. Han, W. Cho, P. J. Sims, and J. D. Bell, *J. Biol. Chem.* **274,** 11494 (1999).
[60] R. Kikuchi-Yanoshita, R. Yanoshita, I. Kudo, H. Arai, T. Takamura, K. Nomoto, and K. Inoue, *J. Biochem. (Tokyo)* **114,** 33 (1993).

and the removal/replacement of calcium, the required cofactor for these enzymes. The use of barium (1 m$M$) as a calcium mimic has been a common method to obtain an inactive, cation-bound sPLA$_2$ that could still bind monomeric and interfacial lipids.[61,62] More recent studies using pancreatic sPLA$_2$, however, have suggested that barium may not form an appropriate enzyme–divalent cation–substrate ternary complex and that two other divalent cations, cadmium (10 $\mu M$) and zinc ($\geq$25 $\mu M$), may allow more complete binding between the membrane phospholipid head group and the active site of the enzyme.[12,63] It should be noted that cation-free sPLA$_2$ does have the capacity to bind a phospholipid interface, depending of the nature of the phospholipids present. One defining character, in this respect, is the presence of a net negative charge in the membrane, which can be accomplished by the introduction of anionic lipids, such as the phosphatidylglycerols or the phosphatidylserines, or free fatty acid to the bilayer.

*Enzyme Fluorescence*

The binding of sPLA$_2$ to liposomes has been measured by instantaneous changes in the tryptophan fluorescence emission spectrum of the enzyme (~1 $\mu M$ sPLA$_2$, 0.01–1 m$M$ DPPC).[13] In general, the intensity of tryptophan emission increases and the spectrum maximum shifts toward shorter wavelengths, suggesting that the residue has been shielded from water.[13] In addition to steady state measurements, changes in binding during hydrolysis time courses can also be assessed from tryptophan fluorescence as shown in Fig. 2.[64] A difficulty in using tryptophan fluorescence to assess binding is that interpretation may be complex depending on the enzyme species. Several of the sPLA$_2$ types contain multiple tryptophan residues, and more than two fluorescent states of the enzyme can therefore exist.

*Fluorescence Resonance Energy Transfer*

The binding of sPLA$_2$ to the surface of liposomes has also been assessed at steady state or during hydrolysis time courses by resonance energy transfer between the tryptophan of the enzyme to $N$-[5-(dimethylamino)-naphthalene-1-sulfonyl]-1,2-dihexadecanoyl-*sn*-glycero-3-phosphoethanola mine (dansyl-DHPE) in the phospholipid bilayer.[14,31] Dansyl-DHPE

[61] M. C. Dam-Mieras, A. J. Slotboom, W. A. Pieterson, and G. H. de Haas, *Biochemistry* **14,** 5387 (1975).

[62] M. F. Roberts, R. A. Deems, and E. A. Dennis, *Proc. Natl. Acad. Sci. USA* **74,** 1950 (1977).

[63] B. Z. Yu, J. Rogers, G. R. Nicol, K. H. Theopold, K. Seshadri, S. Vishweshwara, and M. K. Jain, *Biochemistry* **37,** 12576 (1998).

[64] J. D. Bell and R. L. Biltonen, *J. Biol. Chem.* **264,** 12194 (1989).

(2 mol%) is mixed with liposome lipids in organic solvent during vesicle formation. During experiments, the intensity of dansyl-DHPE fluorescence emission is measured at 510 nm with excitation at both 280 nm (for energy transfer) and at 340 nm (for measurement of direct dansyl-DHPE fluorescence) either before and after addition of sPLA$_2$ or simultaneously during time course experiments. When possible, it is helpful also to measure the intrinsic sPLA$_2$ fluorescence (excitation, 280 nm; emission, 340 nm) simultaneously. During resonance energy transfer, the enzyme tryptophan fluorescence should decrease and the dansyl-DHPE fluorescence should increase. Figure 3A shows an example of such as fatty acid is added to liposomes. Interestingly, in some cases, both sPLA$_2$ and dansyl-DHPE fluorescence increase together (Fig. 3B). This phenomenon has been interpreted to indicate that sPLA$_2$ binds to at least two separate sites on the liposomes,

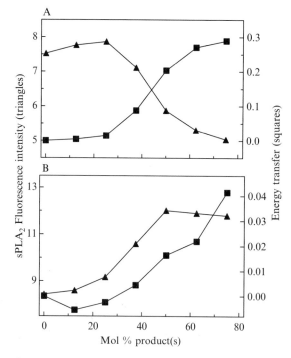

Fig. 3. Intensity of sPLA$_2$ tryptophan fluorescence (triangles) and resonance energy transfer (squares) to dansyl-DHPE (2 mol%) in DPPC LUVs as a function of added palmitic acid (A) or equimolar palmitic acid and lysophosphatidylcholine (B). Reaction conditions were 100 $\mu M$ DPPC, 360 n$M$ sPLA$_2$ (from $A.\ p.\ piscivorus$ venom), 44$^\circ$, and 10 m$M$ EDTA instead of calcium. Energy transfer was calculated by Eq. (1). Excitation was at 280 nm, and emission was at 340 nm (triangles) or 510 nm (squares). Adapted with permission from Ref. 14. Copyright 1998, $Am.\ Chem.\ Soc.$

one rich in phospholipid (i.e., the increase in dansyl-DHPE fluorescence), and one poor in phospholipid (i.e., perhaps rich in fatty acid, indicated by the increased tryptophan fluorescence).

In calculating the amount of energy transfer, it is helpful to normalize the data carefully to account for artifacts. For example, dansyl-DHPE fluorescence is environment sensitive and can therefore be affected by changes in the phase state of the bilayer that can occur as sPLA$_2$ binds, or in the presence of additives such as the fatty acid used in Fig. 3A. We have therefore calculated the amount of energy transfer according to Eq. (1):

$$ ET = \frac{F_{280,\ max} - F_{280,\ i}}{F_{280,\ 0}F_{PLA_2}} \tag{1} $$

$F_{280,\ max}$ is the steady state intensity of dansyl fluorescence (excitation, 280 nm; emission, 510 nm) in the presence of sPLA$_2$ and $F_{280,\ i}$ is the interpolated intrinsic dansyl fluorescence excited at 280 nm.[31] The initial fluorescence of sPLA$_2$ ($F_{PLA_2}$) and dansyl-DHPE ($F_{280,\ 0}$) in the absence of binding or energy transfer is included in the calculation to normalize the data for comparison among samples. The interpolated intrinsic dansyl fluorescence ($F_{280,\ i}$) is calculated from a calibration curve relating dansyl-DHPE fluorescence excited at 340 nm and at 280 nm obtained from a series of excitation spectra at varied intensities in the absence of sPLA$_2$.[31] The parameters of the calibration curve are dependent on the optics of the spectrofluorometer and should be obtained separately for each instrument used.

Similar calculations should be applied to quantify changes in the magnitude of energy transfer that may occur during a reaction time course such as that shown in Fig. 4 (curve c). In this case, $F_{280,\ max}$ is the highest intensity of dansyl-DHPE fluorescence excited at 280 nm during the time course. $F_{280,\ i}$ is obtained by interpolation using the calibration curve described above and dansyl-DHPE fluorescence intensity excited at 340 nm. This calculation allows $F_{280,\ i}$ to be corrected for the abrupt decrease in the intrinsic dansyl-DHPE fluorescence that occurs as hydrolysis products accumulate in the bilayer (see Fig. 4, curve a). This correction is valid only if the shape or wavelength of the dansyl-DHPE excitation spectrum does not change during the experiment. For the experiment shown in Fig. 4, this assumption was verified by control experiments evaluating the excitation spectrum before and after the addition of lysophosphatidylcholine and palmitic acid. Furthermore, the data were also normalized to the intensity of both the dansyl-DHPE fluorescence (excitation at 280 nm) before addition of enzyme ($F_{280,\ 0}$) and the sPLA$_2$ tryptophan fluorescence (Fig. 4, curve b) immediately before the abrupt increase in energy transfer ($F_{PLA_2}$)

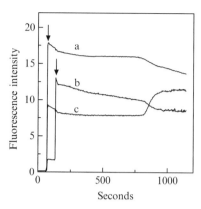

FIG. 4. Time course of intrinsic dansyl-DHPE fluorescence (curve a), sPLA$_2$ tryptophan fluorescence (curve b), and energy transfer from tryptophan to dansyl-DHPE (curve c) during hydrolysis of LUVs (100 $\mu M$ DPPC). Temperature was 38°, and calcium concentration was 10 m$M$. Liposomes (containing 2 mol% dansyl-DHPE) were added at the first arrow, and 360 n$M$ sPLA$_2$ (from *A. p. piscivorus* venom) was added at the second arrow. Excitation was at 280 nm (curves b and c) or 340 nm (curve a). Emission was at 340 nm (curve b) or 510 nm (curves a and c). Adapted with permission from Ref. 31. Copyright 1996, *Am. Chem. Soc.*

to account for experimental variation in the concentrations of these molecules and for adsorption of the enzyme to the cuvette walls.

## Assays of Changes in Membrane Properties during Hydrolysis

One of the difficulties in studying sPLA$_2$ is due to the change imposed on the substrate during sPLA$_2$-dependent hydrolysis. As the hydrolysis products are formed, the membrane must accommodate both changes in volume of the membrane and the introduction of new species, lyso-phospholipid and fatty acid (depending on partition coefficients). One method to examine these membrane changes, and their effect of sPLA$_2$, is through the introduction of membrane probes that are sensitive to subtle changes in the membrane. We will examine the use of two types of probes. These examples should provide a sense of the opportunities and the challenges that can be applied to future efforts.

### Probes of Membrane Phase and Order

Prodan and laurdan are two probes sensitive to the physical properties of their membrane environment. The two differ in the degree of hydropho-bicity and therefore in the position they occupy in the membrane. Prodan partitions weakly into the bilayer and provides information about the head group region of the membrane.[65,66] Laurdan binds more tightly and occupies the region of the glycerol backbones.[66] Both probes report information on the order and fluidity of the membrane.[54,67,68] Their sensitivity to these properties originates in the large solvent relaxation effect on their emission spectrum. When prodan or laurdan are photoexcited, water molecules in the vicinity orient in the field generated by the dipole of the fluorophore-excited state. In the membrane, this reorientation of water molecules is a time-dependent process resulting in a slow shift (on the nanosecond time scale) in the emission spectrum toward longer wavelengths. The presence and mobility of water molecules in different depths of the membrane depends on the order and mobility of lipid molecules in the membrane. Membrane changes detected by prodan and laurdan have been used to make inferences about the membrane environment conducive to attack by sPLA$_2$.[14,25,31,42]

Prodan and laurdan are usually incorporated into the membrane during liposome preparation by mixing them at a concentration of 0.1–0.3 mol% of the total lipid concentration in the chloroform solution before drying. Liposomes should be kept dark during both the preparation steps and storage. Alternatively, they can be incorporated after liposome formation by mixing a small volume of a concentrated stock of the probe in dimethyl sulfoxide with a suspension of liposomes at the mole ratio described above. In such cases, it is usually preferable to equilibrate the probe with the liposomes for 1 h or more at a temperature above the phase transition temperature. The emission spectrum of the probe can be checked repeatedly and vesicles used once the spectrum becomes stable. During experiments, the intensity of prodan or laurdan fluorescence emission is measured at two or three wavelengths (usually 435–440 nm and 490–500 nm for laurdan and 420, 480, and 530 nm for prodan) with excitation at about 350 nm for both. When possible, it is helpful also to assess the hydrolysis reaction simultaneously by the pH-stat method.[31,42]

Changes in the laurdan emission spectrum are quantified by calculating the generalized polarization (GP).[54]

$$GP = \frac{I_B - GI_G}{I_B + GI_G} \qquad (2)$$

$I_B$ is the emission intensity at the shorter wavelength and $I_G$ is the intensity at the longer wavelength, and $G$ is the correction factor for the

[65] P. L. Chong, S. Capes, and P. T. Wong, *Biochemistry* **28**, 8358 (1989).

[66] P. L. Chong and P. T. Wong, *Biochim. Biophys. Acta* **1149**, 260 (1993).

[67] T. Parasassi, E. Gratton, W. M. Yu, P. Wilson, and M. Levi, *Biophys. J.* **72**, 2413 (1997).

[68] E. K. Krasnowska, E. Gratton, and T. Parasassi, *Biophys. J.* **74**, 1984 (1998).

wavelength bias of the instrument. The value of GP varies between −1 and 1. An increase in the value of GP represents a decrease in bilayer polarity and generally corresponds to an increase in order and/or microviscosity.[54,68,69]

Because of the tendency of prodan to bind weakly to the bilayer, a more complex calculation called "three-wavelength GP" is more suitable for that probe.[68] The advantage of that calculation is that it deconvolutes the contributions of bulk water and the membrane to the fluorescence spectrum. Furthermore, it allows estimation of changes in the prodan partition coefficient, a parameter that can be relevant to understanding of membrane properties.[68]

## Pyrene-Labeled Probes for Assessing Changes in Lipid Distribution

Several lines of evidence have suggested that the activity of sPLA$_2$ toward liposome membranes depends on the lateral distribution of lipid components in the bilayer.[14,15,18,31,70–72] Pyrene-labeled lipid probes have been useful in assessing that relationship.[31,72] Pyrene is excited at 344 nm and emits at 396 nm when in a monomer state. However, at high density, pyrene molecules interact and form excimers that emit optimally at 480 nm.[73] This phenomenon allows pyrene-labeled lipids to be used as probes of the state of aggregation of those lipids by examining the relative contributions of monomer and excimer fluorescence under various sets of experimental conditions. For example, pyrene-labeled fatty acid and lyso-phosphatidylcholine have both been used to hypothesize that latent rapid hydrolysis of PC LUVs by sPLA$_2$ (i.e., such as shown in Fig. 2) is initiated by lateral segregation of these lipids within the membrane.[31,72] As illustrated in Fig. 5, segregation is suggested by the sudden rise in excimer fluorescence of the probe (which immediately precedes the onset of rapid hydrolysis[31]). The subsequent decrease in excimer fluorescence is due to dilution of the probe by nascent reaction product generated by hydrolysis.

Pyrene-labeled fatty (pyrene-FA) acid is available commercially. How-ever, pyrene-labeled lysophospholipid (pyrene-lyso) must be synthesized.

[69] T. Parasassi, M. Di Stefano, M. Loiero, G. Ravagnan, and E. Gratton, *Biophys. J.* **66,** 763 (1994).

[70] W. R. Burack, A. R. Dibble, M. M. Allietta, and R. L. Biltonen, *Biochemistry* **36,** 10551 (1997).

[71] W. R. Burack, A. R. Dibble, and R. L. Biltonen, *Chem. Phys. Lipids* **90,** 87 (1997).

[72] W. R. Burack, Q. Yuan, and R. L. Biltonen, *Biochemistry* **32,** 583 (1993).

[73] R. L. Melnick, H. C. Haspel, M. Goldenberg, L. M. Greenbaum, and S. Weinstein, *Biophys. J.* **34,** 499 (1981).

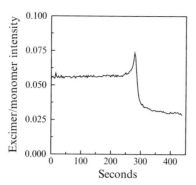

FIG. 5. Time course of pyrene-lyso ($\sim$2 mol%) excimer-to-monomer fluorescence ratio during hydrolysis of DPPC LUVs (100 $\mu M$) by sPLA$_2$ (360 n$M$ from *A. p. piscivorus* venom) at 38° with 10 m$M$ calcium. Excitation was at 344 nm, and emission was at 397 nm (monomer) and 480 nm (excimer). Adapted with permission from Ref. 31. Copyright 1996, *Am. Chem. Soc.*

We have had success with the following procedure. Pyrene-lysophosphatidylcholine is generated by hydrolysis of 1,2-bis-(1-pyrenedecanoyl)-*sn*-glycero-3-phosphocholine (Molecular Probes) with sPLA$_2$. Pyrene-labeled phospholipid is suspended (1 mg/ml) in 35 m$M$ KCl, 10 m$M$ CaCl$_2$, 3 m$M$ NaN$_3$, 55 $\mu M$ phenol red, and 0.6% Lubrol PX. Secretory PLA$_2$ is added (43 $\mu$g/ml) and the pH is adjusted to approximately pH 8 on the basis of the phenol red color. The sample is then incubated at 43° until the color changes to yellow, indicating hydrolysis of the sample (about 20 min). The suspension is lyophilized, and the lipids are extracted from the powder with methanol. Pyrene-lyso and pyrene-FA are separated by thin-layer chromatography on K6 silica gel (Whatman, Clifton, NJ) using chloroform–methanol–water (65:35:4, v/v/v) as the mobile phase. The resolved probes are visualized under ultraviolet light, scraped from the chromatography plates, extracted with methanol, and stored at $-20°$.

In experiments, the probes (either pyrene-lyso or pyrene-FA) are added directly to liposomes in the spectrofluorometer sample cell (final concentration should be about 2 mol%). Experiments are initiated after the probe intensity becomes relatively stable, indicating that steady state has been reached between the probe and the vesicles. Fluorescence is monitored during experiments at both 396 nm (monomer) and 480 nm (excimer), respectively (excitation, 344 nm). The data can be analyzed either by determining the ratio between excimer and monomer fluorescence or by calculating GP values (using wavelengths relevant to pyrene) as in Eq. (2).

Fluorescence Imaging of Giant Unilamellar Vesicles

Giant unilamellar vesicles (GUVs) offer the researcher a model membrane system with several important features. GUVs have a minimum bilayer curvature in comparison with other commonly used liposome preparations, SUVs and LUVs, that may better model cellular membranes. A second, and more important, advantage of the GUV system is the ability to study a single vesicle. Bulk liposome studies generate information about a population of enzymes and vesicles, and calculations based on these data can only reveal patterns and properties of the ensemble. As explained above, sPLA$_2$ is extremely sensitive to liposome structure and will likely be able to detect variations among vesicles within a given population. Being able to examine single vesicles directly allows one to investigate the variety of behaviors that make up the average response and potentially gain new insights into the behavior of the system.

The ability to image a single vesicle and visually inspect the interface, using GUVs, is unique in the liposome area. An image is a particularly powerful medium to convey information about a bilayer and to monitor the changes in the interface as the environment is perturbed with temperature, ionic strength, or the addition of membrane-interacting species (e.g., sPLA$_2$). Unlike electron microscopy, the sample is not fixed, allowing real-time observations, assessment of physical properties with fluorescent probes, and freedom from artifacts associated with fixation of the sample. In Fig. 6 two GUV images are presented to illustrate the impact images can have on our physical understanding of a system and to highlight vesicle heterogeneity that can occur as a result of treatment. In Fig. 6, the simple effect of cooling DPPC GUVs to below their phase transition temperature is striking. The two vesicles were cooled from 55° to below their phase

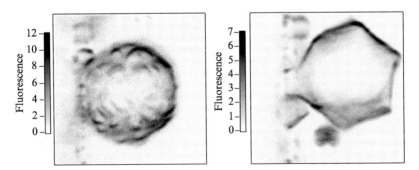

FIG. 6. Fluorescence images of two DPPC GUVs (containing laurdan) after reduction of the sample temperature from 55° to below the main lipid phase transition temperature of 42°. Temperature was decreased from 55° at moderate rates, approximately 1°/min.

transition (42°) at a moderate rate (~1°/min). The formation of lobes and flat panes in the vesicle is a result of the fluid-to-gel volume changes, the initial shape of the vesicle, and the transient presence of liquid/gel-phase coexistence. At slower cooling rates, 0.2°/min, vesicles accommodate the stresses and the nearly spherical shapes are mostly retained. It would seem reasonable to speculate that LUVs on which much sPLA$_2$ work has been carried out, also show such behavior. Surely, given the sensitivity of this enzyme to phospholipid packing, the changes in the organization of phospholipids caused by even mild changes in environmental conditions will greatly influence sPLA$_2$ attack of the surface.

The disadvantage of single vesicle work is the possibility that the event examined is a relatively rare occurrence. This can create interpretive difficulties for the researcher who is looking for the average effect, but is likely encountering a result that contradicts the intuition built through bulk study investigations. With this in mind, one must consider that an observation, no matter how rare, must follow the physics of the system and must be part of the bulk solution distribution of possible events.

The next few sections focus on the use of GUVs as bilayer models for sPLA$_2$. The application of GUVs to sPLA$_2$ is, admittedly, still in an exploratory stage of development. A more complete article on the interaction between GUVs and sPLA$_2$ is available.[74]

*Visualization of Binding*

*Labeling of Secretory PLA$_2$ with Fluorescein.* The dimeric sPLA$_2$ from *Crotalus atrox* venom (Miami Serpentarium, Punta Gorda, FL) is purified according to published procedures.[44] Fluorescein conjugates of the *C. atrox* sPLA$_2$ are prepared with fluorescein succinimidyl ester (Molecular Probes, Eugene, OR) as previously described.[75] Conjugates routinely label with an average of one fluorescein ($\varepsilon_{499} = 70,000\ M^{-1}\ cm^{-1}$ per sPLA$_2$ monomer[76]). There appear to be no significant differences in enzyme activity between the fluorescein-labeled and unlabeled sPLA$_2$ species.

*Preparation of Giant Unilamellar Vesicles.* Giant unilamellar vesicles are prepared by the electroformation procedure developed by Angelova, Dimitrov, and others.[77–79] In this technique, GUVs are formed through

[74] S. A. Sanchez, L. A. Bagatolli, E. Gratton, and T. L. Hazlett, *Biophys. J.* **82**, 2232 (2002).

[75] S. A. Sanchez, Y. Chen, J. D. Muller, E. Gratton, and T. L. Hazlett, *Biochemistry* **40**, 6903 (2001).

[76] E. G. Jablonski, L. Brand, and S. Roseman, *J. Biol. Chem.* **258**, 9690 (1983).

[77] D. S. Dimitrov and M. I. Angelova, *Bioelectrochem. Bioenerg.* **19**, 323 (1988).

[78] M. I. Angelova and D. S. Dimitrov, *Faraday Discuss. Chem. Soc.* **81**, 303 (1986).

[79] J. C. Andre, M. Niclaus, and W. R. Ware, *Chem. Phys.* **28**, 371 (1978).

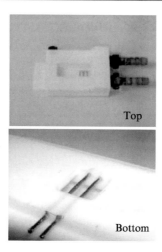

Top

Bottom

FIG. 7. GUV sample preparation chamber.

the introduction of a modulating electric field across aligned platinum wires coated with the appropriate lipid. The wires are approximately 0.5 cm apart and are immersed in water or buffer at low ionic strength. A Teflon, temperature-controlled chamber has been designed (Fig. 7) that allows a working temperature range from approximately 9 to 80°.[80] The general protocol for forming GUVs in this chamber is relatively straightforward. Lipid stock solutions, approximately 0.2 mg/ml in chloroform, are spread on each of the two platinum wires under a stream of dry $N_2$, and the chamber is placed under a vacuum for approximately 1 h to remove any remaining solvent. The chamber and the buffer (routinely in 0.5 m$M$ Tris, pH 8.0) are equilibrated to temperatures above the lipid mixture phase transition(s), typically 10° above the highest phase transition temperature for the given lipid mixture. Once equilibrated, buffer is added to cover the platinum wires and a low-frequency AC field (a 2-V sinusoidal wave function with a frequency of 10 Hz) is applied for approximately 90 min. Experiments on GUVs are conducted with an inverted fluorescence microscope (Axiovert 35; Zeiss, Thornwood, NY) (see below). A charge-coupled device (CCD) video camera (CCD-Iris; Sony) is used to monitor GUV formation and to select the target vesicle. For experiments involving laurdan and prodan, fluorophore is added to the sample chamber after the vesicles are formed. The final bilayer phospholipid:fluorophore ratio is kept greater than 100:1. In the case of rhodamine-labeled phosphatidylethanolamine,

[80] L. A. Bagatolli and E. Gratton, *Biophys. J.* **78,** 290 (2000).

the fluorescent phospholipid is premixed in chloroform with the primary lipid(s) at 0.5 mol% and spread onto the sample chamber wires before GUVs are formed. A more detailed description of this technique can be found in Bagatolli.[81]

*Two-Photon Fluorescence.* Images are collected on a scanning two-photon fluorescence microscope[82] using an LD-Achroplan × 20 long working distance air objective with an NA of 0.4 (Zeiss, Homedale, NJ). In this case, an air objective is useful to thermally isolate the sample. The long working distance is also necessary to provide proper focus at the platinum wires that are positioned slightly above the bottom cover slip in the present chamber design. A mode-locked titanium-sapphire laser (Mira 900; Coherent, Palo Alto, CA) set to 780 nm, pumped by a frequency-doubled Nd:vanadate laser (Verdi; Coherent), is used as the two-photon excitation light source. A galvanometer-driven $x$–$y$ scanner is positioned in the excitation path (Cambridge Technology, Watertown, MA). The samples receive from 5 to 12 mW of 780-nm excitation light, and a frame rate of 9 s/frame is used to acquire the 256 × 256 pixel images. Images are collected as single frames or, for increased image quality, as averages of multiple scans. The fluorescence emission is observed between 350 and 600 nm using a broad, bandpass filter (BG39 filter; Chroma Technology, Brattleboro, VT). A miniature photomultiplier (R5600-P; Hamamatsu, Bridgewater, NJ) is used for light detection in the photon-counting mode.

*Binding Images.* A dramatic image illustrating the preference of *C. atrox* sPLA$_2$ for liquid crystalline-phase phospholipids is shown in Fig. 8. In this case, a GUV is formed with coexisting gel and liquid crystalline phases by using an equimolar mixture of PC species with differing chain lengths (see legend to Fig. 8 for details). Fluorescein-labeled sPLA$_2$ is added and decorates the surface of the GUVs. Immediate addition of prodan, which quickly associates with liquid crystalline phospholipids,[68] demonstrates that sPLA$_2$ decorates regions in the liquid crystalline phase. Unlike many biochemical studies, interpretation of the images is unambiguous.

One last point that should be made is that not every vesicle in a given preparation shows obvious sPLA$_2$ binding, an observation that is likely associated with subtle differences among GUVs. Certainly, the binding of sPLA$_2$ to the GUV membrane surface under the low lipid concentrations used in these studies is relatively weak, and we may be able to detect binding only under optimal conditions of the bilayer structure. The vesicles

[81] L. A. Bagatolli, S. A. Sanchez, T. Hazlett, and E. Gratton, *Methods Enzymol.* **360,** 481, (2003).
[82] P. T. C. So, T. French, W. M. Yu, K. M. Berland, C. Y. Dong, and E. Gratton, *Bioimaging* **3,** 49 (1995).

must be scanned many times in order to see clearly the association above the fluorescence of the free enzyme in solution, and even moderate binding of the enzyme is difficult to detect.

## Visualization of Hydrolysis

The primary effect observed during hydrolysis of GUVs is a steady shrinking of the vesicle as hydrolysis proceeds. The reduction of vesicle size indicates lipid loss from the bilayer. The hydrolysis products have limited solubilities in the aqueous phase, as given by their individual partition co-efficients, and it might be assumed that the products essentially remain in the bilayer. However, GUVs are formed in 4 ml of buffer from 1.2 nmol of lipid. The volume of the aqueous phase relative to the lipid phase is extremely large, which results in a large desorption of products from the bilayer. In studies involving SUVs or LUVs, the lipid concentrations are generally three orders of magnitude greater than in the GUV measurements, and a sensible concentration of the hydrolysis products would be expected to remain in the bilayer. It is important to recognize this difference between the more standard liposome studies and experiments with GUVs when interpreting results.

An interesting issue in sPLA$_2$ research has been the relationship between the interface lipid packing, the quality of the interface, and sPLA$_2$ binding and hydrolysis. As shown in Fig. 8, *C. atrox* sPLA$_2$ appears to bind

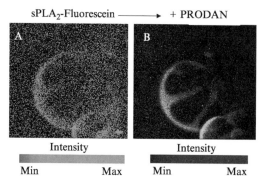

FIG. 8. Association of sPLA$_2$ (from *C. atrox* venom) with GUVs composed of coexisting liquid crystalline- and gel-phase domains. GUVs were formed from 1,2-dilauroyl-*sn*-glycero-3-phosphocholine and diarachidoyl-*sn*-glycero-3-phosphocholine (1:1, mole ratio) equilibrated at 48.1° in the presence of 0.4 m*M* barium (no calcium). (A) Binding of fluorescein-labeled sPLA$_2$ (from *C. atrox* venom, 30 n*M*) to GUVs; (B) same vesicles after prodan addition. Prodan selectively labels liquid crystalline domains. Redrawn from Ref. 74 with permission from the Biophysical Society.

preferentially to fluid domains. An obvious question is whether these liquid domains are where hydrolysis also occurs, or whether the enzyme shows preferential activity at the borders between domains or in the solid-phase domains at concentrations below our detection limits. In Fig. 9, hydrolysis of a mixed lipid GUV, equilibrated at a temperature at which gel and liquid crystalline phases coexist, is displayed. Rhodamine-PE has been added to highlight liquid crystalline domains in this binary lipid mixture.[80] On addition of active sPLA$_2$, the vesicle shrinks but phospholipids within the liquid crystalline (light) regions are lost preferentially. This point can be demonstrated more clearly in a plot of the observed GUV surface area, the solid domain, and the liquid domain as a function of time (Fig. 10). The observation that the liquid domain is hydrolyzed while the gel domain is relatively unchanged is clear. The decrease in the liquid domain appears to be, under these conditions, zero order and matches the kinetics observed with pure liquid GUVs under similar conditions. Combined with the binding results, it appears reasonable to conclude that the borders between these large domains do not appear to create unusually active regions for sPLA$_2$ attack in GUVs.

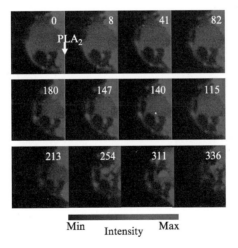

FIG. 9. Fluorescence images of GUVs with coexisting liquid and gel domains (same lipids as in Fig. 8) with 0.5 mol% Lissamine-rhodamine B-1,2-dihexanoyl-*sn*-glycero-3-phosphoethanolamine at (selectively labeling liquid crystalline domains) 46° in the presence of 0.75 m$M$ calcium. Secretory PLA$_2$ (from *C. atrox* venom, 750 n$M$) was added where shown. Numbers refer to the time in seconds after sPLA$_2$ addition. Redrawn from Ref. 74 with permission from the Biophysical Society.

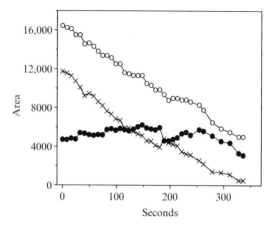

FIG. 10. Time course of changes in the visible surface area of a GUV (same lipids as in Fig. 8) after sPLA$_2$ addition (obtained from images such as those shown in Fig. 9). (○) Total area; (●) gel-phase area; (×) liquid crystalline area. Redrawn from Ref. 74 with permission from the Biophysical Society.

## Visualization of Changes in Membrane Properties

As in the bulk studies described previously, it is useful with single vesicle images to assess changes in membrane properties either as a consequence of experimental manipulations or during liposome hydrolysis. In choosing which probe to use with GUVs, laurdan has several advantages over prodan. By virtue of its acyl chain, laurdan is aligned along the phospholipid acyl chain and has been found to label solid and liquid lipid phases equally. Background fluorescence using laurdan is minimal because it has a greater partitioning into the lipid phase and a lower solution fluorescence than prodan.

Matched images at the two wavelengths required for calculation of GP are collected simultaneously, using a two-channel emission attachment. To acquire images such as those shown in Fig. 11, the sample fluorescence is split into green and blue channels with a Chroma Technology 470DCXR-BS dichroic beam splitter in the emission path. Interference filters, an Ealing 490 or an Ealing 440, are placed in the appropriate emission channels to further isolate the green and blue parts of the emission spectrum. The correction factor, $G$ [see Eq. (2)], is determined at each pixel through the use of a known GP standard (laurdan in dimethyl sulfoxide). The separate green and blue images are then recombined using Eq. (2) to form the sample GP.

In Fig. 11, images of a GUV (DMPC, 26°) in the absence of sPLA$_2$ (Fig. 11A), an early stage of sPLA$_2$-dependent hydrolysis (Fig. 11B), and

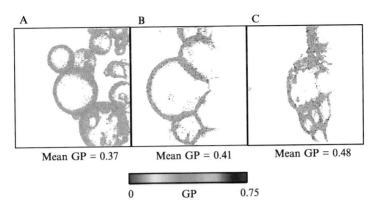

FIG. 11. Laurdan GP images of DMPC GUVs at $26°$. (A) Image collected before sPLA₂ addition (from *C. atrox* venom, 750 n$M$) in the presence of 0.1 m$M$ calcium. (B and C) Images illustrate the same vesicles at early and late stages of hydrolysis, respectively. The mean GP value for each image is noted.

a late hydrolysis stage (Fig. 11C) are shown. The average membrane GP increased from 0.37 to 0.48, indicating an overall decrease in membrane polarity. These data agree with the bulk solution studies.[14,31,59] Moreover, in the GUV system, we can visualize specific membrane changes at the vesicle surface. The GP images in Fig. 11 show the progressive formation of small high GP of, presumably, small product-rich regions formed as the hydrolysis proceeds. The elevated GP probably represents an increase in lipid order due to domain formation and/or interdigitation of hydrolysis products among the surrounding phospholipids. These effects in GUVs are probably due mostly to fatty acid rather than lysophosphatidylcholine, given the partition coefficients of these species and the low lipid concentrations used (see above).

The appearance of fatty acid domains would agree with work reported on monolayers by Grainger *et al.*[83] In their studies the authors used surface pressure to create mixed domain monolayers and explored the interaction, binding, and hydrolysis with *N. naja* sPLA₂. They concluded that the fatty acid released through hydrolysis interacts with the sPLA₂ to form protein-associated fatty acid-rich domains. The GP domains are consistent with this hypothesis, but we have not been able to identify preferential binding of the *C. atrox* sPLA₂ to these regions.[74]

[83] D. W. Grainger, A. Reichert, H. Ringsdorf, and C. Salesse, *Biochim. Biophys. Acta* **1023,** 365 (1990).

## Concluding Remarks

Several of the techniques described here have also been applied to studies of the relationship between membrane structure and susceptibility to $sPLA_2$ in human erythrocytes.[84] Assays of hydrolysis by thin-layer chromatography or the fatty acid-binding proteins and the use of laurdan fluorescence and the two-photon microscopy technique have provided promising results. The simplicity of erythrocytes compared with nucleated cells and the considerable background knowledge available make erythrocytes a good choice for efforts to apply lessons from liposomes to biological membranes. The future of combined approaches such as those described here with liposome membranes of various compositions and simple biological membranes should provide answers to important questions regarding the selective action of $sPLA_2$ toward some, but not all, cell membranes. Furthermore, these techniques should also prove helpful toward studies with other proteins that bind reversibly to membranes.

## Acknowledgments

This work was supported by NSF Grant MCB 9904597 to Brigham Young University and by NIH Resource Grant RR03155 to the Laboratory for Fluorescence Dynamics. We appreciate the assistance of Rebekah Vest in preparing the manuscript.

[84] S. K. Smith, A. R. Farnbach, F. M. Harris, A. C. Hawes, L. R. Jackson, A. M. Judd, R. S. Vest, S. Sanchez, and J. D. Bell, *J. Biol. Chem.* **276,** 22732 (2001).

## [3] Interaction of Proteins with Liposomes as Detected by Microelectrophoresis and Fluorescence

*By* OLAF ZSCHÖRNIG, FRANK OPITZ, and JENS PITTLER

### Introduction

Protein-induced fusion is an important event in intra- and intercellular processes such as viral entry, and exo- and endocytotic transport. In this context the binding of proteins to phospholipid (PL) membranes is of crucial interest. Several water-soluble proteins were used to investigate the mechanisms of their binding to PL vesicles under different circumstances, subsequently leading to phospholipid vesicle aggregation and fusion. Lysozyme,[1–4] cytochrome *c*,[5] annexins,[6] albumin,[7–9] myelin basic protein,[10,11] and clathrin[12] belong to this group of proteins. Arvinte *et al.*[13,14] coupled lysozyme to neutral phospholipid vesicles and measured their fusion with erythrocytes at low pH.

In this chapter we describe selected methods to monitor the interaction of annexin V with phospholipid membranes. Annexin–phospholipid bilayer interactions have been considered a good example of the electrostatic binding of peripheral proteins to biological membranes. As a consequence the mechanism of these associations has been studied intensively *in vitro*, using both lipid monolayers and lipid vesicles as model systems.

The annexins are a family of highly conserved $Ca^{2+}$-dependent membrane-binding proteins that are present in a variety of species and cell types (for review see Refs. 15–17). They share structural homology in a conserved core consisting of 4 domains of about 70 amino acids each, containing the annexin consensus sequence. On the other hand, the N-terminal tails of the individual annexins differ strongly in length and amino acid sequence.[18] The crystal structure of annexin V and its $Ca^{2+}$-binding sites has been described.[19] Annexins are clearly distinguished from the well-known "EF-hand" proteins, such as calmodulin and troponin G, in the sequence and geometry of their calcium-binding sites as well as in their affinity for calcium.

[1] E. Posse, B. F. de Arcuri, and R. D. Morero, *Biochim. Biophys. Acta* **1193,** 101 (1994).

[2] E. Posse, V. A. Lopez, B. F. de Arcuri, R. N. Farias, and R. D. Morero, *Biochim. Biophys. Acta* **1024,** 390 (1990).

[3] G. F. Vechetti, B. F. de Arcuri, E. Posse, J. L. Arrondo, and R. D. Morero, *Mol. Membr. Biol.* **14,** 137 (1997).

[4] K. Arnold, D. Hoekstra, and S. Ohki, *Biochim. Biophys. Acta* **1124,** 88 (1992).

[5] T. J. Pinheiro and A. Watts, *Biochemistry* **33,** 2451 (1994).

[6] G. Kohler, U. Hering, O. Zschornig, and K. Arnold, *Biochemistry* **36,** 8189 (1997).

[7] S. Schenkman, P. S. Araujo, R. Dijkman, F. H. Quina, and H. Chaimovich, *Biochim. Biophys. Acta* **649,** 633 (1981).

[8] S. Schenkman, D. A. Soares, A. Sesso, F. H. Quina, and H. Chaimovich, *Chem. Phys. Lipids* **28,** 165 (1981).

[9] L. A. Garcia, S. Schenkman, P. S. Araujo, and H. Chaimovich, *Braz. J. Med. Biol. Res.* **16,** 89 (1983).

[10] Y. Cajal, J. M. Boggs, and M. K. Jain, *Biochemistry* **36,** 2566 (1997).

[11] M. B. ter Beest and D. Hoekstra, *Eur. J. Biochem.* **211,** 689 (1993).

[12] S. Maezawa, T. Yoshimura, K. Hong, N. Duzgunes, and D. Papahadjopoulos, *Biochemistry* **28,** 1422 (1989).

[13] T. Arvinte, K. Hildenbrand, P. Wahl, and C. Nicolau, *Proc. Natl. Acad. Sci. USA* **83,** 962 (1986).

[14] T. Arvinte, P. Wahl, and C. Nicolau, *Biochim. Biophys. Acta* **899,** 143 (1987).

[15] V. Gerke and S. E. Moss, *Physiol. Rev.* **82,** 331 (2002).

[16] P. Raynal and H. B. Pollard, *Biochim. Biophys. Acta* **1197,** 63 (1994).

[17] R. D. Burgoyne and M. J. Geisow, *Cell Calcium* **10,** (1989).

[18] M. J. Geisow, *FEBS Lett.* **203,** 99 (1986).

[19] R. Huber, M. Schneider, I. Mayr, J. Romisch, and E. P. Paques, *FEBS Lett.* **275,** 15 (1990).

For the annexins several functions are discussed (for review see Moss *et al.*[20]). All these functions are based on the ability of the annexins to bind to negatively charged phospholipid membranes. Annexin V participates in the calcification of cartilage. It is incorporated in the membrane of the matrix vesicles, where it exhibits $Ca^{2+}$ channel activity. The structure of the channel is as yet unclear. $Ca^{2+}$ influx induced by several annexins has been studied.[21–23] Annexin V was found to behave like an integral protein.[24,25] For explanation it has been discussed whether one annexin molecule spans the membrane after conformational change, electroporates the membrane,[26] or may form hexamers.[27]

Annexin V has been shown to bind strongly to negatively charged membrane surfaces in the presence of multivalent cations via electrostatic interactions, and it has been established that such binding may have both a structural and a functional role, mediated by effects on the conformation of the protein as well as on the structure of the membrane surface. Spectroscopic studies have shown that the binding of annexin V induces both protein structural changes involving the immediate environment, as well as extensive alterations in the conformation of the protein backbone, including loosening and destabilizing of the overall protein structure. It has also been demonstrated that annexin V binding alters the structure of the lipid phase of negatively charged membranes. Although the electrostatic interactions between negatively charged membranes and annexin V are well documented and have been discussed extensively within the context of

[20] S. E. Moss, H. T. Haigler, D. D. Schlaepfer, N. D. Horseman, F. Russo Marie, V. Gerke, K. Weber, R. D. Burgoyne, C. E. Creutz, C. Comera, M. Junker, N. G. Kambouris, J. R. Klein, M. R. Nelson, P. Rock, S. L. Snyder, W. Wang, H. B. Pollard, E. Rojas, N. Merezhinskaya, G. A. Kuijpers, M. Srivastava, Z. Y. Zhang-Keck, A. Shirvan, A. L. Burns, R. Huber, R. Berendes, A. Burger, H. Luecke, A. Karshikov, J. R. Dedman, M. A. Kaetzel, R. Hauptmann, C. P. Reutelingsperger, M. Maki, Y. Shidara, and H. Sato, "The Annexins," 1st ed. Portland Press, London, 1992.
[21] M. Diaz Munoz, S. L. Hamilton, M. A. Kaetzel, P. Hazarika, and J. R. Dedman, *J. Biol. Chem.* **265**, 15894 (1990).
[22] E. Rojas, H. B. Pollard, H. T. Haigler, C. Parra, and A. L. Burns, *J. Biol. Chem.* **265**, 21207 (1990).
[23] H. B. Pollard, H. R. Guy, N. Arispe, M. de la Fuente, G. Lee, E. M. Rojas, J. R. Pollard, M. Srivastava, Z. Y. Zhang-Keck, and N. Merezhinskaya, *Biophys. J.* **62**, 15 (1992).
[24] B. R. Genge, L. N. Wu, H. D. Adkisson, and R. E. Wuthier, *J. Biol. Chem.* **266**, 10678 (1991).
[25] R. Bianchi, I. Giambanco, P. Ceccarelli, G. Pula, and R. Donato, *FEBS Lett.* **296**, 158 (1992).
[26] A. Karshikov, R. Berendes, A. Burger, A. Cavalie, H. D. Lux, and R. Huber, *Eur. Biophys. J.* **20**, 337 (1992).
[27] H. Luecke, B. T. Chang, W. S. Mailliard, D. D. Schlaepfer, and H. T. Haigler, *Nature* **378**, 512 (1995).

specific mechanisms of annexin V binding, there are results that indicate the existence of hydrophobic interactions of the protein with a lipid bilayer. Strong experimental support for a hydrophobic component in the interaction has been provided by the finding that all bound annexin V cannot dissociate completely from such membranes, either by increasing the ionic strength of the buffer solution, or by dilution of the bulk protein. This effect occurs despite the fact that the initial binding itself is sensitive to both the ionic strength of the solution and the amount of charge in the lipid membrane, consistent with the electrostatic nature of the interaction.

In summary, there appears to be a general consensus on two points:

1. The initial binding step of annexin V to the surface of a lipid membrane is governed by electrostatics.

2. Subsequent to the initial stage of binding, changes in the structure of the protein and lipid components of the membrane occur, leading to a new and more complex situation involving penetration of annexin V into the lipid phase.

## Preparation of Annexin V

Annexin V is isolated from fresh or frozen bovine liver by the modified procedure of Bandorowicz et al.[28] The tissue of the bovine liver is cut into small pieces, which are homogenized in washing buffer solution $A_1$ [20 m$M$ Tris-HCl (pH 7.5), 100 m$M$ NaCl, 0.2 m$M$ phenylmethylsulfonyl fluoride, aprotinin (5 TIU/liter), 1 m$M$ CaCl$_2$, 1 m$M$ NaN$_3$, and 1% (v/v) Triton X-100] with an UltraTurrax T25 homogenizer (Janke & Kunkel, IKA Labortechnik, Staufen, Germany). For disruption of cells a Potter S-homogenizer (B. Braun Melsungen, Melsungen, Germany) at 0° and 900 min$^{-1}$ is used. Triton X-100 is necessary to bring the membrane-bound annexin fraction into solution. The homogenate is centrifuged at 15,000 g and 4° for 15 min. This step is followed by resuspension of the sediment and centrifugation of the resuspension at 15,000 g (4°, 10 min) three times repeated (washing steps). After the last washing step the suspension is centrifuged at 50,000 g (4°, 20 min). The obtained sediment is gently stirred overnight in washing buffer solution $A_2$ [1 m$M$ CaCl$_2$ is replaced by 5 m$M$ ethylene glycol-bis($\beta$-aminoethyl ether)-$N,N,N',N'$-tetraacetic acid]. The samples are centrifuged for 1 h at 80,000 g (Optima L-60 preparative ultracentrifuge with rotor type 80Ti; Beckman, Fullerton, CA) at 4°. After these procedures proteins are separated by pH or ammonium sulfate

[28] J. Bandorowicz, S. Pikula, and A. Sobota, *Biochim. Biophys. Acta* **1105,** 201 (1992).

precipitation and by ion-exchange column chromatography (NaCl gradient, 0 to 0.5 $M$; flow rate, 1 ml/min) with fast protein liquid chromatography (FPLC) equipment (Pharmacia, Uppsala, Sweden).

## Identification of Annexin V

The purity of the resulted protein fraction after the ion-exchange chromatography is checked by sodium dodecyl sulfate–polyacrylamide gel electrophoresis according to Schagger and von Jagow[29] in the presence of 2% (v/v) 2-mercaptoethanol, stained with Coomassie blue. With monoclonal antibody probes from ICN Biomedicals (Costa Mesa, CA) against annexin V the identity is checked by Western blot.

## Buffers

Experiments are performed in buffer solutions of 10 m$M$ citrate (pH 4.0), 10 m$M$ 2-($N$-morpholino)ethansulfonic acid (MES; pH 5.5), or 10 m$M$ $N$-(2-hydroxyethyl)piperazine-$N'$-(2-ethansulfonic acid) (HEPES; pH 7.4). All solutions contain 0.1 $M$ NaCl.

## Phospholipid Vesicle Preparation

Phosphatidylserine (PS) and phosphatidic acid (PA) are purchased from Avanti Polar Lipids (Alabaster, AL). The purity of the phospholipids is checked by thin-layer chromatography. 3-Palmitoyl-2-(1-pyrenedecanoyl)-L-α-phosphatidylcholine (Pyr-PC) is obtained from Molecular Probes (Eugene, OR). Multilamellar phospholipid vesicles (MLVs) are prepared by the method of Bangham.[29a] The lipid is initially dried from chloroform, subsequently dispersed in buffer solution (10 m$M$ HEPES and 100 m$M$ NaCl, adjusted to pH 7.4), and shaken at a temperature above the gel-to-liquid crystalline transition temperature of the phospholipids for 10 min. Large unilamellar vesicles (LUVs) are prepared by five freeze–thawing cycles of MLVs and subsequent extrusion (five times) through 0.1-$\mu$m pore size filter membranes (Nuclepore, Pleasanton, CA), using an extruder (Lipex Biomembranes, Vancouver, BC, Canada) at 30°. Small unilamellar vesicles are obtained by sonication of the multilamellar lipid suspension in buffer. The phospholipid concentration in the vesicle suspension is measured according to Chen et al.[30]

---

[29] H. Schagger and G. von Jagow, *Anal. Biochem.* **166**, 368 (1987).
[29a] A. D. Bangham, M. W. Hill, and N. G. A. Miller, *Methods Membrane Biol.* **1**, 1–68.
[30] P. S. Chen, T. Y. Toribara, and H. Warner, *Anal. Chem.* **28**, 1756 (1956).

Microelectrophoresis

Electrophoretic mobilities of liposomes can be determined by a variety of methods. The oldest, but still used, equipment consists of microscope-based electrophoresis systems. The basis of these machines is the microscopic observation of the migration of phospholipid vesicles (giant or multilamellar large vesicles). The time $t$ is measured manually or automatically for separate vesicles passing a distinct distance $s$, when an electric field $E$ is applied. The field strength depends on the quotient of the applied voltage $U$ and the distance $d$ between the electrodes.

The electrophoretic mobility $u$ is calculated by

$$u = v/E \tag{1}$$

where $v$ is the quotient $s/t$ for each vesicle.

The $\zeta$ potential as the electrical potential of the hydrodynamic plane of shear can be calculated from

$$\zeta = u\eta/\varepsilon_0\varepsilon_r \tag{2}$$

where $\eta$ is the viscosity and $\varepsilon_0$ and $\varepsilon_r$ are the absolute and relative dielectric constants of the medium, respectively.

These formulas can be used for the determination of large phospholipid vesicles (diameter $> 500$ nm). For determination of the electrophoretic mobilities of smaller particles, light-scattering methods are used. In these cases, calculation of the zeta potentials must be modified because of relaxation effects.

The electrophoretic mobility of PS LUVs is measured with a Zetasizer 4 device (Malvern Instruments, Malvern, UK). The modulation frequency is 250 Hz and the electrode current is 7.5 mA. The surface potential of the vesicles is estimated by means of

$$\Psi_0 = (2RT/F) \cdot \ln\{[1 + q \exp(r/I_D)]/[1 - q \exp(r/I_D)]\} \tag{3}$$

with $r \approx 0.2$ nm and

$$q = [\exp(F\zeta/2RT) - 1]/[\exp(F\zeta/2RT) + 1] \tag{4}$$

$F$, $R$, and $T$ denote the gas constant, the Faraday constant, and the temperature, respectively. The Debye length and the ionic strength are given by $I_D = (\varepsilon_0\varepsilon_r RT/2IF^2)^{0.5}$ and $I = 0.5 \sum z_i^2 c_i$ (summation runs over the charge, $z_i$, and the concentration, $c_i$, of all ions present in the solution), respectively. Finally, the apparent surface charge density is given by

$$\sigma_{ap} = \{2\varepsilon_0\varepsilon_r RT\ c_i[\exp(-z_i F\Psi_0/RT) - 1]\}^{0.5} \tag{5}$$

In a first-order approximation the apparent charge density can be written as the sum of the contribution of the PS moieties, the bound $Ca^{2+}$ ions, and the bound annexin V molecules:

$$\sigma_{ap} = \sigma_{PS} + \sigma_{av} + \sigma_{Ca}$$

$\sigma_{PS}$ was determined from the $\zeta$ potential of PS LUVs in the absence of $Ca^{2+}$ ions.

Consequently, the increment of surface charge density after addition of $Ca^{2+}$ is $\sigma_{Ca} = \sigma_{ap} - \sigma_{PS} = 2e[Ca^{2+}]_{bound}/(0.5[PS]A_L)$. After rearrangement one obtains an estimate of the amount of calcium cations that bind to the outer surface of the vesicles, $[Ca^{2+}]_{bound}$. The area per lipid in the membrane plane is set to $A_L = 0.7$ $nm^2$. The increment of surface charge due to the protein is related to the protein concentration by $\sigma_{av} \approx e\, z_{ax}$ [annexin V]/(0.5[PS]$A_L$), where $z_{ax}$ denotes its effective charge. The analysis in terms of the Helmholtz–Smoluchowski equation predicts that the electrophoretic mobility of a particle should be independent on its size and shape for all charge densities. This is approximately the case for phospholipid vesicles in decimal molar salt solutions.[31] Consequently, our analysis is expected to be independent of aggregation and fusion of vesicles.

In the absence of annexin V the surface charge of the negative PS vesicles increases with increasing $Ca^{2+}$ concentration owing to the binding of cations. The change of the surface charge density yields the amount of calcium that binds to the lipid.

At neutral pH, annexin V has no significant effect on the electrophoretic mobility of PS LUVs in the absence of $Ca^{2+}$. We conclude that there is no annexin V binding to PS vesicles, because the protein is effectively negatively charged at neutral conditions, and thus its interaction with the lipid would decrease the apparent surface charge, in contrast to the observations (Fig. 1). On the contrary, continuous increase of $\sigma_s$ with increasing annexin V concentration is measured under acidic conditions (pH 4.0). This tendency reflects the binding of the positively charged protein to the vesicles, owing to the hydrophobic effect. Addition of $Ca^{2+}$ causes a steeper increase in the $\zeta$ potential in the presence of the protein when compared with the sample without annexin V (at neutral pH) (Fig. 2). If one interprets this effect exclusively in terms of surface electrostatics, then it indicates that annexin V promotes $Ca^{2+}$ binding to PS LUVs. Note that annexin V carries about 13 negative charges at pH 7.4 and potentially binds 5–12 $Ca^{2+}$ ions.[19] After maximum $Ca^{2+}$ binding, annexin V becomes

[31] M. Eisenberg, T. Gresalfi, T. Riccio, and S. McLaughlin, *Biochemistry* **18**, 5213 (1979).

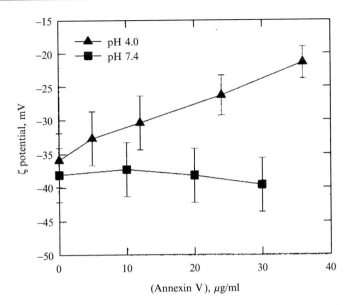

Fig. 1. Influence of pH (pH 4.0, ▲; pH 7.4, ■) and annexin V concentration on the $\zeta$ potential of PS LUVs as determined by laser electrophoresis.

positively charged ($z \approx +10$) and thus complete binding of the protein to the vesicles would increase their apparent surface charge density by $\sigma_{av} = 0.01$ A·s/m$^2$. This value agrees with the measured difference of $\sigma_{ap}$ between PS LUVs in the presence and absence of annexin V. Hence, the respective change in electrophoretic mobility can be explained by complete binding of the protein to the lipid.

## Fluorescence Measurements

All fluorescence assays are carried out on an LS-50B spectrofluorimeter (PerkinElmer, Beaconsfield, UK) or on a Spex FluoroMax-2 spectrofluorimeter (Jobin Yvon, Edison, NJ). Quartz cuvettes equipped with a magnetic stirrer are used. Vesicles are suspended in 1.0 ml of buffer solution, reaching a final concentration of 80 $\mu M$ lipid. The recording starts after the system reaches equilibrium (about 3 to 5 min). Appropriate aliquots of aqueous protein stock solutions are added to the vesicle suspension and stirred continuously. The experiments are performed at 37°. The reproducibility of different runs is within an error of 5 to 10%.

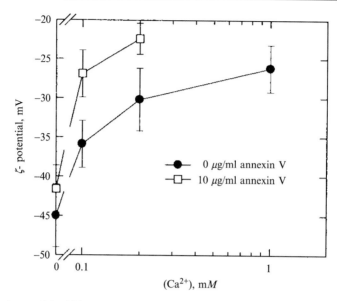

Fig. 2. $\zeta$ potentials of PS LUVs at pH 7.4 in the absence (●) and presence (□) of annexin V (10 $\mu$g/ml) and in the presence of increasing $Ca^{2+}$ concentration.

## Protein Binding to Vesicles

Pyr-PC is mixed with phospholipids before evaporation at a concentration of 10% (mol/mol). Protein adsorption is evaluated by measuring the fluorescence resonance energy transfer (FRET) from protein tryptophan (Trp) residues to the pyrene fluorophore of Pyr-PC.[4] Emission spectra are recorded from 300 to 560 nm. The extent of the FRET is calculated as the pyrene fluorescence intensity integrated over the wavelength range 370–550 nm, corrected as follows:

$$\mathrm{FRET} = \frac{I_{\mathrm{Pyr}} - I_{\mathrm{Pyr}}^0}{I_{\mathrm{Pyr}}^0} \qquad (6)$$

where $I_{\mathrm{Pyr}}$ is the pyrene fluorescence intensity and $I_{\mathrm{Pyr}}^0$ is the initial pyrene fluorescence in the absence of any tryptophan. Tryptophan fluorescence is not excited at its absorption maximum, but at 290 nm in order to minimize initial pyrene fluorescence.

Annexin V contains only one tryptophan residue in the peptide chain. Trp-187 is situated close to the $Ca^{2+}$-binding face of the molecule.[16] We use the chain-labeled Pyr-PC, the fluorescent moiety of which is

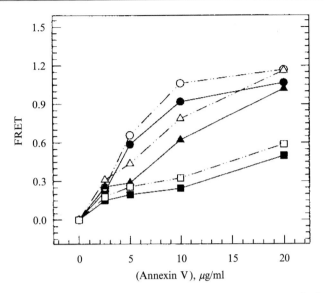

FIG. 3. Influence of annexin V on the fluorescence resonance energy transfer from annexin V Trp-187 to Pyr-PC at different pH values (pH 4.0, ○; pH 5.0, △; pH 7.4, □) in the absence (solid symbols) and in the presence of 40 $\mu M$ Ca$^{2+}$ (open symbols). The liposomes were composed of Pyr-PC and PA (1:9, mol/mol); the total lipid concentration was 100 $\mu M$.

preferentially located in the hydrophobic region of the membrane.[32] The FRET from Trp-187 to Pyr-PC provides a measure of the binding of annexin V to the bilayer because the corresponding Förster distance is 27–31 Å.[33]

At pH 7.4 only a small FRET occurrs after addition of annexin V to PA LUVs in the absence of Ca$^{2+}$ (Fig. 3). However, the subsequent addition of 40 $\mu M$ CaCl$_2$ increases the FRET about twice. Blackwood and Ernst[34] have reported a Ca$^{2+}$ concentration of 0.95 $\mu M$ for half-maximum binding of annexin V to PA. Therefore, 40 $\mu M$ CaCl$_2$ is sufficient to induce the binding. Lowering the pH to the isoelectric point of annexin V (pH 4.8) leads to a significantly stronger FRET, which likewise is increased by addition of 40 $\mu M$ CaCl$_2$. At pH 4.0 the FRET is far stronger than at pH 7.4 and pH 5.0 in the absence of Ca$^{2+}$. Addition of 40 $\mu M$ CaCl$_2$ has no increasing effect at pH 4.0. The binding of annexin V to PS LUVs is comparable to the binding of annexin V to PA LUVs at pH 7.4. At pH 4.0 the

[32] J. Vanderkooi, S. Fischkoff, B. Chance, and R. A. Cooper, *Biochemistry* **13**, 1589 (1974).

[33] G. E. Dobretsov, N. K. Kurek, V. N. Machov, T. I. Syrejshchikova, and M. N. Yakimenko, *J. Biochem. Biophys. Methods* **19**, 259 (1989).

[34] R. A. Blackwood and J. D. Ernst, *Biochem. J.* **266**, 195 (1990).

FRET is much stronger than at pH 7.4; however, it is not as strong as in the case of PA, and is slightly $Ca^{2+}$ dependent.

## Fluorescence Correlation Spectroscopy

For the fluorescence correlation measurements annexin V is labeled with tetramethylrhodamine (FluoReporter F-6163; Molecular Probes). Bovine brain phosphatidylserine (PS), egg phosphatidylcholine (PC), and Lissamine B sulfonyl phosphatidylethanolamine (Rh-PE) are purchased from Avanti Polar Lipids.

### Preparation of Tetramethylrhodamine-Labeled Annexin V

Before the labeling is performed the buffer solution of annexin V must be exchanged for phosphate-buffered saline (pH 7.4) by dialysis. The labeling is achieved by using the information included with the labeling kit F-6163. The protein concentration after chromatography is measured according to Bradford.[35]

### Principles of Fluorescence Correlation Spectroscopy

Fluorescence correlation spectroscopy (FCS) is based on a fluctuation analysis of fluorescence intensity to detect and characterize fluorophores at the single-molecule level in solution. FCS facilitates the measurement of two important physical parameters: the average number of molecules in the detection volume and the translational diffusion time of the molecules through the open volume of detection. Detailed theoretical explanations and potential applications have been discussed.[36–38] The observed fluorescence fluctuations $\delta F(t)$ when correlated with fluorescence fluctuations at time $t + \tau$ yield the normalized autocorrelation function $G(\tau)$ of fluctuations in the signal measured, which is given by

$$G(\tau) = \frac{\langle \delta F(t) \cdot \delta F(t + \tau) \rangle}{\langle F \rangle^2} \tag{7}$$

where the brackets describe the time average and $\langle F \rangle$ is the mean fluorescence signal with $\langle \delta F(t) \rangle = 0$. The autocorrelation function $G(\tau)$ for Brownian motion of fluorescent molecules in the volume of detection is described as follows:

---

[35] M. M. Bradford, *Anal. Biochem.* **72**, 248 (1976).
[36] J. Widengren and R. Rigler, *Cell Mol. Biol. (Noisy le grand)* **44**, 857 (1998).
[37] R. Rigler, *J. Biotechnol.* **41**, 177 (1995).
[38] T. Wohland, R. Rigler, and H. Vogel, *Biophys. J.* **80**, 2987 (2001).

$$G(\tau) = \frac{1}{N} \left( \frac{1}{1 + \frac{\tau}{\tau_D}} \cdot \frac{1}{\sqrt{1 + \frac{\omega_0^2 \cdot \tau}{z_0^2 \cdot \tau_D}}} \right) \qquad (8)$$

where $N$ is the number of fluorescent molecules in the detection volume element of the laser beam defined by a radius $\omega_0$ and a length $2z_0$, and $\tau_D$ is the diffusion time ($\tau_D = \omega_0^2/4D$, where $D$ is the diffusion coefficient). The autocorrelation function $G(\tau)$ mentioned above is valid when the particle size is smaller than the volume element. For the calculation of the average number of molecules per volume element and the diffusion coefficients of free tetramethylrhodamine (TMR) ($\tau_1$), free TMR-labeled annexin V ($\tau_2$), and bound labeled annexin V ($\tau_3$), the system investigated is analyzed by a three-component autocorrelation function following:

$$G(\tau) = \frac{\left(1 - T + T \cdot e^{-\frac{\tau}{\tau_T}}\right)}{N} \cdot \left[ \begin{array}{c} \dfrac{Y_1}{1 + \frac{\tau}{\tau_1}} \cdot \dfrac{1}{\sqrt{1 + \frac{\tau}{K^2 \cdot \tau_1}}} + \dfrac{Y_2}{1 + \frac{\tau}{\tau_2}} \cdot \dfrac{1}{\sqrt{1 + \frac{\tau}{K^2 \cdot \tau_2}}} \\[3ex] + \dfrac{Y_3}{1 + \frac{\tau}{\tau_3}} \cdot \dfrac{1}{\sqrt{1 + \frac{\tau}{K^2 \cdot \tau_3}}} \end{array} \right] \qquad (9)$$

where $Y_1$ denotes the ratio of free TMR, $Y_2$ is the ratio of unbound TMR-labeled annexin V, $Y_3$ is the ratio of vesicle-bound labeled annexin V, $\tau$ is the correlation time, $K$ is the structure parameter (radius/length $= \omega_0/z_0$ of the volume element of the laser beam), $T$ is the ratio of triplet states, and $\tau_T$ is the decay time of triplet states.

## FCS Instrumentation

All FCS measurements are carried out with a ConfoCor2 spectrometer (Carl Zeiss, Jena, Germany) equipped with an $Ar^+$ (458, 488, and 514 nm; 25 mW) and two HeNe (543 and 633 nm; 1 mW) lasers. Tetramethylrhodamine is excited by the HeNe laser beam at 543 nm and a power of 50 $\mu$W (5%). The laser beam is focused by a water immersion objective (C-Apochromat, $\times 40$, 1.2 W; Carl Zeiss). From the focused detection volume, fluorescence is separated by a dichroic mirror (HFT 543 nm) and an emission filter (LP 560 nm) and is detected by an avalanche photodiode. The pinhole diameter is adjusted to 78 $\mu$m. The structure parameter $K$ is determined and held constant for all measurements at 6.0. All measurements are performed on Lab-Tek chambered coverglass slides with eight wells (Nalge Nunc International, Naperville, IL). The fluorescence autocorrelation functions $G(\tau)$ are determined with a digital correlator and are fitted with ConfoCor2 software (version 2.5) by a three-component model according to Eq. (9). The radius of the laser focus is determined with rhodamine 6G (Rh6G), which shows a diffusion time of $\tau = 34 \pm 1 \ \mu s$ at 543 nm. Solving equation $\tau_D = \omega_0^2/4D$, using a radius of confocal laser volume of $\omega_0 = 0.2 \pm 0.03 \ \mu m$ and the diffusion constant of Rh6G with $D_{Rh6G} = 2.8 \times 10^{-10} \ m^2 \ s^{-1}$, which was determined by Magde et al.[39], results in a detection volume of $V = 0.30 \pm 0.09$ fl.

## FCS Experiments

All FCS experiments are performed at room temperature. The time is usually chosen to be 30 s for all measurements and the volume of the solution used is 300 $\mu l$. The condition for binding, obtained by $Ca^{2+}$ titration, is achieved in the binding buffer. The incubation time is 1 min. The concentration of labeled annexin V is 100 nM and that of lipid is 10 $\mu M$. For all measurements, the repetition number is chosen to be 20. The diffusion times of free TMR and of unbound and bound TMR-labeled annexin V are determined separately. For the calibration of the instrument, the reference measurements are performed with free TMR solution. This solution used for the determination of diffusion time $\tau_1$ is diluted in the same buffer [10 mM HEPES (pH 7.4) or citrate (pH 4.0), 100 mM NaCl] as that used to prepare the liposomes. The result $\bar{\tau}_1$ is fixed as a fit parameter in two-component measurements. The diffusion time $\tau_2$ of unbound marked annexin V and the ratio $Y_1$ of free TMR is diluted in the relevant buffer solution that would be used for the investigation of the protein lipid interaction. The results $\bar{\tau}_2$ and $\bar{Y}_1$ are fixed for three-component measurements also. The diffusion time of LUVs is determined with liposomes labeled with 1 mol% Rh-PE in the membrane and is employed as an authentic value for the determination of binding of annexin V on vesicles. These studies show for fluorescence-labeled PC LUVs an average diffusion time of $\bar{\tau} = 3.7 \pm 1.1$ ms at pH 7.4 and of $\bar{\tau} = 3.3 \pm 0.7$ ms at pH 4.0 (data not shown). This result corresponds to a diffusion constant of $D_{PC \ LUVs} = (2.7 \pm 1.6) \times 10^{-12} \ m^2 \ s^{-1}$ at pH 7.4 and of $D_{PC \ LUVs} = (3.0 \pm 1.5) \times 10^{-12} \ m^2 \ s^{-1}$ at pH 4.0. Respectively, the binding constant $K_{ass}$ for a fixed $Ca^{2+}$ value is determined as follows:

$$K_{ass}\left[Ca^{2+}\right] = \frac{Y_3}{(1 - Y_3) \cdot N_{mol} \cdot C_{lipid}} \tag{10}$$

where $C_{lipid}$ denotes the concentration of lipid in solution, $Y_3$ is the ratio of bound marked annexin V, and $N_{mol}$ is the ratio of the binding relevant

[39] D. Magde, E. L. Elson, and W. W. Webb, *Biopolymers* **13**, 29 (1974).

lipid. The calculation of the associated lipids per bound annexin V molecule is made in the context of the following relations. The radius of the outside of the vesicles is taken as $r_a = 50$ nm, and that of the inside as $r_i = 45$ nm, the thickness of bilayer membrane $d = 5$ nm, the overall surface of the vesicle $A_g = A_a + A_i = 5.6 \times 10^{-14}$ m$^2$, and the amount of space per lipid $A_1 = 0.7$ nm$^2$. This supplies a value of $N_1 = 80 \times 10^3$ lipids per vesicle and for the surface a value of $N_2 = 45 \times 10^3$ lipids per vesicle. Ten micromolar lipid represents a value of $C_1 = 6 \times 10^{12}$ lipids/$\mu$l or $C_2 = 75 \times 10^6$ vesicles/$\mu$l. Assuming that bound annexin V per vesicle is evenly distributed, we can calculate the number of associated lipids per bound protein by

$$N_{\text{lipid/protein}} = \frac{N_2 \cdot C_2}{N_a \cdot 10^{-6} \cdot C_{\text{protein}} \cdot Y_3} \tag{11}$$

where $N_a$ is the Avogadro constant, $C_{\text{protein}}$ is the concentration of protein in the binding solution, and $Y_3$ is the ratio of bound protein.

*Binding of Annexin V on PC, PS, and PA Liposomes at pH 7.4 and 4.0*

The binding of annexin V on membranes occurs by electrostatic interaction. The investigation of protein membrane binding of annexin V on PS, PA, and PC LUVs has been carried out to study the influence on the protein fluorescence label of the interaction with lipid. Figure 4 shows the autocorrelation function $G(\tau)$ of free TMR, unbound TMR-labeled annexin V, and $\tau_{\text{annexin V}} = 154 \pm 23$ $\mu$s for PS LUV-bound TMR-labeled annexin V at pH 7.4. For binding to PS LUVs, the Ca$^{2+}$ concentration was set to be 100 $\mu$M. The autocorrelation function confirms a separation of the investigated binding molecules by fluorescence signal correlation analysis. We have determined by FCS for unbound fluorescence-labeled annexin V a diffusion time of $\tau_{\text{annexin V}} = 154 \pm 23$ $\mu$s at pH 7.4. This results in a diffusion constant of $D_{\text{annexin V}} = (65 \pm 29) \times 10^{-12}$ m$^2$ s$^{-1}$. At pH 4.0 the diffusion time of $\tau_{\text{annexin V}} = 180 \pm 54$ $\mu$s has been determined for unbound fluorescence-marked annexin V. This denotes a diffusion constant of $D_{\text{annexin V}} = (56 \pm 33) \times 10^{-12}$ m$^2$ s$^{-1}$. At neutral pH in the absence of Ca$^{2+}$ TMR-annexin V binds to PA liposomes ($K_{\text{ass}} = 5.3 \times 10^5$ $M^{-1}$), whereas PS liposomes show a low binding affinity ($K_{\text{ass}} = 2.1 \times 10^4$ $M^{-1}$) and PC liposomes show no binding affinities. The subsequent addition of 100 $\mu$M Ca$^{2+}$ results in a stronger binding of annexin V to pure PA LUVs ($K_{\text{ass}} = 9.5 \times 10^5$ $M^{-1}$); also, PS LUVs bind TMR-annexin V ($K_{\text{ass}} = 9.9 \times 10^6$ $M^{-1}$). Under these experimental conditions no binding of annexin V to PC liposomes is detected. The data denote about 57 and 62 associated lipids per bound annexin V to PS LUVs and PA LUVs at

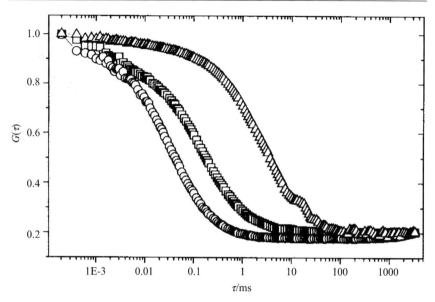

Fig. 4. Intensity autocorrelation function $G(\tau)$ of free TMR ($\bigcirc$), free TMR-labeled annexin V ($\square$), and TMR-labeled annexin V bound to PA LUVs ($\triangle$) in the presence of 100 $\mu M$ $Ca^{2+}$ at pH 7.4 ($\tau_1 = 35$ $\mu s$, $\tau_2 = 190$ $\mu s$, $Y_1 = 18\%$).

100 $\mu M$ $Ca^{2+}$, respectively. Measurements at pH 4.0 indicate a pH-induced binding of annexin V to PS LUVs and PA LUVs despite the absence of $Ca^{2+}$. This alteration of the protein–membrane interaction is caused by modification of the net charge of annexin V at pH 4.0. The binding constants calculated using Eq. (10) are $K_{ass} = 4.9 \times 10^5$ $M^{-1}$ for PS LUVs and $K_{ass} = 8.6 \times 10^5$ $M^{-1}$ for PA LUVs. PC liposomes do not bind TMR-annexin V at pH 4.0. Despite the fact that the presence of PC in PS liposomes (50 mol%) diminishes the surface charge, annexin V binds to these vesicles at pH 4.0 in the absence of $Ca^{2+}$. It is noteworthy that in a sensitive and suitable way, FCS has confirmed the earlier studies of annexin V binding on lipid membranes performed by other techniques.

Discussion

All proposed functions of the annexins depend directly on their ability to bind to phospholipid bilayers.[40] Because biological membranes are negatively charged we studied the interaction of annexin V with vesicles

[40] J. Benz and A. Hofmann, *Biol. Chem.* **378,** 177 (1997).

made from negatively charged phospholipids. PS is the most common negatively charged phospholipid at those membrane faces that are relevant for
fusion. PA is not common in membranes; however, PA is produced by
phospholipase D after stimulation of the cell and may therefore play a role
in processes such as degranulation or exocytosis, which are expected to
proceed under involvement of annexins.[41,42]

Annexin V binding to PA LUVs was studied in the absence and presence of $Ca^{2+}$. Only weak binding of annexin V to PA LUVs occurs in the
absence of $Ca^{2+}$ at pH 7.4. Under these conditions annexin V has a negative net charge, like the phospholipid. Thus, annexin V feels a repulsive
force when it approaches the bilayer. Addition of $Ca^{2+}$ induces and mediates a strong binding between annexin V and the phospholipids. This is due
to electrostatic attraction forces. However, the binding is effective because
of the high affinity of the annexin–phospholipid system for $Ca^{2+}$ ions. This
affinity originates from the special design of the conserved type II $Ca^{2+}$-
binding sites of the annexin, which, together with the phosphoryl moiety
of phospholipids, form a coordinative $Ca^{2+}$ complex.[19]

When the pH falls below the isoelectric point of annexin V ($\sim$4.8)[43]
electrostatic attraction between annexin V and PA or PS occurs. Consequently, there is no need for $Ca^{2+}$ to mediate the binding at such acidic
pH. Interestingly, at pH 4.0 the FRET from annexin V Trp-187 to bi-
layer-residing pyrene is 6- to 8-fold that at pH 7.4 and 40 $\mu M$ $Ca^{2+}$. Because > 80% of the added annexin V should already be membrane bound
at pH 7.4 with $Ca^{2+}$ [34] we must rule out a binding of a 6- to 8-fold higher
amount of annexin V at pH 4.0, compared with pH 7.4. Instead of a higher
binding we assume that this finding monitors the existence of two different
binding modes of annexin V, characterized by different environments and/
or locations (with respect to the bilayer) of Trp-187. The environment and
location of Trp-187 at pH 7.4 are known: Although Trp-187 is a part of an
interhelical loop at the $Ca^{2+}$-binding face of the protein, its fluorescent
moiety is buried in the hydrophobic interior of the protein.[44] In membrane-bound annexin V Trp-187 is located roughly in the depth of the glycerol backbone of the membrane.[44] Obviously, at pH 4.0 the distance
between Trp and the pyrene moiety of Pyr-PC (situated near the center
of the bilayer) is reduced, leading to the observed increase in FRET. In

[41] R. A. Blackwood, J. E. Smolen, A. Transue, R. J. Hessler, D. M. Harsh, R. C. Brower, and
S. French, *Am. J. Physiol* **272,** C1279 (1997).

[42] N. Düzgüneş, *in* "Trafficking of Intracellular Membranes: From Molecular Sorting to
Membrane Fusion" (M. C. Pedrosa de Lima, N. Düzgüneş and D. Hoekstra, eds.), p. 97.
Springer-Verlag, Berlin, 1995.

[43] M. J. Geisow, J. H. Walker, C. Boustead, and W. Taylor, *Biosci. Rep.* **7,** 289 (1987).

[44] P. Meers, *Biochemistry* **29,** 3325 (1990).

addition, the partial quenching of Trp fluorescence occurring in the native protein[44] may be reduced. From the largely increased FRET we therefore deduce a different conformation of annexin V at pH 4.0 compared with pH 7.4.

To summarize, using different methods such as microelectrophoresis and special fluorescence techniques it is possible to monitor protein interactions with liposomes. For a comprehensive calculation of binding constants and the number of bound molecules to liposome surfaces, several data derived from other methods are needed. For interpretation of microelectrophoresis, data information about the average charge of the proteins is needed, which may be solved by developments in electrophoresis instrumentation. Otherwise, material-intensive titration experiments for determination of protein charge are necessary. Laser microelectrophoresis measurements using liposomes are in practice not trivial; practical experience is absolutely necessary. The use of FRET techniques provides a high sensitivity for monitoring protein–liposome interactions. The method is limited to the existence of well-defined fluorescence donors such as tryptophan or tyrosine molecules (the best case would be the existence of one molecule in the protein). External fluorescence labeling of the protein is possible, but this method renders less certain the extent of labeling. The demonstrated FRET measurements indicate the inherent problems; for instance, it is not possible to distinguish whether the amount of bound protein is increased or the distance between donor and acceptor is reduced (penetration of the protein into the liposome bilayer). Several FCS applications using liposomes are reported.[45,46] The nonvaluable advantages of this method are the small volumes ($\sim$50 $\mu$l) necessary for measurements and the extremely low detection limit. Despite the fact that FCS machines are not widespread, FCS measurements and especially the data processing require an experienced user.

[45] J. Korlach, P. Schwille, W. W. Webb, and G. W. Feigenson, *Proc. Natl. Acad. Sci. USA* **96,** 8461 (1999).
[46] N. Kahya, E. I. Pecheur, W. P. de Boeij, D. A. Wiersma, and D. Hoekstra, *Biophys. J.* **81,** 1464 (2001).

## [4] Reconstitution of Membrane Proteins into Liposomes

*By* JEAN-LOUIS RIGAUD and DANIEL LÉVY

### Introduction

Screening of the genomes of various organisms demonstrated that about 25% of the sequenced genes encoded strongly hydrophobic proteins, which are integrated into cell membranes.[1] This observation emphasizes the importance of membrane proteins in many biological processes essential for life. However, the complexity of most biological membranes makes it difficult to study these membrane proteins *in situ*. Therefore, purification from the native membrane and further reincorporation of a purified membrane protein into an artificial membrane continue to be crucial steps in studying the function and structure of these molecules. The necessity for reconstitution arises because many membrane proteins express their full activity only when correctly oriented and inserted in a lipid bilayer. In particular, reconstitution has played a central role in identifying and characterizing the mechanisms of action of membrane proteins with a vectorial transport function.[2-4] More generally, through biochemical and biophysical approaches, it has led to important information about lipid–protein and protein–protein interactions as well as topological and topographical features of different classes of membrane proteins. The reconstitution of membrane proteins to form two-dimensional crystals confined in a membrane has led to important high-resolution structural information by electron crystallography.[5,6]

In many instances the ability to investigate membrane proteins through the use of reconstituted systems has long been limited by the fact that methods for producing high-quality proteoliposomes have not advanced in step with biochemical, biophysical, and molecular biology techniques. Thus, one of the limiting factors in obtaining molecular information is related to the lack of reproducible methods of reconstitution. Therefore, enormous efforts have been required to understand the mechanisms of reconstitution and for new approaches to be evaluated, refined, and applied

[1] D. T. Jones, *FEBS Lett.* **43**, 281 (1998).
[2] G. D. Eytan, *Biochim. Biophys. Acta* **694**, 185 (1981).
[3] E. Racker, *Methods Enzymol.* **55**, 699 (1982).
[4] J. L. Rigaud, B. Pitard, and D. Lévy, *Biochim. Biophys. Acta* **1231**, 223 (1995).
[5] W. Kühlbrandt, *O. Rev. Biophys.* **25**, 1 (1992).
[6] J. L. Rigaud, M. Chami, O. Lambert, D. Lévy, and J. L. Ranck, *Biochim. Biophys. Acta* **1508**, 112 (2000).

to available proteins in order to make reconstitution even more important as a tool for further structure–function relationship studies on membrane proteins.

This chapter deals with the various strategies commonly used to reconstitute proteoliposomes and focuses on approaches that have led to the production of highly functional proteoliposomes. General guidelines and rules are proposed in this area of study, which has long been viewed as more art than science.

## Strategies for Reconstitution of Membrane Proteins into Liposomes

From the abundant literature concerning the insertion of membrane proteins into liposomes, four basic strategies can be outlined: mechanical means, freeze-thawing, organic solvents, and detergents. Although these reconstitution strategies have proved useful to prepare pure phospholipidic vesicles,[7] the additional insertion of a membrane protein during the reconstitution process has imposed many constraints that have hampered seriously their efficiency and applicability for proteoliposome reconstitution. Indeed, besides the need for conditions that preserve the activity of the protein, many criteria must be considered to fully optimize the reconstitution of a membrane protein: the homogeneity of protein insertion and its final orientation, the morphology and size of the reconstituted proteoliposomes, as well as their residual permeability.

For example, proteoliposome reconstitution by mechanical means, such as sonication, has been drastically limited because of local probe heating, which is difficult to control and leads to degradation and denaturation of many membrane proteins. In addition, the small size (10–20 nm) of the resulting proteoliposomes limits the internal volume in which transport membrane proteins can accumulate ions or solutes. In the same framework, organic solvents have been widely used to prepare liposomes with a high capture efficiency, using procedures such as solvent injection[8] and reverse-phase evaporation.[9] However, the usefulness of such strategies for reincorporating membrane proteins has been limited because organic solvents denatured most amphiphilic membrane proteins. Thus, the strategy has been limited to the reconstitution of only a few hydrophobic membrane proteins.[10,11] Even more drastic, many methods for preparing pure

---

[7] F. Szoka and D. Papahadjopoulos, *Annu. Rev. Biophys. Bioenerg.* **9,** 467 (1980)

[8] D. Deamer and A. D. Bangham, *Biochim. Biophys. Acta* **443,** 629 (1976).

[9] F. Szoka and D. Papahadjopoulos, *Proc. Natl. Acad. Sci. USA* **75,** 4194 (1978).

[10] A. Darszon, A. C. Vanderberg, M. H. Ellisman, and M. Montal, *J. Cell Biol.* **81,** 446 (1979).

[11] J. L. Rigaud, A. Bluzat, and S. Büschlen, *Biochem. Biophys. Res. Commun.* **111,** 373 (1983).

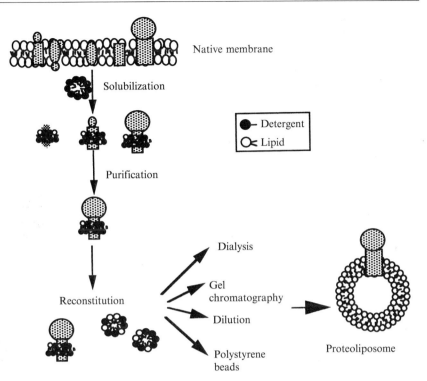

Native membrane

Solubilization

● Detergent
Oc Lipid

Purification

Dialysis

Gel chromatography

Dilution

Reconstitution

Polystyrene beads

Proteoliposome

FIG. 1. Detergent-mediated reconstitution. Most membrane proteins are extracted from native membranes by solubilizing detergent concentrations. After purification, the solubilized protein is supplemented with an excess of lipids and detergent, leading to a solution of mixed lipid–protein–detergent and lipid–detergent micelles. For proteoliposome reconstitution, detergent is removed from these micellar solutions, using various strategies.

liposomes have been difficult to apply successfully to proteoliposome reconstitution because most membrane proteins are purified through the use of detergents, which interfere with the process of vesicle formation. Thus, the vast majority of membrane protein reconstitution procedures involve the use of detergents[2,4] (Fig. 1). Indeed, because of their amphiphilic character, most membrane proteins require detergents, not only as a means of disintegrating the structure of native membranes in the initial step of their solubilization, but also as a means of keeping the protein in a nondenaturing environment during further purification.[12]

The standard procedure in such reconstitutions involves comicellization of the purified membrane protein in an excess of phospholipids and

[12] M. LeMaire, P. Champeil, and J. Möller, *Biochim. Biophys. Acta* **1508,** 86 (2000).

appropriate detergent, to form a solution of mixed lipid–protein–detergent and lipid–detergent micelles. Next, the detergent is removed from these micellar solutions, resulting in the progressive formation of closed lipid bilayers in which the proteins eventually incorporate. All detergent-mediated reconstitutions described in the literature rely on the same standard procedure, differing only in the techniques used to remove the detergent.

*Dialysis*

Dialysis has been the most widely used method for detergent removal. Dialysis of mixed lipid–protein–detergent micelles against a detergent-free aqueous medium is based on the selective retention by a porous membrane of vesicles and micelles compared with detergent monomers. In dialysis experiments, only the detergent monomers diffuse through the dialysis bag and the rate of removal depends on the monomer detergent concentration across the bag. Detergents with high critical micelle concentrations (CMCs) such as octylglucoside, ionic detergents (sodium cholate, sodium deoxycholate), or 3-[(cholamidopropyl)-dimethyl-ammonio]-1-propane-sulfonate (CHAPS) are easily removed by dialysis in 1–2 days; the technique is not appropriate for detergents with low CMCs such as polyoxyethylene glycols, which require 1–2 weeks for complete removal.[13]

In the simplest form of the method, the lipid–protein–detergent micellar solution is placed in a cellulose membrane bag with a cutoff of 14 kDa and dialyzed against a large volume of detergent-free aqueous buffer. Because of the necessity to scale down the amount of material for membrane proteins difficult to produce in large quantities, microdialysis devices have been used in the form of small compartments, allowing dialysis of 50- to 100-$\mu$l samples.[5,6]

The main advantages of dialysis are its simplicity and the low cost of the materials used. Also, the dialysis procedure generally leads to homogeneously sized vesicles. However, in its simplest setup, it suffers from numerous drawbacks, including (1) poor reproducibility, (2) an uncontrolled rate of dialysis, (3) possible retention of molecules on the dialysis membrane, and (4) duration of the experiments, which can be deleterious for many membrane proteins. Using a flowthrough dialysis cell can be advantageous as the rate of removal can be controlled carefully and the dialysis time decreased significantly.[14] Addition of polystyrene beads outside the

[13] T. M. Allen, A. Y. Romans, H. Kercret, and J. P. Segrest, *Biochim. Biophys. Acta* **601,** 328 (1980).
[14] M. H. W. Miloman, R. A. Schwender, and H. Weber, *Biochim. Biophys. Acta* **512,** 47 (1978).

bags, to maintain the external concentration of dialyzed detergent at zero, reduces the number of changes of buffer required during dialysis and can decrease the time of dialysis.[15] In conclusion, although the dialysis method has been applied successfully to many membrane proteins, it is unfortunately not well suited for detergents with low CMCs that require a long time for dialysis, which in turn is not always well tolerated by membrane proteins.

*Gel Filtration*

To avoid prolonged contact between proteins and detergent during reconstitutions, gel filtration has been used as a method for detergent removal. This strategy takes advantage of the different accessibility to the pores of a gel by mixed micelles as compared with liposomes. As the lipid–protein–detergent micellar mixture is eluted through the resin, the dilution resulting from both molecular diffusion and faster access of the smaller aggregates (monomers, mixed micelles) to the gel pore results in the formation of vesicles and their extrusion from the gel pores. The technique is simple, and gels of different pore size selectivity have been used in reconstitution experiments.[13]

The main advantage of this technique is its rapidity (5–10 min), avoiding long times of contact between detergent and protein. This, however, turns into a serious disadvantage in terms of incomplete and/or inhomogeneous protein incorporation and heterogeneous size distribution of the reconstituted preparations.[16] Thus, although it represents an interesting technique for preparation of pure liposomes, gel chromatography is no longer used in detergent-mediated reconstitution of membrane proteins.

*Dilution*

The dilution strategy is based on dilution by a detergent-free buffer of a lipid–protein–detergent micellar solution. The dilution must lower the initial detergent concentration to below its CMC, allowing spontaneous proteoliposome formation.

Cholate dilution was first employed for reconstitution. Lipids were mixed with membrane proteins at a final concentration of about 0.5% cholate, followed by a 25-fold dilution in the appropriate detergent-free buffer.[17] Other membrane proteins have been reconstituted by dilution

[15] J. Philippot, S. Mutaftschiev, and J. P. Liautard, *Biochim. Biophys. Acta* **734,** 137 (1983).
[16] M. Y. Abeywardena, T. M. Allen, and J. S. Charnock, *Biochim. Biophys. Acta* **729,** 62 (1983).
[17] E. Racker, T. F. Chien, and A. Kandrach, *FEBS Lett.* **57,** 1 (1975).

of octylglucoside micellar solutions.[18] The simplicity of the technique is evident, and the strategy has two main advantages that are related to (1) the short times required to decrease detergent concentrations and (2) the possibility of mastering the rate of dilution by progressive addition of the dilution buffer by syringe pumps powered either step by step or by synchronous motors. However, the technique suffers from numerous drawbacks that have made it unsuitable in many cases. Among them is the necessity to use detergents with high CMCs, because this limits the concomitant lipid dilution. Even with a high-CMC detergent, an additional centrifugation step is required to concentrate the diluted proteoliposomes. Another important drawback is that full detergent removal cannot be attained, because this would correspond to infinite dilution and, thus, the residual detergent still must be removed by other procedures.

*Polystyrene Beads*

Detergents with low CMCs that are not readily removed by dialysis or dilution can be removed efficiently through hydrophobic adsorption onto polystyrene bead resins such as Amberlite XAD and Bio-Beads SM-2.

The batch procedure, in which Bio-Beads SM-2 are added directly to the protein–lipid–detergent mixtures, has been demonstrated to be a powerful alternative to conventional dialysis for reconstitution trials.[19–21] Using radioactive detergents, the method has been calibrated precisely in terms of adsorptive capacity of beads and rates of detergent removal. The mechanisms underlying detergent adsorption onto beads have been analyzed, and general rules for the use of polystyrene beads have been proposed. Sufficient reproducibility can now be obtained with knowledge, experience, and careful handling, avoiding the main limitation of this strategy, which is lipid adsorption onto the beads.

This strategy is general and can be used whatever the nature of the detergent. In particular, it has been possible through this strategy to efficiently produce proteoliposomes by Triton X-100- or octaethylene glycol monododecyl ether ($C_{12}E_8$)-mediated reconstitutions. This has been an important step forward, because these low-CMC detergents, although mild for membrane proteins, were avoided in reconstitution trials because of

---

[18] E. Racker, B. Violand, S. O'Neal, M. Alfonzo, and J. Telford, *Biochem. Biophys. Res. Commun.* **198,** 470 (1979).

[19] D. Lévy, A. Bluzat, M. Seigneuret, and J. L. Rigaud, *Biochim. Biophys. Acta* **1025,** 179 (1990).

[20] J. L. Rigaud, D. Lévy, G. Mosser, and O. Lambert, *Eur. Biophys. J.* **27,** 305 (1998).

[21] J. L. Rigaud, G. Mosser, J. J. Lacapère, D. Lévy, A. Olofsson, and J. L. Ranck, *J. Struct. Biol.* **118,** 226 (1997).

their difficult removal by the previous conventional techniques. Another originality in the use of polystyrene beads is the possibility to master accurately the rate of detergent removal, by simply controlling the amount of beads added to the lipid–protein–detergent mixtures. Finally, another important benefit in using Bio-Beads is to remove almost all the detergent from a micellar solution, which allows the production of proteoliposomes with low ionic permeability, a crucial parameter in the study of membrane transport proteins.

From all these considerations, it appears that the use of Bio-Beads SM-2 satisfies all the criteria that make the procedure a powerful and better alternative to dialysis or dilution for proteoliposome reconstitution. This procedure is now used widely and has been demonstrated to be successful for the reconstitution of various classes of membrane proteins solubilized in all types of detergents.[4,20]

Mechanisms of Proteoliposome Formation and Efficiency
of Reconstitution

Despite extensive studies and diverse applications of proteoliposomes, the mechanism of their formation has long been surprisingly ill defined. Reconstitutions from detergent micellar mixtures yielded proteoliposomes of various compositions depending on the nature of the detergent, the particular procedure used to remove it, as well as the nature of the protein and the lipid composition. Therefore, not surprisingly, each membrane protein responded differently to the various reconstitution procedures and the approach has long been entirely empirical.

In the 1990s, important knowledge of the mechanisms of liposome formation,[22] as well as understanding of the physical behavior of lipid–detergent systems,[23,24] resulted in a set of basic principles that has limited the number of experimental variables and the empirical approach of proteoliposome reconstitution.

The first basic concepts to be taken into account are those developed in the model proposed by Lasic[22] for bilayer formation by detergent depletion techniques. As detergent is removed from micellar solutions a series of micelle–micelle interactions is initiated, resulting in the formation of large mixed disklike structures. When they have grown past a critical radius, a subsequent bending of these large micelles occurs and, at a critical micelle size, the amplitude of the bending is sufficient to cause bilayer closure and

[22] D. D. Lasic, *Biochem. J.* **256,** 1 (1988).
[23] D. Lichtenberg, *Biochim. Biophys. Acta* **821,** 470 (1985).
[24] J. R. Silvius, *Annu. Rev. Biomol. Struct.* **21,** 323 (1992).

vesicle formation. Ultimately, these vesicles still undergo size transformation, as long as the level of residual detergent remains high. According to this model, the size and the morphology of the final products of a detergent-mediated reconstitution are related (1) to the size, the morphology, and the composition of the initial micelles, which are closely linked to the properties of the detergents, and (2) to the morphologies of the mixed amphiphilic structures formed during the micelle-to-lipid bilayer transition, which depend on the nature of the detergent and its rate of removal.[25–27]

Additional concepts in detergent-mediated reconstitution are related to the mechanisms that trigger protein insertion into bilayers. Two main mechanisms were initially proposed[2]: (1) detergent removal results in the simultaneous coalescence of initial lipid–detergent and lipid–detergent–protein micelles, and the protein molecules simply participate in the membrane formation process; and (2) detergent removal results in the separate dissociation of lipid–detergent and lipid–detergent–protein micelles, and the protein molecules must insert into preformed detergent-doped bilayers. The nature of the detergent used, as well as the rate of detergent removal, are critical in determining one of the mechanisms of lipid–protein association and consequently in determining the efficiency of the final proteoliposomes.[2,4,28]

To allow realistic experimental monitoring of the mechanisms by which proteins may associate with lipids during detergent-mediated reconstitutions, we have developed a strategy based on the idea that reconstitution by detergent removal from a micellar solution is the mirror image of the solubilization of liposomes by a detergent.[4,29–31] To this end, detergent was first added to preformed liposomes through the range of concentration that causes the transformation of lamellar structure into mixed micelles. This allows a "snapshot" of all the lipid–detergent structures that can be formed during a reconstitution process. The protein is then added at each well-defined step of the solubilization process, allowing easy determination of the optimal conditions under which a protein can associate with lipids in the presence of detergent.

Besides providing original information about the mechanisms of lipid–protein association in the presence of detergent, this "step-by-step" reconstitution strategy was revealed to be a powerful reconstitution procedure,

[25] P. K. Vinson, Y. Talmon, and A. Walter, *Biophys. J.* **56**, 669 (1989).

[26] A. Walter, P. K. Vinson, J. Kaplun, and Y. Talmon, *Biophys. J.* **60**, 1315 (1991).

[27] O. Lambert, D. Lévy, J. L. Ranck, G. Leblanc, and J. L. Rigaud, *Biophys. J.* **74**, 918 (1998).

[28] A. Helenius, M. Sarvas, and K. Simons, *Eur. J. Biochem.* **116**, 27 (1981).

[29] M. T. Paternostre, M. Roux, and J. L. Rigaud, *Biochemistry* **27**, 2668 (1988).

[30] D. Lévy, A. Gulik, M. Seigneuret, and J. L. Rigaud, *Biochemistry* **29**, 9480 (1990).

[31] J. L. Rigaud, M. T. Paternostre, and A. Bluzat, *Biochemistry* **27**, 2677 (1988).

more suitable than the usual methods. This strategy has produced proteoliposomes, which satisfy most of the criteria for efficient reconstitution and sustain activities comparable to those measured in the native membrane[27,32-34]

New Method for Membrane Protein Reconstitution:
Step-by-Step Procedure

The new reconstitution strategy proceeds in four stages (Fig. 2): (1) preparation of large, homogeneous, and unilamellar liposomes, (2) addition of detergent to the preformed liposomes, through all the range of the solubilization process, (3) addition of solubilized protein at each well-defined step of the solubilization process, and (4) detergent removal and characterization of the reconstituted products.

*Preparation of Preformed Pure Liposomes*

The first stage in the reconstitution strategy is to prepare unilamellar and homogeneous preformed liposomes. Importantly, to avoid significant fusion processes on addition of subsolubilizing detergent concentrations, these liposomes may have a mean diameter exceeding 150 nm. This can be achieved using the reverse-phase evaporation method.[9,11]

A typical preparation contains 50 mg of phospholipids, usually solubilized in chloroform and dried under high vacuum. This thin lipidic film is dissolved in 3 ml of diethyl ether. Then, 1 ml of the desired aqueous buffer is added and the resulting two-phase system is sonicated, using a tip sonicator for 2 min at 4°, leading to the formation of a stable water-in-oil emulsion. The organic solvent is then removed, at room temperature, by rotary evaporation under reduced pressure (300–400 mmHg), using a nitrogen gas bleed to regulate the vacuum produced by a water aspirator. After about 15 min, a viscous gel forms that, on further organic solvent removal, collapses into a smooth suspension of large multilamellar vesicles. At this stage, an additional 2 ml of buffer is added to dilute the vesicle suspension, and evaporation (700 mmHg) proceeds for a further 30 min to remove all traces of organic solvent. The liposome suspension obtained is finally sequentially extruded through 0.4- and 0.2-$\mu$m pore size polycarbonate filters by pushing the liposomal suspension with a syringe through a filter holder.

[32] J. Cladera, J. L. Rigaud, and M. Dunach, *Eur. J. Biochem.* **243,** 798 (1997).

[33] D. Lévy, A. Gulik, A. Bluzat, and J. L. Rigaud, *Biochim. Biophys. Acta* **1107,** 283 (1992).

[34] B. Pitard, P. Richard, M. Dunach, G. Girault, and J. L. Rigaud, *Eur. J. Biochem.* **235,** 769 (1996).

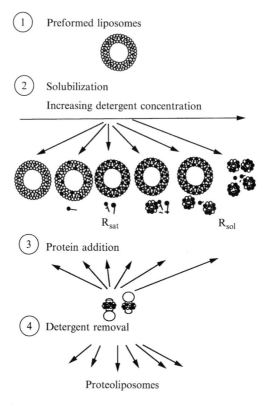

FIG. 2. The step-by-step reconstitution strategy. The standard procedure for reconstituting membrane proteins into proteoliposome is carried out in four stages: (1) preparation of large, homogeneous, and unilamellar liposomes; (2) step-by-step addition of detergent to the preformed liposomes. Some steps in the solubilization process are drawn schematically. $R_{sol}$ corresponds to the detergent-to-lipid ratio in liposomes at the onset of solubilization, whereas $R_{sat}$ corresponds to the detergent-to-lipid ratio in micelles when total solubilization is reached; (3) addition of the solubilized protein at each well-defined step of the solubilization process; and (4) detergent removal.

Although all our studies have been performed with liposomes produced by the reverse-phase evaporation technique, other methods that produce large unilamellar and homogeneous preformed liposomes should work as well. In particular, the Microfluidizer, a specialized apparatus commercially available from Microfluidics (Newton, MA), allows large-scale production of uniformly sized liposomes by extruding a multilamellar lipid suspension, up to 400 mg of lipid per milliliter. However, cryoelectron microscopy studies in our laboratory indicate that a significant proportion of small liposomes is encapsulated into large liposomes. Keeping this limitation in mind, this sequential extrusion procedure can be useful when dealing with specific lipid compositions for which reverse-phase evaporation fails, because of poor lipid solubility in diethyl ether or for lipids with a high temperature transition necessitating the evaporation of organic solvent at high temperature. Other strategies, such as dialysis or gel chromatography of lipid–detergent micelles, can also be used, but care should be taken to ensure complete detergent removal.

### Solubilization of Preformed Liposomes

In the second stage of the strategy, liposomes prepared by reverse-phase evaporation are diluted at the desired concentration and aliquoted to the desired volume. Solubilization is carried out by adding increasing amounts of detergent to aliquoted liposome suspensions. To monitor the solubilization process, the turbidity of phospholipid vesicle suspension (1 to 10 mg of lipid per milliliter) is measured between 400 and 700 nm as a function of detergent concentration. For lipid concentrations below 1 mg/ml, light-scattering changes are monitored with a fluorimeter set at 400 nm in both excitation and emission monochromators in order to measure the relative changes in light at $90°$.

Previous studies have indicated that the solubilization process can be described by a "three-stage" model and quantitatively visualized through changes in turbidity of the lipid–detergent suspensions.[23,24,27,29] In stage I, the detergent partitions into the lipid bilayer liposomes until it saturates the liposomes. During this stage, detergent addition to preformed liposomes does not disrupt the liposomes and induces slight changes in turbidity. Stage II corresponds to the gradual solubilization of detergent-saturated liposomes into small lipid–detergent micelles, inducing a large decrease in turbidity. Stage III is characterized by the complete solubilization of all liposomes into lipid–detergent mixed micelles and the solution becomes optically transparent (Fig. 3).

Table I summarizes the results obtained in the solubilization process with the most commonly used detergents. From Table I, the amount of any detergent to be added to a liposome suspension to reach any step in the solubilization process can be calculated easily.

### Addition of Protein

After the various detergent–phopholipid mixtures have been equilibrated for at least 1 h, the selected membrane protein is added. It is essential to add the protein as a monodisperse detergent–protein solution. This implies that, for any new membrane protein, the monodispersity of the

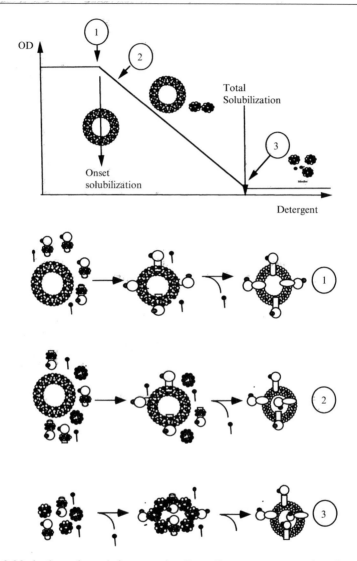

FIG. 3. Mechanisms of protein incorporation. Proteoliposomes are reconstituted according to the step-by-step strategy described in Fig. 2. *Top:* Steps in the lamellar-to-micellar transition at which optimal reconstitution has been observed, depending on the nature of the detergent. The lamellar-to-micellar transition can be analyzed qualitatively by turbidimetry as depicted schematically. *Bottom:* Depending on the nature of the detergent, proteins can either be incorporated directly into detergent-saturated liposomes at the onset of solubilization (mechanism 1), transferred from mixed micelles to detergent-saturated liposomes (mechanism 2), or participate in proteoliposome formation by micellar coalescence (mechanism 3). Note that the final orientation of the protein depends on the mechanism of association.

TABLE I
PARAMETERS DESCRIBING SOLUBILIZATION OF LIPOSOMES BY VARIOUS DETERGENTS[a]

| Detergent | $D_{water}$ | | $R_{sat}$ | | $R_{sol}$ | |
|---|---|---|---|---|---|---|
| | m$M$ | mg/ml | mol/mol | w/w | mol/mol | w/w |
| Triton X-100 | 0.18 | 0.12 | 0.64 | 0.5 | 2.5 | 2.0 |
| $C_{12}E_8$ | 0.20 | 0.11 | 0.66 | 0.45 | 2.2 | 1.5 |
| Octylglucoside | 17 | 4.9 | 1.3 | 0.48 | 3.0 | 1.1 |
| Dodecylmaltoside | 0.3 | 0.15 | 1 | 0.65 | 1.6 | 1.0 |
| Cholate | 3 | 1.29 | 0.3 | 0.16 | 0.9 | 0.5 |
| CHAPS | 3.15 | 1.94 | 0.4 | 0.31 | 1.04 | 0.8 |
| CHAPSO | 1.6 | 1.1 | 0.21 | 0.17 | 0.74 | 0.6 |

*Abbreviations:* CHAPS, 3-[(cholamidopropyl)-dimethyl-ammonio]-1-propanesulfonate; CHAPSO, 3-[(cholamidopropyl)-dimethyl-ammonio]-2-hydroxy-1-propanesulfonate.

[a] The process of solubilization can be described quantitatively by the general equation $D_{total} = D_{water} + R_{eff}$ [lipid], in which $D_{total}$ is the concentration of the detergent to be added to reach any step in the solubilization process; [lipid] is the lipid concentration; $D_{water}$ is the aqueous monomeric detergent concentration, that is, the CMC determined in the presence of lipids; $R_{eff}$ is the effective detergent-to-lipid ratio in mixed lipid–detergent aggregates ($R_{eff} = R_{sat}$ at the onset of solubilization represents the detergent-to-lipid ratio in detergent-saturated liposomes; $R_{eff} = R_{sol}$ at total solubilization represents the detergent-to-lipid ratio in lipid–detergent micelles).

solubilized preparation should be checked by gel chromatography or gel electrophoresis before reconstitution.

The association of a protein with lipid–detergent mixtures has been shown to be a time-dependent process, which depends on the nature and the concentration of the detergent used for reconstitution. It is generally recommended that the protein be incubated for 1 h with the lipid–detergent mixture before detergent removal. However, this time must be decreased to about 5–10 min when dealing with detergents that can be deleterious.

Another factor is related to the concentration of the presolubilized protein. The higher this initial concentration is, the lower is the amount of detergent brought with the protein into the equilibrated lipid–detergent mixtures. A protein concentration above 1 mg/ml allows the amount of detergent brought with the protein to be minimized. As a typical example, Ca-ATPase is solubilized at 1 mg/ml in $C_{12}E_8$ at 2 mg/ml. For standard reconstitution at a lipid-to-protein ratio of 80 (w/w), 50 $\mu$l of protein solution is added to 1 ml of detergent–lipid suspension containing lipid at 4 mg/ml. Thus, the amount of detergent brought with the protein corresponds to 0.1 mg of $C_{12}E_8$, that is, to a detergent-to-lipid ratio of 0.025 (w/w).

This ratio must be compared with the detergent-to-lipid ratios between 1 and 2 (w/w) generally needed for total lipid solubilization by other detergents. A consequence of this low amount of detergent added with the protein is the possibility to perform reconstitutions of any membrane protein with any detergent different from that used for initial purification of the protein.

*Detergent Removal*

The last step of reconstitution is related to the removal of detergent from the equilibrated lipid–detergent–protein mixtures. Although there are various methods for detergent removal, adsorption on Bio-Beads SM-2 has been demonstrated to be the most efficient for removal of all kind of detergents and convenient for a rapid screening of reconstitution trials.[20]

Before use, Bio-Beads are washed several times in methanol followed by several more washes in distilled water and stored in water at 4°. For use in reconstitution excess water is aspirated away from Bio-Beads and a small quantity is deposited on a Kim-Wipes filter paper. After adsorption of most of the water, wet Bio-Beads are weighed quickly to prevent total drying. For detergent removal at room temperature, the standard procedure is the following: Bio-Beads are added directly to each lipid–protein–detergent solution, at a Bio-Bead-to-detergent ratio of 10 (w/w), and stirred for 1 h. A second portion of beads is then added for an additional 1 h of incubation, followed by a third addition for about 2 h to ensure complete detergent removal. For less hydrophobic detergents, such as ionic detergents or CHAPS, a fourth addition of beads is performed because of the lower adsorptive capacity of these detergents.

The most accurate method to analyze detergent removal is based on the use of radioactive detergents.[19–21] However, turbidity measurements provide a convenient way to quantitatively monitor the reconstitution process[30,34]: starting from an optically transparent micellar solution, the turbidity increases on detergent removal, reaching a plateau value at a time depending on the amount of beads added to the detergent solution. This steady state absorbance has been shown to be a good index of the end of the micelle to detergent–saturated liposome transition. However, to ensure total detergent removal from the detergent-saturated vesicles, it is recommended that an extra amount of beads be added, once the steady state has been reached.

*Parameters to Be Varied*

*Choice of Detergent.* The main lesson to be learned from past studies is that no one detergent is likely to serve equally well for the reconstitution of all membrane proteins, and the experimental approach must be kept as broad as possible. Importantly, the choice should not be limited to the detergent used for the purification of a membrane protein. Indeed, other detergents reported deleterious when measuring the activity of a detergent-solubilized protein can be efficient, because, in reconstitution trials, phospholipids are present in excess and may have a protective effect. Thus, the optimal detergent that allows optimal protein incorporation, while avoiding its denaturation, must be found experimentally.

Regarding optimal protein incorporation, we have identified three mechanisms by which membrane proteins can associate with lipids to give proteoliposomes[4,27,32–34] (see Fig. 3). Depending on the nature of the detergent, proteins can be either incorporated directly into detergent-saturated liposomes (octylglucoside or dodecylmaltoside-mediated reconstitutions), transferred from mixed micelles to detergent-saturated liposomes (Triton X-100-mediated reconstitutions), or involved in proteoliposome formation during the micellar-to-lamellar transition (cholate, CHAPS, CHAPSC, and $C_{12}E_8$). Because reconstitution of a membrane protein into preformed liposomes ensures a unidirectional insertion of the protein in the membrane of unilamellar, homogeneous liposomes, it generally produces the most efficient proteoliposomes.[4] Thus, octylglucoside, dodecylmaltoside, and Triton X-100 are important detergents to check in reconstitution trials of any new membrane proteins. However, despite their efficiency in terms of protein incorporation, these detergents can be deleterious and/or induce protein aggregation. In this case, it is recommended that other classes of detergents also be analyzed.

As a general experimental scheme, it can be proposed that various prototypical detergents, including glycosylated detergents (octylglucoside, dodecylmaltoside), polyoxyethylenic detergents (Triton X-100 and $C_{12}E_8$), as well as ionic detergents (cholate), be tested. In preliminary trials, reconstitution should be analyzed at a few characteristic steps in the solubilization process by each of these detergents: onset of solubilization, half solubilization, and total solubilization. This allows rapid determination of the best detergent for reconstitution of any membrane protein. Once the best detergent has been determined, other parameters such as lipid composition, temperature, buffer composition, and rate of detergent removal may be optimized.

*Nature of Lipids.* Although all our experiments, using membrane proteins isolated from different organisms, have been performed with egg phosphatidylcholine–egg phosphatidic acid mixtures (9/1, mol/mol), preformed liposomes can be prepared from a variety of phopholipids and phospholipid mixtures for further reconstitution trials. The most commonly used phospholipids are phosphatidylcholine molecules derived from natural sources. They are often used because of their low cost relative to other phospholipids and because of their neutral charge and chemical inertness. It is recommended that charged species such as egg phosphatidic acid or egg phosphatidylglycerol (5–20%) be added. First, these negatively charged lipids avoid liposome fusion and/or aggregation, which enables the use of a liposome preparation for 1 week. Second, full activity of many membrane proteins depends on the presence of negatively charged lipids. Cholesterol can also be included into phospholipid mixtures to provide greater stability and better impermeability of reconstituted proteoliposomes to ions and small polar molecules.

An important basic point for the preparation of well-defined liposomes is to use well-characterized lipids. Be aware that these products vary widely in purity depending on the manufacturer and even the lot number. Also important is to use the purest quality in order to avoid lipid oxidation or impurities, such as lysoderivatives, which have been shown to induce large changes in permeability of reconstituted bilayers.

When selecting a liposome composition with synthetic saturated lipids (such as dimyristoyl or dipalmitoyl derivatives), it is best to keep in mind that phospholipids form smectic mesophases that undergo a characteristic gel–liquid crystalline phase transition. This transition is a function of the chain length and it is necessary, during all the reconstitution process, to work at a temperature above the transition of the higher melting component. In the same framework, care should be taken when using non-bilayer-forming lipids (phosphatidylethanolamine) or detergent-resistant lipidic compositions (sphingomyelin, cholesterol).[35]

*Lipid-to-Protein Ratio in Reconstitution Experiments.* The standard lipid-to-protein ratio used in proteoliposome reconstitutions is about 80 (w/w), which is roughly equivalent to a lipid-to-protein molar ratio of 8000, considering a membrane protein of 100 kDa. At this ratio, assuming a homogeneous protein distribution, it can be calculated that a proteoliposome 200 nm in diameter contains about 25 protein molecules of 100 kDa.

For many membrane proteins, the step-by-step method allows efficient reconstitutions for lipid-to-protein ratios ranging from 800 to 10 (w/w), with the final activity increasing proportionally with the amount of protein present initially. For lipid-to-protein ratios below 10 (w/w), although all the protein could be incorporated, limitations in reconstitution efficiency are, nevertheless, related to protein aggregation, and to a drastic increase in proteoliposome permeability due to the high protein concentration in the reconstituted bilayer.

---

[35] R. E. Brown, *J. Cell Sci.* **111,** 1 (1998).

*Rate of Detergent Removal.* When reconstitutions are performed from lipid–detergent–protein micellar solutions, an additional parameter to be analyzed is the rate of detergent removal. Generally, slow detergent removal is recommended to ensure homogeneous and efficient reconstitution when starting from micellar solutions. Fast detergent removal, however, has been demonstrated to be more efficient for those membrane proteins with a high tendency for aggregation during detergent removal; for example, reconstitution of Ca-ATPase, starting from Triton X-100 or $C_{12}E_8$ micellar solutions, has been demonstrated to require fast detergent removal to avoid insertion and aggregation of the protein in a low percentage of liposomes.[33] On the other hand, fast detergent removal is required when reconstituting a protein with a detergent that can be deleterious on long exposure, as demonstrated for octylglucoside-mediated reconstitution of Ca-ATPase.[33] Finally, fast detergent removal will also be required in those detergent-mediated reconstitutions that lead to multilamellar structures, as reported for dodecylmaltoside-mediated reconstitutions.[27]

The rate of removal can be controlled easily by the regimen and the amount of beads added.[19–21] For fast detergent removal, addition at once of an amount of beads equivalent to a Bio-Beads-to-detergent ratio of 30 (w/w) is sufficient to remove all the detergent from a lipid–protein–detergent mixture in about 1 h. For slow detergent removal, successive additions, every 30 min, of small amounts of beads at Bio-Beads-to-detergent ratios of 1 to 2 (w/w) will allow the removal of the detergent in times ranging from 5 to 12 h.

When temperature must be optimized in a detergent-mediated reconstitution, it must be kept in mind that the rate of detergent adsorption onto Bio-Beads depends drastically on the temperature, doubling every 12°. This implies that either the time of detergent removal or the amount of beads must be adapted to the temperature at which the reconstitution is performed.[19,21]

### Characterization of Reconstituted Proteoliposomes

Besides the need for measuring the activity of the protein, any method of membrane protein reconstitution should fulfill a number of important criteria that must be analyzed to characterize unequivocally the efficiency of the reconstitution (Fig. 4).

*Functional Activity*

The first parameter to analyze in order to check the efficiency of a reconstitution trial is the activity of the protein after reconstitution. Functional assays depend on the specific function of the protein under study.

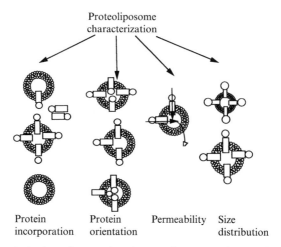

Fig. 4. Characterization of reconstituted proteoliposomes. Any method of membrane protein reconstitution should fulfill a number of important criteria. Besides the need to measure the activity of the protein, the following parameters must be analyzed: protein incorporation and protein distribution among the proteoliposomes, final protein orientation, size distribution, and permeability.

For each protein, activity measurements are set up to allow not only a measurement consuming small amounts of material, but also a rapid measurement, facilitating a rapid screening of the samples after reconstitution.

*Protein Incorporation*

Because nonincorporated and/or aggregated membrane proteins may still be active, it is essential to characterize a reconstitution trial by determining the amount of protein incorporated in the reconstituted proteoliposomes. The most accurate method to analyze the efficiency of membrane protein incorporation is a density gradient. The nonincorporated protein is separated easily from liposomes containing different amounts of proteins after centrifugation on a 0–30% sucrose gradient.[31–33] The reconstituted material is loaded on top of a continuous or discontinuous 0–30% sucrose gradient made in the desired buffer and centrifuged at 100,000 *g* at 4° [TLS55, 34,000 to 35,000 rpm (Beckman, Fullerton, CA) or other Beckman centrifuge with SW-type rotor and appropriate speed to achieve 100,000 *g*]. Proteoliposomes can also be subjected to discontinuous flotation gradients. In this case, the reconstituted sample is first deposited in a 30% sucrose layer. Successive layers of 20, 15, 10, 5, and 2.5% sucrose (w/w) are then deposited on top of this layer. After centrifugation the various sucrose layers are analyzed for protein and lipid content, using colorimetric determinations or more sensitive radioactive or fluorescent labeling.

Incorporation of a membrane protein can also be monitored by freeze–fracture electron microscopy.[32,36] The presence and distribution of intramembrane particles in the fracture faces demonstrate efficiency and homogeneity of protein incorporation. Furthermore, examination of convex and concave fracture faces can reveal a significant difference in particle density related to a preferential orientation of the protein in the reconstituted membranes.

*Protein Orientation*

As a general approach, sidedness of membrane proteins after reconstitution can be determined from gel electrophoresis patterns before and after proteolytic treatment of the reconstituted samples.[11,32] To this end proteoliposomes (protein at 100 $\mu$g/ml) are incubated in the presence of a specific proteolytic enzyme (e.g., trypsin, chymotrypsin, Pronase, and/or papain). After the proteolytic reaction is stopped, the samples are delipidated[37] before sodium dodecyl sulfate–polyacrylamide and gel electrophoresis (SDS–PAGE). The percentage of protein orientation is determined from the ratio of the intensities of the protein band before and after proteolytic digestion, using a laser densitometer.

Functional tests, using one-sided inhibitors, must be specifically adapted to the membrane protein under study. For membrane proteins with ATPase activity, the orientation after reconstitution in the bilayer is assessed by determining the fraction of molecules accessible to one-sided inhibitors that react with the cytoplasmic domains, thus abolishing the ATPase activity [e.g., fluorescein isothiocyanate (FITC)[38] for Ca-ATPase or ouabain for Na/K-ATPase[39]]. Accordingly, the residual ATPase activity after full reaction with a one-sided inhibitor represents the fraction of ATPase molecules facing the inside of intact reconstituted vesicles. Detergent-solubilized reconstituted vesicles are generally taken as controls in which all molecules are accessible to inhibitors. For light-activated membrane proteins that function whatever the final protein orientation, measurements are simply performed in the presence or absence of a one-sided nonpermeant inhibitor (e.g., millimolar concentration of lanthanides that

[36] T. Gulik-Krzywicki, M. Seigneuret, and J. L. Rigaud, *J. Biol. Chem.* **262**, 15580 (1987).

[37] D. Wessel and U. I. Flügge, *Anal. Biochem.* **138**, 141 (1984).

[38] H. S. Young, J. L. Rigaud, J. J. Lacapère, L. G. Reddy, and D. L. Stokes, *Biophys. J.* **72**, 2545 (1997).

[39] F. Cornelius, *Biochim. Biophys. Acta* **1071**, 19 (1991).

inhibit bacteriorhodopsin proton-pumping activity when facing the carboxy terminus of the protein[40]).

## Size Distribution

Various methods, already developed for liposomes, have been applied to determine not only the average size but also the size distribution of reconstituted proteoliposomes.

Light-scattering procedures, particularly laser-based quasi-elastic light-scattering methods, are popular to obtain information about the size and polydispersity of a reconstituted proteoliposome preparation.[41] These techniques are based on the time-dependent intensity fluctuations of scattered laser light due to Brownian motion of particles in solution. Analysis of fluctuation as a function of time yields lateral diffusion values that can be related to the Stokes radius of the particles. The advantage of light-scattering methods is that information can be obtained in a few minutes. However, caution must be taken because misleading results can be obtained for heterogeneous systems exhibiting bimodal or more complex size distributions.[42]

If only an approximate idea of the size range is required, gel chromatography can also be recommended, because it is a quick and convenient method for both fractionation and average size determination of proteoliposomes. Different gels can be used, depending on the mean size of the proteoliposomes. Keeping in mind that most detergent-mediated reconstitutions produce proteoliposomes less than 300 nm in diameter, Sephacryl S-1000 columns are especially appropriate for separating vesicles.[40,43]

Electron microscopy is the most precise method available to measure vesicle size, whatever the size range. Negative staining, although the simplest electron microscopic approach, must be interpreted with caution because of stain artifacts and significant distortions of large proteoliposomes that collapse on carbon-coated grids. Freeze–fracture electron microscopy is also useful for size and morphology analysis of a proteoliposome preparation.[36] However, care must be taken when measuring size distribution, because the random cleavage plane does not necessarily go through the midplane and thus can reveal a smaller diameter vesicle. Finally, cryoelectron microscopy is the best technique in characterizing the size distributions and morphology of reconstituted liposomes. It is

[40] M. Seigneuret and J. L. Rigaud, *FEBS Lett.* **228,** 79 (1988).

[41] H. Ruf, Y. Georgalis, and E. Grell, *Method Enzymol.* **172,** 364 (1986).

[42] O. Lopez, M. Cocera, E. Wehrli, J. L. Parra, and A. de la Maza, *Arch. Biochem. Biophys.* **367,** 153 (1999).

[43] J. A. Reynolds, Y. Nozaki, and C. Tanford, *Anal. Biochem.* **130,** 471 (1983).

useful in discerning changes in size and volume distributions, because this technique avoids the artifacts of staining and drying procedures and permits the observation of undistorted samples (see Chapter 29 in Volume 373[44]).

Determining the volume of the aqueous compartment of proteoliposomes is important when dealing with the activity of membrane transport proteins. Most of the methods involve measurements of radioactive or fluorescent markers trapped inside liposomes after removal of the untrapped marker by dialysis, gel chromatography, or centrifugation. It is usually also necessary to measure the amount of residual marker present outside the liposomes, because removal is often incomplete and leakage may occur during washing procedures. Fluorescence techniques have been developed that avoid the need for removal of external marker.[45] Using calcein ($10^{-4}$ $M$) as a fluorescent marker, the addition of cobalt or copper cations ($10^{-2}$ $M$) outside the liposomes is sufficient to quench the fluorescence of the external marker. The fraction of the total volume that is within the proteoliposomes is obtained as the fraction of the fluorescence that remains after adding cations that, when chelated by calcein, quench its fluorescence. Although a rapid and simple method, care must be taken about time-dependent aggregation of proteoliposomes containing negatively charged phospholipids.

*Permeability*

An important aspect in reconstitution is the residual detergent, which can alter the passive permeability of proteoliposomes. The most precise method to analyze residual permeability consists of measuring proton and counterion flux generated by an external acid pulse.[46] To this end, proteoliposomes are prepared in the presence of 200 $\mu M$ pyranine, a fluorescent pH-sensitive probe ($\lambda_{ex}$ = 460 nm; $\lambda_{em}$ = 510 nm). After reconstitution, external pyranine can be removed by passing the proteoliposomes through a prepacked G-25 Sephadex column. The resulting pyranine-containing proteoliposomes are subjected to an external acidic pulse of 0.5 pH unit, and changes in internal fluorescence are monitored as a function of time. The changes in the rate of internal acidification, following the acidic pulse, can be related to proton and counterion permeability coefficients.

[44] O. Lambert and J. L. Rigaud, *Methods Enzymol.* **373,** 29 (2003).
[45] N. Oku, D. A. Kendall, and R. C. Macdonald, *Biochim. Biophys. Acta* **691,** 332 (1982).
[46] M. Seigneuret and J. L. Rigaud, *Biochemistry* **26,** 6723 (1986).

Conclusion

Reconstitution of membrane proteins into liposomes is a powerful tool that can be used to identify the mechanism of action of membrane proteins. As shown in this chapter, it appears that reconstitution is no longer "black magic," and the prospects of achieving optimal proteoliposome reconstitution are obviously good when using reliable methods and systematic experimental analysis.

The future of membrane protein reconstitution appears bright in the light of the steadily expanding number of membrane proteins that have been identified in the sequencing of the genomes of different organisms. Powerful methods of overexpression of genes are now available that will allow the production of new, interesting membrane proteins. In this context, although not largely exploited up to now, the reconstitution of membrane proteins into liposomes should be an efficient strategy for the purpose of pharmaceutical, cosmetic, and chemical applications (e.g., specific targeting, drug delivery, and antigenicity or gene therapy).

# [5] Reconstitution of Purified Bacterial Preprotein Translocase in Liposomes

By CHRIS VAN DER DOES, JEANINE DE KEYZER, MARTIN VAN DER LAAN, and ARNOLD J. M. DRIESSEN

Introduction

The last decade has seen a major advance in the study of bacterial protein translocation (for reviews see Refs. 1 and 2). Secretory and membrane proteins are synthesized at the ribosome with an N-terminal signal sequence or hydrophobic transmembrane segments that direct the proteins to the bacterial inner (cytoplasmic) membrane. Many inner membrane proteins are targeted to the membrane as a ribosome-bound nascent chain via the signal recognition particle (SRP) and FtsY, the SRP receptor. The targeting of most periplasmic and outer membrane proteins occurs posttranslationally as a completely synthesized polypeptide, often in association with the molecular chaperone SecB.[1] At the membrane, both

[1] E. H. Manting and A. J. M. Driessen, *Mol. Microbiol.* **37,** 226 (2000).

[2] A. J. M. Driessen, E. H. Manting, and C. van der Does, *Nat. Struct. Biol.* **8,** 492 (2001).

targeting routes converge at the translocase.[3] Translocase consists of a complex of the ATP-driven molecular motor SecA[4] and a heterotrimeric membrane protein complex composed of the SecY, SecE, and SecG subunits.[5] The SecYEG complex forms a protein-conducting channel in the membrane.[6] SecA is a dissociable subunit of the translocase that associates with high affinity with the SecYEG complex.[7] SecA[8] serves both as a receptor for precursor proteins (preproteins) and as an ATP-driven molecular motor that fuels the translocation reaction. Multiple cycles of ATP binding and hydrolysis of ATP by SecA[9] drive the stepwise translocation of preprotein segments across the membrane.[10] For the cotranslational insertion of membrane proteins into the inner membrane, nascent membrane proteins are first released from SRP in a GTP-dependent and FtsY-dependent reaction. During this process, the nascent chain is transferred to the SecY subunit of the translocase and further membrane assembly can be driven by chain elongation at the ribosome, whereas large, hydrophilic periplasmic loops may require the action of SecA (for review see Herskovits *et al.*[11]). The inner membrane protein YidC is a translocase-associated component that at an early stage interacts with membrane-inserted transmembrane segments of nascent membrane proteins.[12,13] YidC is homologous to Oxa1p, a mitochondrial inner membrane protein that facilitates the insertion of some membrane proteins into the inner mitochondrial membrane.[14]

All known components involved in protein translocation have been purified, and functionally reconstituted in liposomes. These proteoliposomes mediate an authentic energy-dependent preprotein translocation

[3] Q. A. Valent, P. A. Scotti, S. High, J. W. de Gier, G. von Heijne, G. Lentzen, W. Wintermeyer, B. Oudega, and J. Luirink, *EMBO J.* **17**, 2504 (1998).

[4] A. Economou and W. Wickner, *Cell* **78**, 835 (1994).

[5] L. Brundage, C. J. Fimmel, S. Mizushima, and W. Wickner, *J. Biol. Chem.* **267**, 4166 (1992).

[6] E. H. Manting, C. van der Does, H. Remigy, A. Engel, and A. J. M. Driessen, *EMBO J.* **19**, 852 (2000).

[7] F. U. Hartl, S. Lecker, E. Schiebel, J. P. Hendrick, and W. Wickner, *Cell* **63**, 269 (1990).

[8] J. P. Hendrick and W. Wickner, *J. Biol. Chem.* **266**, 24596 (1991).

[9] E. Schiebel, A. J. M. Driessen, F. U. Hartl, and W. Wickner, *Cell* **64**, 927 (1991).

[10] J. P. van der Wolk, J. G. de Wit, and A. J. M. Driessen, *EMBO J.* **16**, 7297 (1997).

[11] A. A. Herskovits, E. S. Bochkareva, and E. Bibi, *Mol. Microbiol.* **38**, 927 (2000).

[12] P. A. Scotti, M. L. Urbanus, J. Brunner, J. W. de Gier, G. von Heijne, C. van der Does, A. J. M. Driessen, B. Oudega, and J. Luirink, *EMBO J.* **19**, 542 (2000).

[13] J. C. Samuelson, M. Chen, F. Jiang, I. Moller, M. Wiedmann, A. Kuhn, G. J. Phillips, and R. E. Dalbey, *Nature* **406**, 637 (2000).

[14] K. Hell, W. Neupert, and R. A. Stuart, *EMBO J.* **20**, 1281 (2001).

reaction *in vitro*.[15,16] In its minimal form the actively reconstituted *Escherichia coli* translocase consists of SecY, SecE, and SecA.[17] SecG co-purifies with SecY and SecE[15,16] and strongly stimulates the translocation activity.[18] Proteoliposomes containing coreconstituted[19,20] SecYEG complex and YidC support the SRP-, FtsY-, and GTP-dependent targeting and transfer of nascent inner membrane proteins to the SecY and YidC subunits,[21] demonstrating that these proteoliposomes catalyze the initial stages of membrane protein insertion.

This chapter describes methods to express, detergent solubilize, purify, and functionally reconstitute preprotein translocation in proteoliposomes, using *E. coli* translocase components. In addition, a coreconstitution proto-col for SecYEG and YidC, and a fluorescence assay to monitor preprotein translocation *in vitro,* are described.

## Growth and Purification of SecYEG and YidC

Using specific combinations of plasmids and strains, it is possible to overexpress SecYEG[16] and YidC[21] to approximately 20% of the total inner membrane protein. The proteins can be solubilized from the membrane and purified to homogeneity (>95%) by anion exchange and nickel-nitrilotriacetic acid (Ni-NTA) affinity chromatography.

### Isolation of Membranes Harboring Overexpressed SecYEG or YidC Protein

*Escherichia coli* SF100[22] is transformed either with plasmid pEH1-hisYidC (overexpression of YidC protein with an amino-terminal histi-dine tag, *kan*)[21] or pET610 (overexpression of SecYEG complex with a

[15] L. Brundage, J. P. Hendrick, E. Schiebel, A. J. M. Driessen, and W. Wickner, *Cell* **62**, 649 (1990).

[16] C. van der Does, E. H. Manting, A. Kaufmann, M. Lutz, and A. J. M. Driessen, *Biochemistry* **37**, 201 (1998).

[17] M. Hanada, K. I. Nishiyama, S. Mizushima, and H. Tokuda, *J. Biol. Chem.* **269**, 23625 (1994).

[18] K. Nishiyama, M. Hanada, and H. Tokuda, *EMBO J.* **13**, 3272 (1994).

[19] C. van der Does, J. Swaving, W. van Klompenburg, and A. J. M. Driessen, *J. Biol. Chem.* **275**, 2472 (2000).

[20] K. Douville, A. Price, J. Eichler, A. Economou, and W. Wickner, *J. Biol. Chem.* **270**, 20106 (1995).

[21] M. van der Laan, E. N. Houben, N. Nouwen, J. Luirink, and A. J. M. Driessen, *EMBO Rep.* **2**, 519 (2001).

[22] F. Baneyx and G. Georgiou, *J. Bacteriol.* **172**, 491 (1990).

histidine tag at the amino terminus of the SecY subunit, amp).[23] Cells are grown overnight at 37° in 100 ml of Luria broth (LB), 0.5% (w/v) glucose, and ampicillin at 0.1 mg/ml (SecYEG) or kanamycin at 0.05 mg/ml (YidC). Of this culture, 25 ml is used to inoculate 1 liter of LB containing the appropriate antibiotic. At an $OD_{660}$ of about 0.6, the culture is induced with 0.5 m$M$ isopropyl-$\beta$-D-thiogalactopyranoside (IPTG). Growth is continued for an additional 2 h. Cells are harvested by centrifugation (5000 g, 10 min, 4°) and resuspended in 50 m$M$ HEPES-KOH (pH 8.0), 20% (w/v) sucrose. Cells are rapidly frozen as small nuggets by slowly dropping the cell suspension into liquid nitrogen. The nuggets can be stored for several months at −20°. Before use, the frozen nuggets are slowly thawed in water. To prevent proteolysis, subsequent steps are performed at 4°. The suspension is supplemented with 1 m$M$ dithiothreitol (DTT) (Roche Diagnostics, Mannheim, Germany), 0.5 m$M$ phenylmethylsulfonyl fluoride (PMSF), DNase (1 mg/ml), and RNase (1 mg/ml) (all from Sigma, St. Louis, MO). The cells are lysed by French press (One Shot cell disrupter; Constant Systems, Daventry, UK) treatment (two times at 8000 lb/in²). After the first lysis step, the DTT and PMSF concentrations are increased to 2 and 1 m$M$, respectively. Unbroken cells are removed by centrifugation (6000 g, 15 min), and the membranes are collected by ultracentrifugation (125,000 g, 90 min). The membrane pellet is resuspended in buffer A [50 m$M$ HEPES-KOH (pH 8.0), 20% (v/v) glycerol] and applied on a sucrose step gradient in 50 m$M$ HEPES-KOH, pH 8.0. The membranes isolated from a 1-liter culture are divided over two tubes (4 ml) containing layers of 55% (0.4 ml), 51% (1 ml), 45% (0.4 ml), and 36% (0.4 ml) (w/v) sucrose, respectively. The gradient is spun for 30 min at 4° at 250,000 g in a TLA100.4 rotor. The brownish inner membrane fraction is collected from the 45% sucrose layer, diluted 5-fold with 50 m$M$ HEPES-KOH, pH 8.0, and recollected by centrifugation (250,000 g, 30 min). The pellet is resuspended in buffer A, frozen in liquid nitrogen, and stored at −80°. One liter of culture yields 10–15 mg of membrane protein.

*Purification of SecYEG Complex*

Membranes are solubilized at a protein concentration of 1 mg/ml in buffer B [10 m$M$ Tris-HCl (pH 8.0), 20% (v/v) glycerol] with 2% (w/v) of the detergent dodecylmaltoside (DDM) (Anatrace, Maumee, OH) for 1 h at 4° under gentle mixing. Nonsolubilized material is removed

[23] A. Kaufmann, E. H. Manting, A. K. J. Veenendaal, A. J. M. Driessen, and C. van der Does, *Biochemistry* **38,** 9115 (1999).

by centrifugation (125,000 $g$, 30 min), and the supernatant is loaded on an anion-exchange column (DEAE, HiTrap Sepharose Q, or MonoQ). The column is washed with 1 column volume of buffer B with 0.05% DDM, and bound protein is eluted in 3 column volumes with a linear gradient of 0–300 m$M$ KCl in buffer B with 0.05% DDM. His-tagged SecYEG complex elutes at a KCl concentration of about 100 m$M$, before most other proteins. Fractions are analyzed by 15% sodium dodecyl sulfate–polyacrylamide gel electrophoresis (SDS–PAGE) without boiling the samples. SecYEG-containing fractions are pooled and diluted with an equal volume of buffer B supplemented with 0.1% DDM, 100 m$M$ KCl, and 10 m$M$ imidazole, pH 8.0. For further purification, a 1- or 5-ml HiTrap chelating HP column (Amersham Pharmacia Biotech, Uppsala, Sweden) is loaded with $Ni^{2+}$, using a 0.1 m$M$ $NiCl_2$ solution, and equilibrated with buffer B with 0.1% DDM, 100 m$M$ KCl, and 10 m$M$ imidazole, pH 8.0. The pooled SecYEG fractions are loaded onto the column at a flow rate of less than 0.5 ml/min. The column is washed with 5 volumes of buffer B with 0.1% DDM, 100 m$M$ KCl, and 10 m$M$ imidazole, pH 8.0, and the SecYEG complex is eluted with 5 volumes of buffer B with 0.1% DDM, 100 m$M$ KCl, and 200 m$M$ imidazole, pH 7.0. Fractions are collected and analyzed by 15% SDS–PAGE. The fractions containing purified SecYEG are rapidly frozen in small aliquots in liquid nitrogen and stored at $-80°$. One liter of culture yields about 0.5 mg of purified SecYEG.

*Purification of YidC*

Membranes are solubilized at a protein concentration of 1 mg/ml in buffer C [10 m$M$ HEPES-KOH (pH 8.0), 100 m$M$ KCl, 20% (v/v) glycerol] with 10 m$M$ imidazole and 2% (w/v) DDM for 1 h at 4° under gentle mixing. Insoluble material is removed by centrifugation (125,000 $g$, 30 min), and the supernatant is incubated under gentle mixing overnight at 4° with washed Ni-NTA agarose beads (100 $\mu$l of beads per milligram of membrane protein). Before use, the Ni-NTA beads are washed according to the instructions of the manufacturer and equilibrated with buffer C with 10 m$M$ imidazole and 0.1% (w/v) DDM. The mixture is poured into a chromatography column and washed with 5 volumes of buffer C with 10 m$M$ imidazole and 0.1% DDM and 5 volumes of buffer C with 30 m$M$ imidazole and 0.1% DDM. Bound YidC protein is eluted with 8 column volumes of buffer C with 400 m$M$ imidazole (pH 7.0) and 0.1% DDM. Fractions are analyzed by 12% SDS–PAGE, pooled, frozen in liquid nitrogen, and stored at $-80°$. One liter of culture yields about 1 mg of purified membrane protein.

Analysis of Purified Components

The purified Sec YEG and YidC can be analyzed by SDS–PAGE, but when the purified proteins are analyzed by blue native polyacrylamide gel electrophoresis (BN–PAGE),[24] information about the oligomeric state can also be obtained. The next section describes the protocol for BN–PAGE, using the commercially available Mini-PROTEAN II gel system (Bio-Rad, Hercules, CA). As shown in Fig. 1, purified SecYEG and YidC are both present in monomeric and dimeric forms after purification.

*Blue Native Polyacrylamide Gel Electrophoresis*

Purified proteins are mixed with a loading dye solution containing Serva Blue G and analyzed by polyacrylamide gel electrophoresis under native conditions. First, the stacking gel and gradient running gel are poured. The volumes and composition of the stacking and running gels for a Bio-Rad Mini-PROTEAN II gel system with 1-mm spacers are given in Table I. The linear 5–15% acrylamide running gel is made with a simple gradient maker. Alternatively, a 5–15% gradient gel can be made by carefully stacking six layers of 15, 13, 11, 9, 7, and 5% acrylamide and decreasing concentrations of glycerol (20, 16, 12, 8, 4, and 0%, respectively) on top of each other. Gels are poured the day before use and stored at 4°. Loading dye [10 × loading dye: 100 m$M$ Bis-Tris (pH 7.0), 500 m$M$ 6-aminocaproic acid, 5% (w/v) Serva Blue G] is added and the samples are directly applied on the gel. To dissociate membrane protein complexes, SDS [1% (w/v) final concentration] can be added to the sample. After

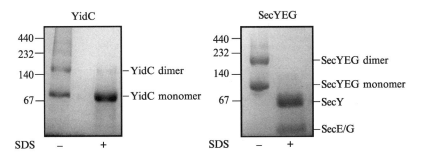

FIG. 1. Blue native–PAGE analysis of YidC and SecYEG. Purified YidC and SecYEG were analyzed directly after purification or after 10 min of incubation in 1% SDS at room temperature.

[24] H. Schagger and G. von Jagow, *Anal. Biochem.* **199,** 223 (1991).

TABLE I
COMPOSITION OF STACKING AND RUNNING GELS FOR
BLUE NATIVE–POLYACRYLAMIDE GEL ELECTROPHORESIS[a]

|  | Stacking gel (4% T) | Gradient running gel | |
| --- | --- | --- | --- |
|  |  | 5% T | 15% T |
| Acrylamide | 0.6 ml | 1.10 ml | 3.35 ml |
| Gel buffer (3×) | 2.5 ml | 3.65 ml | 3.65 ml |
| Glycerol | — | — | 2.20 ml |
| $H_2O$ | 4.4 ml | 6.25 ml | 1.80 ml |
| APS (10%, w/v) | 30 µl | 40 µl | 40 µl |
| TEMED | 3 µl | 4 µl | 4 µl |
| Total volume: | 7.5 ml | 11.0 ml | 11.0 ml |

*Abbreviations:* APS, Ammonium persulfate; TEMED, $N, N,$
$N', N'$-tetramethylethylenediamine.
[a] Acrylamide mix: 49.5% T, 3% C (48 g acrylamide per
100 ml and 1.5 g bisacrylamide per 100 ml). Gel buffer
(3×): 150 m$M$ Bis-Tris (pH 7.0), 200 m$M$ 6-aminocaproic
acid.

loading, the gel is placed in the electrophoresis apparatus and the cathode
[50 m$M$ Tricine, 15 m$M$ Bis-Tris (pH 7.0), 0.02% Serva Blue G] and anode
(50 m$M$ Bis-Tris, pH 7.0) reservoirs are filled with the indicated buffers.
Electrophoresis is started at 50 V until the protein sample is just within
the running gel. The voltage is then set to 150 V (running time,
about 6 h). When the blue dye has reached the second half of the gradient
gel, the cathode buffer is replaced with buffer without Serva Blue G. This
makes the proteins become visible in the gel and prevents excess dye from
interfering with Western blotting and subsequent detection. During
the whole procedure the temperature of the gel is maintained at about 4°
by using prechilled buffers and by performing the electrophoresis in a
cold room or box. For calibration purposes a high molecular weight
(HMW) calibration kit for native electrophoresis (Amersham, Bucking-
hamshire, UK) is used. However, these proteins can be used only to obtain
a rough estimate of the molecular mass range as proteins differ in their
ability to bind Serva Blue G.

Reconstitution of Membrane Proteins into Proteoliposomes

A multitude of methods have been described for the reconstitution of
membrane proteins into proteoliposomes. The methods of hydrophobic
absorption or rapid dilution for detergent removal have proved to be

applicable for the functional reconstitution of SecYEG complex and YidC. When the solubilized proteins are mixed before removal of detergent, both methods result in incorporation of two different proteins in the same liposomes.[21,25]

### Bio-Beads

The Bio-Bead method is useful when a detergent with a low critical micelle concentration (CMC) is used in the purification. Acetone/ether-washed *E. coli* phospholipids (Avanti Polar Lipids, Alabaster, AL) are dried under vacuum and the lipid film is hydrated with water to yield a final concentration of 20 mg/ml. The suspension is sonicated at low power to yield an opalescent solution. About 200 $\mu$l of a suspension containing purified SecYEG complex (0.2 mg/ml) and/or YidC (0.1–0.75 mg/ml) is mixed with 200 $\mu$l of a solution containing acetone/ether-washed *E. coli* phospholipids diluted to 4 mg/ml in 0.5% (v/v) Triton X-100. The suspension is incubated for 30 min at 4° under gentle mixing. Before use, Bio-Beads SM-2 adsorbents (Bio-Rad) are washed in a series of solutions: (1) twice in methanol, (2) twice in ethanol, (3) twice in demineralized water, and (4) twice in buffer D [50 m$M$ HEPES-KOH (pH 8.0), 50 m$M$ KCl, 1 m$M$ DTT]. The Bio-Beads are resuspended in 1 ml of buffer D to a final concentration of about 80 mg/ml, and 500 $\mu$l is added to the protein–detergent–phospholipid suspension. After 2 h of incubation at 4°, the Bio-Beads are sedimented by a brief centrifugation in an Eppendorf centrifuge and discarded. The overlaying solution is mixed with 40 mg of fresh Bio-Beads. After incubation at 4° for 4 h, the Bio-Beads are sedimented and the supernatant is mixed with 60 mg of fresh Bio-Beads. The mixture is finally incubated overnight at 4°. After sedimentation, the supernatant is collected, the Bio-Bead fraction is washed separately with 1 ml of buffer D, and the supernatants of both steps are pooled. Proteoliposomes are collected by ultracentrifugation (200,000 $g$, 30 min), washed twice with buffer D, and resuspended in 200 $\mu$l of buffer D. Proteoliposomes are aliquoted, frozen in liquid nitrogen, and stored at −80°.

### Rapid Dilution

Rapid dilution is a fast and efficient method to obtain functionally reconstituted membrane proteins with high recovery of activity. A disadvantage of the method is that it readily yields protein-free liposomes.

[25] A. J. M. Driessen, L. Brundage, J. P. Hendrick, E. Schiebel, and W. Wickner, *Methods Cell Biol.* **34,** 147 (1991).

To 40 $\mu$l of the acetone/ether-washed *E. coli* phospholipids (prepared as described previously), 200 $\mu$l of solubilized SecYEG complex (0.2–0.6 mg/ml) is added and the mixture is incubated for 5 min on ice. Next, the sample is diluted with 4 ml of buffer E (50 m$M$ KCl, 50 m$M$ Tris-HCl, pH 8.0). After 30 min, proteoliposomes are collected by ultracentrifugation (125,000 g, 30 min) and resuspended in 200 $\mu$l of buffer E. Samples are frozen in liquid nitrogen and stored at $-80°$. Before use, samples are thawed, followed by brief sonication (3 times, 10 s each) in a bath sonicator. This improves the homogeneity of the proteoliposomes and reduces the content of protein-free liposomes.

Preprotein Translocation Assays

To monitor the activity of SecYEG proteoliposomes, the SecA-dependent translocation of a model preprotein can be followed by a protease-accessibility assay.[26] Conventional assays make use of a radiolabeled precursor protein synthesized by means of an *in vitro* transcription/translation system.[26] This section describes a highly sensitive and widely applicable translocation assay that utilizes a fluorescently labeled preprotein. In addition, the translocation activity of SecYEG proteoliposomes can be determined by measuring the preprotein-stimulated ATPase activity of SecA,[27] using an optimized microtiter plate assay. The assay of YidC-mediated membrane protein insertion in proteoliposomes containing coreconstituted SecYEG complex and YidC by means of a direct cross-linking approach is described elsewhere.[21]

*Purification of SecA*

The purification of SecA is performed by a method modified from Cabelli: *et al.*[28] *Escherichia coli* DH5$\alpha$[29] is transformed with plasmid pMKL18 [unpublished, gift of R. Freudl (Institut Für Biotechnologie 1, Forschungszentrum Jülich, Jülich, Germany); SecA gene cloned in pUC19 vector, expression of SecA, *amp*], grown overnight at 37° in LB, 0.5% (w/v) glucose, and ampicillin (0.1 mg/ml), and 20 ml is used to inoculate 2 liters of LB with ampicillin (0.1 mg/ml). Growth is continued at 37° until an OD$_{660}$ of about 0.6, whereupon overexpression is induced by the addition of 0.5 m$M$ IPTG. To facilitate further growth, 0.4% arabinose can be added. After 2 h of growth, the cells are collected by centrifugation

---

[26] K. Cunningham, R. Lill, E. Crooke, M. Rice, K. Moore, W. Wickner, and D. Oliver, *EMBO J.* **8,** 955 (1989).

[27] R. Lill, K. Cunningham, L. A. Brundage, K. Ito, D. Oliver, and W. Wickner, *EMBO J.* **8,** 961 (1989).

and resuspended in 50 m$M$ Tris-HCl (pH 7.6) and 20% sucrose. The suspension is supplemented with 1 m$M$ DTT, 0.5 m$M$ PMSF, DNase (1 mg/ml), and RNase (1 mg/ml). Cells are lysed by three passes through a French pressure cell at 8000 lb/in$^2$. Unbroken cells and membranes are removed by subsequent centrifugation steps for 15 min at 30,000 $g$ and for 60 min at 125,000 $g$, respectively. The supernatant is diluted to 50 ml with buffer F [50 m$M$ Tris-HCl (pH 7.6), 10% (v/v) glycerol, and 1 m$M$ DTT] and loaded on a MonoQ anion-exchange column equilibrated with buffer F supplemented with 150 m$M$ NaCl. The column is washed with 8 column volumes of buffer F with 180 m$M$ NaCl, and the SecA protein is eluted with 14 column volumes of a linear gradient of 200 to 400 m$M$ NaCl in buffer F. Fractions are analyzed by 10% SDS–PAGE, pooled, and concentrated by Amicon ultrafiltration to 5 ml, using 30-kDa cutoff filters. The concentrated fraction is loaded on a gel-filtration column (HiLoad 26/60 Superdex 200, Amersham Pharmacia Biotech, Uppsala, Sweden) equilibrated with buffer F, and the SecA is eluted from the column with the same buffer. Fractions containing the purified SecA are pooled and frozen in aliquots in liquid nitrogen. The purified SecA is stored at $-80°$. One liter of culture yields about 6 mg of purified protein.

### Production and Purification of proOmpA

For protein translocation assays, the precursor of outer membrane protein A (proOmpA) is an ideal substrate as it can be obtained easily in large amounts as a precursor harboring the signal sequence. For this purpose, proOmpA is expressed in temperature-sensitive *E. coli* MM52,[30] which bears a conditional lethal mutation in the *secA* gene. At the submissible temperature of 37°, this strain exhibits a major secretion defect, which prevents the cleavage of the signal sequence and causes the accumulation of the precursor form of OmpA in the cytosol. *Escherichia coli* MM52 is transformed with pET503 (overexpression of proOmpAC290S, *amp*)[6] and grown overnight in LB with 0.5% glucose and ampicillin (0.1 mg/ml) at 30°. From this culture, 10 ml is used to inoculate 140 ml of LB supplemented with 0.5% glucose and ampicillin (0.1 mg/ml). Growth is continued at 37° to an $OD_{660}$ of about 1. To reduce the remaining concentration of glucose, the culture is diluted with 750 ml of LB with ampicillin (0.1 mg/ml) that is prewarmed at 37°. After 30 min of growth at 37°, high-level expression of the *ompA* gene is induced by the

[28] R. J. Cabelli, K. M. Dolan, L. P. Qian, and D. B. Oliver, *J. Biol. Chem.* **266**, 24420 (1991).
[29] D. Hanahan, *J. Mol. Biol.* **166**, 557 (1983).
[30] D. B. Oliver and J. Beckwith, *Cell* **25**, 765 (1981).

addition of 1 m$M$ IPTG. After 2 h, the cells are harvested, washed with 10 m$M$ Tris-HCl, pH 7.0, and resuspended in about 5 ml of the same buffer. Cells are lysed by sonication, using a tip sonicator for 10 min with pulses of 10 s on a 50% duty cycle. The inclusion bodies and cell debris are collected by low-speed centrifugation (1800 $g$, 10 min), and dissolved in 10 ml of 8 $M$ urea, 50 m$M$ Tris-HCl (pH 7.0). Nonsoluble material is removed by centrifugation (200,000 $g$, 60 min). The supernatant is loaded on a HiTrap Q Sepharose column (Amersham Pharmacia Biotech) equilibrated with 8 $M$ urea, 50 m$M$ Tris-HCl (pH 7.0), and the column is washed with the same buffer. ProOmpA elutes with the nonbound protein fraction, whereas the contaminating proteins remain bound to the column. The urea-dissolved proOmpA is frozen in liquid nitrogen and stored at −20°. One liter of culture yields about 40 mg of purified protein.

*Fluorescein Labeling of Purified proOmpA*

OmpA contains two cysteine residues at positions 290 and 302. These can be fluorescently labeled with fluorescein maleimide. To obtain a proOmpA molecule that is labeled at a single position, the cysteine at position 290 is mutated through site-directed mutagenesis, yielding proOmpA(C290S). The protocol, however, also works with the wild-type proOmpA molecule bearing the two cysteine residues, in which case both cysteines are labeled.

Urea-dissolved proOmpA (1–3.5 mg/ml) is reduced with a 10-fold molar excess of tri(2-carboxyethyl)phosphine (TCEP; Molecular Probes, Leiden, The Netherlands) for 30 min at room temperature. Next, a 20-fold molar excess of fluorescein maleimide is added. After 30 min, the reaction is terminated by the addition of a 10-fold molar excess of DTT. To remove the nonreacted fluorescein, 75 $\mu$l of the labeled proOmpA is spun through a Micro BioSpin 6 chromatography column (Bio-Rad) equilibrated with 8 $M$ urea and 50 m$M$ Tris-HCl, pH 7.0. To remove the remaining nonreacted fluorescein, proOmpA is precipitated with 5% (w/v) trichloroacetic acid. After 60 min of incubation on ice, the precipitated proOmpA is collected by centrifugation for 10 min in an Eppendorf centrifuge at maximum speed. The pellet is washed twice with ice-cold acetone and redissolved in 75 $\mu$l of 8 $M$ urea and 50 m$M$ Tris-HCl, pH 7.0.

*Translocation of Fluorescently Labeled proOmpA*

*In vitro* translocation of fluorescently labeled proOmpA is assayed by its accessibility to added proteinase K.[26] ProOmpA (5 $\mu$g/ml) is incubated at 37° in 50 $\mu$l of a buffer containing 50 m$M$ HEPES-KOH (pH 7.5),

30 m$M$ KCl, bovine serum albumin (BSA, 0.5 mg/ml), 10 m$M$ DTT, 2 m$M$ magnesium acetate, 10 m$M$ phosphocreatine, creatine kinase (50 $\mu$g/ml), SecYEG proteoliposomes (20 $\mu$g/ml), and SecA (20 $\mu$g/ml). The translocation reaction is started by the addition of 2 m$M$ ATP and terminated after 5–30 min by chilling the samples on ice. Samples are treated with proteinase K (0.1 mg/ml) for 15 min on ice, and the protease-protected proOmpA is precipitated by the addition of 150 $\mu$l of 10% (w/v) trichloroacetic acid. After a further incubation for 30 min on ice, the proOmpA is pelleted by centrifugation for 10 min in an Eppendorf centrifuge. Pellets are washed with ice-cold acetone and redissolved in 10 $\mu$l of 2-fold concentrated SDS–PAGE sample buffer. Samples are loaded on 10% SDS–polyacrylamide gels and the fluorescence of the protease-protected proOmpA is, without fixation, visualized *in gel* in a highly sensitive Roche Applied Science Lumi-Imager F1 with the emission filters set at 520 nm. An example of a translocation reaction imaged by fluorescence is shown in Fig. 2. Alternatively, imaging can be done with a regular gel documentation system equipped with a charge-coupled device (CCD) camera using a general ultraviolet (UV) tray for excitation. However, care should be taken that the excitation light is distributed evenly over the slab gel.

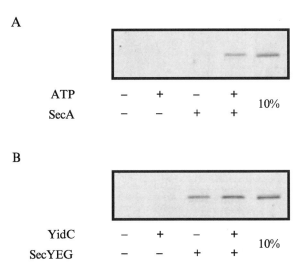

FIG. 2. *In vitro* translocation of proOmpA–fluorescein. (A) Translocation of proOmpA–fluorescein into proteoliposomes containing purified SecYEG in the presence or absence of ATP and SecA. (B) Translocation of proOmpA–fluorescein into liposomes or proteoliposomes containing purified YidC, SecYEG, or SecYEG and YidC. 10%, 10% of the total added proOmpA–fluorescein.

*ProOmpA-Stimulated SecA ATPase Activity Assay*

The preprotein-stimulated SecA ATPase activity in the presence of SecYEG proteoliposomes (SecA translocation ATPase) can be performed essentially as described by Lill et al.[27] To monitor the release of inorganic phosphate on the hydrolysis of ATP, a malachite green molybdate reagent is prepared as follows: malachite green (340 mg) is dissolved in 75 ml of water and mixed with a solution containing 10.5 g of ammonium molybdate dissolved in 250 ml of 4 $M$ HCl. The volume is adjusted to 1 liter by the addition of demineralized water and kept on ice for at least 1 h. The solution is filtered through Whatman (No. 3) paper, and it can be used for at least 3 months when stored at 4°. Translocation ATPase is measured in a 96-well microtiter plate according to the following procedure: a solution containing 50 m$M$ Tris-HCl (pH 8.0), 50 m$M$ KCl, 5 m$M$ MgCl$_2$, 1.0 m$M$ DTT, and BSA (5 $\mu$g/ml) is preheated at 37° and, just before use, supplemented with 1 m$M$ ATP and purified SecA (10 $\mu$g/ml). Next, SecYEG proteoliposomes (25 $\mu$l, 5–25 $\mu$g of protein) are premixed with 150 $\mu$l of this solution to yield the reaction mixture. For background ATPase activity, 1 $\mu$l of an 8 $M$ urea solution in 50 m$M$ Tris-HCl, pH 7.0, is added to the wells of a microtiter plate. For pro-OmpA-stimulated SecA activity, wells are supplemented with 1 $\mu$l of proOmpA (2 mg/ml). A multichannel pipette is used to add 30 $\mu$l of the reaction mixture to each well containing urea or proOmpA. The microtiter plate is incubated at 37° for 30 min. Finally, 200 $\mu$l of the malachite green solution is added and the plates are incubated for 2–5 min at room temperature. The activity is monitored by measurements of OD$_{660}$ with zero adjustment using malachite green.

Conclusion

Protein translocation is one of the examples of a complex membrane-bound biological process that has been functionally reconstituted into proteoliposomes. This process relies on several membrane proteins that need to be coreconstituted in order to yield a functional system. The functional membrane integration of newly synthesized inner membrane proteins, using the purified translocase and associated components, remains a challenge for future research. Other challenges lie in the reconstitution of the complete assembly of multisubunit membrane proteins.

Acknowledgment

The authors thank Nico Nouwen for careful reading of the manuscript.

# [6]   Liposomes in the Study of Pore-Forming Toxins

*By* MAURO DALLA SERRA and GIANFRANCO MENESTRINA

## Introduction

Membrane-interacting toxins are secreted as soluble peptides or proteins by a variety of organisms such as bacteria, plants, and animals.[1,2] To express their pathogenic effects they are able to incorporate into the cell plasma membrane and interfere with its integrity. They can either permit the translocation to the cytosol of a toxic moiety with enzymatic activity or directly increase the permeability of the membrane to small molecules. Interestingly, both these routes lead to the formation of new exogenous channels in the plasma membrane. Toxins in the first group are referred to as the A-B superfamily,[3,4] whereas toxins of the second kind belong to the so-called membrane-damaging/pore-forming supergroup.[5-7] Because of the similarities in membrane interaction, we deal in this chapter with both of them, although the term pore-forming toxins would normally be reserved for members of the second group. Our aim is to show the importance and significance of liposomes as a model of the cell plasma membrane to better understand cell–toxin interaction.

## Pore-Forming Toxins

### General Aspects

Pore-forming toxins (PFTs) are biological weapons improved during evolution. For example, in the case of bacteria, they effectively enhance the performance of the invading microbe by lowering the host defense

[1] A. L. Harvey, *in* "Handbook of Toxicology" (W. T. Shier and D. Mebs, eds.), p. 1. Marcel Dekker, New York, 1990.

[2] J. E. Alouf and J. H. Freer, "The Comprehensive Sourcebook of Bacterial Protein Toxins," 2nd ed. Academic Press, London, 1999.

[3] O. Rossetto, M. de Bernard, R. Pellizzari, G. Vitale, P. Caccin, G. Schiavo, and C. Montecucco, *Clin. Chim. Acta* **291,** 189 (2000).

[4] S. Olsnes, H. Stenmark, J. O. Moskaug, S. McGill, I. H. Madshus, and K. Sandvig, *Microb. Pathog.* **8,** 163 (1990).

[5] G. Menestrina, G. Schiavo, and C. Montecucco, *Mol. Aspects Med.* **15,** 79 (1994).

[6] G. Menestrina and M. Dalla Serra, *in* "Encyclopedia of Life Sciences," at: http://www.els.net. Nature Publishing Group, Macmillan Reference, London, 2000.

[7] F. G. van der Goot, "Pore-Forming Toxins," Vol. 257. Springer-Verlag, Berlin, 2001.

system and by providing access to nutrients. They exert their action by the formation of pores in the plasma membrane of the target cell, which has turned out to be one of the most common killing mechanisms adopted by toxins.[6-8] When the number of pores in the cell membrane becomes large enough, the intra- and extracellular concentrations of ions and small solutes equilibrate. An osmotic imbalance, due to the presence of internal macromolecules that are too large to permeate through the new toxin pores, is thereby created. This causes an uncontrollable water influx into the cell, which responds by swelling and eventually lysing. Under these conditions, red blood cells release their hemoglobin content. It was proposed that freed hemoglobin could become a source of iron required for bacterial growth. Indeed, in many cases, the synthesis and expression of pore-forming toxins are tightly regulated by iron, that is, conditions of iron deprivation induce toxin expression. The number of toxins known to belong to this group is already large, yet it is continuously growing.

*Proposed Classifications*

Because pore formation is a general mechanism, PFTs have been subdivided into a number of classes according to different criteria, for example, the molecular size, the structure, or the producing organism.

*Subdivision according to Size*

On the basis of polypeptide size, PFTs are usually divided between small pore formers and large pore formers. The first class includes peptides up to approximately 4–5 kDa, which can be either secreted by microorganisms (e.g., *Staphylococcus aureus* δ-lysin,[9] or amoebapore[10]) or produced by animals (e.g., bee venom melittin[11] and amphibian skin magainin[12]), plants (e.g., thionins[13] and plant defensins[14]), and fungi (e.g., alamethicin[12]). Although their secondary structure may be different, they have in common an amphipathic organization, often stabilized by disulfide bridges, and generally also by the presence of an excess of positive charges in the amphiphilic portion.[15] Large pore formers are, instead, proteins

[8] J. I. Kourie and A. A. Shorthouse, *Am. J. Physiol.* **278,** C1063 (2000).
[9] J. H. Freer and T. H. Birkbeck, *J. Theor. Biol.* **94,** 535 (1982).
[10] M. Leippe, *Parasitol. Today* **13,** 178 (1997).
[11] C. E. Dempsey, *Biochim. Biophys. Acta* **1031,** 143 (1990).
[12] B. Bechinger, *J. Membr. Biol.* **156,** 197 (1997).
[13] P. Hughes, E. Dennis, M. Whitecross, D. Llewellyn, and P. Gage, *J. Biol. Chem.* **275,** 823 (2000).
[14] K. Thevissen, F. R. G. Terras, and W. F. Broekaert, *Appl. Environ. Microbiol.* **65,** 5451 (1999).

ranging between 10 and 200 kDa, which can also be produced by plants, animals, and microorganisms.[6] They do not normally have disulfide bridges, nor a strictly common structural motif, but simply share a mechanism of action. This classification does not refer, and remains unrelated, to the size of the pore formed, the radius of which may vary broadly (typically remaining within a range of 0.5 to 3 nm), but in some cases extending to huge holes up to 30 nm in diameter.

### Subdivision according to Structure

With regard to the secondary structure, PFTs are usually classified in two groups: $\alpha$-PFTs if they are predicted to form pores via membrane-inserted $\alpha$ helices, and $\beta$-PFTs if they form transmembrane $\beta$ barrels.[16]

$\alpha$-*PFTs.* High-resolution three-dimensional (3-D) structures [either from X-ray or nuclear magnetic resonance (NMR) studies] are available only for the monomeric, water-soluble form of $\alpha$-PFTs, for example, colicins and *Bacillus thuringiensis* $\delta$-endotoxins as protein toxins,[16,17] or for alamethicin and magainin as peptides.[18,19] All of them contain hydrophobic, or amphipathic, helices long enough to span the bilayer (typically more than 18 residues). The pore-forming domain, in the soluble form, comprises a helical bundle, with the hydrophobic helices placed in the interior and the amphipathic helices at the surface. On interaction with the membrane, an inside-out transition might lead to the exposure, and penetration into the lipid, of the hydrophobic helices, whereas the amphipathic helices remain flat (or only partially inserted) on the bilayer surface. Aggregation of inserted helices may thereafter originate the transmembrane barrel pore. In one exception, *Escherichia coli* ClyA monomers form the transmembrane barrel without an inside-out transition, using only a small hydrophobic patch present on their surface.[20]

$\beta$-*PFTs.* $\beta$-PFTs have predominantly hydrophilic amino acid sequences, with no evidence of long hydrophobic segments.[21] The high-resolution structure of an assembled $\beta$-PFT pore, *S. aureus* $\alpha$-toxin, has been solved.[22]

[15] R. E. W. Hancock, *Expert Opin. Investig. Drugs* **9,** 1723 (2000).

[16] D. B. Lacy and R. C. Stevens, *Curr. Opin. Struct. Biol.* **8,** 778 (1998).

[17] J. Li, *Curr. Opin. Struct. Biol.* **2,** 545 (1992).

[18] P. M. Hwang and H. J. Vogel, *Biochem. Cell Biol.* **76,** 235 (1998).

[19] M. Zasloff, *Nature* **415,** 389 (2002).

[20] A. J. Wallace, T. J. Stillman, A. Atkins, S. J. Jamieson, P. A. Bullough, J. Green, and P. J. Artymiuk, *Cell* **100,** 265 (2000).

[21] A. P. Heuck, R. K. Tweten, and A. E. Johnson, *Biochemistry* **40,** 9065 (2001).

[22] L. Song, M. R. Hobaugh, C. Shustak, S. Cheley, H. Bayley, and J. E. Gouaux, *Science* **274,** 1859 (1996).

This heptameric pore results from the cooperative transition of a flexible, glycine-rich, $\beta$ strand, which unfolds from each monomer and inserts into the bilayer in the form of a hairpin, thus generating a 14-stranded $\beta$ barrel with some similarity to bacterial porins. Besides staphylococcal $\alpha$-toxin, a number of other bacterial toxins are supposed to adopt the same $\beta$-barrel topology,[16,21,23] such as the cholesterol-dependent cytolysins pneumolysin, perfringolysin, and streptolysin (PLO, PFO, and SLO),[24] and the protective antigen of anthrax toxin.[25]

### Subdivision according to Producing Organism and Function

*Microbial Hemolysins and Leukotoxins.* Many microbes produce PFTs that lyse erythrocytes and are toxic for other cells, such as epithelial and connective tissue cells or the various leukocyte species.[5–7] These leukotoxins can inhibit the mobility, chemotaxis, and respiratory burst of neutrophils, induce apoptosis of T lymphocytes, release inflammatory mediators from granulocytes or macrophages, prevent phagocytosis by polymorphonuclear cells, and disrupt the phagosome, allowing invasion of the phagocyte. Altogether these effects reduce strongly the immune response of the host against the microorganism, thus inducing a state of immunodepression. Almost all pathogens produce at least one PFT, suggesting that such toxins are important, if not indispensable, for microbial virulence.

*Antimicrobial PFTs.* Microorganisms also produce PFTs that are directed against other microbes with which they are in competition, thus obtaining an advantage. Examples include antibiotics such as nisin[26] or gramicidin,[27] and proteins such as the large group of colicins.[28]

*Venom Components.* Many poisoning animals and plants utilize one or more PFTs as components in their venom. Examples include peptides such as melittin[11] or pardaxin,[29] small proteins such as cobra venom cardiotoxin,[30] and larger proteins such as cnidarian[31] and *Cerebratulus* cytolysins.[32]

[23] C. Lesieur, B. Vécsey-Semjén, L. Abrami, M. Fivaz, and F. G. van der Goot, *Mol. Membr. Biol.* **14,** 45 (1997).

[24] M. Palmer, *Toxicon* **39,** 1681 (2001).

[25] C. Petosa, R. J. Collier, K. R. Klimpel, S. H. Leppla, and R. C. Liddington, *Nature* **385,** 833 (1997).

[26] E. Breukink and B. de Kruijff, *Biochim. Biophys. Acta* **1462,** 223 (1999).

[27] B. A. Wallace, *Bioessays* **22,** 227 (2000).

[28] K. S. Jakes, P. K. Kienker, and A. Finkelstein, *Q. Rev. Biophys.* **32,** 189 (1999).

[29] D. Rapaport, R. Peled, S. Nir, and Y. Shai, *Biophys. J.* **70,** 2502 (1996).

[30] T. K. S. Kumar, G. Jayaraman, C. S. Lee, A. I. Arunkumar, T. Sivaraman, D. Samuel, and C. Yu, *J. Biomol. Struct. Dyn.* **15,** 431 (1997).

[31] G. Anderluh and P. Macek, *Toxicon* **40,** 111 (2002).

[32] J. Liu and K. M. Blumenthal, *J. Biol. Chem.* **263,** 6619 (1988).

*Effectors of Immunity.* Although normally not considered toxins, some molecules share the same purpose and the same mechanism of action of PFTs, and therefore they are worth mentioning here. They include proteins, such as complement and perforin,[33] or smaller peptides with antimicrobial activity.[18,19] They may either occur freely in the body fluids (e.g., in the hemolymph of insects and annelids[34,35] or in the saliva of mammal[36]), or be confined inside dischargeable organelles of specialized cells (e.g., granular defensins[37] and cathelicidins[38]).

## Mode of Action: Multiple Steps Leading to Pore Formation

In addition to their ability to form ion channels, PFTs may also exert on cell membranes a range of other effects related to their nature as cell-targeted amphipathic molecules. The relative importance of these effects depends on the toxin concentration and/or on the distance from the site where the PFT is produced or injected.[5]

At sublytic concentrations (such as those occurring a long distance from the site of bacterial infection or venom injection) they can elicit a number of effects, possibly related to the existence of specific protein receptors on some cells. When engaged by the PFT these receptors may become altered in their normal function. As an example, bacterial leukotoxins may induce several effects in sensitive cells (usually leukocytes): cell activation, release of inflammatory mediators and cytokines, stimulation of $Ca^{2+}$ channels, and initiation of apoptosis.

At higher concentrations, such as those found within a short distance of the PFT source, the number of monomers adsorbed becomes large enough to form pores. When enough of these are present in a cell, the intra- and extracellular concentrations of small solutes (ionic or neutral) equilibrate, whereby internal impermeant macromolecules create an osmotic imbalance. This causes an uncontrollable influx of water through the plasma membrane and the pores themselves, ensuing cell swelling, and eventually cell disruption.

[33] J. D. Young, Z. A. Cohn, and E. R. Podack, *Science* **233,** 184 (1986).

[34] R. I. Lehrer and T. Ganz, *Curr. Opin. Immunol.* **11,** 23 (1999).

[35] S. Lange, E. Kauschke, W. Mohrig, and E. L. Cooper, *Eur. J. Biochem.* **262,** 547 (1999).

[36] D. M. Rothstein, P. Spacciapoli, L. T. Tran, T. Xu, F. D. Roberts, M. Dalla Serra, D. K. Buxton, F. G. Oppenheim, and P. Friden, *Antimicrob. Agents Chemother.* **45,** 1367 (2001).

[37] R. I. Lehrer and T. Ganz, *Curr. Opin. Immunol.* **14,** 96 (2002).

[38] M. Scocchi, S. L. Wang, and M. Zanetti, *FEBS Lett.* **417,** 311 (1997).

In strict contact with the site of production, the PFT concentration may be so great that it can produce not only cell disruption but also extensive tissue damage, facilitating the propagation of the venom, or the microorganism, through the epithelia and into the tissues. This effect requires large amounts of bound toxin and ensues from its amphipathic nature destabilizing the bilayer.

When we consider channel assembly we must admit that in most cases the exact mechanism of the conformational changes leading to pore opening is still elusive. Nevertheless, the following scheme may be considered: (1) binding of the toxin monomers on the external surface of the cell membrane, (2) oligomerization of the membrane-adsorbed toxin monomers, (3) transmembrane insertion of the multimeric complex, and (4) reversible transition from a closed to an open pore configuration. In some cases steps 3 and 4 are virtually indistinguishable.

As stated above, PFTs may have high-affinity binding sites, for example, protein receptors, on some specific target cells. However, in most cases, they simply use lipids (or certain subclasses of lipids) as low-affinity acceptors. The reason for this lies in the nature of the lipids themselves. In fact, lipids are fundamental and abundant components of all cell membranes, they are widespread on different types of cells, and they are not easily mutated. This maximizes the probability that a PFT will induce damage and minimizes the risk of it suddenly becoming ineffective because of a mutation appearing in the receptor. It has been shown that specialized lipid regions of the membrane (e.g., raft domains) may be the preferred site of action of PFTs.[39] They may facilitate PFT action in at least two ways: (1) by promoting a high local concentration desirable for oligomer assembly and (2) by favoring interaction with elements of the signal transduction chain, for example, G proteins, that are colocalized in the raft domains.

Although, in general, the insertion of proteins in membranes is a complex process that requires the assistance of a number of proteins besides the one to be inserted,[40] in the case of PFT insertion is a completely spontaneous, or at most self-assisted, process. This is apparently achieved through interaction with specific lipids, but, more importantly, by an aggregation mechanism involving multiple copies of the same toxin or of some strictly related toxin components.[21,41] In fact, whereas the conformational change occurring within one monomer is generally small, the

[39] M. Fivaz, L. Abrami, and F. G. van der Goot, *Protoplasma* **212,** 8 (2000).

[40] F. A. Agarraberes and J. F. Dice, *Biochim. Biophys. Acta* **1513,** 1 (2001).

[41] M. Ferreras, F. Höper, M. Dalla Serra, D. A. Colin, G. Prévost, and G. Menestrina, *Biochim. Biophys. Acta* **1414,** 108 (1998).

cooperative transition of *n* monomers, and the favorable noncovalent interactions occurring in the oligomer, may concur to generate a stable membrane-inserted protein complex.

Hence, PFTs are a unique model system, the study of which is important not only to understand the toxin function but also to shed light on more general mechanisms of action of proteins, such as the conformational transitions leading to aggregation (protein–protein interaction), those leading to membrane binding and insertion (protein–lipid interaction), and those leading to opening and closing of the pores (gating mechanisms).

## Liposomes

We only briefly address here the most practical issues related to the use of liposomes as a convenient model to study PFT activity. More general and exhaustive reviews can be found in other chapters of this series and elsewhere (e.g., New[42]).

### Liposomes as Convenient Model System

Liposomes are synthetic lipid vesicles with an internal aqueous volume separated from the environment by a tight lipid membrane. Their main components are usually phospholipids. They form spontaneously after the dispersion of these amphiphiles in water. According to the method of preparation, it is possible to vary their dimension, entrapped volume, and lamellarity. Liposomes can be classified according to their dimensions (ranging from a diameter of 20–30 nm for sonicated vesicles to several microns for giant liposomes), to the number of lamellae in the lipid phase, and to the method of preparation.

One of the advantages offered by these vesicles is that their lipid composition can be varied, almost at will, without compromising their stability. This provides a useful system with which to investigate the lipid requirements, or preferences, of PFTs. Another advantage is that the information they provide is averaged over a large population of equivalent entities (typically $10^9$–$10^{11}$) and is therefore less prone to the artifacts that may occur when just a few pore copies are studied, for example, in electrophysiological studies performed with planar lipid membranes or single-cell patches.[43]

[42] R. R. C. New, "Liposomes: A Practical Approach." IRL Press, Oxford, 1990.

[43] M. Dalla Serra and G. Menestrina, *in* "Bacterial Toxins, Methods and Protocols" (O. Holst, ed.), p. 171. Humana Press, Totowa, NJ, 2000.

During the process of formation, liposomes entrap the molecules that are present in the aqueous solution (fluorescent markers, quenchers, sugars, ions, etc.) with variable efficiency depending on the lipid composition and the protocol of preparation. Labeled lipids and lipophilic drugs can also be mixed with the phospholipids and become incorporated into the membrane bilayer. The insertion of bacterial PFTs into a vesicle is normally achieved in a simple way. Soluble toxins are mixed with preformed vesicles in the water phase, where they spontaneously undergo the transition from hydrophilic to amphiphilic that renders them competent for insertion into the membrane. In some cases, the presence of a specific, suitable lipid component may be necessary.

*Preparations Suitable to Study of PFTs*

The easiest liposome preparation, that is, the dispersion of lipid in the liquid phase by thorough stirring, leads to a suspension of multilamellar vesicles (MLVs). Despite their heterogeneous size and low entrapment capacity, which might be a disadvantage, they can be successfully employed, for example, to investigate toxin-induced changes in the lipid phase. They are also useful to study toxin binding, because they are easily precipitated by centrifugation and thereby separated from unbound toxin.

For characterization of permeabilizing activity, however, a unilamellar system with well-defined and homogeneous diameter is to be preferred. Small unilamellar vesicles (SUVs), as obtained by ultrasonic treatment of MLVs, can be used.[41] These are normally 20–30 nm in diameter, with a small internal aqueous volume and a high radius of curvature, which stresses the asymmetry between the internal and external leaflets. Alternatively, it is advisable to use large unilamellar vesicles (LUVs) obtained, for example, by peeling the MLVs via their extrusion through polycarbonate filters with a well-defined pore diameter (normally 100–200 nm).[44,45] Bigger dimensions are also possible, but usually at the expense of unilamellarity.

*Characterization of Liposome Systems*

The interaction of PFTs with liposomes may depend strongly on some physical parameters of the lipid system such as lamellarity, size, internal

[44] M. Tejuca, M. Dalla Serra, M. Ferreras, M. E. Lanio, and G. Menestrina, *Biochemistry* **35,** 14947 (1996).
[45] M. Tejuca, M. Dalla Serra, C. Alvarez, C. Potrich, and G. Menestrina, *J. Membr. Biol.* **183,** 125 (2001).

volume, and membrane integrity.[46] The liposome diameter regulates the membrane curvature, which often affects lipid–toxin interaction. A common way to determine liposome diameter, under physiological conditions, is offered by dynamic or quasi-elastic light scattering (QLS).[47,48]

Often an accurate estimate of phospholipid concentration is also important. The most common methods, based on total phosphate measurement, are generally tedious and require a multistep and time-consuming treatment of the samples with the preparation of elaborate reagent solutions. Alternatively, in the case of phosphatidylcholine (PC) or mixtures with a fixed proportion of PC, we have successfully used a simple, rapid, and inexpensive method based on a commercial kit for the enzymic determination of the choline head group (Menagent phospholipid kit; Menarini Diagnostics, Florence, Italy), which is normally employed in clinical analysis.[44]

Quantitative determination of vesicle lamellarity can be achieved by measuring the proportion of lipids in the outer and inner layers of liposomes. A direct, noninvasive, spectroscopic method consists of measuring the phosphorus NMR signal before and after the addition, in the external medium, of impermeant reagents causing either broadening (e.g., $Mn^{2+}$) or shift (e.g., $Pr^{3+}$) of the NMR signal.[49]

## Pore Formation Studies with Liposomes

Liposomes may be useful models to study all steps of the general mechanism of PFT–membrane interaction. Here we describe some simple experimental procedures with which we have direct experience and that are suitable to characterize PFT action on liposomes. For the sake of simplicity we focus separately on each step leading to pore opening, and in the natural order in which they occur, even though the property usually checked first in an experimental test is probably the ability of the toxin to induce liposome permeabilization.

[46] J. A. Gazzara, M. C. Phillips, S. LundKatz, M. N. Palgunachari, J. P. Segrest, G. M. Anantharamaiah, W. V. Rodrigueza, and J. W. Snow, *J. Lipid Res.* **38**, 2147 (1997).

[47] R. Pecora, "Dynamic Light Scattering." Plenum Press, New York, 1985.

[48] C. Alvarez, M. Dalla Serra, C. Potrich, I. Bernhart, M. Tejuca, D. Martinez, I. F. Pazos, M. E. Lanio, and G. Menestrina, *Biophys. J.* **80**, 2761 (2001).

[49] N. Duzgunes, J. Wilschut, K. Hong, R. Fraley, C. Perry, D. S. Friend, T. L. James, and D. Papahadjopoulos, *Biochim. Biophys. Acta* **732**, 289 (1983).

## Binding and Insertion of Toxin into Lipid Matrix

*Detecting Toxin Binding.* Spectroscopic techniques offer the opportunity to measure the extent and rate of toxin binding to liposomes. For this it is necessary to have a signal that changes, depending on whether the toxin is in the lipid phase or in the water phase. The most used indicator is tryptophan, an intrinsic chromophore whose fluorescence is environment dependent. Because the variations in quantum yield and emission maximum induced by polarity changes are not strong, in some cases suitable quenchers have been used to enhance the differences.[50] If only one tryptophan residue is present in the toxin (either naturally or after specific mutagenesis), the use of quenchers with a precise location in the lipid bilayer, for example, lipids with bromine ions or electron paramagnetic resonance (EPR) spin labels at different positions along the chain,[51] can provide information about the membrane topology of the fluorophore.

Alternatively, the toxin can be labeled with an environment-sensitive fluorophore at precise positions, for example, by the modification of specific cysteines with acrylodan, and the emission of the label during interaction can be measured. Coupled with the single-cysteine substitution technique, which introduces unique cysteines at well-defined positions along the protein sequence, this has allowed the precise determination of which part (or parts) of the PFT molecule interacts with the membrane. Examples include PFO,[52] SLO,[53] *S. aureus* α-toxin,[54] and sea anemone toxin.[55] Fluorescence resonance energy transfer (FRET), between suitable donor and acceptor fluorophores, has also been used for precise estimation of the lipid–toxin molecular interaction both with peptide toxins[56] and with protein toxins,[57] allowing more distant evaluations.[58,59] In some cases,

[50] P. Macek, M. Zecchini, C. Pederzolli, M. Dalla Serra, and G. Menestrina, *Eur. J. Biochem.* **234,** 329 (1995).

[51] F. S. Abrams and E. London, *Biochemistry* **31,** 5312 (1992).

[52] R. W. Harris, P. J. Sims, and R. K. Tweten, *J. Biol. Chem.* **266,** 6936 (1991).

[53] M. Palmer, P. Saweljew, I. Vulicevic, A. Valeva, M. Kehoe, and S. Bhakdi, *J. Biol. Chem.* **271,** 26664 (1996).

[54] A. Valeva, A. Weisser, B. Walker, M. Kehoe, H. Bayley, S. Bhakdi, and M. Palmer, *EMBO J.* **15,** 1857 (1996).

[55] G. Anderluh, A. Barlic, Z. Podlesek, P. Macek, J. Pungercar, F. Gubensek, M. Zecchini, M. Dalla Serra, and G. Menestrina, *Eur. J. Biochem.* **263,** 128 (1999).

[56] E. Gazit, W. J. Lee, P. T. Brey, and Y. Shai, *Biochemistry* **33,** 10681 (1994).

[57] J. H. Lakey, D. Duche, J. M. Gonzalez-Manas, D. Baty, and F. Pattus, *J. Mol. Biol.* **230,** 1055 (1993).

[58] J. H. Lakey, D. Baty, and F. Pattus, *J. Mol. Biol.* **218,** 639 (1991).

[59] B. A. Steer and A. R. Merrill, *Biochemistry* **33,** 1108 (1994).

more sophisticated environment-sensitive fluorescence lifetime techniques have been applied with success.[60]

Topological information about protein folding can be obtained with an extrinsic fluorophore such as anilinonaphthalene sulfonate (ANS), which can stick to exposed apolar regions of proteins, thereby increasing its fluorescence. Any variation of protein structure increasing its hydrophobic surface, as is necessary for membrane interaction, can be reported by a change in the fluorescence of ANS.[61,62]

Toxin–liposome interactions are often characterized by the partition coefficient of the toxin between the lipid and water phases. Determination of this physical parameter requires measuring the fraction of toxin bound to the liposomes and the fraction free in solution. Circular dichroic (CD) spectroscopy can be used when the toxin undergoes extensive changes in its hydrophilic to lipophilic transition.[63] Alternatively, an ultracentrifugation step can be employed to precipitate the liposomes together with the bound toxin (2–4 h at 100,000 $g$ is normally enough), thus separating free from bound toxin. From the pellet it is possible to determine the lipid-to-bound toxin ratio, for example, by polarized Fourier transform infrared (FTIR) spectroscopy, as reported by Menestrina,[64] concentration of free toxin can be estimated, for instance, by determining the permeabilizing activity remaining in the supernatant.[44]

The partition coefficient ($K$) is defined as

$$K = \frac{T_b}{T_f \cdot L} = \frac{T_b}{(T_0 - T_b) \cdot L} \tag{1}$$

where $T_0$ is the total concentration of toxin applied, $L$ is the total concentration of lipid present, and $T_f$ and $T_b$ are the concentrations of free and bound toxin, respectively.

*Oligomerization and Prepore Formation*

*Infrared and Ultraviolet Spectroscopy to Detect Changes in Secondary Structure and Folding.* Information about the secondary and three-dimensional structure of PFT is paramount to understand the change from a

[60] O. Shatursky, A. P. Heuck, L. A. Shepard, J. Rossjohn, M. W. Parker, A. E. Johnson, and R. K. Tweten, *Cell* **99**, 293 (1999).

[61] M. Dalla Serra, J. M. Sutton, F. Höper, J. A. Downie, and G. Menestrina, *Biochem. Biophys. Res. Commun.* **263**, 516 (1999).

[62] K. Sandvig and J. O. Moskaug, *Biochem. J.* **245**, 899 (1987).

[63] G. Schwarz and G. Beschiaschvili, *Biochim. Biophys. Acta* **979**, 82 (1989).

[64] G. Menestrina, *in* "Bacterial Toxins, Methods and Protocols" (O. Holst, ed.), p. 115. Humana Press, Totowa, NJ, 2000.

water-soluble to a membrane-embedded form, which is often monitored by the aggregation of toxin monomers. Accurate X-ray structures of membrane proteins are becoming available, but, because of intrinsic crystallization problems, only at a slow pace. Besides the photosynthetic reaction center,[65] they refer mainly to channels such as porins[66] $K^+$,[67] $Cl^-$,[68] and glycerol channels[69]; and aquaporin.[70] Similarly, although the X-ray structures of several PFTs have now been established,[22,25,71–76] only one of these, staphylococcal $\alpha$-toxin,[22] is in the putative pore-forming conformation (obtained with a detergent rather than with lipids). All the others are in the soluble state.

FTIR and far-UV CD spectroscopy provide an alternative means to estimate the secondary structure of proteins simply and quickly,[77–79] although with limited precision compared with crystallography. Even if they do not provide information about 3-D folding, such techniques have the advantage of being applicable to membrane-adsorbed proteins as well as to protein in solution.[64,80,81] This is extremely important in the case of

[65] J. Deisenhofer and H. Michel, *EMBO J.* **8,** 2149 (1989).

[66] M. S. Weiss, A. Kreusch, E. Schiltz, U. Nestel, W. Welte, J. Weckesser, and G. E. Schulz, *FEBS Lett.* **280,** 379 (1991).

[67] D. A. Doyle, J. H. Morais Cabral, R. A. Pfuetzner, A. Kuo, J. M. Gulbis, S. L. Cohen, B. T. Chait, and R. MacKinnon, *Science* **280,** 69 (1998).

[68] R. Dutzler, E. B. Campbell, M. Cadene, B. T. Chalt, and R. MacKinnon, *Nature* **415,** 287 (2002).

[69] D. Fu, A. Libson, L. J. Miercke, C. Weitzman, P. Nollert, J. Krucinski, and R. M. Stroud, *Science* **290,** 481 (2000).

[70] H. Sui, B. G. Han, J. K. Lee, P. Walian, and B. K. Jap, *Nature* **414,** 872 (2001).

[71] M. W. Parker, J. T. Buckley, J. P. M. Postma, A. D. Tucker, K. Leonard, F. Pattus, and D. Tsernoglou, *Nature* **367,** 292 (1994).

[72] J. Rossjohn, S. C. Feil, W. J. McKinstry, R. K. Tweten, and M. W. Parker, *Cell* **89,** 685 (1997).

[73] J.-D. Pédelacq, L. Maveyraud, G. Prévost, L. Baba-Moussa, A. Gonzalez, E. Courcelle, W. Shepard, H. Monteil, J.-P. Samama, and L. Mourey, *Structure* **7,** 277 (1999).

[74] R. Olson, H. Nariya, K. Yokota, Y. Kamio, and J. E. Gouaux, *Nat. Struct. Biol.* **6,** 134 (1999).

[75] M. W. Parker, J. P. M. Postma, F. Pattus, A. D. Tucker, and D. Tsernoglou, *J. Mol. Biol.* **224,** 639 (1992).

[76] A. Athanasiadis, G. Anderluh, P. Macek, and D. Turk, *Structure* **9,** 341 (2001).

[77] J. L. Arrondo, A. Muga, J. Castresana, and F. M. Goñi, *Prog. Biophys. Mol. Biol.* **59,** 23 (1993).

[78] M. Jackson and H. H. Mantsch, *Crit. Rev. Biochem. Mol. Biol.* **30,** 95 (1995).

[79] J. T. Yang, C.-S. C. Wu, and H. M. Martinez, *Methods Enzymol.* **130,** 208 (1986).

[80] J. Corbin, N. Methot, H. H. Wang, J. E. Baenziger, and M. P. Blanton, *J. Biol. Chem.* **273,** 771 (1998).

[81] L. K. Tamm and S. A. Tatulian, *Q. Rev. Biophys.* **30,** 365 (1997).

PFTs, which are proteins with two stable conformations, soluble and lipid bound. Both techniques can thus provide insight into the molecular changes occurring on membrane insertion. In the case of CD, because of the high light-scattering contribution of liposomes in the UV region, SUVs are preferred because of their smaller size. The amount of scattered light, in fact, increases with vesicle radius. Alternatively, lipid-mimetic systems, such as trifluoroacetic acid (TFA) and detergents, may be used. Both FTIR and CD spectroscopy require considerable elaboration to derive meaningful structural information. FTIR spectra are deconvoluted to enhance resolution and then curve fitted to derive the amplitudes of all the different signal components representing each secondary structure element (i.e., $\alpha$ helix, $\beta$ sheet, $\beta$ turn, and random coil). Similarly, CD spectra are normally fit as the sum of pure secondary structure spectral contributions,[79] although some more sophisticated mathematical models have also been developed[82] (see also http://www2.umdnj.edu/cdrwjweb/).

Polarized FTIR and CD spectroscopy can also provide information about toxin orientation in the lipid matrix. In a typical attenuated total reflectance (ATR)-FTIR experiment[64,81] multilayers of lipid, or of lipid with adsorbed toxin, are deposited on a crystal, the internal reflecting element (IRE), starting from a LUV preparation. The layers maintain a parallel orientation with respect to the IRE surface, and with respect to each other, the extent of which can be estimated by the differential absorbance of light that is polarized either parallel or perpendicular to the plane of the IRE (Fig. 1). The relative orientation of specific secondary structure elements of the toxin with respect to the IRE and to the lipid can also be estimated in the same way.[64,81,83] Similar information can also be obtained by CD in a polarized membrane environment.[84]

*Methods to Detect Formation of Oligomers.* The presence and molecularity on the lipid membrane of sufficiently stable PFT oligomers can be demonstrated via sodium dodecyl sulfate–polyacrylamide gel electrophoresis (SDS–PAGE). To do this, the toxin is preincubated with suitable lipid vesicles (normally LUVs). Unbound toxin is then washed out by ultrafiltration through filters, the pores of which allow protein release but retain the much bigger vesicles, for example, polysulfone filters with a molecular weight cutoff of 300 kDa [nominal molecular weight limit (NMWL); Millipore, Bedford, MA].[41,85] The filtrates are tested for the presence of free

[82] A. Perczel, K. Park, and G. D. Fasman, *Anal. Biochem.* **203,** 83 (1992).

[83] S. A. Tatulian and L. K. Tamm, *Biochemistry* **39,** 496 (2000).

[84] W. T. Heller, K. He, S. J. Ludtke, T. A. Harroun, and H. W. Huang, *Biophys. J.* **73,** 239 (1997).

[85] C. Pederzolli, L. Cescatti, and G. Menestrina, *J. Membr. Biol.* **119,** 41 (1991).

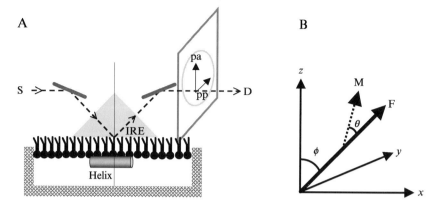

Fig. 1. Determination by ATR-IR spectroscopy of the orientation of peptides inserted in thin lipid films. (A) The IR beam emitted by the source (S) is reflected to the internal reflecting element (IRE), which is represented here as a single-reflection 45°-cut germanium crystal. After total reflection at the interface the beam exits the IRE and is reflected through the polarizer to the detector (D). The polarizer selects photons with orientation either parallel (pa) or perpendicular (pp) to the plane of the internal reflections. At the interface, part of the beam (the evanescent wave) is absorbed differently by the sample depending on the polarization. The dichroic ratio is calculated as the ratio $R = A_{pa}/A_{pp}$, where $A_{pa}$ and $A_{pp}$ are the integrated absorption bands in the parallel and perpendicular configuration, respectively. (B) The average tilt angle $\phi$ of the molecular fiber axis (F) with respect to the $z$ axis (i.e., the perpendicular to the plane of the membrane) can be derived knowing the refractive index of the IRE and the angle $\theta$ between the transition moment (M) and the long axis of the molecule. The orientation of both lipid and peptide molecules can be obtained independently, by looking at different regions of the spectrum (CH stretching around 3000 cm$^{-1}$ and amide I band around 1650 cm$^{-1}$, respectively). The fiber axis is either the axis defined by the lipid chains of the film or the helical axis of the attached peptide in these two cases. The relative orientation of lipid and peptide can finally be derived.

toxins, for example, by measuring the hemolytic activity, or by another sensitive technique, and the washing process is repeated until free toxin is not detected any more. Thereafter, the retentates, which consist of permeabilized vesicles and bound toxin, are subjected to SDS–PAGE. In some cases, oligomers are not stable in SDS and the presence of lipid-inserted oligomers can be demonstrated only after stabilization by chemical cross-linking.[86]

An alternative, powerful approach for checking oligomer formation exploits the FRET effect. This effect occurs between two fluorophores with overlapping excitation and emission spectra, if they are located within a short distance, (typically in the range of 2 to 5 nm[87]). Mixing

[86] G. Belmonte, C. Pederzolli, P. Macek, and G. Menestrina, *J. Membr. Biol.* **131,** 11 (1993).
[87] J. Yguerabide, *Biophys. J.* **66,** 683 (1994).

vesicles with a population of toxin monomers comprising molecules labeled with two different, and suitable, fluorophores leads to FRET only if the monomers achieve close contact on the membrane, for example, by oligomerization.[88]

*Pore Opening: Membrane Permeabilization*

*Release of Fluorescent Molecules.* The capacity to open nonselective, or poorly selective, pores through the lipid membrane is the main characteristic of PFTs and can be exploited to obtain insight into the mechanism of pore formation. Vesicles loaded with fluorescent molecules that have a different quantum yield depending on whether they are inside or outside the vesicles, and that have a size smaller than the cutoff of the channel, are often used to detect pore opening. Examples are dyes such as calcein and carboxyfluorescein (CF),[89] which undergo a marked self-quenching when they are highly concentrated in the inner compartment (typically 50–200 m$M$), or ion couples, such as 8-aminonaphthafene-1,3,6-trisulfonic acid (ANTS)–$P$-xylene-bis-pyridinium bromide (DPX),[90] which will emit fluorescence only when they are dissociated, usually in the outer solution. Other fluorescent couples often used are terbium ion (Tb$^{3+}$)–dipicolinic acid (DPA$^-$) and 6-methoxy-$N$-(3-sulfopropyl)-quinolinium (SPQ$^+$)–SO$_4^{2-}$.[91] In this case the presence of a second binding molecule in the outer solution, acting as a quencher of the fluorescence (respectively, EDTA and Cl$^-$ in the two previous cases), is usually exploited. When using such a configuration, two additional aspects should be considered. First, the pore permeability of the second binder must also be taken into account. Dequenching, in fact, may result either from the exit of the dye from the vesicle, or from the entry of the external binder. Second, if one of the two ions forming the couple is too big to pass through the pore, a membrane potential can be established across the vesicle membrane (see Measurements of Membrane Potential, p.117). The formation of such potential can stop ion transport and prevent the full dissociation of the couple and the development of the related fluorescence change.

Loading of these dyes into the vesicles is normally achieved by preparing the liposomes (lipid typically 1 to 10 mg/ml, suspended in 0.5–1.0 ml of

[88] E. M. Hotze, A. P. Heuck, D. M. Czajkowsky, Z. Shao, A. E. Johnson, and R. K. Tweten, *J. Biol. Chem.* **277,** 11597 (2002).

[89] C. Kayalar and N. Düzgüneş, *Biochim. Biophys. Acta* **860,** 51 (1986).

[90] R. A. Parente, S. Nir, and F. C. Szoka, Jr., *Biochemistry* **29,** 8720 (1990).

[91] B. Vécsey-Semjén, S. Knapp, R. Mollby, and F. G. van der Goot, *Biochemistry* **38,** 4296 (1999).

aqueous solution) in the presence of the dye (80 m$M$ calcein-NaOH at pH 7.0) and thereafter removing the external, nonentrapped, pool. A useful way to do this is to spin the suspension through minicolumns filled with a desalting gel (e.g., Sephadex G-50 medium; Pharmacia, Uppsala, Sweden) using a minicentrifuge. This technique is fast, avoids sample dilution (a dilution factor less than 2-fold is obtained easily), and permits the simultaneous washing of many samples.[42,48,92]

The vesicles are finally placed in a stirred fluorometer cuvette and exposed to the PFT, and the time course of the fluorescence increase due to marker release is measured. This reflects the kinetics of disappearance of intact vesicles from the solution, caused by the formation of pores, because the time course of marker release through a single pore is normally too fast to be detected (in the millisecond range). The toxin concentration applied depends strongly on the PFT used and on the experimental conditions (e.g., lipid composition, liposome size, pH, and ionic strength). Normally, lipid-to-toxin ratios range from 5 to 5000. One hundred percent release is obtained at the end of the kinetic phase by destroying the vesicles either with detergent (e.g., 1 m$M$ Triton X-100) or by sonication.

This approach has a number of advantages: it is easy, straightforward, and, in the case of concentration-dependent dyes (e.g., CF), it allows easy discrimination between two conceptually different mechanisms of release, all-or-none or graded. All-or-none release indicates the formation of stable pores with an open lifetime long enough to completely empty the vesicle (which usually takes less than 1 s). Graded release instead indicates the presence of pores that are only transiently open.[93,94] Discriminating the two mechanisms, after an incomplete release, is achieved by determining the degree of self-quenching of the dye retained in the vesicles at steady state after the released dye has been removed again.[95] If the all-or-none mechanism applies, a population of intact vesicles, with a maximum degree of self-quenching, will be present. In the other case, all vesicles will have a degree of self-quenching intermediate between maximum and nil. A less straightforward method for the case in which a quenching couple is used, for example, ANTS–DPX, has also been described,[96] which does

[92] M. Dalla Serra, G. Fagiuoli, P. Nordera, I. Bernhart, C. Della Volpe, D. Di Giorgio, A. Ballio, and G. Menestrina, *Mol. Plant Microbe Interact.* **12**, 391 (1999).

[93] G. Schwarz and A. Arbuzova, *Biochim. Biophys. Acta* **1239**, 51 (1995).

[94] A. S. Ladokhin, M. E. Selsted, and S. H. White, *Biophys. J.* **72**, 794 (1997).

[95] J. N. Weinstein, R. D. Klausner, T. Innerarity, E. Ralston, and R. Blumenthal, *Biochim. Biophys. Acta* **647**, 270 (1981).

[96] A. S. Ladokhin, W. C. Wimley, and S. H. White, *Biophys. J.* **69**, 1964 (1995).

not require separation of the liposomes from the released dye(s) and, furthermore, takes into account even the possibility that one or the other of the two ions is preferentially released.[97]

Particularly noteworthy is the possibility of performing experiments with a microplate reader instead of a conventional fluorometer. This allows for the simultaneous determination of the reaction kinetics from a large number of samples (typically 96, but conceivably also 384 or 1536) with a smaller volume.[41,44,48,92,98] It also reduces the adverse influence of unavoidable variations in experimental conditions ($1/f$ noise) such as temperature fluctuations and age and concentration of liposomes and toxins. For a 96-well microplate a lipid concentration as low as 6 to 20 $\mu M$ has been used. The PFT will be serially 2-fold diluted through the wells of a single row (total well volume is 200 $\mu$l). A pretreatment of the plastic microplates for 30 min with Prionex (0.1 mg/ml; Pentapharm, Basel, Switzerland) or poly(vinyl alcohol) strongly reduces nonspecific adsorption of the PFT to the plastic hydrophobic sites, which would otherwise artifactually reduce the amount of toxin available for the vesicles. This high-throughput approach could be used, for example, for the screening of samples obtained during the purification of PFT[99] or during the search for a lipid or protein receptor.

*Pore Diameter Estimation and Passage of Neutral Molecules.* Determination of the diameter of the channel is obviously important for understanding the mechanism of PFT action. Not surprisingly, a number of methods have also been devised with liposomes, although the results are not yet entirely satisfactory. In a conceptually simple experiment, fluorescently labeled molecules of two different sizes (e.g., dextran of 5 and 50 kDa) are coencapsulated inside liposomes.[100] The pore diameter is obtained by measuring the relative release of the two markers, after they have been separated through a gel-filtration column and quantitatively monitored with a flow-through cuvette in a fluorometer. Although effective, the experiment is not single-step. Alternatively, it is possible to perform experiments in which vesicles are preloaded, each with a different single-size fluorescently labeled dextran. Outside these vesicles a large quenching molecule is added, for example, an antibody against the fluorophore, which cannot enter the vesicle via the

[97] A. S. Ladokhin, W. C. Wimley, K. Hristova, and S. H. White, *Methods Enzymol.* **278,** 474 (1997).

[98] S. Anzlovar, M. Dalla Serra, M. Dermastia, and G. Menestrina, *Mol. Plant Microbe Interact.* **11,** 610 (1998).

[99] J. M. Rausch and W. C. Wimley, *Anal. Biochem.* **293,** 258 (2001).

[100] A. S. Ladokhin, M. E. Selsted, and S. H. White, *Biophys. J.* **72,** 1762 (1997).

toxin pore.[101] If the dextran can escape through the pore, a decrease in the fluorescent signal is detected when it contacts the quencher. Exploiting a similar approach, we loaded LUVs of PC and phosphatidylserine (PS), (1:1 molar ratio; nominal diameter, 400 nm) at the same time with the couple $Tb^{3+}$–$DPA^-$ and with a fluorescein isothiocyanate-labeled dextran of 20 kDa (FD-20). These fluorophores have similar fluorescence emission maxima (543 and 525 nm, respectively), but widely different excitation maxima (280 and 490 nm, respectively). Therefore, when these vesicles are transferred, at a concentration of 50 $\mu M$, into a buffer solution (0.1 $M$ NaCl, 2 m$M$ NaN, 10 m$M$ HEPES, pH 6.0) containing a suitable amount of the respective quenchers (EDTA and rabbit anti-fluorescein antibodies[102]), it is possible to determine the degree of quenching of each fluorophore with a single excitation spectrum. In fact, scanning the excitation from 250 to 500 nm and monitoring the emission through a high-pass band filter (>520 nm), we could observe two distinct peaks, one at about 280 nm for $Tb^{3+}$ emission and one at about 490 nm for FD-20 emission. With this experiment we could demonstrate that *E. coli* hemolysin A induces the release of $Tb^{3+}$, but not of FD-20, confirming its fixed-size channel nature, whereas melittin (at a concentration of 10 $\mu M$, at which it has a strong destabilizing action) immediately releases both molecules, as does Triton X-100 (M. Dalla Serra, S. Angeretti, L. Conci, and G. Menestrina, 1988).

To estimate the pore diameter of preformed channels, unlabeled, neutral molecules can also be used in an osmotic swelling assay. This is based on the fact that LUVs (but not SUVs) are good osmometers and have a high intrinsic permeability to water.[103,104] When the vesicles are placed in a hyperosmotic medium, they first rapidly shrink, as water flows out along the osmotic gradient, and then reswell as the permeable osmolite [e.g., sugars, dextrans, or polyethylene glycols (PEGs)] flows in, together with water. The ensuing volume changes can be observed by measuring the light scattering, which increases as vesicles shrink because of the variation in the form factor,[47] and decreases when they swell again. Because the intrinsic permeability of the lipid membrane to most solutes is low, the rate of reswelling reports the permeation of the osmolite through the channels, which are already present. Typically, shrinking is finished within 200 ms, whereas swelling takes from a few to several seconds, depending on the osmolite dimension. Therefore, a stopped-flow

---

[101] J. C. Sharpe and E. London, *J. Membr. Biol.* **171,** 209 (1999).

[102] A. Stutzin, *FEBS Lett.* **197,** 274 (1986).

[103] J. P. Reeves and R. M. Dowben, *J. Membr. Biol.* **3,** 123 (1970).

[104] B.-S. Mui, P. R. Cullis, E. A. Evans, and T. D. Madden, *Biophys. J.* **64,** 443 (1993).

apparatus ensuring rapid and complete mixing within a few milliseconds is required to satisfactorily monitor the time course of the event. By choosing osmolites with different sizes, it is possible to evaluate the cutoff of the pores.[45,105]

*Measurements of Membrane Potential.* The creation and/or dissipation of a transmembrane potential through the vesicles can be used to detect the insertion of ion channels and to define their selectivity. The potential is measured with a fluorescent lipophilic dye whose partitioning in the lipid phase depends on membrane potential and whose quantum yield changes with the environment. Examples are ANS, cyanine, styryl, and oxonol dyes.[106] Because liposomes do not have a transmembrane potential by themselves, this must be generated, for example, by forming a gradient of one ion across the vesicle membrane and thereafter applying a specific carrier for that ion (e.g., $K^+$ and valinomycin). A Nernst potential is generated in this way through the membrane, which prevents further movement of the ions and dissipation of the gradient. Insertion of a nonselective (or poorly selective) toxin channel will dissipate such a potential, producing a detectable fluorescence change.

Interestingly, the dependence of toxin insertion on the transmembrane voltage can be studied with this technique simply by creating a variable voltage, using different $K^+$ gradients. The voltage-dependent depolarizing effects of proteins such as diphtheria[107] and tetanus toxin[108] or smaller peptides such as insect cecropin[56] and staphylococcal $\delta$-lysin[108] have been studied in this way. Alternatively, if the toxin channels are by themselves selective enough, they may be used directly to establish the transmembrane potential with suitable ion gradients. From the magnitude and the sign of the established voltage it is possible to determine the type and extent of channel selectivity. In this way we have been able to establish the cation selectivity of *E. coli* hemolysin-A pores in a liposome system,[109] confirming our earlier observations in planar lipid membranes (PLMs).[110] Creation of a transmembrane potential in LUVs composed of pure PC or asolectin (170 $\mu M$) is indicated by a change in ANS fluorescence. LUVs were prepared in a buffer containing $TEA^+$ (200 m$M$, pH 7) and transferred to a solution containing ANS (6 $\mu M$) and a different cation (e.g., $Gdn^+$, $Na^+$, $K^+$, or $TBA^+$). The creation of an inner positive potential, after the formation

[105] J. Carroll and D. J. Ellar, *Eur. J. Biochem.* **214**, 771 (1993).

[106] J. Plasek and K. Sigler, *J. Photochem. Photobiol. B* **33**, 101 (1996).

[107] J. W. Shiver and J. J. Donovan, *Biochim. Biophys. Acta* **903**, 48 (1987).

[108] G. Menestrina, S. Forti, and F. Gambale, *Biophys. J.* **55**, 393 (1989).

[109] G. Menestrina, C. Pederzolli, M. Dalla Serra, M. Bregante, and F. Gambale, *J. Membr. Biol.* **149**, 113 (1996).

[110] M. Ropele and G. Menestrina, *Biochim. Biophys. Acta* **985**, 9 (1989).

of transmembrane pores by *E. coli* hemolysin-A (40 n$M$), is indicated by an increase in fluorescent signal. With this approach we demonstrated that hemolysin-A channels permit a higher rate of influx of most cations tested against $TEA^+$, in the order $Na^+ > K^+ > Gdn^+$. $TBA^+$, instead, is less permeant than $TEA^+$, and therefore induces a slow efflux of $TEA^+$ with the consequent development of an internal negative potential. Some attention must be paid to the fact that, being hydrophobic ions, these dyes will in general respond not only to the transmembrane potential, but also to the surface potential generated by charged lipids. In addition, they may act as carriers of counterions, thus possibly interfering with the establishment of the diffusion potential.

Similarly, the creation of a Donnan potential has been used to detect the formation of alamethicin channels. The potential was established in permeabilized vesicles by supplementing the external medium with a few millimoles of a large charged polymer such as polyacrylate or polyethyleneimide.[111] Using this approach the effects of a preformed electric potential on oligomer formation and pore conformation have been studied.

Other techniques have been used to directly measure the release of ions from vesicles. A simple technique involves the use of ion-selective electrodes to monitor the PFT-induced increase in the outer compartment concentration of an ion, for example, $K^+$ or $Ca^{2+}$, which was initially present only in the inner compartment. Although straightforward, this technique is not widely adopted because of the inherently low signal-to-noise ratio and possible artifacts. Another, conceptually similar technique is that of using entrapped radioactive tracers, for example, $Rb^{+86}$. Once relatively common, this approach is now seldom used because of the associated security problems, which render fluorescence experiments much preferred.

An elegant way to study the transport properties of PFT ion channels incorporated in lipid vesicles is to apply the patch-clamp technique. Because traditional LUVs are too small for this technique, giant proteoliposomes, produced by repeatedly fusing normal vesicles one with another, are used.[112] Such a configuration is similar to the BLM system,[43,113] with an even higher time resolution, and with all the pros and cons of such an electrophysiological technique.

[111] V. Brumfeld and I. R. Miller, *Biochim. Biophys. Acta* **1024,** 49 (1990).
[112] O. Moran, O. Zegarra-Moran, C. Virginio, and G. Rottini, *FEBS Lett.* **283,** 317 (1991).
[113] G. Menestrina, *in* "Sourcebook of Bacterial Protein Toxins" (J. E. Alouf and J. H. Freer, eds.), p. 215. Academic Press, London, 1991.

*Theoretical Models*

*Demonstrating Formation of Toxin Oligomers.* Because almost all PFTs act by making oligomeric channels, some general questions that arise concern whether and where oligomerization takes place, the size of the oligomer, and whether this size is fixed or not. Information about the occurrence and extent of aggregation is embodied in the kinetics of pore formation, which can be easily assessed by the release experiments. It is therefore useful to have models for the interpretation of such kinetics, allowing the extrapolation of some basic notions such as equilibrium and rate constants for aggregation, and size of the conducting oligomer.

A thorough and useful treatment of such a general problem has been provided by Schwarz and co-workers in a series of articles on the action of various pore-forming peptides including alamethicin, melittin, and mastoparan.[114–116] This modeling has the advantage of offering an exact analytical solution, but has the disadvantage that it requires precise hypotheses on all the involved steps, which have therefore to be kept to a minimum number. Usually, exact solutions are obtained only up to the formation of a trimer.

Alternatively, an approach of a more statistical nature was presented by Parente *et al.*[90] and later improved by Rapaport *et al.*[29] In this model the process of permeabilization is divided into three steps (Fig. 2): (1) partitioning of toxin monomers into the lipid bilayer (which is assumed to be fast), (2) aggregation of membrane-inserted monomers (assumed to be rate limiting), and (3) formation of a conducting unit and marker release, as soon as the aggregate has reached a critical size comprising at least $M$ monomers.

The incorporation process (first step) is governed by a partition coefficient [Eq. (1)] related to the kinetic parameters by

$$K = k_{-1}/k_1 \tag{2}$$

where $k_1$ and $k_{-1}$ are the forward and backward rate constants for toxin binding.

The aggregation process, instead, is characterized by $k_2$ and $k_{-2}$, the rate constants for adding or removing a monomer from the oligomer. For simplicity they are assumed to be the same at each step of the aggregation. The percentage of release is given by

$$R_\% = 100 \sum_{i=M}^{N} A_i \cdot Z(M, t, i, c, d) \tag{3}$$

[114] G. Schwarz and C. H. Robert, *Biophys. J.* **58**, 577 (1990).
[115] G. Schwarz and C. H. Robert, *Biophys. Chem.* **42**, 291 (1992).
[116] G. Schwarz, *Biophys. Chem.* **86**, 119 (2000).

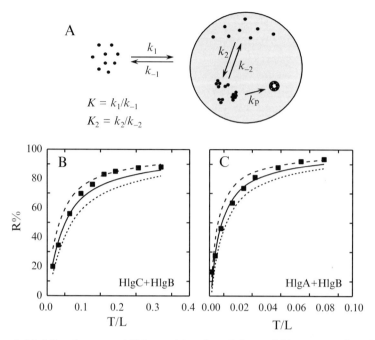

FIG. 2. Modeling the permeabilizing activity of staphylococcal bicomponent hemolysins (Hlg). (A) Scheme of the Parente–Rapaport (P–R) model discussed in text. (B and C) Steady state calcein release induced by the couples HlgC+HlgB (B) and HlgA+HlgB (C) as a function of the toxin-to-lipid (T/L) ratio. Results are expressed as $R_\%$, the percent fraction of the maximal release obtained with Triton X-100 (adapted from Ferreras $et\ al.$[41]). SUVs were composed of PC–cholesterol (1:1 molar ratio). The solid lines are the best fit of the P–R model, whereas the two dotted lines define the confidence limit for the parameter $M$ (an integer representing the size of the oligomeric pore). With both toxins the best fit provides $M = 6 \pm 1$ (for the upper and lower limits shown). Best fit values and confidence limits for the other two parameters of the model, that is, the equilibrium constants for the partitioning from buffer to lipid, $K$, and for the aggregation within the lipid phase, $K_2$, were obtained similarly. $K$ was $(9 \pm 4) \times 10^4\ M^{-1}$ for both toxins, whereas $K_2$ was $0.04 \pm 0.01$ for HlgC+HlgB and $0.25 \pm 0.1$ for HlgA+HlgB. By fitting also the time courses of the fluorescence increase it was possible to obtain separately the forward and backward rates for adding or removing a monomer from the aggregate ($k_2$ and $k_{-2}$, respectively). For HlgC+HlgB and HlgA+HlgB we obtained, respectively, $k_2 = (3 \pm 2) \times 10^{-4}\ s^{-1}$ and $(7 \pm 2) \times 10^{-4}\ s^{-1}$, and $k_2 = (7.5 \pm 5) \times 10^{-3}\ s^{-1}$ and $(2.8 \pm 1) \times 10^{-3}\ s^{-1}$ (not shown).

where $A_i$ is the fraction of vesicles with $i$ monomers bound, which is time independent because it results from the fast partitioning step, $N$ is the maximum number of toxin molecules that can be bound to one vesicle, and $Z(M, t, i, c, d)$ is the probability that a vesicle containing $i$ bound monomers will also have an aggregate of order not smaller than $M$ at time $t$. Complete expressions for $A_i$ and the probability factors $Z(i)$ have been reported.[29,90]

This model has the disadvantage that many unwarranted assumptions are taken (e.g., step-independent reaction rates) and that the details of the interaction are not apparent in the model parameters. On the other hand, it has the advantage of being applicable to any system, independent of its complexity, and of providing simple estimates of important parameters such as $M$ and $K$. In addition to peptides,[29,90] we found this model useful for several PFTs, including *S. aureus* leukotoxins,[41] the plant defense protein linusitin,[98] some toxic metabolites of *Pseudomonas syringae*,[92] and even detergents.[92] The permeabilizing activity of leukotoxins could be conveniently measured either at constant lipid concentration (4 $\mu M$ LUVs, PC–cholesterol, 1:1), dispensing through the microplate different amounts of leukotoxins (ranging from 5 to 500 n$M$); or in the opposite way, that is, varying the lipid concentration (1 to 100 $\mu M$) and maintaining the toxin constant (i.e., at 0.2 $\mu M$). The whole set of steady state release data was then plotted against the toxin-to-lipid ratio and fitted with the Parente–Rapaport model described previously.

*Effects of Toxin on Lipid Organization*

As we have mentioned, PFTs, usually at higher concentrations, can also induce notable changes in lipid organization. Many such effects have been described ranging from aggregation/fusion of bilayers, phase changes, and detergent-like solubilization. Once again, lipid vesicles are of paramount importance in investigating such changes in the lipid bilayer structure, and are usually coupled with spectroscopic techniques. We have used static and dynamic light scattering to monitor the change in size ensuing from agent-induced aggregation and fusion of LUVs[117–120] (Fig. 3), or from liposome micellization after amphipathic detergent-like attack.[118] We have described more subtle changes, detected by an RET fluorescence approach, demonstrating that one PFT, an actinoporin, is able to enhance lipid flip–flop rates[48] during pore formation. This last mechanism is compatible with the induction by the PFT of toroidal lipid pores, described by Yang *et al.* for the action of magainin[121] and of melittin, under certain circumstances.[122]

[117] G. Menestrina, C. Pederzolli, S. Forti, and F. Gambale, *Biophys. J.* **60**, 1388 (1991).

[118] M. Dalla Serra and G. Menestrina, *Life Sci. Adv. Biophys.* **12**, 109 (1993).

[119] R. Chignola, C. Anselmi, M. Dalla Serra, A. Franceschi, G. Fracasso, M. Pasti, E. Chiesa, J. M. Lord, G. Tridente, and M. Colombatti, *J. Biol. Chem.* **270**, 23345 (1995).

[120] E. de Leeuw, K. te Kaat, C. Moser, G. Menestrina, R. A. Demel, B. de Kruijff, B. Oudega, J. Luirink, and I. Sinning, *EMBO J.* **19**, 531 (2000).

[121] L. Yang, T. M. Weiss, R. I. Lehrer, and H. W. Huang, *Biophys. J.* **79**, 2002 (2000).

[122] L. Yang, T. A. Harroun, T. M. Weiss, L. Ding, and H. W. Huang, *Biophys. J.* **81**, 1475 (2001).

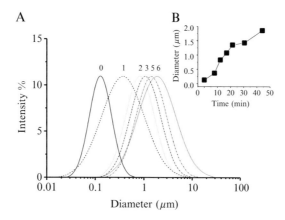

Fig. 3. Determination of the change in size of lipid vesicles exposed to a fusing agent by quasi-elastic light scattering (QLS) measurements. (A) The size of extruded vesicles was determined before adding the agent (curve 0) and at successive times after its addition (curves numbered from 1 to 6). LUVs had an initial average diameter of 110 nm; they then grew over time to about 1.4 $\mu$m in diameter. The diameter was directly determined with a particle size analyzer (ZetaSizer 3; Malvern). The involved agent is FtsY,[120] a bacterial protein involved in protein trafficking and associated with lipid membranes, with some similarity to PFT. (B) Time course of the size increase as obtained from the mean of the distributions in (A).

## Conclusions and Perspectives

As we have tried to substantiate in this chapter, liposomes are probably the most practical and useful model system with which to study the effects of PFTs and similar agents. Compared with other systems, like BLMs, they may not provide the same level of molecular detail about some specific aspects of the interaction, but have certainly the advantage of offering unsurpassed versatility for studying different aspects of the toxin–membrane relation. All the steps leading to pore formation, for example, binding, aggregation, conformational changes, transmembrane insertion, pore opening, and lipid reorganization, can be covered. Intrinsic to this system is also the possibility of applying simultaneously different, or complementary, techniques to observe the various aspects of interaction and their temporal development and sequence. Although one of the most interesting aspects, this potentiality has not yet been explored thoroughly. We have described previously such an approach in some detail.[118] The assay allowed continuous and simultaneous monitoring of two different effects (vesicle aggregation and permeabilization) induced by the

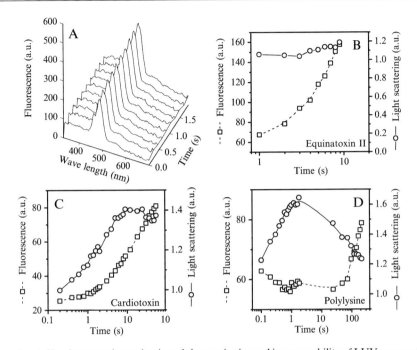

FIG. 4. Simultaneous determination of changes in size and in permeability of LUVs exposed to toxins. (A) Fluorescence excitation spectra of calcein-loaded extruded vesicles taken at consecutive times after their exposure to a PFT: cardiotoxin from the venom of the cobra snake, *Naja naja kaouthia*. Fluorescence excitation was scanned by a spinning monochromator in 8 ms, and emission was collected at 90° without filtration at a maximum rate of 10 spectra per second. Unfiltered spectra contain contributions from both calcein fluorescence ($C_f$), the bell-shaped region around 490 nm, and light scattering ($L_s$), the last contributing significantly only at shorter wavelengths.[124] Both $L_s$ and $C_f$ increased with time for the effect of the toxin, indicating the occurrence of aggregation and permeabilization. LUVs were composed of pure PS: the lipid concentration was 100 $\mu M$, and the cardiotoxin concentration was 24 $\mu M$. (B–D) Similar experiments allowed to obtain separately the time course of $C_f$ (squares) and $L_s$ (circles) with three different agents: sea anemone equinatoxin II (1 $\mu M$; B), cardiotoxin (24 $\mu M$; C), and polylysine (0.12 $\mu M$; D). LUVs were as in (A) except in (B), where they were smaller and composed of PC–sphingomyelin (a 1:1 mixture at a final concentration of 125 $\mu M$). $C_f$ was calculated by subtracting from the peak value at 488 nm the $L_s$ contribution at the same wavelength (extrapolated assuming a second power variation between 425 and 525 nm[103]). $L_s$ was calculated as the average diffused intensity between 375 and 415 nm and was normalized by dividing by the initial value. Full details are in Dalla Serra and Menestrina.[118]

same agent (Fig. 4). Another interesting effort, also using a stopped-flow apparatus, has appeared.[123] Intrinsic tryptophan fluorescence and RET to membrane-bound lipid pores were used to differentiate, in the same experiment, the time course of lipid binding and transmembrane insertion of staphylococcal $\delta$-lysin. This kind of approach is certainly worthy of further investigation. Another promising and fast-expanding area of study is the determination of precise 3-D models of PFT–membrane interaction through the use of enhanced scattering techniques (e.g., low-angle X-ray and neutron scattering) as well as new microscopies such as cryoelectron microscopy and atomic force microscopy. Fluorescence will also continue to hold the stage, particularly through microscope-coupled techniques such as single-molecule tracking and multiphoton excitation.

### Acknowledgments

The authors were supported by the Italian Consiglio Nazionale delle Ricerche (CNR) and Istituto Trentino di Cultura (ITC), and in part also by the Provincia Autonoma di Trento (PAT) Fondo Progetti, through the project StaWars.

---

[123] A. Pokorny, *Biophys. J.* **80,** 548a (2001).

[124] V. K. Miyamoto and W. Stoekenius, *J. Membr. Biol.* **4,** 252 (1971).

## [7] Liposomes as Models for Antimicrobial Peptides

*By* RICHARD M. EPAND and RAQUEL F. EPAND

### Questions to Be Asked when Studying Antimicrobial Peptides with Liposomes

There is considerable interest in developing new approaches to designing antimicrobial peptides. A major reason for the increased interest is that traditional antibiotics are becoming less effective because of the emergence of resistant organisms. The potential of this approach is enhanced by the fact that there are numerous antimicrobial peptides with a variety of chemical structures and properties.[1]

Many of the known antimicrobial peptides interact with membrane bilayers. There are many aspects of cytotoxicity in which membrane interactions may play a role. Liposomes can play an important role in evaluating the relationship between these interactions and the biological activities of the peptides.

---

[1] R. M. Epand and H. J. Vogel, *Biochim. Biophys. Acta* **1462,** 11 (1999).

*Access of Peptide to Target*

Regardless of whether the membrane is the final target for the cytotoxic action of the peptide, for the peptide to affect the cell it must either bind to the membrane or pass through it. Despite this fact, the potency of antimicrobial peptides has not been correlated with their effects on model membranes.[2] There are several possible reasons for this. The peptides used in the study by Zhong *et al.*[2] encompassed a range of different types of peptides; it is likely that they did not all function by the same mechanism. It is also likely that some antimicrobial peptides have intracellular targets and may therefore be less dependent on membrane interactions. In addition, the model liposomes did not contain all the features of the biological membrane, some of which may be important for the action of certain peptides. For example, it has been shown that nicin binds specifically to the peptidoglycan precursor lipid II.[3–6] The chemical composition of biological membranes is complex and also includes membrane proteins. It would be difficult to completely mimic this environment with model liposomes. In addition, biological membranes have transmembrane asymmetry that is difficult to duplicate in a model system. Nevertheless, for several antimicrobial peptides, the importance of membrane interaction and its relevance to the mechanism of action of the peptide have been established.[7,8]

*Selectivity of Toxicity*

Independent of the role of membranes in the cytotoxic action of antimicrobial peptides is the question of the specificity of these agents for microbes. For these peptides to be useful in human and veterinary applications, the peptides must have low toxicity for mammalian cells while maintaining high toxicity against microbes. There is evidence that the basis of this selectivity is through interactions with the membrane. There are several factors that can lead to this specificity. It is known that bacterial and mammalian membranes differ in several respects. The relative importance of several of these factors can be studied by using liposomes as simplified

[2] L. Zhang, A. Rozek, and R. E. Hancock, *J. Biol. Chem.* **276**, 35714 (2001).

[3] E. Breukink, B. B. Bonev, I. Wiedemann, H. G. Sahl, A. Watts, and B. de Kruijff, *Biophys. J.* **80**, 2A (2001).

[4] E. Breukink, P. Ganz, B. de Kruijff, and J. Seelig, *Biochemistry* **39**, 10247 (2000).

[5] E. Breukink, I. Wiedemann, C. van Kraaij, O. P. Kuipers, H. G. Sahl, and B. de Kruijff, *Science* **286**, 2361 (1999).

[6] I. Wiedemann, E. Breukink, C. van Kraaij, O. P. Kuipers, G. Bierbaum, B. de Kruijff, and H. A. Sahl., *J. Biol. Chem.* **276**, 1772 (2001).

[7] K. Matsuzaki, *Biochem. Soc. Trans.* **29**, 598 (2001).

[8] K. Matsuzaki, *Biochim. Biophys. Acta* **1376**, 391 (1998).

model systems. An aspect that appears to be particularly important in this regard is the electrostatic charge on the membrane surface. The major anionic lipids of the membranes of mammalian cells are sequestered on the cytoplasmic side of the surface membrane and are not accessible to interaction with antimicrobial peptides. In gram-positive bacteria, phosphatidylglycerol, as well as other anionic lipids, are exposed on the cell surface, whereas gram-negative bacteria have anionic lipopolysaccharides exposed in the cell wall. Most antimicrobial peptides are cationic, and the sequestering of these peptides to a negatively charged membrane surface results in making these peptides more effective against bacteria. Another difference is the absence of cholesterol from bacterial membranes and its presence in mammalian cell membranes. This difference contributes to antimicrobial activity but is not as important as the electrostatic effect.[9] In addition, for some antimicrobial peptides there is an interaction between the peptide and a specific molecular component of the bacterial membrane.[6]

### Nature and Extent of Membrane Damage

The issue of the importance of a direct effect of antimicrobial peptides on membranes has come into question because many of the studies demonstrating effects of these peptides on leakage of dyes from liposomes have used much higher peptide concentrations than are required for antimicrobial action. Antimicrobial peptides encompass a large variety of chemical structures with different physical properties and different relative potencies against different bacteria. Almost certainly they do not all have the same mechanism of action. It is likely that some act by damaging the surface membrane either by complete lysis of the cell or by altering the permeability of the membrane. It has been suggested that dissipation of the electrical or pH gradient may be important for the action of certain antimicrobial peptides.[10,11]

### Properties

### Lipid Binding

Liposomes are useful for determining the affinity of antimicrobial peptides for membrane bilayers and the chemical or physical features of the membrane that modulate the binding affinity. There are several

---

[9] K. Matsuzaki, K. Sugishita, N. Fujii, and K. Miyajima. *Biochemistry* **34,** 3423 (1995).
[10] O. Aguilera, H. Ostolaza, L. M. Quiros, and J. F. Fierro, *FEBS Lett.* **462,** 273 (1989).
[11] H. J. Kim, S. K. Han, J. B. Park, H. J. Baek, B. J. Lee, and P. D. Ryu, *J. Peptide Res.* **53,** 241 (1999).

methods that can be used to study the binding of peptides to membranes. The following are some that have been applied more extensively in the study of antimicrobial peptides.

*Centrifugation.* Centrifugation is a simple and direct method for studying the binding of peptides to membranes, and it has general applicability. The basis of the method is simple. It requires the physical separation of lipid-bound peptide from free peptide. However, there are several considerations that must be taken into account in applying the method. Lipid has a partial specific volume close to that of water, so that there is a large buoyant force slowing the sedimentation of liposomes. There are two ways around this problem. The simplest one is to use multilamellar vesicles (MLVs). Because MLVs are large aggregates with a high molecular mass, they can be sedimented more easily. However, MLVs are formed by concentric rings of bilayers and only the outer ring, a small fraction of the total lipid, is exposed to the external environment. Thus, to ensure equilibration between the peptide and all of the lipid, the sample must be taken through several freeze–thaw cycles in order to achieve penetration of the peptide into the inner lamellae.

Instead of MLVs, it is also possible to use large unilamellar vesicles (LUVs) made by extrusion; see chapter 1 of volume 367.[11a] An advantage of LUVs is that if the binding process is slow they can be incubated with the peptide for a determined amount of time at a fixed temperature. These LUVs are sufficiently large not to have any strain due to curvature. However, it should be taken into account that only half of the lipid of the LUVs is exposed to the outside and this is therefore the effective concentration that can interact with peptide. LUVs are much smaller than MLVs and are therefore more difficult to sediment. One way around this is to use sucrose-loaded vesicles,[12] which have a higher density of entrapped sucrose-containing buffer compared with the density of the surrounding isosmotic medium.

A precaution in using centrifugation methods is to ensure that the peptide does not sediment without binding to lipid. This can be ascertained by running a control without lipid, that is, centrifuging the peptide solution by itself. When there is some loss of peptide because of aggregation, the binding can still be studied by centrifugation if a density gradient is used. Because of their different densities, liposomes, liposomes with bound lipid, and aggregated peptide can, in principle, all be separated from each other. Another concern, particularly when using low concentrations of peptides or hydrophobic peptides, is the loss of the peptide on the walls of vessels,

---

[11a] M. J. Hope, *Methods Enzymol.* **367A,** 3 (2003).
[12] C. A. Buser and S. McLaughlin, *Methods Mol. Biol.* **84,** 267 (1998).

pipettes, and so on. This can be averted by silanization procedures applied to all containers coming in contact with the peptides.

*Fluorescence.* In principle, any spectral technique can be used, in which a difference between the free peptide in solution and the lipid-bound peptide can be measured. By adding increments of lipid to a solution of peptide the binding affinity can be determined. Fluorescence is a particularly convenient method to use for peptides that contain tryptophan (Trp). The method is sensitive to low concentrations of peptide and precautions can be taken to diminish light-scattering artifacts.[13] In general, there is an increase in the intensity of Trp fluorescence as well as a shift of the emission spectrum to the blue on insertion of Trp into the more hydrophobic environment of a membrane. Even when such a spectral shift does not occur, fluorescence can still be used by covalently attaching to the lipid a fluorophore that can undergo energy transfer with Trp. Energy transfer will take place only when the two fluorophores are sufficiently close to each other, that is, when the peptide is bound to the membrane. Peptides without Trp can also be studied by chemically modifying them with a fluorophore, of course, with the caveat that the probe does not alter the properties of the system. 7-Nitro-2,1,3-benzoxadiazol-4-yl (NBD)-labeled peptides have been used in a variety of cases. The other application of labeled peptides is to study their self-association by energy transfer, using two different labels: one to serve as a fluorescent donor and the other an acceptor of the excitation energy from the donor.[14,15]

*Titration Calorimetry.* Improvements in the performance of calorimeters have allowed for greater sensitivity. Many studies have applied isothermal titration calorimetry to the study of the binding of antimicrobial peptides to membranes. The calorimeter is filled with a solution, usually containing peptide, which is continually stirred and maintained at constant temperature. Into this solution is added small aliquots of a second solution, usually containing LUVs. The heat evolved or absorbed as a consequence of the reaction of the two components is then recorded as a function of injection number. The only limitations of the method is that the peptide must be sufficiently soluble and the rate of reaction between lipid and peptide must be rapid and result in a heat change. The method has been applied successfully to several antimicrobial peptides.[4,16–20]

[13] A. S. Ladokhin, S. Jayasinghe, and S. H. White, *Anal. Biochem.* **285**, 235 (2000).
[14] Y. Pouny, D. Rapaport, A. Mor, P. Nicolas, and Y. Shai, *Biochemistry* **31**, 12416 (1992).
[15] E. Gazit, W. J. Lee, P. T. Brey, and Y. Shai. *Biochemistry* **33**, 10681 (1994).
[16] H. Heerklotz, and J. Seelig, *Biophys. J.* **81**, 1547 (2001).
[17] G. Machaidze, A. Ziegler, and J. Seelig *Biochemistry* **41**, 1965 (2002).
[18] T. Wieprecht, O. Apostolov, and J. Seelig. *Biophys. Chem.* **85**, 187 (2000).

*Structural Analysis.* A variety of methods can be used to assess the conformational properties of an antimicrobial peptide, as well as the manner in which it inserts into membranes. One aspect of the interaction of the peptide with the membrane is its orientation with respect to the plane of the bilayer. This is an important issue relating to the mechanism of action of antimicrobial peptides. Many peptides have been found to be located parallel to the surface of the membrane without inserting too deeply, stabilized by noncovalent interactions with the lipid headgoups, whereas others insert into the bilayer perpendicularly, with their axis parallel to the acyl chains. Circular dichroism (CD) is a convenient tool for measuring changes in secondary structure that may occur on binding of the peptide to lipid. In addition, CD of oriented samples allows for the determination of the orientation of a helical peptide with respect to the plane of the membrane.[21–24] Other methods that have been used include equilibrium dialysis[25] and electrophoretic mobility,[26] and Fourier transform infrared spectroscopy, including the use of linear dichroism.[27,28] A more detailed analysis of peptide conformation can be achieved by nuclear magnetic resonance (NMR),[29] including the use of solid-state NMR.[30] Diffraction methods have also been applied. This has shown the effect of several antimicrobial peptides, causing a thinning of the membrane.[31–33] In addition, neutron scattering has been used to attempt to model a leakage pore with magainin, albeit at high peptide concentrations and with an equal

[19] T. Wieprecht, O. Apostolov, M. Beyermann, and J. Seelig, *Biochemistry* **39,** 442 (2000).

[20] T. Wieprecht, M. Beyermann, and J. Seelig, *Biochemistry* **38,** 10377 (1999).

[21] S. J. Ludtke, K. He, Y. Wu, and H. W. Huang, *Biochim. Biophys. Acta* **1190,** 181 (1994).

[22] S. J. Ludtke, K. He, W. T. Heller, T. A. Harroun, L. Yang, and H. W. Huang, *Biochemistry* **35,** 13723 (1996).

[23] Y. Wu, H. W. Huang, and G. A. Olah, *Biophys. J.* **57,** 797 (1990).

[24] L. Yang, T. A. Harroun, T. M. Weiss, L. Ding, and H. W. Huang, *Biophys. J.* **81,** 1475 (2001).

[25] G. Montich, S. Scarlata, S. McLaughlin, R. Lehrmann, and J. Seelig. *Biochim. Biophys. Acta* **1146,** 17 (1993).

[26] R. M. Peitzsch, and S. McLaughlin, *Biochemistry* **32,** 10436 (1993).

[27] E. Goormaghtigh, V. Raussens, and J. M. Ruysschaert, *Biochim. Biophys. Acta* **1422,** 105 (1999).

[28] E. Gazit, I. R. Miller, P. C. Biggin, M. S. Sansom, and Y. Shai, *J. Mol. Biol.* **258,** 860 (1996).

[29] D. J. Schibli, H. N. Hunter, V. Aseyev, T. D. Starner, J. M. Wiencek, P. B. McCray, Jr., B. F. Tack, and H. J. Vogel, *J. Biol. Chem.* **277,** 8279 (2002).

[30] B. Bechinger, *Biochim. Biophys. Acta* **1462,** 157 (1999).

[31] F. Y. Chen, M. T. Lee, and H. W. Huang, *Biophys. J.* **82,** 908 (2002).

[32] T. A. Harroun, W. T. Heller, T. M. Weiss, L. Yang, and H. W. Huang, *Biophys. J.* **76,** 3176 (1999).

[33] W. T. Heller, A. J. Waring, R. I. Lehrer, T. A. Harroun, T. M. Weiss, L. Yang and H. W. Huang, *Biochemistry* **39,** 139 (2000).

concentration of peptide on both sides of the membrane.[22] In addition to differential scanning calorimetry (DSC), X-ray diffraction methods can determine the effect of antimicrobial peptides on lipid polymorphism, phase separation, and segregation of lipids into domains.[34,35]

## Leakage

*Dye Leakage.* The simplest application of dye leakage studies is to determine whether an antimicrobial peptide can rupture a lipid bilayer membrane and whether this process is dependent on the nature of the lipid. These assays are generally designed so that a fluorophore, such as calcein or carboxyfluorescence, is entrapped within an LUV at sufficiently high concentration that the dye is self-quenched.[35a] Another variation uses the entrapment of a fluorophore such as 8-aminonaphthalene-1,3,6-trisulfonic acid (ANTS) together with a quenching agent such as p-xylene-bis-pyridinium bromide (DPX). The dye outside of the liposome is then removed by gel filtration. After addition of a lytic peptide the dye can leak out into the larger extravesicular volume, where it can become fluorescent by the relief of collisional quenching as a consequence of dilution.

There are several variations of this method that allow probing of specific aspects of the mechanism of membrane lysis by a peptide. One variation is to entrap a dye that is covalently linked to a larger molecule. A series of fluorescently labeled dextran molecules is commercially available. The purpose of doing this is to assess the size of the pore that is formed by the peptide.[36] If the pore is too small, larger compounds will not leak out of the membrane. In such studies it is difficult to have the fluorophore at self-quenching concentrations. Therefore the dye remaining in the liposome must be separated from the leaked dye by gel filtration, making the assay more cumbersome and preventing continuous monitoring of the rate of leakage.

A gradual release of liposome contents can be distinguished from a complete release of contents by a fraction of the liposomes. This is done with an entrapped quenched fluorophore. A certain time after peptide-induced leakage, the liposomes are separated from the external medium and their fluorescence determined. If the dye leaked out in an all-or-none fashion, then fluorescence of the dye remaining in the liposomes should be just as quenched as it was before peptide addition. However, if the leakage

[34] K. Lohner, and E. J. Prenner, *Biochim. Biophys. Acta* **1462,** 141 (1999).
[35] E. J. Prenner, R. N. Lewis, L. H. Kondejewski, R. S. Hodges, and R. N. McElhaney, *Biochim. Biophys. Acta* **1417,** 211 (1999).
[35a] C. Kayalar and N. Düzgüneş, *Biochim. Biophys. Acta* **860,** 51 (1986).
[36] A. S. Ladokhin, M. E. Selsted, and S. H. White, *Biophys. J.* **72,** 1762 (1997).

is gradual, then the dye remaining in the liposome should be partially dequenched. The extent of dequenching can be calibrated as a function of the extent of dilution or by fluorescence-requenching methods.[37]

In the case of the gradual release of fluorophore it is also possible to measure the kinetics of dye release. The kinetics have been modeled in terms of the establishment of a transient pore whose lifetime can be estimated from the kinetics of dye release.[38] The synergistic actions of magainin and PGLa have been explained as being a consequence of fast pore formation and moderate pore stability.[39]

*Small Ion Leakage.* In addition to lysis of the membrane, leading to the formation of large pores or to complete rupture, it is also possible that the antimicrobial peptide causes damage to the membrane that allows for leakage of ions. Such a process could result in toxicity to a cell by dissipating ion or pH gradients. This could provide a mechanism for inducing membrane damage by some antimicrobial peptides that would allow the peptides to be toxic to bacteria even at concentrations below that which is required for membrane lysis.

Several techniques have been applied to study the dissipation of ion gradients by antimicrobial peptides. These methods include the use of ion-selective electrodes to monitor the release of entrapped ions,[40] ion-specific fluorescent probes,[41] quenching of fluorophores by ions,[41a] and NMR.[42]

## Lipid Flip–Flop

One observation about the effects of antimicrobial peptides on membranes is that they promote phospholipid flip–flop.[43] This phenomenon, along with other evidence, has been used to establish the pore model for the action of magainin.[8] This can be distinguished from channel-forming peptides in which the pore is composed essentially only of peptide, whereas in the case of the magainin pore it has both peptide and lipid.

[37] A. S. Ladokhin, W. C. Wimley, K. Hristova, and S. H. White, *Methods Enzymol.* **278,** 474 (1997).

[38] K. Matsuzaki, O. Murase, and K. Miyajima, *Biochemistry* **34,** 12553 (1995).

[39] K. Matsuzaki, Y. Mitani, K. Y. Akada, O. Murase, S. Yoneyama, M. Zasloff, and K. Miyajima, *Biochemistry* **37,** 15144 (1998).

[40] Y. Sokolov, T. Mirazabekov, D. W. Martin, R. I. Lehrer, and B. L. Kagan, *Biochim. Biophys. Acta* **1420,** 23 (1999).

[41] A. S. Verkman, *Am. J. Physiol.* **259,** C375 (1990).

[41a] E. P. Bruggemann and C. Kayalar, *Proc. Natl. Acad. Sci. USA* **83,** 4273 (1986).

[42] M. M. Pike, S. R. Simon, J. A. Balschi, and C. S. Springer, Jr. *Proc. Natl. Acad. Sci. USA* **79,** 810 (1982).

[43] E. Fattal, S. Nir, R. A. Parente, and F. C. Szoka, Jr., *Biochemistry* **33,** 6721 (1994).

## Peptide Translocation

Peptide translocation is another feature of the pore mechanism of membrane leakage. Of course, any membrane rupture that is sufficiently large will allow the passage of an antimicrobial peptide. However, there are aspects of peptide translocation that suggest a more specific type of pore mechanism, at least for some antimicrobial peptides. One feature is that the time course of peptide translocation is similar to the time courses for lipid flip–flop and for leakage of internal aqueous contents of the liposome.[8] This is particularly suggestive of a mechanism, because the leakage of vesicle contents does not follow a simple kinetic expression; rather, the leakage stops before completion. The observation that the peptide can translocate across the membrane may be important both to understand the mechanism of interaction as well as to estimate how efficiently the peptide might enter cells. One antimicrobial peptide, buforin, is efficient in its ability to cross membranes.[44,45]

## Lipid Exchange

Assays of lipid mixing are frequently used to monitor membrane fusion processes.[46] This phenomenon also is a measure of the instability of a membrane. In addition, there is evidence of certain antimicrobial peptides, such as polymyxin, promoting lipid exchange in bacteria.[47] This process could have profound effects on gram-negative bacteria by promoting the exchange of lipid between the inner and outer membranes.

## Planar Bilayers

The presence of voltage-dependent ion channels can be detected with planar bilayer membranes. Asymmetric bilayers can be made by this method, thus simulating natural membranes; however, because of their instability, only a limited number of lipids can be used with this technique.[48]

The properties at the hydrophobic–hydrophilic interface and the mechanism of insertion of antimicrobial peptides into monolayers have been extensively reviewed.[49]

[44] S. Kobayashi, K. Takeshima, C. B. Park, S. C. Kim, and K. Matsuzaki, *Biochemistry* **39**, 8648 (2000).

[45] C. B. Park, K. S. Yi, K. Matsuzaki, M. S. Kim, and S. C. Kim *Proc. Natl. Acad. Sci. USA* **97**, 8245 (2000).

[46] D. K. Struck, D. Hoekstra, and R. E. Pagano, *Biochemistry* **20**, 4093 (1981).

[47] J. T. Oh, Y. Cajal, E. M. Skowronska, S. Belkin, J. Chen, T. K. Van Dyk, M. Sasser, and M. K. Jain, *Biochim. Biophys. Acta* **1463**, 43 (2000).

[48] D. S. Cafiso, *Annu. Rev. Biophys. Biomol. Struct.* **23**, 141 (1994).

[49] R. Maget-Dana, *Biochim. Biophys. Acta* **1462**, 109 (1999).

Concluding Remarks

Liposomes are a useful tool with which to identify the mechanism of action of antimicrobial peptides, and they provide a simple system to complement cell studies for optimization of the potency and specificity of these agents.

## [8]  Preparation of Vesicles from Nonphospholipid Amphiphiles

*By* PIERRE-ALAIN MONNARD and DAVID W. DEAMER

Introduction

Liposomes are defined as vesicular structures bounded by lipid bilayers. The concept of liposomes as model membrane systems was introduced by Bangham and co-workers (see Bangham[1] for review). As technology progressed, specific terminology came to be associated with the physical and chemical properties of various preparations.[2] For instance, the original liposomes produced by Bangham and co-workers are now referred to as multilamellar vesicles (MLVs) because each vesicle is bounded by multiple lipid bilayers. The heterogeneity of MLV preparations is often a limitation, and it was later found that small unilamellar vesicles (SUVs) having dimensions in the range of 25–100 nm could be prepared by sonication of MLV preparations.[3] Large unilamellar vesicles (LUVs) ranging from 100 nm to 1 $\mu$m in diameter were first prepared in the mid-1970s[4–8] and are standard preparations today. A widely used method for LUV preparation is by extrusion through polycarbonate filters.[9]

[1] A. D. Bangham, *in* "Liposomes: From Physical Studies to Therapeutic Applications" (G. Knight, ed.), p. 1. Elsevier/North-Holland, Amsterdam, 1981.
[2] M. C. Woodle and D. Papahadjopoulos, *Methods Enzymol.* **171,** 193 (1989).
[3] C. Huang, *Biochemistry* **8,** 344 (1969).
[4] D. Deamer and A. D. Bangham, *Biochim. Biophys. Acta* **443,** 629 (1976).
[5] F. Szoka and D. Papahadjopoulos, *Proc. Natl. Acad. Sci. USA* **75,** 4194 (1978).
[6] M. Kasahara and P. C. Hinkle, *Proc. Natl. Acad. Sci. USA* **73,** 396 (1976).
[7] U. Pick, *Arch. Biochem. Biophys.* **212,** 186 (1981).
[8] O. Zumbuehl and H. G. Weder, *Biochim. Biophys. Acta* **640,** 252 (1981).
[9] M. J. Hope, M. B. Bally, G. Webb, and P. R. Cullis, *Biochim. Biophys. Acta* **812,** 55 (1985).

Modern liposomes are typically prepared from phospholipids, sometimes with admixtures of cholesterol to improve properties such as stability and permeability. However, it has been known since 1973 that much simpler amphiphilic compounds such as oleic acid also form vesicular structures.[10,11] For certain applications such liposomes are superior to those prepared from ordinary phospholipids. For instance, vesicles composed of fatty acids can be used to define the minimal properties required to produce stable vesicles.[11,12] Furthermore, the effect of chain length and admixtures of other amphiphilic compounds on stability and permeability can also be more easily investigated in fatty acid vesicles than in ordinary liposomes. Another example is that whereas phospholipid vesicles cannot undergo any form of growth, vesicles composed of oleic acid can grow by the hydrolysis of a nonmembranous precursor, oleic anhydride.[13–17]

These properties are particularly relevant to the origin of lipid bilayer membranes that would be required for the first forms of cellular life. It is highly improbable that phospholipids were available in the prebiotic environment, so that simpler molecules such as fatty acids, fatty alcohols, and isoprenoid derivatives[18] were more likely to be the primary membrane components.

In this chapter, we describe properties and preparation of lipid vesicles from the simplest kinds of amphiphilic compounds. We address issues of stability, permeability, and encapsulation of macromolecules, and show how such vesicles can be used to encapsulate functional enzymes. We also discuss how these vesicles are related to naturally occurring substances, particularly the vesicular membranes that are produced from ancient organic compounds present in carbonaceous meteorites.

[10] J. M. Gebicki and M. Hicks, *Nature* **243**, 232 (1973).

[11] W. R. Hargreaves and D. W. Deamer, *Biochemistry* **17**, 3759 (1978).

[12] C. L. Apel, D. W. Deamer, and M. N. Mautner, *Biochim. Biophys. Acta* **1559**, 1 (2002).

[13] P. Walde, *in* "Self-Production of Supramolecular Structures: From Synthetic Structures to Models of Minimal Living Systems" (G. R. Fleischaker, S. Colonna, and P. L. Luisi, eds.), Vol. 1, p. 209. Kluwer Academic, Dordrecht, The Netherlands, 1994.

[14] P. Walde, A. Goto, P.-A. Monnard, M. Wessicken, and P. L. Luisi, *J. Am. Chem. Soc.* **116**, 7541 (1994).

[15] F. Mavelli and P. L. Luisi, *J. Phys. Chem.* **100**, 16600 (1996).

[16] K. Morigaki, S. Dallavalle, P. Walde, S. Colonna, and P. L. Luisi, *J. Am. Chem. Soc.* **119**, 292 (1997).

[17] E. Blochliger, M. Blocher, P. Walde, and P. L. Luisi, *J. Phys. Chem. B* **102**, 10383 (1998).

[18] S.-I. M. Nomura, K. Yoshikawa, O. Dannenmuller, S. Chasserot-Golaz, G. Ourisson, and Y. Nakatani, *Chembiochem* **2**, 457 (2001).

Formation of Vesicles by Self-Assembly of Fatty Acids, and Fatty Acid–Alcohol Mixtures

## Pure Fatty Acid Vesicles

We use the term *critical bilayer concentration* (CBC), by analogy to the *critical micelle concentration* (CMC), to define an amphiphile concentration at which bilayer vesicles are formed. Above their CBC, monocarboxylic acids, also called fatty acids, with a carbon chain ranging from $C_8$ to $C_{18}$ with various degrees of unsaturation (Table I), spontaneously form bilayer vesicles when the acids are suspended in an aqueous medium at a pH close to their apparent $pK_a$. These structures are stabilized by hydrogen bonding between deprotonated and protonated carboxylic molecules,[19–21] and are in dynamic equilibrium with single molecules and micelles dissolved in the aqueous medium.

The fatty acid can be suspended either by adding a solution of pure fatty acid to the aqueous medium[11,12] or by swelling a fatty acid film with

TABLE I
FATTY ACID PROPERTIES

| Name of fatty acid | Length:unsaturation of hydrocarbon chains | State of fatty acid at $25°$ | Critical bilayer concentration (m$M$) | $pK_a$ | Ref. |
|---|---|---|---|---|---|
| Heptanoic | 7:0 | Liquid | $a$ | $a$ | |
| Octanoic | 8:0 | Liquid | 130 | 6.3–6.9 | b, c |
| Nonanoic | 9:0 | Liquid | 85 | 7.0 | c |
| Decanoic | 10:0 | Crystalline$^d$ | 43 | 7.1–7.3 | b, c |
| Dodecanoic (lauric) | 12:0 | Crystalline$^e$ | 22.5 | 7.5 | b |
| Tetradecanoic (myristic) | 14:0 | Crystalline$^f$ | | 8.1–8.2 | b |
| 9,10-Octadecenoic (oleic) | 18:1 (9-*cis*) | Liquid | 0.7–1.4 | 8.0–9.0 | g, h |

[a] Heptanoic acid alone does not form stable bilayers.
[b] J. R. Kanicky, A. F. Poniatowski, N. R. Mehta, and D. O. Shah, *Langmuir* **16,** 172 (2000).
[c] C. L. Apel, D. W. Deamer, and M. N. Mautner, *Biochim. Biophys. Acta* **1559,** 1 (2002).
[d] Melts at $32°$.
[e] Melts at $43°$.
[f] Melts at $58°$.
[g] D. M. Small, "The Physical Chemistry of Lipids: Handbook of Lipid Research 4." Plenum Press, New York, 1986.
[h] H. Fukuda, A. Goto, H. Yoshioka, R. Goto, K. Morigaki, and P. Walde, *Langmuir* **17,** 4223 (2001).

the aqueous medium. Fatty acid films are prepared by dissolving the fatty acid in an organic solvent, for example, oleic acid in chloroform–methanol (9:1, v/v), followed by evaporation to produce a thin film on the interior of a glass vessel. In the case of $C_{10}$–$C_{12}$ fatty acids, it is essential to maintain the aqueous medium at a temperature above the bulk melting point during vesicle formation to avoid crystallization (see Table I).

To optimize spontaneous self-assembly, it is convenient first to completely deprotonate the acid by adding a strong base (1 $M$ NaOH). At pH 9–11, ordinary fatty acids typically form micelles, and these suspensions are totally clear. The pH of a micellar solution is then adjusted with acid addition to a pH range near the apparent $pK_a$ of the fatty acid. At this pH the solution appears slightly opalescent. Microscopic examination of the suspension, as well as a gradual increase in light scattering at 500 nm on acid addition, can be used to monitor vesicle formation. Further lowering of the pH below the $pK_a$ value results in protonation of the fatty acid molecules, which then become insoluble and form droplets.

*Procedure.* To prepare 5.0 ml of decanoic acid (DA) vesicles, the fatty acid crystals are first melted at 43°. A 72-$\mu$l aliquot of DA is added to 4.5 ml of 10 m$M$ Tris, pH 7.4, heated at 43°. This aliquot represents 370 $\mu$mol of liquid DA having a density of 0.886 g/ml at 40°.[22] The DA forms droplets on top of the aqueous medium. NaOH (1.0 $M$) is then added to the solution in 20-$\mu$l aliquots, with vortexing between additions, until the acid is completely dissolved near pH 11.0 (approximately 330 $\mu$l of total NaOH). The suspension is transparent with abundant foam. HCl (1.0 $M$) is then added in 10-$\mu$l aliquots (approximately 120 $\mu$l of total HCl), with vortexing and checking of the pH after each addition, until pH 7.4 is reached, at which point an opalescent suspension of DA vesicles is present. The suspension volume is then adjusted to 5 ml for a final DA concentration of 74 m$M$.

## Mixed Fatty Acid–Alcohol Vesicles

The inclusion of a cosurfactant into fatty acid bilayers changes the physical properties of these membranes.[11,12,23] For example, an aliphatic alcohol having the same chain length as the fatty acid will lower its CBC and extend the pH range over which bilayers are stable under alkaline

[19] H. L. Rosano, A. P. Christodolou, and M. E. Feinstein, *J. Colloid Interface Sci.* **29,** 335 (1969).

[20] T. H. Haines, *Proc. Natl. Acad. Sci. USA* **80,** 160 (1983).

[21] D. P. Cistola, J. A. Hamilton, D. Jackson, and D. M. Small, *Biochemistry* **27,** 1881 (1988).

[22] "Handbook of Chemistry and Physics," Weast, R. C. 70th Ed. CRC Press, Boca Raton, FL, 1989.

[23] P.-A. Monnard, C. L. Apel, A. Kanavarioti, and D. W. Deamer, *Astrobiology* **2,** 139 (2002).

conditions. This alcohol effect is observed in a narrow range of molar ratios of fatty alcohol to fatty acid ($C_8$–$C_{11}$) of approximately 1:10. Below this ratio, the stabilization effect of the alcohol cosurfactant lessens, and above this ratio droplets are formed by the excess cosurfactant.

To form mixed bilayers, the acid is added to the desired aqueous medium heated to a temperature above the bulk melting point (see Table I), and titrated with NaOH as described earlier to produce micelles. When the aliphatic alcohol is added under constant vortexing to the micellar solution at alcohol-to-acid molar ratios of approximately 1:10, formation of vesicles is triggered in the alkaline medium. The pH can then be adjusted to a pH close to the apparent $pK_a$ of the fatty acid to maximize the number of vesicles. Such suspensions are typically incubated overnight to allow equilibrium of the bilayer composition. In the case of mixed vesicles containing decanoic acid/decanoate and decyl alcohol, the CBC decreases from 43 m$M$ [CBC of pure decanoic acid (DA)] to approximately 20 m$M$ total amphiphiles (18.5/1.9 m$M$).

*Procedure.* To prepare 5 ml of 40 m$M$ DA–decyl alcohol (37 m$M$/ 3 m$M$) vesicles, the fatty acid crystals are first melted and 36 $\mu$l of DA (185 $\mu$mol) is added to 4.5 ml of 10 m$M$ Tris, pH 7.4, heated at 43°. The acid forms droplets at the surface of the aqueous medium. NaOH (1.0 M; 10-$\mu$l aliquots) is added to the solution, with vortexing between additions, until the acid is completely dissolved, typically near pH 10.0. Decyl alcohol (2.9 $\mu$l) is added, and on vortexing for 30 s an opalescent dispersion of vesicles is produced. HCl (1.0 $M$; 5-$\mu$l aliquots) is then added, with vortexing and checking of the pH after each addition, until the $pK_a$ of the acid (approximately pH 7.4) is reached. The total suspension volume is then adjusted to 5 ml.

### Mixed Fatty Acid–Polyol Vesicles

Monoglycerides such as glycerol monooleate have been previously used to form planar bilayers.[24] We have reported that they can also form mixed vesicles with enhanced stability compared with pure fatty acid vesicles.[23] Mixtures of glycerol monodecanoate and DA (GMD/DA) form stable vesicles at concentrations up to 67.5 m$M$ (22.5 m$M$ GMD/45 m$M$ DA). GMD has a CBC of approximately 1 m$M$ in 10 m$M$ Tris, pH 7.4. However, mixing both fatty acids and GMD molecules modifies the GMD characteristics so that preformed GMD vesicles at low GMD concentrations (1 to 2.5 m$M$) are solubilized by the added fatty acid molecules, when the fatty acid:GMD ratio exceeds 3:1. This phenomenon has been observed

[24] S. B. Hladky and D. A. Haydon, *Biochim. Biophys. Acta* **274,** 294 (1972).

previously in mixtures of phosphatidylcholine and detergents, such as cholic acid.[25]

To prepare the vesicles, the procedure described for pure fatty acid preparation is used, but because of the lower solubility of these compounds it is necessary to optimize the mixing of the membrane components within the mixed bilayers by freeze-thawing the vesicle suspensions. The suspension container is immersed in dry ice–acetone until the aqueous mixture is completely frozen, and then thawed at room temperature and vortexed. This procedure is repeated 4–10 times until a stable turbid suspension is obtained in which no aggregation occurs.

*Procedure.* To prepare 5.0 ml of vesicles composed of 7.5 m$M$ GMD/ 15 m$M$ DA, 9.25 mg of solid GMD (37.5 $\mu$mol) and 14.7 $\mu$l of melted decanoic acid are added to 4.5 ml of 10 m$M$ Tris, pH 7.4, preheated at 43°. At this point the GMD has already formed vesicles and the DA is present as droplets at the surface. After vortexing and pH adjustment to pH 10.0, the fatty acid partitions into the GMD vesicles, resulting in an increase in turbidity and the complete disappearance of the DA droplets. The pH is then adjusted to pH 7.4 with HCl to maximize vesicle formation. During HCl addition, the turbidity of the suspension increases further. The suspension volume is now adjusted to 5.0 ml. To stabilize the vesicles, which tend to aggregate, this suspension is freeze-thawed 4–10 times. The resulting suspension is stable for several weeks at room temperature.

*Microscopy*

Fatty acid membranes can be visualized by phase-contrast and epifluorescence light microscopy. Under phase contrast, emulsions of fatty acid and alcohol droplets appear as phase-bright spheres. Vesicles are revealed as spheres having obvious unilamellar or oligolamellar membranes, whereas multilamellar structures appear as phase-dark spheres due to differences in refractive index of the amphiphile structures and water (Fig. 1).

Epifluorescence examination of vesicular structures requires a fluorescent dye marker. For instance, water-soluble impermeant dyes such as pyranine (1-hydroxypyrene-3,6,8-trisulfonic acid) provide information about internal volume, whereas hydrophobic or surface-binding dyes allow vesicle membranes to be visualized.[12] Water-soluble dyes must be added during the vesicle formation step, whereas membrane marker dyes such as rhodamine 6G and B can be added after vesicles are present, typically at molar lipid:dye ratios lower than 1000:1.

[25] P. Schurtenberger, N. Mazer, S. Waldvogel, and W. Känzig, *Biochim. Biophys. Acta* **775,** 111 (1984).

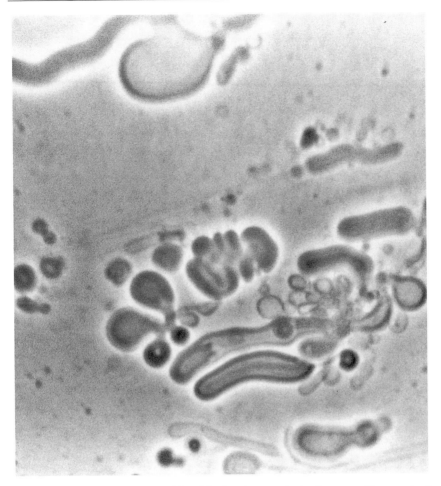

Fig. 1. Visualization of fatty acid vesicles by phase-contrast microscopy. Phase-contrast micrographs of 125 m$M$ nonanoic acid suspensions at pH 6.9. The sample was prepared by pH vesiculation. The bilayer membranes are clearly darker than the background, and the intensity of shading is directly proportional to the number of membranes stacked together in the vesicle boundary.

*Procedure.* An aliquot of decanoic acid/decanoate vesicle preparation (5 $\mu$l, 74 m$M$) is pipetted onto a glass slide and then mixed with 0.5 $\mu$l of 100 $\mu M$ rhodamine 6G placed on the coverslip. Unilamellar or oligolamellar vesicles having diameters in the micron range or larger appear as intensely fluorescent circles with dark interiors (Fig. 2). If fluid fatty acid

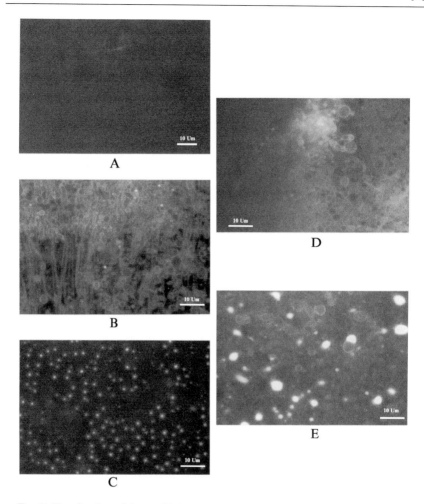

FIG. 2. Visualization of fatty acid–fatty alcohol vesicles by epifluorescence microscopy. Epifluorescence micrographs of 74 m$M$ DA suspension (CBC, 43 m$M$) at various pH values, stained with rhodamine 6G, are shown: (A) pH 11.0; (B) pH 7.34; (C) pH 5.5. At high pH (A), fatty acid molecules form micelles with an average size of several nanometers (size range below the resolution of a light microscope). At intermediate pH (B), vesicles are clearly visible together with droplets (the bright spots). At low pH (C), droplets form that adhere to the glass slide. Additional micrographs (D and E) show transformations occurring on addition of NaOH (D) or HCl (E) to the suspension in (B). (D) On the left side of the picture, 1 $M$ NaOH was added to the DA vesicle suspension, and as DA molecules deprotonate, vesicles are disrupted. (E) HCl addition to DA vesicles promotes the protonation of DA molecules, followed by vesicle disruption and droplet formation (bright spots).

or alcohol droplets are present, they appear as uniformly stained spherical structures having no apparent boundary membranes.

## Critical Issues

The stability of fatty acid vesicles depends on several factors. For instance, fatty acid vesicles are in dynamic equilibrium with their monomeric species, so that dilution to concentrations below the CBC of the component fatty acid results in disruption of the vesicles. The ionic strength of the aqueous medium can also influence the stability of fatty acid vesicles. If the NaCl concentration is raised above a threshold ratio, vesicles collapse completely to form crystals or sheetlike aggregates that can be observed by epifluorescence microscopy. Divalent cations such as magnesium and calcium profoundly inhibit the ability of ordinary fatty acids to form stable vesicles. The aggregation of the decanoic acid/decanoate vesicles can be demonstrated with those cations at a lipid:cation ratio of approximately 65:1.[23] In general, the sensitivity of the vesicles to salts follows the sequence decanoic acid (DA) > decanoic acid–decyl alcohol (DA/DOH) $\gg$ glycerol monodecanoate/decanoic acid (GMD/DA) $\approx$ glycerol monodecanoate (GMD). Chelating agents such as EDTA can remove $Ca^{2+}$ and $Mg^{2+}$ from the divalent cation–fatty acid complexes and permit re-formation of vesicles.

## Encapsulation of Macromolecules

### Methods of Encapsulation

Macromolecules can be entrapped in fatty acid vesicles by pH-induced vesiculation and dehydration–rehydration.[26,27] The pH-induced vesiculation is based on the procedure used to prepare fatty acid vesicles, except that the solute to be encapsulated is added to the micellar solution before the pH is adjusted to a value close to the apparent $pK_a$ of the fatty acid. Fatty acid vesicles readily trap solutes during the pH adjustment.[14,28]

The dehydration–rehydration method involves mixing preformed vesicles with macromolecules such as proteins or nucleic acids, followed by dehydration under a constant $N_2$ flow over 2–3 h at 30° for a 200-$\mu$l sample. Under these conditions the vesicles fuse and the macromolecules

[26] D. W. Deamer and G. L. Barchfeld, *J. Mol. Evol.* **18**, 203 (1982).

[27] R. L. Shew and D. W. Deamer, *Biochim. Biophys. Acta* **816**, 1 (1985).

[28] T. Oberholzer, R. Wick, P. L. Luisi, and C. K. Biebricher, *Biochem. Biophys. Res. Commun.* **207**, 250 (1995).

TABLE II
METHODS TO SEPARATE LIPOSOMES FROM UNENCAPSULATED SOLUTE

| Liposome size (nm) | Solute size | |
|---|---|---|
| | Small | Large |
| <800 | Dialysis, gel chromatography | Dialysis, gel chromatography; with additional preparation |
| ≥800 | Dialysis, centrifugation | Dialysis, centrifugation; with additional preparation |

are captured between layers of the lipid deposited on the vial walls. The solute–fatty acid film is carefully rehydrated, first using a moist cotton plug at 37° for 20 min to provide 100% relative humidity, and then with additions of the desired aqueous phase. The vesicles that form during rehydration encapsulate up to approximately half the macromolecules,[12,27,29] depending on the molar ratio of lipid to solute molecule, the salt concentration,[30] and the liposome size.[29] Some examples of encapsulation using these methods are described in detail in the section Enzymatic Reactions Captured in Liposomes (see p.146).

*Removal of Nonentrapped Material*

It is often desirable to remove unencapsulated solutes from a liposome preparation, and dialysis, centrifugation, and gel-filtration chromatography are commonly used for this purpose.[31] The main criterion in selecting the type of separation is the liposome size (Table II). In some instances partial enzymatic hydrolysis of large solute molecules is required before removal steps can be undertaken.[29]

The dialysis method can be employed for any liposome size, because even the smallest vesicles will not pass through the pores of a dialysis membrane having a molecular weight cutoff value of 20,000. However, separation is typically slow, requiring several hours, and a permeant solute with relatively high diffusion rates across bilayers will be released from the vesicles.

[29] P.-A. Monnard, T. Oberholzer, and P. L. Luisi, *Biochim. Biophys. Acta* **1329**, 39 (1997).

[30] P.-A. Monnard, N. Berclaz, K. Conde-Frieboes, and T. Oberholzer, *Langmuir* **15**, 7504 (1999).

[31] A. Chonn, S. C. Semple, and P. R. Cullis, *Biochim. Biophys. Acta* **1070**, 215 (1991).

Large liposomes (MLVs and LUVs) can be pelleted by centrifugation in an Eppendorf centrifuge (10,000 g, 30–45 min) to separate vesicles from unencapsulated solutes. Gel chromatography, also called size-exclusion chromatography, can be applied to SUV or LUV suspensions containing vesicles up to 800 nm in diameter. Larger vesicles and aggregates tend to adhere to the column material. The column size varies from spin columns in a 1-ml syringe[31] to standing columns filled with Sephadex, Sephacryl, or Bio-Gel media, depending on the size of the solute to be excluded.

A spin column packed with Bio-Gel A 15 m, mesh 200–400 efficiently removes small solutes such as nucleotide triphosphates (NTPs), tRNA,[29] or 120-bp DNA fragments,[32] as long as no direct interaction between the membranes and the solutes exist. Larger polymers of nucleic acids or proteins may require enzymatic degradation before the separation.[29]

*Procedure.* A spin column can be prepared from a 1-ml syringe plugged with glass wool.[31] A suspension of packing material (e.g., Bio-Gel A-15 m, 200–400 mesh size; Bio-Rad, Hercules, CA) is first added to completely fill the syringe. As the gel settles in the tube, further additions are made until the bed volume reaches the syringe top. Once the syringe is filled with Bio-Gel, the column is centrifuged once at 165 g for 2 min to pack the bed. The column is then equilibrated by adding 200 μl of buffer to the column, followed by centrifugation at 165 g for 2 min. The operation is repeated with 50-μl buffer aliquots until the bed volume stabilizes. If it falls below 1.0 ml, the column should be topped with fresh Bio-Gel, and the equilibration repeated. For optimal separation, sample volumes containing up to 120 mM fatty acid should be in the range of 150 to 250 μl. The bed should appear dry after the first centrifugation step, and elution can then be carried out by adding 50-μl buffer aliquots followed by centrifugation. The vesicles are recovered in fractions 2–8 (the fraction volume usually is 50–70 μl).

## Critical Issues during Separation

Fatty acid vesicles are sensitive to dilution during removal of non-entrapped solute. To prevent vesicle disruption by dilution, columns should be preequilibrated with a suspension containing a fatty acid concentration equivalent to the CBC. Aqueous media used for dialysis, elution, and centrifugation should also contain empty vesicles at or above the CBC for maximal recovery of the vesicles with entrapped solute.

[32] T. Oberholzer, E. Meyer, I. Amato, A. Lustig, and P.-A. Monnard, *Biochim. Biophys. Acta* **1416**, 57 (1999).

*Determination of Encapsulation Efficiency*

Encapsulation efficiency can be determined by capturing a relatively impermeant dye marker, such as 10 $\mu M$ pyranine (1-hydroxypyrene-3,6,8-trisulfonic acid) or 6-carboxyfluorescein. Once the external solute is removed, the amount of encapsulated solute is determined by first increasing the pH with NaOH to disperse the vesicles, and then measuring the absorbance of the pyranine at 415 nm. Because the original concentration of dye is known, the encapsulated volume can be calculated.

Permeability

The lipid moiety of cellular membranes represents the main barrier to the free diffusion of solutes. However, lipid bilayers are not completely impermeable, and permeability of bilayers to solutes ranging from water to ions is therefore of interest. Permeability to a given solute is a function of molecular size, polarity, and ionic charge of the solute. In general, solutes with relatively high partition coefficients permeate by dissolving in the bilayer and then diffusing across, whereas ionic solutes such as protons and sodium and potassium ions apparently use transient defects in the bilayer to cross the permeability barrier.[33] It follows that relatively fragile bilayers, such as those composed of fatty acids, will have large numbers of transient defects and for that reason will be more permeable to ions than bilayers composed of phospholipids.

*Uptake of Externally Added Solutes*

The relative permeability of a bilayer to a given solute can be investigated by monitoring light-scattering changes.[11,12,34] Vesicle suspensions are placed in a spectrophotometer cuvette and absorbance is measured at 500 nm. Osmotically active solutes are then added (e.g., 0.1 $M$ sucrose). Osmotic flow of water out of the vesicles causes them to decrease in volume over a few seconds, thereby increasing light scattering (measured here as absorbance) to a maximum. This is followed by a slower decrease in absorbance as the solute crosses the membrane and the gradient decays (Fig. 3). The rate at which the vesicles return to their previous size and shape is a function of the bilayer permeability to a given solute.[32]

The permeability can also be assessed by incubating empty liposomes in the presence of a slowly permeating solute and removing the

[33] S. Paula, A. G. Volkov, A. N. Van Hoek, T. H. Haines, and D. W. Deamer, *Biophys. J.* **70**, 339 (1996).
[34] B. E. Cohen and A. D. Bangham, *Nature* **236**, 173 (1972).

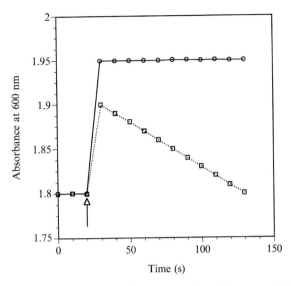

Fig. 3. Permeability of nonanoic acid/nonoate vesicles. The permeability of 120 m$M$ nonanoic acid/nonoate bilayers to 0.1 $M$ KCl (open circles) and 0.1 $M$ glycerol (open squares) is investigated by monitoring the absorbance changes at 600 nm.[12] The addition of solute (arrow) after 20 s causes light scattering to increase. After 30 s, the light scattering decreases because of the permeation of glycerol, whereas that of KCl samples does not because this solute does not diffuse across the bilayers, even after several hours.

non-entrapped material by size-exclusion chromatography at regular time intervals. The passive diffusion of AMP, ADP, and ATP across oleic acid bilayers was established in this way.[14]

*Efflux of Encapsulated Solutes*

Permeability can be measured directly by monitoring changes in the internal solute concentration after removal of external solute molecules. The solute concentration can be then spectroscopically determined. Marker dyes such as 1.0 m$M$ carboxyfluorescein or pyranine can also be used and are described here in detail.

*Procedure.* DA vesicles (125 m$M$) containing 2 m$M$ carboxyfluorescein are prepared as previously described. A separation column (25 cm in length, 1 cm in diameter) is half-filled with Sephadex G-50 and equilibrated with a suspension of empty vesicles (45 m$M$ DA), and then loaded with a 0.5-ml aliquot of dye-containing vesicles. A typical encapsulation yield is 2.75% of the initial dye. If elution is stopped once the free dye has separated from the vesicles, so that two clear bands are visible on the column, solute concentrations can then equilibrate across the bilayers. After 140 min on the column, only 1.75% of the dye coelutes with the vesicles when elution is completed. In this case, the dye permeation rate is 0.26%/min.

For a more conventional use of carboxyfluorescein for permeability measurements in phospholipid liposomes, see the procedure described by New.[35]

### Enzymatic Reactions Captured in Liposomes

Fatty acid bilayers are more permeable to large ionic solutes than their phospholipid counterparts having similar hydrocarbon chain lengths.[14,36] This property can be useful in designing vesicle-encapsulated catalytic reactions. For instance, Chakrabarti et al.[36] showed that only phospholipid vesicles with hydrocarbon chains 14 carbons in length (dimyristoyl-*sn*-glycerol-3-phosphocholine) allow adenosine diphosphate (ADP) to permeate at a rate sufficient to support an enzyme-catalyzed polymerization reaction. Walde et al.[14] reported similar results with vesicles composed of oleic acid, which has an 18-carbon chain ($C_{18:1}$). From these results it is clear that both phospholipid liposomes and fatty acid vesicles can capture macromolecules with catalytic activity, and the bilayer permeability is sufficient to permit the inward flow of substrates, but prevents the release of encapsulated macromolecular catalysts and products.

The preparation of catalytic systems entrapped in vesicles can be divided into four stages (Fig. 4).

Step I. All components of the catalytic system are entrapped within vesicles. The suspensions can be further processed at this stage to obtain specific vesicle sizes.

Step II. The nonentrapped catalytic units must be removed from the external medium or inactivated to prevent any reaction in the outer medium.

Step III. Substrate solutes are added to the external medium, and incubation is initiated. At this stage, vesicle suspensions can be subjected to physical alterations such as temperature-induced phase transitions to assure a sustained substrate supply.

Step IV. The reaction is stopped by inactivation of the entrapped enzyme, and the product is analyzed. In this step, particular care should be given to vesicle solubilization in order to recover all products.

[35] R. R. C. New, *in* "Liposomes: A Practical Approach" (R. R. C. New, ed.), p. 105. IRL Press, Oxford, 1992.

[36] A. C. Chakrabarti, R. R. Breaker, G. F. Joyce, and D. W. Deamer, *J. Mol. Evol.* **39,** 555 (1994).

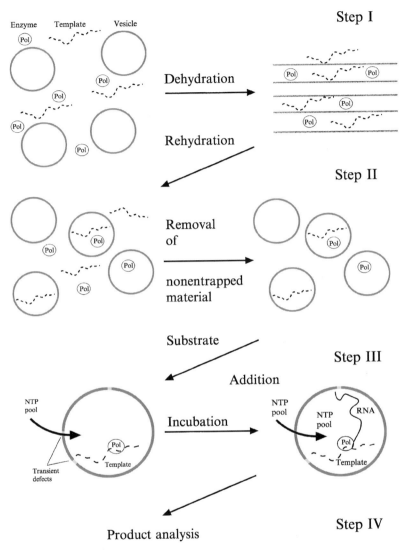

FIG. 4. Preparation of an enzymatic reaction entrapped in fatty acid vesicles. This schematic representation illustrates a template-directed RNA polymerization in fatty acid vesicles. Step I: The enzymatic system (the enzyme, its template, and its metal catalyst) is entrapped according to the dehydration–rehydration method.[26] The encapsulation is random, and therefore three populations of liposomes are obtained: empty vesicles, those containing one of the system components, and those with the complete system. Step II: External reaction is prevented by removing the nonentrapped enzymes and templates after enzymatic digestion. Step III: Substrates, NTPs, are added externally, and begin to permeate through bilayer transient defects, so that the encapsulated reaction proceeds. Step IV: The reaction is quenched, and the product is analyzed.

*Procedures*

The following three examples illustrate preparation and analysis of vesicle-entrapped catalytic systems.

*Catalase Activity within Fatty Acid–Fatty Alcohol Vesicles.*[12] Catalase is an enzyme that decomposes hydrogen peroxide to oxygen. The substrate has a high permeability across fatty acid bilayers, making it an ideal candidate for encapsulated reactions.

Decanoic acid (200 $\mu$mol) is added to 5.0 ml of water to a concentration of 40 m$M$ and the pH is raised to pH 7.5 by the dropwise addition of NaOH to induce micellization; at pH 7.5 catalase is stable. Beef liver catalase (12.5 $\mu$l of a 1-mg/ml solution) is then added to the decanoic acid solution. The subsequent addition of 3.0 $\mu$l of decyl alcohol with vortex mixing initiates vesicle formation, which is completed by adjusting to pH 7.2 with HCl, an optimum pH for both enzyme function and membrane stability. The suspension of catalase and vesicles is equilibrated overnight at 25° (step I). To inhibit unencapsulated enzyme activity, 10 $\mu$l of a 10.0-mg/ml solution of *streptomyces griseus* protease can be added to the catalase–vesicle solution, which is then incubated for 6 h (step II). One milliliter of the digested solution is then placed in a standard quartz cuvette (1.0-cm path length). To start the encapsulated reaction (step III), 10 $\mu$l of 30% hydrogen peroxide is added to the external medium. Hydrogen peroxide has an ultraviolet (UV) absorbance peak at 240 nm. Over time, the degradation of hydrogen peroxide to oxygen and water is observed as a decrease in absorbance at 240 nm.

In control experiments, the same amount of catalase is added to a 5-ml decanoic acid suspension after vesicle formation. In this case, the protease digests the enzyme completely, so that no enzymatic activity is detected after a 6-h incubation. In this experiment, the fourth step described in the overview is omitted because the reaction can be monitored *in situ*. Alternatively, oxygen production may be monitored with an oxygen electrode.

*Polynucleotide Phosphorylase (PNPase)*[14]*/Q$\beta$ Replicase*[28] *Reaction within Fatty Acid Vesicles.* These two polymerization reactions can be carried out with extruded oleic acid–oleate vesicles formed by pH vesiculation,[9] and yield RNA oligomers. The PNPase incorporates NDPs into nascent RNA molecules, and the Q$\beta$ replicase is an RNA template-directed enzyme that uses NTPs as substrates. The entrapped enzymes must be removed from the fatty acid vesicle preparation before RNA product analysis can be carried out.

The encapsulation of the PNPase (PNPase, 4.3 mg/ml; 80 m$M$ oleic acid–oleate) is carried out as described for catalase (step I). Nonentrapped enzymes are removed by gel filtration on a Sepharose 4B column (step II).

The vesicle fractions (turbid fractions) are then pooled, and ADP is added externally (step III). In the experiments described by Walde et al.,[14] oleic acid–oleate vesicles (2 ml of 8 m$M$ oleic acid in 50 m$M$ Tris-HCl buffer, pH 9) are mixed with 0.2 ml of 100 m$M$ ADP. Note that the pH of the ADP stock solution must be adjusted to pH 9 before this solution can be used. After incubation at 25°, the reaction is stopped by directly injecting the liposome suspension into an HPLC/DEAE 4000–7 anion-exchange column, using 4 $M$ urea in the elution buffer, which ensures complete denaturation of the enzyme. The RNA product is measured by UV spectroscopy (step IV).

Q$\beta$ replicase is an RNA template-directed enzyme active at 25°. For this reason the enzyme, cofactor (magnesium), and template must all be present within the same vesicle. Oberholzer et al.[28] chose to trap the four substrates and NTPs (step I), so that replication begins as soon as the compounds are mixed. An excess of EDTA in the external medium totally inhibits the reaction (step II). Vesicular systems are prepared as described above for catalase with 200 n$M$ Q$\beta$ replicase; 200 n$M$ midivariant RNA; 500 $\mu M$ CTP, GTP, and UTP; 250 $\mu M$ ATP; 25 $\mu$Ci of [$^{35}$S]ATP$\alpha$S; 4 m$M$ MgCl$_2$; 0.25 m$M$ dithiothreital (DTT); 200 m$M$ Bicine buffer (pH 8.5); and 50 m$M$ oleic acid–oleate.

After incubation at 25° (step III), the reaction is stopped by isolating the RNA fragments by phenol–chloroform extraction and subsequent ethanol–salt precipitation (step IV). Phenol–chloroform not only removes the entrapped protein, but also the fatty acids. Products can be analyzed by gel electrophoresis and Phosphorimager techniques. Interestingly, after incubation, the addition of RNase A to the external medium of intact vesicles does not reduce product yields, clearly demonstrating that fatty acid vesicles protect the encapsulated product.

Each of the above-described protocols demonstrates how fatty acid vesicles can be used as nanoscale bioreactors. The results with PNPase and catalase show that external substrate can supply encapsulated enzymes by passive diffusion. The fatty acid bilayers are stable enough to encapsulate the catalytic species, and they protect both enzymes and RNA products. This essential property permits accumulation of products. The Q$\beta$ replicase experiment further demonstrates the possibility of simultaneous entrapment of a complex metabolic system (eight components) in fatty acid vesicles.

### Relevance to Early Forms of Cellular Life

The pathway by which prebiotic molecular systems undergoing chemical evolution assembled into the earliest forms of cellular life remains an unsolved problem. However, at some point simple self-replicating systems of polymeric molecules essential for life must have been captured in membrane-bounded compartments.[37–41] Compartmentalization is a necessary prerequisite for maintaining the integrity of such interdependent molecular systems and for permitting variations required for speciation.[42]

Early self-assembling amphiphilic molecules could have been provided either by delivery of extraterrestrial organic material during late accretion or by abiotic synthesis. Short-chain monocarboxylic acids ranging from $C_8$ to $C_{11}$ have been detected in organic material extracted from the Murchison carbonaceous meteorite.[43–47] Furthermore, substantial yields of both monocarboxylic acids and alcohols in this range of chain lengths have been synthesized by Fischer–Tropsch-type reactions that simulate plausible geochemical conditions.[48,49] Monoglycerides are readily synthesized by drying a mixture of glycerol and fatty acid under simulated prebiotic conditions as reported by Hargreaves *et al.*[50] If the early membranes formed by self-assembly of organic compounds, then the prebiotically plausible mixed fatty acid vesicles represent an attractive model, even though early membranes must have been composed of an even broader mixture of amphiphiles that included fatty acids, their derivatives, aliphatic hydrocarbons, as well as polycyclic aromatic hydrocarbons.

As the amphiphile concentration rose above the threshold value needed for self-assembly (CBC), spontaneous vesicle formation would have taken place with concomitant trapping of any solutes present in the aqueous medium. Vesicles that happened to encapsulate catalytically active solutes, such as ribozyme precursors, would have represented a first step toward a primitive form of cellular life. To survive, evolve, and self-replicate, these early cells had to interact efficiently with their environment, their only source for nutrients and energy at this early stage. Because they presumably lacked any complex protein transport systems, passive

[37] D. W. Deamer and J. Oro, *Biosystems* **12,** 167 (1980).

[38] D. W. Deamer, *Microbiol. Mol. Biol. Rev.* **61,** 230 (1997).

[39] A. Lazcano, in "Early Life on Earth" (S. Bengtson, ed.), Nobel Symposium No. 84, p. 60. Columbia University Press, New York, 1994.

[40] P. L. Luisi, P. Walde, and T. Oberholzer, *Curr. Opin. Colloid Interface Sci.* **4,** 33 (1999).

[41] J. W. Szostak, D. P. Bartel, and P. L. Luisi, *Nature* **409,** 387 (2001).

[42] D. S. Tawfik and A. D. Griffiths, *Nat. Biotechnol.* **16,** 652 (1998).

[43] J. G. Lawless and G. U. Yuen, *Nature* **282,** 431 (1979).

[44] D. W. Deamer, *Nature* **317,** 792 (1985).

[45] D. W. Deamer and R. M. Pasley, *Orig. Life Evol. Biosph.* **19,** 21 (1989).

[46] M. Komiya, A. Shimoyama, and K. Harada, *Geochim. Cosmochim. Acta* **57,** 907 (1993).

[47] M. Mautner, R. L. Leonard, and D. W. Deamer, *Planetary Space Sci.* **43,** 139 (1995).

[48] T. M. McCollom, G. Ritter, and B. R. T. Simoneit, *Orig. Life Evol. Biosph.* **29,** 153 (1999).

[49] A. I. Rushdi and B. R. T. Simoneit, *Orig. Life Evol. Biosph.* **31,** 103 (2001).

[50] W. R. Hargreaves, S. J. Mulvihill, and D. W. Deamer, *Nature* **266,** 78 (1977).

diffusion of solutes would have been the most plausible exchange mechanism.[14,23,51] Again, fatty acid vesicles represent an attractive model because they can spontaneously trap complex catalytic molecular assemblies that remain functional within their small aqueous compartment and can therefore process substrates that passively diffuse across the fatty acid bilayers.

Self-reproduction is an essential property of living systems. Once a primitive cell was able to replicate its encapsulated macromolecular components, an increase of its membrane area would be necessary to accommodate the internal growth. To model this evolutionary step, fatty acid vesicle systems have been designed successfully, using anhydride derivatives as a source of additional membrane amphiphiles. In their study, Luisi and co-workers[13–17] have demonstrated that the hydrolysis of oleic anhydride, a water-immiscible precursor of oleic acid, could supply additional amphiphile molecules to be incorporated into preexisting oleic acid–oleate vesicles, which in turn accelerated further the hydrolysis rates. Both membrane growth and a simple metabolism (PNPase-catalyzed RNA polymerization) could occur simultaneously in this system.

The prebiotic availability of fatty acids or their precursors, and their properties of encapsulation, high permeability, and membrane growth, make them ideal model systems for investigating primitive forms of life.

## Acknowledgments

The authors thank C. L. Apel for help during manuscript preparation.

[51] P.-A. Monnard and D. W. Deamer, *Orig. Life Evol. Biosph.* **31,** 147 (2001).

# [9] Liposomes in the Study of GDP/GTP Cycle of Arf and Related Small G Proteins

*By* KARINE ROBBE and BRUNO ANTONNY

## Introduction

Most G proteins are localized at the cytosolic surface of lipid membranes. The Ras-like fold, which forms the core domain of all G proteins including small GTP-binding proteins, has no ability to interact directly with membrane lipids. However, this domain is frequently flanked by N- or C-terminal extensions that function as anchors to permit membrane attachment.[1] Small G proteins of the Ras, Rab, and Rho families have a

15- to 40-amino acid C-terminal extension, which points outward from the core domain and ends with a CAAX sequence.[2] The cysteine residue is the site of attachment of an isoprenyl group, which, together with basic residues or additional lipid modifications, makes the C terminus a strong membrane anchor. The membrane partitioning of Rab and Rho proteins is controlled by GDP dissociation inhibitor (GDI) proteins, which prevent the lipid segment from inserting into the lipid bilayer.[3] GDI proteins interact preferentially with the GDP-bound form of Rab and Rho to form a cytosolic complex. ADP ribosylation factors (Arfs) and the related small G protein Sar interact with membrane lipids through their N terminus. Here, membrane attachment is mediated by an amphipathic N-terminal helix and also (except for Sar1 and some Arf-like proteins) by the myristoylation of the N-terminal glycine. The remarkable feature of this N-terminal extension is that its availability for membrane interaction is directly controlled by the binary GDP/GTP conformational switch of the core domain.[4,5] Thus, Arf and Sar cycle between a soluble GDP-bound form and a membrane-associated GTP-bound form without the help of an accessory protein.

The Ras-like fold contains most regions that mediate the interactions of small G proteins with other proteins, including effectors of the GTP-bound form, exchange factors (GEFs) that catalyze the replacement of GDP by GTP, and GTPase-activating proteins (GAPs) that catalyze GTP hydrolysis.[1] Minimalist studies in solution with soluble forms obtained by N- or C-terminal truncation are therefore generally justified to explore the molecular basis of small G protein interactions. However, many aspects of the functioning of small G proteins can be understood only in the presence of lipid membranes. Restriction of motion at the membrane surface strongly influences the interaction of small G proteins with their partners. Moreover, in the case of Arf and Sar, the lipid membrane cannot be viewed solely as a planar surface where proteins diffuse and interact. In the GTP-bound state, Arf and Sar1 interact with proteins that deeply remodel lipid membranes. Arf and Sar1 promote the recruitment of protein coats, which shape membranes into buds and small vesicles. In addition, Arf activates lipid-modifying enzymes such as phospholipase D and phosphatidylinositol (PI) kinases.[6–8]

[1] I. R. Vetter and A. Wittinghofer, *Science* **294**, 1299 (2001).

[2] M. C. Seabra, *Cell Signal.* **10**, 167 (1998).

[3] B. Olofsson, *Cell Signal.* **11**, 545 (1999).

[4] B. Antonny, S. Beraud-Dufour, P. Chardin, and M. Chabre, *Biochemistry* **36**, 4675 (1997).

[5] J. Goldberg, *Cell* **95**, 237 (1998).

[6] J. G. Donaldson and C. L. Jackson, *Curr. Opin. Cell Biol.* **12**, 475 (2000).

[7] J. E. Rothman and F. T. Wieland, *Science* **272**, 227 (1996).

[8] R. Schekman and L. Orci, *Science* **271**, 1526 (1996).

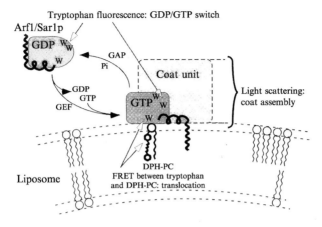

FIG. 1. Schematic representation of the spectroscopic methods used to study the GDP/GTP cycle of Arf1 or Sar1 on liposomes.

Liposomes of submicrometer size and defined composition provide a minimal system by which the membrane confinement of proteins through specific and nonspecific interactions with lipids can be mimicked in the test tube. Several steps of the Arf/Sar cycle have been reconstituted, using liposomes and purified proteins. By a combination of biochemical and morphological approaches, the GTP-dependent translocation of Arf and Sar1 to membrane lipids, the subsequent recruitment of coat complexes, and the deformation of membranes induced by these polymeric structures can be analyzed.[9–12] In addition, several spectroscopic methods have been developed to monitor some reactions in real time (Fig. 1). Tryptophan fluorescence allows monitoring of the conversion of Arf or Sar1 between GDP- and GTP-bound forms.[4] Fluorescence resonance energy transfer (FRET) between tryptophan residues and a fluorescent lipid analog such as diphenyl hexatriene-phosphatidylcholine (DPH-PC) allows monitoring of the translocation of Arf1 on liposomes.[4] Finally, light scattering allows monitoring of the assembly of protein coats on liposomes.[13] These methods,

[9] A. Spang, K. Matsuoka, S. Hamamoto, R. Schekman, and L. Orci, *Proc. Natl. Acad. Sci. USA* **95,** 11199 (1998).

[10] K. Matsuoka, L. Orci, M. Amherdt, S. Y. Bednarek, S. Hamamoto, R. Schekman, and T. Yeung, *Cell* **93,** 263 (1998).

[11] K. Matsuoka, Y. Morimitsu, K. Uchida, and R. Schekman, *Mol. Cell.* **2,** 703 (1998).

[12] M. Bremser, W. Nickel, M. Schweikert, M. Ravazzola, M. Amherdt, C. A. Hughes, T. H. Sollner, J. E. Rothman, and F. T. Wieland, *Cell* **96,** 495 (1999).

[13] B. Antonny, D. Madden, S. Hamamoto, L. Orci, and R. Schekman, *Nat. Cell Biol.* **3,** 531 (2001).

together with a liposome sedimentation assay, are described in this chapter. The emphasis is placed on Arf and Sar, although some experiments with Rac, a member of the Rho family, are also presented.

## Liposome Preparation

### Azolectin Liposomes

The reversed phase method[14] (see also Düzgüneş[14a]) is used to prepare liposomes from unpurified soybean lipids (azolectin type IIS: Sigma, St. Louis, MO). Azolectin lipids are stored at $-20°$ under argon. Lipids are weighed and dissolved in diethylether (6 ml/20 mg azolectin lipids) in a 100-ml round-bottom flask. An aqueous buffer (1 ml) containing sucrose or monovalent salts (see below) is then added. The mixture is sonicated at $4°$ for 1–2 min in a bath sonicator to obtain a homogeneous opalescent dispersion in which lipids dissolved in the organic phase form inverted micelles encapsulating water. The solvent is slowly evaporated in a rotary evaporator (100 rpm) under moderate vacuum and at $20–25°$. After 5 min, a viscous gel forms at the balloon surface. When most solvent has evaporated, the gel falls off to form a liposome suspension. After 45 min, the liposome suspension is collected in an Eppendorf tube and is further incubated in a vacuum chamber for 15 min to eliminate traces of ether. When necessary, the solution is completed up to 1 ml in buffer. As checked by dynamic light scattering (DynaPro apparatus; Proterion, Piscataway, NJ; www. proterion.com), these liposomes display a hydrodynamic radius ranging from 0.2 to 1 $\mu$m. To eliminate lipid aggregates, the liposome suspension can be filtered through a 0.8-$\mu$m pore size ester cellulose filter (Millipore, Billerica, MA). The suspension can also be extruded to obtain liposomes of the desired diameter.

### Liposomes of Defined Lipid Composition

Most synthetic and natural lipids are purchased from Avanti Polar Lipids (Alabbster, AL; www.avantilipids.com) as chloroform solutions (1–20 m$M$). DPH-PC is from Molecular Probes (Eugene, OR; www. probes.com). Phosphatidylinositol 3,4,5-trisphosphate is from Matreya (State College, PA; www.matreya.com). Others phosphoinositides are from Echelon (Salt Lake City, UT; www.echelon-inc.com). Phosphoinositol bis- and trisphosphates are solubilized in chloroform–methanol–$H_2O$

---

[14] F. Szoka, Jr. and D. Papahadjopoulos, *Proc. Natl. Acad. Sci. USA* **75**, 4194 (1978).
[14a] N. Düzgüneş, *Methods Enzymol.* **367**, 23 (2003).

(1:1:0.3, v/v/v). Sonication and addition of 10 m$M$ HCl facilitate the solubilization of these highly charged and polar lipids. Lipid solutions (0.5–10 m$M$) are stored in glass vials at $-20°$ under argon.

To prepare liposomes of defined composition, lipids are mixed at the desired molar ratio in a pear-shaped glass flask, using glass syringes (Hamilton, Reno, NV). Classically, the total amount of lipids is between 1 and 5 mg and the volume of solvent, mostly chloroform, is 1 ml. For mixtures containing phosphoinositides (up to 5%), a cloudy suspension is observed, reflecting the poor solubility of these lipids in chloroform. Adding an additional half-volume of methanol leads to a clear solution. The solvent is evaporated in a rotary evaporator under vacuum at 20–25° and at 500 rpm. A lipid film forms rapidly on the glass surface. After 30 min, a connection between the evaporator and an argon bottle is opened to break the vacuum. The remaining solvent is evaporated in a vacuum chamber. The lipid film is gently resuspended at 1–4 mg/ml in an aqueous buffer containing sucrose or monovalent salts (see below) to give a suspension of multilamellar liposomes. The suspension is submitted to five freeze–thaw cycles in liquid nitrogen and in water (40°). After the last freezing step, the suspension is either stored at $-20°$ or extruded immediately.

*Liposome Extrusion*

The extrusion method combined with the freeze–thaw protocol allows the preparation of homogeneous liposomes[15] (see Mui *et al.*[15a]). The pressure forces the liposome suspension through a polycarbonate filter of defined pore size. The concentric layers of multilamellar liposomes are broken and form smaller liposomes. The miniextruder from Avanti Polar Lipids or a custom-made device similar to the one described by MacDonald *et al.*[16] is used to extrude, 19 times, the suspension of multilamellar liposomes through 0.1- or 0.4-$\mu$m pore size polycarbonate filters [Corning Life Sciences, Corning, NY (www.corning.com) or Millipore]. As checked by dynamic light scattering (DynaPro apparatus; Proterion), extruded liposomes display a hydrodynamic radius close to the pore size of the polycarbonate filter (see application note number 103 at www.protein-solutions.com/appinfo.htm). Extruded liposomes are kept at room temperature under argon and used within 1 week of extrusion.

[15] L. D. Mayer, M. J. Hope, and P. R. Cullis, *Biochim. Biophys. Acta* **858,** 161 (1986).

[15a] B. Mui, L. Chow, and M. Hope, *Methods Enzymol.* **367,** 3 (2003).

[16] R. C. MacDonald, R. I. MacDonald, B. P. Menco, K. Takeshita, N. K. Subbarao, and L. R. Hu, *Biochim. Biophys. Acta* **1061,** 297 (1991).

Buffers, Proteins, and Reagents

*Buffers Used for Preparation of Liposomes*

As a general rule, the buffer used to prepare liposomes should be isosmotic with the "assay" buffer in which proteins and liposomes are finally mixed. In addition, it is preferable to omit divalent salt ($MgCl_2$) from the liposome buffer to prevent long-term fusion. As an example, the buffer used to prepare liposomes for the coat protein complex II (COPII) light-scattering assay is 20 m$M$ HEPES-KOH (pH 7.0), 160 m$M$ potassium acetate. For sedimentation experiments, the buffer used for the preparation of liposomes should contain a concentration of sucrose isosmotic with the monovalent salt present in the final buffer. Thus, for sedimentation experiments at 100 m$M$ NaCl, the corresponding sucrose concentration is 180 m$M$ (for a complete conversion table, see Ref. 17).

*Buffers Used in Sedimentation, Fluorescence, and*
*Light-Scattering Experiments*

Sedimentation experiments are performed in 20 m$M$ HEPES-KOH (pH 7.5), 100 m$M$ NaCl, 1 m$M$ $MgCl_2$, and 1 m$M$ dithiothreitol (DTT) (buffer A). Fluorescence measurements on Arf1 are performed in 50 m$M$ HEPES-KOH (pH 7.5), 100 m$M$ KCl, 1 m$M$ $MgCl_2$, 1 m$M$ DTT (buffer B). Because the COPII complex Sec23/24p aggregates in solution below 160 m$M$ potassium acetate, experiments on COPII proteins are performed in 20 m$M$ HEPES-KOH (pH 7.0), 160 m$M$ potassium acetate, 1 m$M$ $MgCl_2$, 1 m$M$ DTT (buffer C). Note that all buffers contain 1 m$M$ $MgCl_2$. Millimolar amounts of $Mg^{2+}$ prevent small G proteins from losing bound GDP. When indicated, 2 m$M$ EDTA is added to obtain 1 $\mu M$ free $Mg^{2+}$. At this concentration, the rate of GDP dissociation is accelerated, thus facilitating its replacement by GTP. It is important to filter and degas buffers before use to prevent artifacts from bubbles and dust particles during fluorescence and light-scattering measurements.

Detailed protocols for the purification of myristoylated Arf1-GDP,[18] Arf nucleotide-binding site opener (ARNO),[19] Arf1-GAP,[20] ASAP-1 (Arf-GAP containing SH3, ankyrin repeat, and pH domain 1),[21] and

[17] "Handbook of Chemistry and Physics," 70th Ed., p. D264. CRC Press, Boca Raton, FL, 1989.

[18] M. Franco, P. Chardin, M. Chabre, and S. Paris, *J. Biol. Chem.* **270**, 1337 (1995).

[19] S. Beraud-Dufour and S. Robineau, *Methods Enzymol.* **329**, 264 (2001).

[20] I. Huber, M. Rotman, E. Pick, V. Makler, L. Rothem, E. Cukierman, and D. Cassel, *Methods Enzymol.* **329**, 307 (2001).

[21] P. A. Randazzo, K. Miura, and T. R. Jackson, *Methods Enzymol.* **329**, 343 (2001).

Sar1-GDP and COPII components[22] can be found elsewhere. Geranyl-geranylated Rac1 is coexpressed with RhoGDI-1 in *Saccharomyces cerevisiae* and purified as described for RhoA/GDI.[23] The spectroscopic methods presented here require relatively pure proteins (>80%) devoid of aggregates. To keep the liposomes intact, detergents should be avoided. Some commercial preparations of GTPγS contain significant amounts of GDP (up to 30%), which competes with GTPγS during the nucleotide-loading step. Therefore, GTP and analogs should be of the highest purity commercially available (e.g., Roche Molecular Biochemicals, Indianapolis, IN; www.biochem.roche.com/).

## Sedimentation Assay of Membrane Translocation of Arf1 and Rac on Liposomes

Ultracentrifugation of sucrose-loaded liposomes is a simple method by which the lipid-binding properties of small proteins or model lipid-modified peptides can be assessed.[24] This method can be used to study the GTP-dependent recruitment of geranyl-geranylated Rac1 and myristoylated Arf1 to lipid membranes.[4,18]

Azolectin liposomes (20 mg/ml) are prepared by the reversed phase method in a 180 m$M$ sucrose buffer. The following mixtures (80-$\mu$l total volume in buffer A) are prepared in 8 × 34 mm polycarbonate ultracentrifuge tubes (Beckman Coulter, Fullerton, CA; www.beckmancoulter.com/): Rac1-GDP/RhoGDI (0.5 $\mu M$) or Arf1-GDP (0.5 $\mu M$), azolectin liposomes (1.5 mg/ml), and GDP or GTPγS (50 $\mu M$). For experiments with GTPγS, 2 m$M$ EDTA is added to accelerate the replacement of GDP by GTP. Mixtures are incubated for 20 min at 30° in a Thermomixer (300 rpm; Brinkmann Instruments, Westbury, NY; www.brinkmann.com). After incubation, MgCl$_2$ (2 m$M$ final concentration) is added to the sample with EDTA before centrifugation. The tubes are centrifuged at 100,000 rpm (360,000 $g$) in a TL-100.1 rotor for 20 min at 25°. Supernatants are recovered with a Hamilton syringe, and liposome pellets are resuspended in the same volume of buffer. Samples are analyzed by sodium dodecyl sulfate–polyacrylamide gel electrophoresis (SDS–PAGE; 13% acrylamide) with Coomassie staining (Fig. 2).

In the presence of GDP, Rac1 and Arf1 partition between the supernatant and the liposomes. This partitioning depends on the amount of

[22] C. Barlowe, L. Orci, T. Yeung, M. Hosobuchi, S. Hamamoto, N. Salama, M. F. Rexach, M. Ravazzola, M. Amherdt, and R. Schekman, *Cell* **77,** 895 (1994).
[23] P. W. Read and R. K. Nakamoto, *Methods Enzymol.* **325,** 15 (2000).
[24] C. A. Buser and S. McLaughlin, *Methods Mol. Biol.* **84,** 267 (1998).

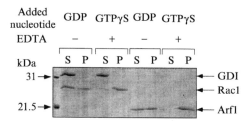

FIG. 2. GTP-dependent membrane translocation of Arf1 and Rac on liposomes. Geranyl-geranylated Rac1 in complex with Rho-GDI-1 (0.5 $\mu M$) or myristoylated Arf1 (0.5 $\mu M$) was incubated with azolectin liposomes (1.5 mg/ml) with the indicated nucleotides (50 $\mu M$) for 20 min. Temperature, 30°. To promote fast GTP$\gamma$S loading, the concentration of free $Mg^{2+}$, initially at 1 m$M$, was set at 1 $\mu M$ free $Mg^{2+}$ during incubation by adding 2 m$M$ EDTA. After centrifugation at 360,000 g, supernatants (S) and pellets (P) were analyzed by SDS–PAGE (13% acrylamide) with Coomassie staining.

liposomes. After GTP$\gamma$S loading, Rac1 and Arf1 are found mainly in the liposome pellet, whereas Rho-GDI remains in the supernatant.

The sedimentation method has been applied successfully for analysis of the lipid-binding properties of Arf1,[4,18]Arf1-GAP,[25] and its exchange factor ARNO.[26] The inverse method, a flotation method, in which liposomes prepared in aqueous buffer are collected over dense sucrose fractions, is more adapted to analysis of the binding of large protein complexes.[9,10,12] The high sedimentation coefficient of protein coats and their tendency to form aggregates preclude the use of sedimentation assays.

### General Requirements for Fluorescence, Fluorescence Resonance Energy Transfer, and Light-Scattering Experiments

In the following sections, we describe three real-time assays based on the measurements of fluorescence, FRET, and light-scattering changes (see Fig. 1). All measurements can be performed in a standard fluorometer (90° format). Importantly, the fluorometer should contain a thermostatted cuvette holder equipped with a magnetic stirrer. Quartz cuvettes (10 × 10 or 4 × 10 mm) adapted to magnetic bars can be purchased from Hellma (Müllhein, Germany; www.hellma-worldwide.de/). For most experiments, custom-made cylindrical quartz cuvettes (internal diameter, 8 mm) inserted in a three-window metal adaptor are used. The shape and size of

[25] B. Antonny, I. Huber, S. Paris, M. Chabre, and D. Cassel, *J. Biol. Chem.* **272**, 30848 (1997).
[26] P. Chardin, S. Paris, B. Antonny, S. Robineau, S. Beraud-Dufour, C. L. Jackson, and M. Chabre, *Nature* **384**, 481 (1996).

these cuvettes give the best compromise for minimizing the sample volume (500 $\mu$l) and allowing continuous mixing with a small magnetic bar (2 × 7 mm; Hellma). Injections from stock solutions are done with Hamilton syringes through a guide in the cover of the fluorometer. The guide is positioned so that the tip of the syringe needle touches the meniscus of the sample without crossing the light beam. With such a device, the recording is not interrupted by the injections, and kinetics in the range of a few seconds can be monitored. To further reduce the sample volume, some experiments are performed in microcuvettes (50–100 $\mu$l; Hellma). In this case, injection and mixing are performed manually with 20- and 100-$\mu$l pipettes, respectively, interrupting the recording for a few seconds.

Experiments on mammalian Arf1 and on yeast Sar1 are performed at 37 and 27°, respectively. Tryptophan fluorescence is recorded at 340 nm with a wide (10- to 30-nm) bandwidth to enhance the signal-to-noise ratio. Because large amounts of nucleotide are used, it is preferable to excite the sample at $\lambda > 290$ nm (e.g., 297.5 nm; bandwidth, 1–5 nm). FRET between tryptophan residues and the fluorescent lipid DPH-PC (at 5% in liposomes) is measured at 460 nm (bandwidth, 20 nm) on excitation at 280 nm (bandwidth, 1.5 nm). For light-scattering measurements, the excitation and emission monochromators are set at 350 nm with small (e.g., 1–2 nm) bandwidths. The gain, the sampling rate, and the time response of the fluorometer should be adjusted to optimize the linearity, the sensitivity, and the time resolution of the measurement. The quality of measurement depends not only on the intrinsic sensitivity of the fluorometer but also on the use of freshly degassed and filtered buffers, correctly aligned and tuned optics, and accurate mixing and injection devices.

## Activation and Deactivation Assays Based on Tryptophan Fluorescence

Two tryptophans at positions 64 and 78 act as intrinsic probes of the conformation of Arf1. On GDP-to-GTP exchange, these two aromatic groups move away from each other, inducing a dramatic (+250%) increase in the intrinsic fluorescence of Arf1.[4,5] Sar1 shows a more modest fluorescence increase (+100%), probably reflecting the fact that only one of these two tryptophans is conserved in Sar1. The two reactions that govern the cycle of these small G proteins, namely GDP to GTP exchange and GTP hydrolysis, can be monitored by tryptophan fluorescence with a much better time resolution than that of conventional assays using radiolabeled nucleotides. In a typical experiment, the tryptophan fluorescence of Arf1 (or Sar1) is monitored continuously. Sequential additions from stock solutions are made to activate and subsequently deactivate the small G protein.

Activation is seen as a fluorescence increase and inactivation as a fluorescence decrease. In these experiments, liposomes are essential, as full-length Arf1 and Sar1 cannot adopt the GTP conformation in the absence of lipids. Moreover, because many GEFs and GAPs contain domains that mediate their translocation to membrane lipids (e.g., PH domain), varying the composition of liposomes can dramatically affect the kinetics of Arf1 activation or deactivation.[25,26]

Figure 3 shows the time course of Arf1 activation as measured by tryptophan fluorescence in the presence of ARNO, an exchange factor for Arf proteins made of three domains: coiled-coil, Sec7, and PH.[26] The Sec7 domain is responsible for the nucleotide exchange activity. The cuvette contains liposomes (final concentration of phospholipids, 0.3 mg/ml) obtained by extrusion through a 0.1-$\mu$m pore size polycarbonate filter and diluted in 600 $\mu$l of buffer B. The percentage of anionic phospholipids (phosphatidylglycerol, PG; phosphatidylinositol 4,5-bisphosphate, PInsP$_2$) in liposomes is indicated. The remaining lipid is egg PC. The initial fluorescence level before the addition of Arf1-GDP is due to light scattering from liposomes and is arbitrarily set at zero. At the times indicated, myristoylated Arf1 complexed with GDP (0.5 $\mu M$), GTP$\gamma$S (10 $\mu M$), or ARNO

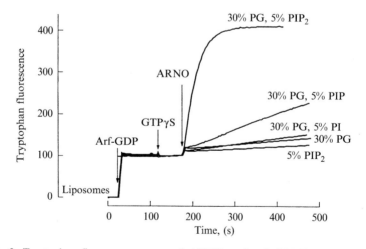

Fig. 3. Tryptophan fluorescence assay of ARNO-catalyzed GDP/GTP exchange on Arf1 in the presence of liposomes of defined composition. Myristoylated Arf1 (0.5 $\mu M$) was injected into a fluorescence cuvette containing extruded liposomes (diameter, ~0.1 $\mu$m; concentration, 0.1 mg/ml). The percentage of phosphatidyl glycerol (PG), phosphatidylinositol (PI), phosphatidylinositol 4-phosphate (PIP), and phosphatidylinositol 4,5-bisphosphate (PIP$_2$) is indicated. The bulk lipid was egg PC. At the indicated time, GTP$\gamma$S (10 $\mu M$) and ARNO (25 n$M$) were added. Temperature, 37°. Note that optimal activation requires both PIP$_2$ and anionic lipids such as PG.

(25 n$M$) is added from stock (20×–100×) solutions. On GTPγS addition, a slow increase in the fluorescence of Arf1 is observed that reflects spontaneous exchange of GDP by GTPγS. The rate of this reaction is limited by the dissociation of GDP, which is slow at millimolar magnesium concentrations. Catalytic amounts of ARNO dramatically accelerate the reaction. However, as shown in Fig. 3, the effect of ARNO is highly dependent on the presence of anionic lipids in liposomes. Optimal activation is observed with mixtures containing small amounts (up to 5%) of phosphatidylinositol 4,5-bisphosphate and large amounts (up to 30%) of phosphatidylglycerol or phosphatidylserine.

Sedimentation experiments with ARNO or its PH domain and sucrose-loaded liposomes (see the protocol described earlier) reveal that the requirement for anionic lipids in the exchange reaction matches the lipid-binding properties of ARNO.[26,27] ARNO interacts with phosphoinositides through its PH domain and with cationic lipids through a patch of basic residues at the C terminus. As nucleotide exchange on full-length Arf1 occurs at the surface of the lipid membrane, the recruitment of ARNO is critical for its exchange activity. It should be noted that a variant of ARNO with a single amino acid insertion in the PH domain is more specific for phosphatidylinositol 3,4,5-trisphosphate than the form used here.[28]

The assay for GAP-catalyzed GTP hydrolysis on Arf1 is based on the associated decrease in tryptophan fluorescence.[25] In the first stage, myristoylated Arf1-GDP (0.2 $\mu M$) is incubated with liposomes (phospholipids at 0.2 mg/ml) of defined lipid composition (Fig. 4). To activate Arf1 without the help of an exchange factor, EDTA (2 m$M$) is added after the addition of GTP (40 $\mu M$). This lowers the concentration of free Mg$^{2+}$ to 1 $\mu M$, allowing fast exchange of the bound GDP for GTP. As shown in Fig. 4, the kinetics of Arf1 activation at micromolar Mg$^{2+}$ is almost independent of the liposome composition. This reflects the fact that the interaction of Arf1 with membrane lipids, although essential for the exchange reaction, is mostly hydrophobic and does not involve specific interactions with lipid polar heads. Thus, provided that sufficient amounts of liposomes are used (in the lipid range of 0.1 mg/ml), a complete activation of Arf1 is observed. After completion of exchange, 2 m$M$ MgCl$_2$ is added to obtain 1 m$M$ free Mg$^{2+}$. No fluorescence decrease is observed, as Arf1 has no intrinsic GTPase activity. GTP hydrolysis on Arf1 is initiated by the addition of GAP proteins. In Fig. 4, two GAPs are used: first, a fragment (amino acids 1–257) encompassing the catalytic domain of the Golgi protein

[27] E. Macia, S. Paris, and M. Chabre, *Biochemistry* **39**, 5893 (2000).
[28] J. K. Klarlund, W. Tsiaras, J. J. Holik, A. Chawla, and M. P. Czech, *J. Biol. Chem.* **275**, 32816 (2000).

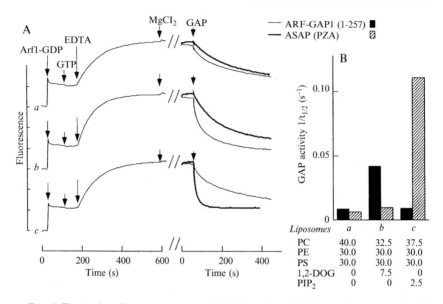

Fig. 4. Tryptophan fluorescence assay of GAP-catalyzed GTP hydrolysis on Arf1 with liposomes of defined composition. In the first stage, myristoylated Arf1-GDP (0.2 $\mu M$) was incubated in buffer B with liposomes of the indicated composition [see (B)] and with 40 $\mu M$ GTP. To promote fast GTP loading on Arf1, the concentration of free $Mg^{2+}$, initially at 1 m$M$, was temporarily set at 1 $\mu M$ by adding 2 m$M$ EDTA. The total volume of the sample in the fluorescence cuvette was 1200 $\mu$l. After the activation stage, $MgCl_2$ (2 m$M$) was added and the sample was divided into two 600-$\mu$l aliquots. Arf1-GAP (30 n$M$) was added to one aliquot, whereas PZA (40 n$M$) was added to a second aliquot. The corresponding fluorescence recordings are shown as thin and bold face traces, respectively, and have been superimposed. GAP activity was determined from the apparent half-time of the fluorescence decrease (B). Temperature, 37°.

Arf-GAP1[20,29]; second, a construct (PZA) containing the PH domain, the catalytic domain with a zinc finger, and the ankyrin repeat domain of ASAP-1, a GAP for Arf that regulates the actin cytoskeleton.[21,30] Notably these two constructs exhibit distinct lipid dependencies (Fig. 4): the association of Arf-GAP1 (1–257) with membrane lipids and its activity toward membrane-bound Arf1-GTP is favored by lipids with conical shape such as 1,2-dioleoylglycerol (1,2-DOG), whereas ASAP-1 interacts through its PH domain with phosphatidylinositol 4,5-bisphosphate.

[29] E. Cukierman, I. Huber, M. Rotman, and D. Cassel, *Science* **270,** 1999 (1995).
[30] P. A. Randazzo, J. Andrade, K. Miura, M. T. Brown, Y. Q. Long, S. Stauffer, P. Roller, and J. A. Cooper, *Proc. Natl. Acad. Sci. USA* **97,** 4011 (2000).

The two assays presented are convenient for studying the GTPase cycle of Arf1 in the presence of catalytic amounts of GAPs or GEFs. When stoichiometric amounts of GAPs or GEFs are used, background fluorescence from these proteins should be determined from control experiments in the absence of Arf1 or in the presence of a nonhydrolyzable analog of GTP. Similar tryptophan fluorescence-based assays can be used to study the GTPase cycle of Sar1.[13] For Sar1, it is important to use liposomes containing unsaturated (e.g., palmitoyl-oleoyl or dioleoyl) lipids, as this favors membrane translocation and stabilization of Sar1 in the GTP-bound state.[10]

## FRET Assay to Monitor Membrane Translocation of Arf1 on Liposomes

In addition to the sedimentation method with sucrose-loaded liposomes, the translocation of Arf1 on artificial liposomes can be monitored in real time by FRET. The measurement is based on an energy transfer signal between tryptophan residues from Arf and the fluorescent lipid DPH-PC.[4] The absorption spectrum of DPH ($\lambda_{max}$ = 350 nm) nicely overlaps with the emission spectrum of tryptophan ($\lambda_{max}$ = 340 nm). With this method, the rate of Arf1-GTP desorption from labeled liposomes can be determined. This rate is a direct index of the strength of the Arf1-GTP/ lipid interaction.

The fluorescence cuvette initially contains PC liposomes (160 $\mu M$ phospholipids) containing 5% DPH-PC, obtained by extrusion through a 0.1-$\mu$m pore size polycarbonate filter. The excitation is set at 280 nm. At this wavelength, tryptophan excitation is maximal, whereas direct excitation of the probe is minimal. Fluorescence of DPH-PC is monitored continuously at 460 nm. Myristoylated Arf1-GDP (0.5 $\mu M$), GTP$\gamma$S (40 $\mu M$), and EDTA (2 m$M$) are sequentially added from stock solutions (Fig. 5). No FRET signal is observed on the addition of Arf1-GDP. The large drop induced by the addition of the GTP analog is due to light absorption by the nucleotide. After the addition of EDTA, an exponential fluorescence increase is observed. This signal correlates with the GDP/GTP exchange reaction as monitored by tryptophan fluorescence or [35S]GTP$\gamma$S binding.[4] This shows that the translocation and the activation of Arf1 are correlated events. After completion of exchange, 2 m$M$ MgCl$_2$ and a 5-fold excess of unlabeled PC liposomes are added (Fig. 5). The observed FRET decrease monitors the exchange of Arf1-GTP from labeled to unlabeled liposomes. The time course of Arf1-GTP desorption from PC liposomes ($\tau_{off}$ = 100 s) is much slower than that observed for model myristoylated peptides, suggesting that some residues of Arf1 interact directly with the lipid bilayer.

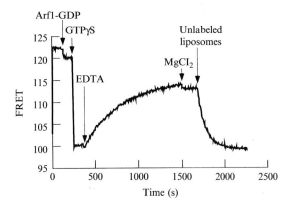

Fig. 5. Translocation/desorption kinetics of Arf1 on liposomes as detected by energy transfer from tryptophan to the fluorescent lipid DPH-PC. The fluorescence cuvette contained PC liposomes (diameter, 0.1 $\mu$m; concentration, 160 $\mu M$) with 5 mol% DPH-PC in buffer B. At the indicated times, myristoylated Arf1-GDP (0.5 $\mu M$), GTP$\gamma$S (40 $\mu M$), EDTA (2 m$M$), MgCl$_2$ (2 m$M$), and unlabeled PC liposomes (800 $\mu M$) were added. Temperature, 37°. Adapted with permission from Antonny et al.[4] Copyright 1997, American Chemical Society.

Similar measurements on Arf1 mutants demonstrate that some N-terminal hydrophobic residues participate in the interaction of Arf1 with membrane lipids.[4]

### Light-Scattering Assay to Monitor Assembly of COPII Coat

In the GTP bound state, Sar1-GTP promotes the assembly of the COPII coat on endoplasmic reticulum (ER) membranes, whereas Arf proteins promote the assembly of various coats (COPI and clathrin/adaptators) on Golgi and plasma membranes.[7,8] These protein coats are made by the assembly of large cytosolic complexes and mediate the budding of transport vesicles from donor compartments. Here, we describe an assay by which the stepwise assembly of the COPII coat on liposomes is monitored in real time by measuring the relative 90° light-scattering signal from liposomes in a standard fluorimeter. With this method, the dynamics of the COPII coat with GTP can be explored.[13]

The rationale underlying the light-scattering measurement is that the contrast of liposomes over water increases as proteins coat the lipid surface. In a solution containing similar amounts of proteins and extruded liposomes, only liposomes and bound proteins contribute to the light-scattering signal. It should be noted that light scattering by large liposomes is a complex phenomenon that can be affected not only by protein coating, but also by morphological changes (induced, e.g., by an osmotic stress). Therefore, proper control experiments (e.g., titration, order of additions) should be considered before assigning a light-scattering change to a protein translocation event.

Figure 6 shows a typical COPII light-scattering experiment performed in a small quartz cuvette (sample volume, 100 $\mu$l). The cuvette contains extruded (0.4-$\mu$m) liposomes (0.1 mg/ml) with an optimal composition for the assembly of the COPII coat (see Fig. 6). For simplicity, the light-scattering intensity is normalized to the initial level observed in the absence of added proteins. On the sequential addition of Sar1-GDP (950 n$M$), GMP-PNP (100 $\mu M$), Sec23/24p (160 n$M$), and Sec23/31p (260 n$M$) from stock solutions, three light-scattering changes are observed: first, a slow and small increase that reflects the translocation of Sar1 on GDP to GMP-PNP exchange; then, two rapid and large increases that reflect the translocation of the large COPII complexes Sec23/24p and Sec13/31p. This experiment, together with titration and order-of-addition experiments, validates light scattering as a simple quantitative assay by which the assembly and disassembly of a protein coat on liposomes can be monitored in real time. Notably, light scattering is useful when studying the transient assembly of the COPII coat with GTP instead of a nonhydrolyzable analog. At first sight, the light-scattering assay may appear redundant with the FRET assay with DPH-PC-labeled liposomes (described above). With both

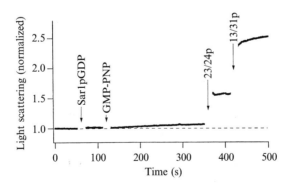

FIG. 6. Light-scattering assay of the assembly of COPII on liposomes. Light scattering of extruded liposomes (diameter, 0.4 $\mu$m; concentration, 0.1 mg/ml) was recorded at a right angle in a fluorometer ($\lambda = 350$ nm). Temperature, 27°. The composition of liposomes was as follows (mol%): PC (50), PE (23), PS (8), phosphatidic acid (5), PIns (9), PIns(4)P (2.2), PIns(4,5)P$_2$ (0.8), CDP-diacylglycerol (2). The lipid mixture was also supplemented with 20% (w/w) ergosterol. PC, PE, PS, and PA were equimolar mixtures of 1,2-dioleoyl and 1-palmitoyl-2-oleoyl derivatives. Other lipids were of biological origin.

assays, it is possible to monitor the recruitment of the small G protein (Arf1 or Sar1) on liposomes. However, light scattering is more adapted to detection of the binding of Sec23/24p and Sec23/31p. Indeed, the distance from the membrane surface increases with the sequence in which COPII components are assembled in a coat,[31] making FRET experiments with DPH-PC liposomes less suitable for the detection of Sec23/24p and Sec13/31p binding.

### Acknowledgments

We thank Dr. R. K. Nakamoto for the yeast expression system used to produce the Rac–GDI complex and Dr P. Chardin for comments on the manuscript. Karine Robbe is a recipient of a Ph.D. fellowship from the Ministère de la Recherche. This work is supported by the Ligue Nationale contre le Cancer. Experiments on COPII proteins were performed during a sabbatical in Dr. Randy Schekman's laboratory (UC Berkeley).

[31] K. Matsuoka, R. Schekman, L. Orci, and J. E. Heuser, *Proc. Natl. Acad. Sci. USA* **98,** 13705 (2001).

## [10] Liposomes with Multiple Fluorophores for Measurement of Ionic Fluxes, Selectivity, and Membrane Potential

*By* RÉMY GIBRAT and CLAUDE GRIGNON

### Introduction

Two main difficulties are encountered in the study of ion transport in liposomes. At the experimental level, a small aqueous lumen (typically 0.1 $\mu$m in diameter) is available to monitor ion electrochemical gradients. In addition, relatively fast filling or emptying kinetics, the deciphering of which forms the basis of our understanding of ion transport,[1] need to be measured. At the theoretical level, the difficulty is due to the dielectric barrier formed by the lipidic bilayer, which isolates the aqueous lumen. Salt diffusion involves fundamentally an electrical coupling of various ion fluxes, and the membrane potential ($E_m$) itself may be affected strongly by the capacitor properties of small vesicles (see Membrane Potential).

[1] W. D. Stein, ed., "Transport and Diffusion across Cell Membranes." Academic Press, New York, 1986.

For these reasons, a comprehensive analysis of ion transport across membrane vesicles requires acquiring information about several net ion fluxes, and about the size of the membrane potential ($E_m$). This is difficult to achieve with radioisotopes, whereas various ion-specific and membrane potential-sensitive fluorescent dyes are available for continuous and sensitive monitoring of electrochemical gradients across membrane vesicles. Methods for determining electrochemical gradients of individual ions across membrane vesicles have been described already in several volumes of this series. Only a few comments on dye methodology, essentially concerning $E_m$ determination and use, are made here; this review focuses on the rules and principles, illustrated by practical examples, for using multilabeled liposomes and modern spectrofluorometers fitted with motorized gratings (or filter wheels, or diode arrays). Indeed, such devices enable the investigator to make use of many advantages offered by liposomes, which are formed easily, are stable, and are handled easily for biochemical treatments or reinsertion of purified membrane proteins. As such, they offer a good avenue by which to decipher transport mechanisms of ion carriers, that is, electroenzymes exhibiting large conformational modification along the transport reaction cycle. Indeed, these transport systems are relatively abundant in biological membranes, but display too low a conductance relative to ion channels forming aqueous pores, to be easily studied by classic electrophysiological methods such as the patch-clamp.

## Designing Multilabeling Experiments

### Buffers and Salts

The number of dyes that can be used simultaneously in multilabeling experiments remains limited (see Multilabeling Experiments) and, generally, one of those must be dedicated to the determination of $E_m$. On the other hand, any transported ionic species should be monitored, ideally, to check the consistency of the procedure of calibration and analysis of signals. Indeed, the experimental value of the sum of net ion fluxes (multiplied by the valency of the species) must be equal to zero to account for the electroneutrality of the compartments. Therefore, a practical rule should be that only permeant ionic species required strictly for the transport study, together with nonpermeant counterions, should be present in the outer and inner media.

Concerning the pH control, various nonpermeant buffers are available, but they can also be used as nonpermeant co-ions of the studied transported ions [e.g., morpholineethanesulfonic acid (MES) or hydroxyethylpiperazine ethane sulfonic acid before (HEPES) as co-ion of $K^+$; Bis-Tris propane (BTP) as co-ion of $Cl^-$; see practical examples in the final section). Extreme pH values may be required, especially in plants or microorganisms, in which the $H^+$ electrochemical gradient plays a central role in bioenergetics. Below pH 5.0, for example, the chelator EDTA can be used as a nonpermeant buffer, or as a co-ion.

Incubation at such extreme pH values may be detrimental when vesicles contain proteins. For this reason, pH jumps generated by the instant diffusion of the freely permeant (electroneutral) species of certain weak acids or amines, for example, $NH_4^+$, are often used. The lumenal pH expected in such experiments, for example, when $NH_4^+$-loaded vesicles are diluted in an $NH_4^+$-free assay medium, can be evaluated in a test experiment in which the pH of the same buffer (except that the $NH_4^+$ salt is replaced by the corresponding $H^+$ salt) is measured with a pH meter. To improve the reliability and the reproducibility of such a pH jump, two nonpermeant buffers of suitable $pK_a$, for instance, BTP and MES, should be used to control both initial (neutral) and final (acidic) internal pH after dilution of the liposomes. No detrimental effect of such an acidic internal pH on integral proteins will occur as long as they have been unidirectionally reinserted into liposomes with their cytoplasmic face exposed to the external medium buffered at neutral pH (see the next section).

Another point to be considered is that $H^+$ ions, naturally present in media, are about 1 million times more permeant across lipid bilayers[2] than usual permeant inorganic ions ($K^+$, $Cl^-$, etc.). Even at neutral pH, a significant net $H^+$ flux is driven by $E_m$ and must be taken into account, even in transport experiments dedicated to other ionic species.[3] In practice, pH jump tests may be used to ascertain that buffers or other ionic species in the assay media can be considered nonpermeant in the time course of the experiments. An acidic-inside pH gradient across vesicles, following the dilution of $NH_4^+$-loaded vesicles, can be detected easily with fluorescent permeant amine dyes such as 9-amino-6-chloro-2-methoxyacridine (ACMA). Paradoxically, because $H^+$ ions are freely permeant, the dissipation of the formed pH gradient is limited strictly by the diffusion of another ionic species electrically compensating the outward diffusion of $H^+$. Therefore, if only nonpermeant ionic species are present in both internal and external media, the pH gradient will remain stable.[4] Finally, addition of $\approx 0.1$ m$M$ EDTA may be suitable to eliminate undesired contamination by polyvalent cations.

[2] M. Rossignol, P. Thomas, and C. Grignon, *Biochim. Biophys. Acta* **684,** 195 (1982).

[3] P. Pouliquin, J. P. Grouzis, and R. Gibrat, *Biophys. J.* **76,** 360 (1999).

[4] G. Borrelly, J. C. Boyer, B. Touraine, W. Szponarski, M. Rambier, and R. Gibrat, *Proc. Natl. Acad. Sci. USA* **98,** 9660 (2001).

*Membrane Vesicles*

A variety of methods[5,6] are available for the preparation of unilamellar liposomes of various lipid compositions and in a broad size range. The so-called small unilamellar vesicles (SUVs) are small liposomes (about 20 nm), the bilayer packing of which is affected by the low radius of curvature relative to cell membranes. Large unilamellar liposomes (LUVs) are stable vesicles of intermediate size (about 200 nm) exhibiting relatively slow filling/emptying kinetics, permitting a description of the (nonfacilitated) ion diffusion and selectivity across the lipid moiety of biological membranes.

Whereas the two leaflets of the lipidic bilayer of biological membranes exhibit asymmetrical compositions,[7] classic methods for the preparation of LUVs (e.g., by detergent dilution or extrusion) give symmetrical bilayers. This facilitates greatly the quantitative analysis of $E_m$ dye signals, which generally interact strongly with both internal and external leaflets of the bilayer (see the example of oxonol VI, described below).

LUVs are used largely to reinsert integral transport proteins. Ion carriers, also called electroenzymes,[8] are subjected to strong electroconformational changes, depending on the size and orientation of the electrical field, associated with a variation in their activity. Moreover, the two opposite faces of transporters may display different sensitivities to inorganic ions and organic molecules. The cytoplasmic face of V-ATPase or P-ATPase of plant or fungi, for instance, exhibits a strongly different sensitivity to $H^+$ ions, relative to the external face. Therefore, the composition of the inner and outer media of transporter-containing vesicles may have to be different to obtain physiologically relevant information about their function. An obvious example is provided by the supply of nonpermeant energizing substrates to the catalytic site of ion pumps.

For these reasons, it is important to reinsert transport proteins unidirectionally in liposomes. Most transporters have highly asymmetrical structures, often with large cytoplasmic moieties with several functional domains, and smaller external moieties. This is the case for P-ATPases, V-ATPases, $K^+$ channels, $Na^+$ or $H^+$ cotransporters, and so on. Spontaneous reinsertion of various integral proteins into preformed bilayers was the first experimental approach that allowed unidirectional

[5] G. Gregoriadis ed., "Liposome Technology," Part 1: "Preparation of Liposomes." CRC Press, Boca Raton, FL, 1984.
[6] M. C. Woodle and D. Papahadjopoulos, *Methods Enzymol.* **171,** 193 (1989).
[7] H. A. Rinia and B. de Kruijff, *FEBS Lett.* **504,** 194 (2001).
[8] D. Gradmann, M. R. Blatt, and G. Thiel, *J. Membr. Biol.* **136,** 327 (1993).

orientation,[9,10] taking advantage of the asymmetrical structure of integral proteins and of the formation of hydrophobic defects at the surface of bilayers. In contrast to slow detergent dilution, which leads to scrambled orientation, some fast detergent depletion methods have been shown also to permit unidirectional reinsertion of proteins,[11] probably because the lipidic bilayer was formed before the spontaneous reinsertion of detergent–protein complexes.[12]

The tightness of liposomes to small polar molecules is another important property that depends essentially on alkyl chain unsaturation of phospholipids and/or on the presence of sterols.[13] The tightness of proteoliposomes appears also to depend on the proportions of the various phospholipid head groups.[3]

### Membrane Potential: Long-Wavelength Oxonol VI Dye

$E_m$ is a thermodynamic pivot for the analysis of ion transport, because measurement of the electrochemical gradients of the transported species is necessary to determine whether the transport process is passive or active. At a kinetic level, determination of the $E_m$ change on activation of a transport system is crucial for the characterization of the transport mechanism (electrogenic versus electroneutral mechanism, electroconformational regulations, etc.).

As pointed out above, integral proteins may be reinserted unidirectionally in liposomes, with their large "cytosolic" moiety exposed toward the outside. In the cell, this face is at a lower (more negative) electrical potential than the opposite face, which corresponds to the imposition of positive-inside $E_m$ across liposomes. Thus, among the variety of dyes usable to monitor negative-inside or positive-inside $E_m$, the latter are generally more suitable for *in vitro* transport studies on liposomes.

Freely permeant dyes are more sensitive to $E_m$ than are electrochromic dyes embedded in the bilayer. Because they are thermodynamically distributed in the two aqueous compartments separated by the membrane, their aqueous concentrations at equilibrium are accounted for by the Nernst law:

[9] D. Zakim and A. W. Scotto, *Methods Enzymol.* **171,** 253 (1989).

[10] F. Simon-Plas, K. Venema, J. P. Grouzis, R. Gibrat, J. Rigaud, and C. Grignon, *J. Membr. Biol.* **120,** 51 (1991).

[11] J. P. Grouzis, P. Pouliquin, J. Rigaud, C. Grignon, and R. Gibrat, *Biochim. Biophys. Acta* **1325,** 329 (1997).

[12] J. L. Rigaud, B. Pitard, and D. Levy, *Biochim. Biophys. Acta* **1231,** 223 (1995).

[13] W. K. Subczynski, A. Wisniewska, J. J. Yin, J. S. Hyde, and A. Kusumi, *Biochemistry* **33,** 7670 (1994).

$$D_{in} = D_{out} \exp\left[(-ZF/RT)E_m\right] \tag{1}$$

where $R$ and $F$ are the classic thermodynamic constants, $T$ is the absolute temperature, $Z$ is the valency ($-S$ for oxonol VI), and $D_{in}$ and $D_{out}$ are the free dye concentrations in the lumen and in the external medium, respectively. Among "nernstian" dyes, oxonol VI offers several advantages: (1) it is a long-wavelength dye with an excitation wavelength near 600 nm, permitting the simultaneous use of other dyes of lower wavelength for specific ion detection; and (2) among the cyanine dyes, the lipophilic anion oxonol VI is the most permeant $E_m$ dye,[14] allowing the investigator to monitor fast positive-inside $E_m$ changes.

Some fluorescent nernstian dyes may be quenched after their passive accumulation inside the vesicle lumen, or even within the internal membrane leaflet. Concentration quenching of fluorophores is generally observed at high concentration in the millimolar range when considering aqueous solutions. To reach such concentrations inside the lumen of vesicles sustaining an $E_m$ of 120 mV, external concentrations of about 10 $\mu M$ for a univalent anionic dye are required. Because of the high permeability coefficient of lipophilic dyes, such concentrations promote a large anionic conductance that decreases the (positive-inside) $E_m$ and, therefore, are unsuitable for quantitative analysis of the electrical polarization of liposomes. Concentrations of oxonol VI lower than 50 n$M$ are required to achieve that goal. In this condition, a red shift of fluorescence and an augmentation of fluorescence intensity (FI) due to an increase in the quantum yield are observed in response to $E_m$. This response does not originate from the free dye passively accumulated in the vesicle lumen, but from the augmentation of the number of dye molecules bound onto hydrophobic sites of the inner leaflet of the vesicle membrane. The second rationale for using a low oxonol VI concentration (50 n$M$) is that $E_m$ can be monitored only as long as internal binding sites remain unsaturated. Accounting for the $K_d$ of binding sites of liposomes (5 $\mu M$) and the depletion of the free form of the dye at high $E_m$, about 60% of the internal binding sites still remain unoccupied at $E_m = 130$ mV, the maximum diffusion $E_m$ that can be reached (see below).

Anionic amphipolar dyes are inserted at the membrane interface within sites constituted of four phosphatidylcholine molecules,[15] the major (neutral) phospholipid in biological membranes. The presence of one phosphatidylethanolamine, the second major phospholipid, within the quadratic sites conformationally hinders the insertion of amphipolar dye molecules.

[14] R. J. Clarke and H. J. Apell, *Biophys. Chem.* **34**, 225 (1989).
[15] D. H. Haynes and H. Staerk, *J. Membr. Biol.* **174**, 313 (1974).

Thus, liposomes made from natural extracts of mixed phospholipids,[16] or liposomes containing at least 20% phosphatidylethanolamine, exhibit low binding site numbers for amphipolar dyes. This is important for two reasons. First, a large mean distance between binding sites minimizes the deactivating energy transfer between fluorescent dyes.[17] The quantum yield of bound dyes remains constant along the binding isotherm and, thus, the variation of FI is proportional to the number of bound molecules. Second, the apparent dissociation constant ($K_d$) of the binding sites, which depends on the electrostatic repulsion of free dye molecules by the field of the membrane surface charge,[3] remains unaffected by the binding of the anionic dye.

The fluorescence ratio $\rho = \Delta F/F_0$, $F_0$ and $\Delta F$ corresponding to the fluorescence of bound dye at $E_m = 0$ and on $E_m$ generation, respectively (Fig. 1A), is generally used to empirically calibrate the dye. This is achieved by plotting $\rho$ versus theoretical (Nernst) $E_m$ values expected for imposed diffusion gradients. However, when increasing $K^+$–valinomycin gradients are used, for example, an increasing discrepancy is observed between the dye response and expected diffusion $E_m$, and $\rho$ reaches a plateau at the theoretical $E_m$ of ~130 mV.[3] This discrepancy originates from the high capacitance ($C_m$) of biological membranes, well recognized in electrophysiological studies. Because of the high surface-to-volume ratio, the maximum value of diffusion $E_m$ across small vesicles is restricted by $C_m$.[18] Indeed, a significant number of diffusive ions, for instance, $K^+$ in the presence of valinomycin, must cross the dielectric to polarize the membrane vesicle capacitor from 0 to $E_m$. This preliminary $K^+$ entry in the small vesicle lumen significantly increases the initial $K^+$ concentration for high expected $E_m$ (i.e., for large imposed gradients). For example, an "infinite" Nernst $E_m$ is expected by imposing an infinite inward $K^+$–valinomycin gradient (200 m$M$ outside and no $K^+$ inside initially) across liposomes with a 28-nm radius. Nevertheless, using classic capacitor equations with $C_m = 1 \ \mu F \cdot cm^{-2}$, the initial $K^+$ entry required to polarize such vesicle capacitors should correspond to an increase in the lumenal $K^+$ concentration from 0 to 1.3 m$M$ $K^+$, which accounts for the plateau of the dye response.[3]

Therefore, it is important to achieve a direct determination of $E_m$ that does not rely on calibration plots with a theoretical diffusion $E_m$. For reasons indicated above, FI can be considered linear with the concentration of bound dye. Therefore, $\rho = \Delta F/F_0$ is an experimental measure of the accumulation of bound dye onto the internal leaflet relative to the external

[16] R. Gibrat, C. Romieu, and C. Grignon, *Biochim. Biophys. Acta* **736,** 196 (1983).
[17] J. Perrin, *C.R. Acad. Sci. Paris* **184,** 1097 (1927).
[18] H. J. Apell and B. Bersch, *Biochim. Biophys. Acta* **903,** 480 (1987).

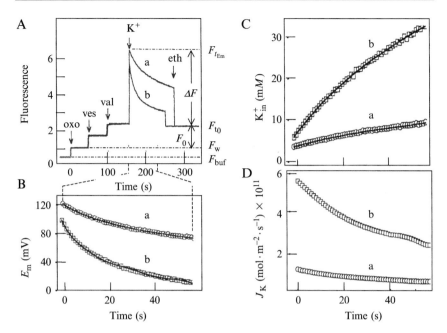

FIG. 1. Monitoring of a positive-inside diffusion potential with oxonol VI and determination of net ion fluxes. (A) The following FIs are recorded: (1) $F_{buf}$, FI of the assay buffer containing 0.1 $M$ HEPES-Li (pH 7.4) and 50 m$M$ Li$_2$SO$_4$; (2) $F_w$, FI of oxonol VI (oxo = 50 n$M$) free in bulk solution; (3) $F_{t_{Em}}$, FI after addition of liposomes with reinserted plant plasma membrane proteins (ves = 50 $\mu$g of phospholipid $\cdot$ ml$^{-1}$) and containing 0.1 $M$ HEPES-Li (pH 7.4), 50 m$M$ Li$_2$SO$_4$, valinomycin (val, 20 n$M$), and K$^+$ at 200 m$M$ to generate a positive-inside $E_m$; K$^+$ is added either as K$_2$SO$_4$ (trace a) or as K$_2$SO$_4$ plus KNO$_3$ to make final concentrations of K$^+$ and NO$_3^-$ equal to 200 and 15 m$M$, respectively (trace b); (4) FI at zero $E_m$ after its short-circuiting by Li$^+$ ionophore (eth, 2 nmol $\cdot$ ml$^{-1}$) ($F_{t_0}$); FIs are corrected for successive dilutions before calculating $E_m$ with Eqs. (3) and (4) (see text); the ratio $\rho = \Delta F / F_0$ is determined from the FI increase ($\Delta F = F_{t_{Em}} - F_{t_0}$) after $E_m$ imposition and the FI increase ($F_0 = F_{t_0} - F_w$) after liposome addition. (B) $E_m$ is calculated from the four preceding FIs and Eqs. (3) and (4). (C) Internal K$^+$ concentration (K$_{in}^+$) at equilibrium with $E_m$, calculated with Eq. (5); the kinetics are fitted with a polynomial regression algorithm (continuous line). (D) Net K$^+$ fluxes ($J_K$) in the presence and in the absence of NO$_3^-$ are calculated from the derivatives of the polynomial fits [Eq. (6)] with an experimental estimate of $r$ = 28 nm. The net NO$_3^-$ flux ($J_N$) is given by the NO$_3^-$-dependent augmentation of $J_K$ (see text). Reproduced from Ref. 3, published in the *Biophysical Journal* with permission of the Biophysical Society (United States).

leaflet ($D_{bound\ in}/D_{bound\ out}$) at a given $E_m$. Assuming, for the time being, that $D_{bound\ in}$ and $D_{bound\ out}$ are linear with the free dye concentrations in the two compartments $D_{in}$ and $D_{out}$, $E_m$ can be calculated directly from $\rho$ and the thermodynamic Eq. (1):

$$E_m = (RT/F) \log (2\rho + 1) \tag{2}$$

accounting for the fact that $D_{in}$ and $D_{out}$ for each leaflet are proportional to $(\Delta F + F_0/2)$ and $F_0/2$, respectively.

Equation (2) is not usable as such because of the high accumulation of dye within the vesicles at high $E_m$. Indeed, the external medium is generally significantly depleted of free dye because a low total dye concentration must be used. Furthermore, $D_{in}$ often reaches values comparable to, or even higher than, the $K_d$. Internal binding sites become saturated and, therefore, $D_{bound\ in}$ is related to $D_{in}$ by a nonlinear binding isotherm. These depletion and saturation effects can be quantitatively accounted for by measuring in each record (Fig. 1A) the FI values, respectively, after dye addition in the absence of vesicles $(F_w)$, after vesicle addition at $E_m = 0$ $(F_{t_0})$, and after $E_m$ generation $(F_{t_{Em}})$. As detailed elsewhere,[3] Eqs. (3) and (4) may be used to determine $E_m$:

$$\rho = F_w[(F_{t_{Em}} - F_{t_0})(\Phi_r - 1)]/[(\Phi_r F_w - F_{t_{Em}})(F_{t_0} - F_w)] \tag{3}$$

$$E_m = (RT/F) \ln [(2\rho + 1)/\{1 - [D_t(2\rho + 1) \\ (F_{t_{Em}} - F_w)]/[N \operatorname{Lip} F_w(\rho + 1)(\Phi_r - 1)]\}] \tag{4}$$

where $\Phi_r$, $N$, Lip, and $D_t$ are the ratio of the fluorescence yield of bound/free dye, the number of sites per mole of lipid, the lipid concentration, and the total dye concentration, respectively. The parameters $\Phi_r$ and $N$ are to be determined once by titrating a fixed dye concentration by liposomes, and a fixed liposome concentration by the dye. Although these parameters depend greatly on the lipid composition of liposomes, they seem remarkably insensitive to variations of ionic conditions.

### Diffusion Potential and Analysis of Filling/Emptying Kinetics

In the preceding section, the oxonol VI dye has been used as a direct and quantitative sensor for continuously monitoring $E_m$. Importantly, $E_m$ dyes can also be used as indirect reporters of passive ion fluxes across liposomes when ion-specific fluorescent dyes for a given ion species are not available. This can be achieved by comparing quantitatively imposed diffusion $E_m$ kinetics in the presence or absence of this species[3] (see below). For instance, typical $K^+$–valinomycin diffusion potential kinetics are shown in Fig. 1B. After reaching "instantaneously" its maximum initial value, $E_m$ slowly decreases, indicating that the $K^+$ diffusion gradient slowly dissipates. This dissipation results actually from $K^+$ filling of the vesicle lumen, and not from $K^+$ emptying of the external compartment; indeed, $K^+_{out}$ remains constant because the size of the external compartment is "infinite" relative to the lumenal compartment. Therefore, the internal concentration $(K_{in}^+)$ during the filling kinetics can be deduced simply from the known imposed value of $K_{out}^+$ and the $E_m$ value measured during the depolarization kinetics (Fig. 1C), considering that $K^+$ is virtually at thermodynamic equilibrium, despite the slow dissipation of its gradient actually observed.[3] Indeed, the slow net transport of $K^+$ occurs at a negligible $K^+$ electrochemical potential gradient (about 15 $J \cdot mol^{-1}$) because of the high permeability conferred by valinomycin.

$$(K_{in}^+) = (K_{out}^+) \exp\left[-(F/RT)E_m\right] \tag{5}$$

A nonlinear regression fit of $(K_{in}^+) = f(t)$ may be used to determine the polynomial equation $K_{in}^+ = \sum_{i=0}^{i=n} a_i t^i$, which accounts for the $K^+$ filling kinetics and which can be derived easily to obtain $J_K$, the net $K^+$ influx (Fig. 1D):

$$J_K = d(K_{in}^+ V)/(A dt) = (r/3) dK_{in}^+/dt \tag{6}$$

where $V$ is the internal volume, $A$ is the surface area, and $r$ is the internal radius of liposomes.

In the experiment corresponding to trace a in Fig. 1, $K^+$ is the only added permeant species. Nevertheless, a significant net entry of $K^+$ (initial $J_K \approx 1.2 \times 10^{-9}$ mol $\cdot$ m$^{-2} \cdot$ s$^{-1}$ at $E_m \approx 0.1$ V) is observed, which indicates that an equivalent counterion transport occurs to electrically counterbalance $J_K$. The latter is actually compensated by an efflux of $H^+$ $(J_H)$, the only strongly permeant ion naturally present in all media. This can be checked with vesicles multilabeled with $K^+$, $H^+$, and $E_m$ dyes (see experiment in Fig. 5). $J_H$ $(= J_K)$ and $E_m$ values may be used to calculate the permeability coefficient of the vesicles to $H^+$ $(P_H)$ by using the Goldman–Hodgkin–Katz (G–H–K) relation,[1] simplified in this case because internal and external pH are the same:

$$P_H = -J_H[RT/(FE_m)]/10^{-pH} \tag{7}$$

In agreement with literature data,[2] a high $P_H$ is obtained $(10^{-6}$ m $\cdot$ s$^{-1})$.

Adding another passively transported ionic species (i.e., by diffusion or by an electrically driven protein uniport) will result in augmentation of $J_K$, equal to the net equivalent flux of this species. This is exemplified, in the experiment corresponding to trace b in Fig. 1, by the 5-fold increase in the initial $J_K$ value after $NO_3^-$ addition to the outside, together with the polarizing concentrated $K^+$ aliquot. The $NO_3^-$-dependent augmentation of $J_K$ could be used as an indirect measurement of the net $NO_3^-$ influx $(J_{NO3} = \Delta J_K = 5 \times 10^{-9}$ mol $\cdot$ m$^{-2} \cdot$ s$^{-1})$. The permeability coefficient of the vesicles to $NO_3^-$ $(P_{NO3} = 9 \times 10^{-11}$ m $\cdot$ s$^{-1})$ may be calculated from $J_{NO3}$ and the corresponding $E_m$ value (0.09 V), using a simplified G–H–K relation (as vesicles were initially deprived of $NO_3^-$):

$$P_{NO3} = -J_{NO3}[RT/(-FE_m)]\{1 - \exp[-(F/RT)E_m]\}/NO_{3\text{ out}}^- \qquad (8)$$

Despite the fact that no specific dyes are available to measure $NO_3^-$ flux across membrane vesicles, this procedure permits the functional characterization *in vitro* of a plant plasma membrane uniporter.[19]

*Multilabeling Experiments*

Analyzing ion transport experiments requires at least the determination of the net fluxes of the studied species and of the co-ion, and the $E_m$ value. This requires that signals from three different dyes be measured at a high rate to resolve emptying or filling kinetics. T-shaped spectrofluorometers allow simultaneous measurement of two fluorescence signals through two distinct data acquisition channels. In addition, modern spectrofluorometers are fitted with fast motor-driven monochromators with slewing rates up to $500 \text{ nm} \cdot \text{s}^{-1}$. Moreover, spectrofluorometers fitted with diode array detectors for almost instantaneous acquisition of whole fluorescence spectra or fluorescence intensity at any wavelength are now available. Such a combination of a fast-slewing excitation monochromator and a diode array detector constitutes an efficient device to obtain rapid acquisition cycles of fluorescence signals from dye mixtures. Another important technical requirement concerns the sample holder, which must be thermostatted and fitted with an efficient stirrer to minimize the mixing time of added aliquots of reactants, as well as kinetic distortions due to unstirred layers at the surface of membrane vesicles. Of course, such experiments generate large data flows that require specific home made computer programs.

The internal concentration of certain ionic species, for example, $H^+$ ions, can be monitored either with permeant or with nonpermeant dyes, but the latter are more suitable for a straightforward quantitative analysis. Moreover, certain ion-specific nonpermeant dyes display changes in fluorescence spectra when they are complexed, and an FI ratio can be obtained from the free and bound forms, providing sensitive, reproducible, and nonarbitrary data. (FI values are arbitrary, depending on the spectrofluorometer settings, whereas FI ratios are accounted for by the Beer–Lambert law, and are therefore not arbitrary.) In addition, they provide isobestic excitation or emission wavelengths, which are useful in analyzing complex fluorescence dye mixtures (see the example on p. 177).

Choosing the dyes for multilabeling experiments is not straightforward and a compromise is to be searched for, taking into account their relative efficiencies (affinity, selectivity, sensitivity, ratio acquisition, etc.) and the

[19] P. Pouliquin, J. C. Boyer, J. P. Grouzis, and R. Gibrat, *Plant Physiol.* **122**, 265 (2000).

compatibility of their excitation/emission bands (band overlapping is to be minimized, whereas excitation or emission bands that are far apart increase the duration of acquisition cycles).

Calibrating mixtures of ion-specific dyes do not differ fundamentally from that described in the literature for individual dyes. High concentrations of ion-specific dyes are to be entrapped for a sensitive labeling of liposomes. Excitation/emission spectra of the dye mixture in liposomes are to be compared with those observed after liposome permeabilization. The similarity of measured spectra, as well as that of calibration curves, ascertains the absence of disturbing interactions between dyes.

### Practical Example: Study of Electrically Coupled $H^+$:$K^+$ Exchange

The aim of this multilabeling experiment is to determine net flux ($J_K$) and permeability coefficient ($P_K$) of liposomes to $K^+$, using the specific $K^+$ dye potassium-binding benzofuran isophthalate (PBFI; Molecular Probes, Eugene, OR). Involving also the pH dye pyranine, this experiment points out that careful attention must be paid to $H^+$ transport even at neutral pH, or even if $H^+$ is not directly involved in a transport mechanism. Moreover, this example illustrates that measuring $H^+$ flux is another way to determine indirectly the permeability coefficient of liposomes and membrane selectivity to various ions when specific dyes are not available.

### Liposomes and Encapsulation of Nonpermeant Dyes

Liposomes are made from mixed soybean phospholipids (L-$\alpha$-phosphatidylcholine, type II-S; Sigma, St. Louis, MO) or an egg phosphatidylcholine–phosphatidylglycerol mixture [typically 10 mg, 8:2 (mol/mol)] in chloroform, dried under argon on a Vortex mixer. Phospholipids are dispersed in 300 $\mu$l of reconstitution buffer (50 m$M$ BTP-SO$_4$ at pH 7.0 or histidine-HEPES at pH 6.75 and 1 $M$ mannitol or glycerol as indicated) by vigorous mixing for 15 min under argon in the presence of glass beads, sonicated for 15 min in a bath sonicator under argon, and diluted 10-fold in the same buffer to which an aliquot of deoxycholate (DOC) is added to obtain an effective ratio of 0.6: $R_e =$ [(detergent) $-$CMC]/(phospholipids), where CMC is the critical micellar concentration. Thereafter, LUVs are prepared by one of the detergent elimination methods described below, depending on the nonpermeant dye that must be encapsulated. Note that the permeant dye oxonol VI is simply added to the assay cuvette containing LUVs before FI measurements.

Inexpensive dye mixtures may be entrapped in LUVs along with their formation as follows: 1 ml of the dye mixture in reconstitution buffer is layered and allowed to run into the top of a Sephadex G-50 column

(35 × 0.5 cm) equilibrated in the buffer indicated above; thereafter, an aliquot of a mixture containing both phospholipid–DOC micelles and the dyes is applied directly to the column; reconstituted LUVs with entrapped dyes are eluted in the cloudy void volume, whereas nonentrapped dyes are retained in the column.

A more sparing method, in which encapsulation follows LUV formation, is usable for expensive nonpermeant dye mixtures, such as those containing PBFI. Plastic syringes filled with 1.5 ml of Sephadex G-50 equilibrated in the above-described buffer are centrifuged for 5 min at 180 g to release extra buffer. Thereafter, 220-$\mu$l aliquots of lipid–detergent micelles are layered onto the columns and centrifuged again to eliminate DOC. Reconstituted LUVs are recovered in the cloudy eluted volumes, which account for at least 90% of the initially applied lipids and volume. This LUV suspension is concentrated 5-fold (Ultrafree-PF, 100,000 NMWL; Millipore, Bedford, MA) before addition of aliquots of PBFI and pyranine (1 and 0.5 m$M$ final concentrations, respectively), and the mixture is sonicated for 30 s in a bath sonicator and freeze–thawed once (liquid nitrogen–room temperature). Thereafter, nonencapsulated dyes are removed from the Sephadex G-50 column (35 × 0.5 cm) as described above.

Addition of the nonpermeant pyranine quencher p-xylene-bis-pyridinium bromide (DPX) allows confirmation that more than 90% of the dyes in the LUV suspension are encapsulated. The specific internal volume $V_{in}$, in liters per mole of phospholipid, can then be estimated from

$$V_{in} = D_a/(D_{lip}Lip) \tag{9}$$

in which Lip is the molar phospholipid concentration in the assay cuvette, $D_{lip}$ is the known concentration of entrapped dye in liposomes, and $D_a$ is the macroscopic dye concentration in the assay cuvette, deduced by comparing the sample FI with a calibration curve performed with known bulk dye concentrations. The mean internal radius ($r$) of LUVs can be calculated from

$$r = 3V_{in}/(2A) = 6V_{in}/(Ns) = 133V_{in} \tag{10}$$

where $A$, the mean vesicle area (in Å$^2$), corresponds to 2Å of phospholipid area, $N$ is Avogadro's number, and $s$ is the estimated mean molecular area of phospholipids (75 Å$^2$).[16] Other methods, such as gel filtration, give similar results.

*Fluorescence Measurements*

FI values are measured with T-shaped multichannel spectrofluorometers (SLM-Aminco 8000C, or Aminco-Bowman; Thermo Spectronic, Madison, WI), allowing the investigator to record two fluorescence signals simultaneously through A and B channels. FI values are ratioed with the signal of a rhodamine reference cell and corrected for intrinsic FI values of buffer at different wavelengths, as well as for dilution due to addition of successive aliquots. Channel A is used to acquire oxonol VI FI at 646 nm through a monchromator after excitation at 614 nm. Channel B is used to acquire successively PBFI and pyranine FI values through a 495-nm long-pass cutoff filter. This design is highly sensitive, and it also circumvents the shift of the maximum FI of PBFI from 550 to 500 nm when $K^+$ concentration increases. Although maximum excitation of PBFI is noted at 340 nm, PBFI is excited at 336 nm, which corresponds to one of the two isobestic wavelengths of pyranine. In this condition, the small contribution of pyranine is independent of the pH. Thereafter, pyranine is excited successively at 410, 421, and 460 nm.

## Calibration of the pH Dye Pyranine

Both protonated (PyrH) and deprotonated (Pyr$^-$) forms of pyranine display the same maximum fluorescence at 511 nm, but their maximum excitation is observed at 410 nm ($FI_{410}$) and 460 nm ($FI_{460}$), respectively. To calibrate pyranine, LUVs ($\sim$80 $\mu g \cdot ml^{-1}$) are diluted in the assay cuvette containing the reconstitution buffer used above, and the electroneutral $K^+/H^+$ exchanger nigericin (0.5 $\mu M$) is added. Thereafter, the external pH is increased step by step between pH 6.75 and pH 9.0 by adding concentrated aliquots of KOH. The nigericin exchanger allows a free equilibration of the internal pH, and the ratio $FI_{460}/FI_{410}$ is noted after each pH jump. The plot log ($FI_{460}/FI_{410}$) versus pH is linear according to the Henderson–Hasselbalch relation (Fig. 2):

$$\log (FI_{460}/FI_{410}) = pH - pK_a + \log (\beta/\alpha) \tag{11}$$

where $\alpha$ and $\beta$, respectively, are the coefficients of the linear relationship between $FI_{410}$ and PyrH, and between $FI_{460}$ and Pyr$^-$. The same calibration curves are obtained when only pyranine is entrapped or free in bulk solution, or when the two other dyes are present.

According to the Debye–Hückel theory, $H^+$ ions are accumulated in the vicinity of pyranine because of electrostatic interactions with this polyanionic dye (valency of $-3$ to $-4$, depending on pH). Therefore, titration of the dye gives access only to an apparent $pK_a$ value and, as expected, the lower the ionic strength, the higher this value is relative to its intrinsic value.[20] Therefore, high lumenal ionic strengths of nonpermeant salts limit

[20] K. Venema, R. Gibrat, J.-P. Grouzis, and C. Grignon, *Biochim. Biophys. Acta* **1146,** 87 (1993).

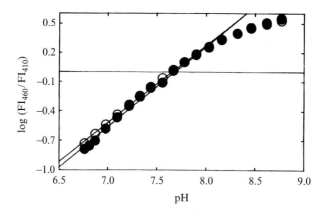

Fig. 2. Calibration of the nonpermeant pH dye pyranine. Liposomes (80 $\mu$g · ml$^{-1}$) containing 50 m$M$ histidine-HEPES (pH 6.75), 10% glycerol (w/v), and 0.5 m$M$ pyranine (see text) are diluted in the same medium without the dye. The K$^+$/H$^+$ exchanger nigericin (0.5 $\mu M$) is then added to allow a free equilibration of the internal pH, the pyranine dye is titrated with concentrated aliquots of KOH, and the FIs at 460 and 410 nm are rationed. Calibration curves when the dye is encapsulated in liposomes (●) or free in solution (○) are similar. No difference is observed when liposomes are multilabeled with pyranine, PBFI, and oxonol VI (not shown). Reprinted from Ref. 20, published in *Biochimica et Biophysica Acta* with permission from Elsevier Science.

distortions of H$^+$–dye interactions during filling/emptying experiments involving high ion concentrations. Ion-specific dyes are generally fluorescent, ionized chelators subjected to electrostatic interactions with their ligand. Certain experimental designs can circumvent this difficulty. For example, addition of a small KOH aliquot to the external medium in the presence of nigericin allows an "instant" equilibration of the internal pH at constant ionic strength when liposomes are buffered with BTP. Indeed, when K$^+$ is the only permeant ion added, the influx of K$^+$ is compensated by an efflux of H$^+$ released by the deprotonation of the amine groups of the BTP buffer, which become electroneutral.

*Calibration of K$^+$ Dye PBFI*

As PBFI dye binds K$^+$, its fluorescence increases. The complex KPBFI is formed according to the mass action law:

$$K_{d\,app} = [(PBFI)(K^+)/(KPBFI)] \qquad (12)$$

The apparent value of the dissociation constant ($K_{d\,app}$) depends on the ionic strength, as discussed above.[20] To calibrate PBFI, LUVs

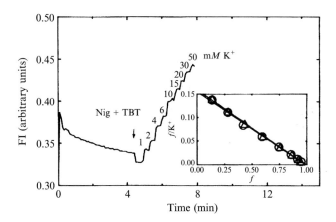

FIG. 3. Calibration of the nonpermeant $K^+$ dye PBFI. Conditions are as in Fig. 2, except that liposomes contain 1 m$M$ PBFI. PBFI is titrated with concentrated aliquots of KCl after addition of nigericin (0.5 $\mu M$) for $K^+/H^+$ exchange and TBT (0.5 m$M$) for $Cl^-/OH^-$ exchange, allowing a free equilibration of the internal KCl concentration. *Inset:* Scatchard plot in which $f = (FI_K - FI_0)/(FI_{max} - FI_0)$ (see text). Similar results were obtained when liposomes were multilabeled with PBFI, pyranine, and oxonol VI. Reprinted from Ref. 20, published in *Biochimica et Biophysica Acta* with permission from Elsevier Science.

($\sim$80 $\mu$g $\cdot$ ml$^{-1}$) are diluted in the assay cuvette containing the reconstitution buffer used above, and the two electroneutral antiporters nigericin (0.5 $\mu M$) and tributyltin chloride (TBT, 0.5 m$M$, mediating a $Cl^-/OH^-$ exchange) are added. Thereafter, PBFI fluorescence is titrated by adding concentrated aliquots of KCl (Fig. 3). Instant KCl equilibration is achieved at each step by the two combined exchangers. Note $FI_0$ and $FI_K$, the fluorescence in the absence and in the presence of $K^+$, respectively:

$$FI_K = FI_0 + \gamma(KPBFI) \tag{13}$$

where $\gamma$ is a constant.[21] The binding fluorescence isotherm is therefore given by

$$[FI_K - FI_0] = [FI_{max} - FI_0](K^+)/[K_{d\,app} + (K^+)] \tag{14}$$

The same $K_{d\,app}$ values are determined for entrapped or free PBFI (inset in Fig. 3). This KCl titration aims essentially at determining the values of $FI_{max}$ and $FI_0$, in addition to the $K_{d\,app}$ value at the initial pH. Indeed, $FI_{max}$ and $FI_0$ remain constant in the pH range used, suggesting that the

[21] P. Jezek, F. Mahdi, and K. D. Garlid, *J. Biol. Chem.* **265,** 10522 (1990).

intrinsic fluorescence properties of the dye are not affected, despite the strong pH dependence of the PBFI $K^+$ response. This dependence actually results from a competition of $H^+$ and $K^+$ ions for the binding site of the dye, described by the mass action law:

$$K_{d\,app} = K_{dK}[1 + (H^+)]/K_{dH} \qquad (15)$$

where $K_{dK}$ and $K_{dH}$ stand for the dissociation constants for $K^+$ and $H^+$, respectively. The titration performed with KOH in the presence of nigericin to calibrate the pyranine dye can be used to take into account this effect of $H^+$ ions (see the preceding section). At each step of this titration, where both $K^+$ and $H^+$ concentrations are known, the $K_{d\,app}$ value can be determined using Eq. (14). Then, the experimental plot of $K_{d\,app}$ versus $H^+$ appears linear, as expected, and $K_{dK}$ and $K_{dH}$ may be determined from a regression plot according to Eq. (15) (Fig. 4). The same $K_{dK}$ and $K_{d\,H}$ values are observed for free or encapsulated dye.

After calibration of the pyranine and PBFI dyes, the lumenal pH during the $K^+$ filling experiment is determined using Eq. (11), and the lumenal $K^+$ concentration is obtained from Eq. (16), taken from Eqs. (14) and (15):

$$(K^+) = \{K_{dK}[1 + (H^+)]/K_{dH}\}[(FI_K - FI_0)/(FI_{max} - FI_K)] \qquad (16)$$

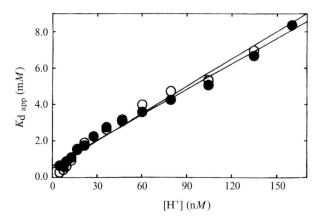

FIG. 4. Competition of $H^+$ ions for the $K^+$-binding site of the nonpermeant dye PBFI. Conditions are as in Fig. 2, except that liposomes contain 1 m$M$ PBFI and 0.5 m$M$ pyranine. Free PBFI in solution is titrated by adding concentrated aliquots of KOH (○); the $K^+/H^+$ exchanger nigericin (0.5 $\mu M$) is added (before the titration) when PBFI is encapsulated in liposomes (●). $K_{d\,app}$ is calculated from Eq. (14), and $K_{d\,K}$ and $K_{d\,H}$ (0.7 × 10$^{-3}$ and 1.5 × 10$^{-8}$ $M$, respectively) are calculated from the linear regression plot of data, using Eq. (15). Reprinted from Ref. 20, published in *Biochimica et Biophysica Acta* with permission from Elsevier Science.

*Ion Fluxes and Permeability Coefficient*

The $K^+$ filling kinetics (Figs. 5 and 6) are initiated by adding a concentrated aliquot of $K_2SO_4$ to the assay medium containing multilabeled LUVs (80 $\mu$g $\cdot$ ml$^{-1}$). At the end of the kinetics, a DOC aliquot (0.05%, w/v) is added to dissipate the transmembrane gradients. Kinetics of internal $H^+$ and $K^+$ concentrations are determined from calibration curves and equations detailed above. The net $K^+$ flux ($J_K$) is then calculated with

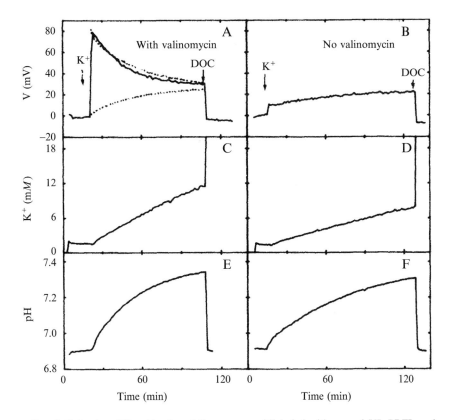

FIG. 5. Potassium filling kinetics of liposomes multilabeled with oxonol VI, PBFI, and pyranine dyes. Conditions and calibrations are as in preceding figures. The filling reaction is started by addition of 20 m$M$ $K_2SO_4$ to the external medium in the presence or in the absence of valinomycin and stopped by addition of deoxycholate (0.05%, w/v). (A and B) $E_m$ determined with oxonol VI (solid line), or calculated from the measured $K^+$ gradient with Eq. (5) (A, upper dotted line) or from the measured $H^+$ gradient (A, lower dotted line). (C and D) Internal $K^+$ concentration measured with PBFI. (E and F) Internal pH measured with pyranine. Reprinted from Ref. 20, published in *Biochimica et Biophysica Acta* with permission from Elsevier Science.

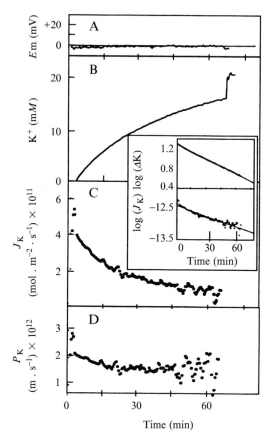

FIG. 6. Potassium filling kinetics of liposomes multilabeled with oxonol VI and PBFI. Liposomes (80 $\mu g \cdot ml^{-1}$) containing 0.5 m$M$ PBFI, 50 m$M$ BTP-SO$_4$ (pH 7.0), 10% glycerol (w/v), 20 m$M$ (NH$_4$)$_2$SO$_4$, and 0.2 $\mu M$ CCCP are diluted in the same medium with 50 n$M$ oxonol VI without PBFI. Dye calibrations are described in text and in the preceding figures. The filling reaction is started by addition of 10 m$M$ K$_2$SO$_4$, and final equilibration is achieved by addition of valinomycin. (A) $E_m$ determined with oxonol VI; (B) internal K$^+$ concentration (K$_{in}^+$) measured with PBFI; (C) net K$^+$ flux ($J_K$) calculated with Eq. (6) from the derivative numerically determined with the Savitzky–Golay algorithm (Aminco-Bowman software) from each (K$_{in}^+$) value measured in (B); (D) permeability coefficient to K$^+$ calculated with Eq. (20) for each value of (K$_{in}^+$ (mean $P_K$; $1.9 \times 10^{-12}$ m $\cdot$ s$^{-1}$). *Top inset:* Log plot of the K$_{in}^+$ kinetics measured in (B), the slope of which gives $P_K = 1.6 \times 10^{-12}$ m $\cdot$ s$^{-1}$ from Eq. (21). *Bottom inset:* Log plot of the $J_K$ kinetics determined in (C), the slope of which gives $P_K = 1.9 \times 10^{-12}$ m $\cdot$ s$^{-1}$.

Eq. (6) from the derivate of $(K_{in}^+) = f(t)$ numerically determined with the Aminco-Bowman software, involving the Savitzky–Golay algorithm. In Fig. 5, the net $H^+$ flux $(J_H)$ is calculated in the same way, taking the buffer strength $(B)$ into account, according to

$$J_H = B(r/3)dpH_{in}/dt \tag{17}$$

The buffer strength is calculated for the two buffer systems (histidine and HEPES) at concentrations $C_1$ and $C_2$, respectively:

$$B = 2.3C_1D_1(1 - D_1) + 2.3C_2D_2(1 - D_2) \tag{18}$$

where $D_1 = [10^{(pH - pK_1)} + 1]^{-1}$ and $D_2 = [10^{(pH - pK_2)} + 1]^{-1}$. Using $pK_1 = 6.0$ and $pK_2 = 7.5$, the buffer strength calculated with Eq. (18) is in good agreement with the measured buffer strength.

In the absence of any ionophore, the net $K^+$ influx into liposomes appears electrically coupled by a small $E_m$ to a net $H^+$ efflux (Fig. 6B), and $P_K$ may be calculated from the G–H–K relation:

$$P_K = J_K(RT/FE_m)[\exp(FE_m/RT) - 1][(K_{in}^+)\exp(FE_m/RT) - (K_{out}^+)]^{-1} \tag{19}$$

$E_m$ slowly increases due to the inward $H^+$ gradient progressively built in response to the $K^+$ entry (highly permeant $H^+$ ions are exchanged with $K^+$ ions). To determine the maximum $J_K$ across liposomes (i.e., not restricted by electrical constraints), $K^+$ must be freely exchanged with $H^+$ ions while $E_m$ is clamped at a negligible value. This can be achieved, as ascertained with oxonol VI dye in the experiment shown in Fig. 6, where $(NH_4)_2SO_4$ is present in both internal and external media to clamp $\Delta pH$ to zero ($\Delta pH$ is dissipated by the freely permeant $NH_3$ form), together with the protonophore carbonylcyanide $m$-chlorophenylhydrazone (CCCP). Despite the fact that $E_m$ is negligible, the $H^+$ efflux remains electrically coupled to the $K^+$ influx because the permeability coefficient of liposomes to $H^+$ is high in the presence of the protonophore. This experiment illustrates two ways to determine $P_K$. Fick's law may be used, because in this case the transport of $K^+$ and $H^+$ can be assimilated to the transport of electroneutral solutes:

$$P_K = -J_K/\Delta K^+ \tag{20}$$

where $\Delta K^+ = (K_{out}^+) - (K_{in}^+)$. This equation allows a determination of a $P_K$ value for each $(K_{in}^+)$ given by one acquisition cycle of fluorescence data. This can be useful to show rapid variations of $P_K$ following, for example, the addition of effectors. Combining and integrating Eqs. (6) and (20) gives a second way to determine $P_K$:

$$\Delta K^+ = \Delta K_{t=0}^+ \exp(-3P_Kt/r) \tag{21}$$

As expected, the plot log $(\Delta K^+) = f(t)$ is linear and its slope permits a determination of $P_K$ from the whole filling kinetics data. This determination is more reliable, because it does not require the derivation of primary $K^+$ filling data.

Finally, the analysis of data in Fig. 5 shows that $J_H$, the net proton efflux determined from pyranine signals, quantitatively mirrors $J_K$, the net $K^+$ influx determined from PBFI signals.[20] The electroneutrality of $K^+/H^+$ exchange fluxes ensures the validity of this multilabeling experiment, including calibration and analysis steps. It ensures also that $K^+$ is really the only significantly permeant species added to the medium, the efflux of $H^+$ ions present naturally in medium being the only way to neutralize $K^+$ entry. In this way, double labeling with pyranine and oxonol VI permits, for example, the indirect determination of the net influx of various alkali cations for which no specific probes are available, and the sequence of the permeability coefficients of various ions in LUVs ($P_K > P_{Na} > P_{Rb} > P_{Cs} \approx P_{Li}$).[20]

### Acknowledgment

The authors are indebted to Dr. Tim Tranbarger for kind revision of the manuscript.

# [11] Liposomes in Assessment of Annexin–Membrane Interactions

By Andreas Hofmann and Robert Huber

### Introduction

The calcium-dependent binding of annexins to membranes makes these proteins an ideal subject for interaction studies with membrane vesicles (liposomes). However, as years of annexin research have revealed, modes of interaction are manifold and the occurrence of different processes at the same time can complicate characterization enormously.[1] Although binding to a membrane surface is an extremely basic event on the scale of biological/cellular processes, it represents already a complicated system from the biophysical point of view. In terms of matter, binding reactions in this context are constituted by ternary systems consisting of the membrane

[1] A. Hofmann and R. Huber, in "Annexins: Biological Importance and Annexin-Related Pathologies" (J. Bandorowicz-Pikula, ed.). Landes Bioscience, Austin, TX, 2003.

(viewed in a simplified manner as a homogeneous participant), the protein, and calcium ions. Thereby, anions are neglected and the solvent is treated as an environmental parameter.

Further complications might arise from changes in these environmental parameters, such as ionic strength and transmembrane voltages, which are frequently occurring phenomena in *in vivo* systems. Another inherent parameter to be considered is the constitution of the membrane. Whereas naturally occurring membranes are mixtures of a variety of different phospholipids, synthetic membranes (phospholipid vesicles, liposomes) usually consist of two phospholipid components in a certain mixture. The nature of phospholipid head groups (spatial extension, charge) impacts possible protein–membrane interactions.

Annexin–membrane interactions occur in different modes. While the peripheral nature of these interactions is still controversially discussed, striking evidence for membrane penetration by annexins has yet to be presented. Moreover, an overwhelming collection of biophysical data supports a peripheral attachment of the protein to membrane surfaces.

The different binding modes range from single annexin molecules adsorbed at a phospholipid layer to clusters of protein molecules and states of two-dimensional crystallization in which the protein covers the phospholipid layer in a protective fashion. From electron micrographs of these different two-dimensional crystalline states one can anticipate that the transition from individually bound molecules to higher order structures might be an interesting subject to investigate. "Building blocks" of annexin oligomers (trimers, dimers of trimers, and trimers of trimers)[2] have been observed frequently with certain annexins, as well as rodlike structures spanning long distances[3] on a membrane surface. It is not yet clear what makes some annexins form highly symmetrical structures on phospholipid layers and what prevents other annexins from doing so.

Early in annexin research it was also reported that several members of this protein family have the ability to permeabilize membranes.[4] The growing number of identified annexin proteins and the apparent differences in their permeabilization behaviors, despite highly conserved three-dimensional structures,[1] called for reliable procedures to investigate binding and permeabilization properties of different annexins.

Biochemical liposome-based assays with annexins concentrate on five phenomena:

[2] F. Oling, W. Bergsma-Schutter, and A. Brisson, *J. Struct. Biol.* **133**, 55 (2001).
[3] A. Hofmann and A. Brisson, unpublished results (2002).
[4] E. Rojas and H. B. Pollard, *FEBS Lett.* **217**, 25 (1987).

1. Binding of annexins to membranes
2. Annexin-induced permeabilization and stabilization/destabilization of liposomes
3. Annexin-induced aggregation and fusion of phospholipid vesicles
4. Annexin effects on membrane properties
5. Folding studies in the presence of liposomes

Within the scope of this chapter we present various methods and approaches that have been used successfully to address these five aspects. To limit the extent of this chapter we do not discuss specific results of these experiments because the motivation is primarily to compile a set of methods rather than to present current results in annexin research.

## Liposome Preparation

An inherent problem with calcium-mediated annexin–membrane interactions is the fact that calcium ions *per se* might cause membrane aggregation.[5] Nelsestuen and Lim[6] found that membranes containing only phosphatidylethanolamine (PE), phosphatidylglycerol (PG), and phosphatidylcholine (PC) are not affected by calcium, at least at concentrations of $Ca^{2+}$ below 10 m$M$. Membrane preparations with a phosphatidylserine (PS) content lower than 30% also did not seem to be affected seriously. However, when using liposomes with a PS content of more than 50%, one should expect $Ca^{2+}$-dependent aggregation. In general, binding of monovalent ions to acidic phospholipids occurs with low affinity, as does binding of $Ca^{2+}$ to liposomes containing zwitterionic phospholipids. With acidic phospholipids, the affinity for binding of $Ca^{2+}$ is not much higher; however, highly increased binding is observed because of electrostatic effects.[7] Certain liposome compositions have been reported to undergo metal-dependent fusion[8,9]; control experiments for assays involving liposomes, annexin, and additional components are therefore required. When designing buffers for liposome experiments one should always check carefully for the balance of osmotic pressure between the intra- and extravesicular spaces of the liposomes to avoid vesicle damage.

[5] D. Papahadjopoulos, W. J. Vail, W. A. Pangborn, and G. Poste, *Biochim. Biophys. Acta* **448,** 265 (1976).
[6] G. L. Nelsestuen and T. K. Lim, *Biochemistry* **16,** 4164 (1977).
[7] R. B. Gennis, "Biomembranes: Molecular Structure and Function," p. 257. Springer-Verlag, New York, 1989.
[8] N. Düzgüneş, R. M. Straubinger, P. A. Baldwin, D. S. Friend, and D. Papahadjopoulos, *Biochemistry* **24,** 3091 (1985).
[9] J. Bentz, D. Alford, J. Cohen, and N. Düzgüneş, *Biophys. J.* **53,** 593 (1988).

TABLE I

PROPERTIES OF VARIOUS UNILAMELLAR VESICLES

| | Average diameter | Trapped volume ($\mu$l/$\mu$mol) | Membrane thickness (nm) | Surface area per PL head group (nm$^2$) | |
| --- | --- | --- | --- | --- | --- |
| | | | | Inner leaflet | Outer leaflet |
| SUVs | 31 nm[a] | | 3.7[a] | 0.61[b] | 0.74[b] |
| LUVs | 120 nm,[a,c] | | 5[a] | | 0.55[e] |
| | 380 nm[a,d] | | | | |
| | 60–100 nm[f] | 1–3[f] | | | |
| HUVs | 260–480 nm[g] | | | | |
| IFVs | 1–6 $\mu$m[h] | 20–25[h] | | | |

*Abbreviations:* HUVs, huge unilamellar vesicles; IFVs, interdigitation fusion vesicles; LUVs, large unilamellar vesicles; PL, phospholipid; SUVs, small unilamellar vesicles.
[a] Y. Lu, M. D. Bassi, and G. L. Nelsestuen, *Biochemistry* **34,** 10777 (1995).
[b] C. Huang and J. T. Mason, *Proc. Natl. Acad. Sci. USA* **75,** 308 (1978).
[c] Freeze–thaw.
[d] Ether injection.
[e] D. Deamer and A. D. Bangham, *Biochim. Biophys. Acta* **443,** 629 (1976).
[f] M. J. Hope, M. B. Bally, G. Webb, and P. R. Cullis, *Biochim. Biophys. Acta* **812,** 55 (1985).
[g] T. C. Evans and G. L. Nelsestuen, *Biochemistry* **33,** 13231 (1994).
[h] P. L. Ahl, L. Chen, W. R. Perkins, S. R. Minchey, L. T. Boni, T. F. Taraschi, and A. S. Janoff, *Biochim. Biophys. Acta* **1195,** 237 (1994).

For an overview of selected properties of various types of liposomes see Table I.

*Huge Unilamellar Vesicles: Injection Method*

A method for production of huge unilamellar vesicles (HUVs) was introduced by Deamer and Bangham[10] and is commonly referred to as the ether injection method. The technique is based on two rationales: First, the efficiency of producing unilamellar structures should be higher if the lipids are not allowed to form multilamellar structures before mixing with the aqueous phase. Second, one should apply gentle conditions in order to ensure the formation of large vesicles (i.e., no sonication). To obtain HUVs it proved successful to prepare a solution of the desired lipids (final concentration, 2 m$M$) in ether and inject it into 4 ml of buffer at a temperature above the boiling point of the ether (55–65°). It is recommended that the injection be carried out with a mechanical drive at a speed of 0.25 ml/min.

[10] D. Deamer and A. D. Bangham, *Biochim. Biophys. Acta* **443,** 629 (1976).

### Large Unilamellar Vesicles by Reeves–Dowben Method

Reeves and Dowben[11] provide a protocol for the preparation of large unilamellar vesicles (LUVs), which is also known as the dehydration/rehydration method (although being different from the liposome preparation protocol described by Gregoriadis et al.[12]) and yields liposome populations with an average diameter of 1.5 $\mu$m. For a 1-$\mu$mol batch of PS/PE liposomes (molar ratio, 3:1) 60 $\mu$l of PS and 20 $\mu$l of PE (each 10 mg/ml in $CHCl_3$) are mixed in a 250-ml Erlenmeyer flask, covered with a few milliliters of $CHCl_3$–methanol (2:1, v/v), and dried under a smooth stream of $N_2$. Uniform distribution of the lipid solution in the flask should be ensured to avoid lump formation. The remaining film is resuspended in a small amount of diethyl ether and dried again under a smooth $N_2$ stream (dehydration). For rehydration, the precipitate is then exposed for 30 min to an $H_2O$-saturated $N_2$ stream. The remaining lipid film is covered with 1–2 ml of the appropriate buffer (see Binding Assays, p.192, for details) and incubated for 2–3 h at 37° (preferably) or overnight at room temperature. Liposomes are collected by centrifugation of the final suspension (12,000 $g$, 10 min, 10°); the use of 1.5-ml Eppendorf tubes eases removal of the supernatant. For a 10-$\mu$mol liposome batch, more incubation buffer (10–20 ml) is used and centrifugation is performed at least two times for 30 min.

### Large Unilamellar Vesicles by Freeze–Thaw Method

LUVs can also be prepared by a freeze–thaw procedure.[13,14] Lipids are dissolved in a few milliliters of $CHCl_3$–methanol (4:1, v/v). The solvent is removed in a rotary evaporator, and the lipid film is suspended in 1 ml of buffer. The suspension is frozen and thawed five times, using liquid nitrogen. To obtain a homogeneous unilamellar vesicle population the suspension is extruded 40 times at 60° through a polycarbonate membrane with an exclusion diameter of 100 nm (LiposoFast; Avestin, Toronto, ON, Canada).

### Large Unilamellar Vesicles by Reversed-Phase Evaporation

An elaborate protocol for preparation of LUVs was introduced by the group of Papahadjopoulos.[15,16] Lipids (10 $\mu$mol) are dried under high vacuum and dissolved in 1 ml of peroxide-free diisopropyl ether. The

---

[11] J. R. Reeves and R. M. Dowben, *J. Cell. Physiol.* **73**, 49 (1969).

[12] G. Gregoriadis, P. D. Leathwood, and B. E. Ryman, *FEBS Lett.* **14**, 95 (1971).

[13] M. J. Hope, M. B. Bally, G. Webb, and P. R. Cullis, *Biochim. Biophys. Acta* **812**, 55 (1985).

[14] C. Jatzke, "Untersuchungen zur Calcium und Phospholipidbindung von Annexin V." Ph.D. thesis, Universität Münster, Munich, Germany, 1999.

appropriate buffer (0.34 ml) is added, and the mixture is sonicated under argon until a stable emulsion is obtained. The ether is removed from the emulsion in a rotary evaporator under vacuum at 45–50°. The resulting gel is broken up by vortex mixing several times during the procedure. An additional 0.66 ml of buffer is added, and high vacuum is applied in the late stage of this step to remove any residual ether. The vesicle suspension is then centrifuged (10,000 g, 20 min, 25°) and the supernatant is extruded under $N_2$ pressure at 45–50° through a polycarbonate membrane with a pore diameter of 0.2 $\mu$m. For a detailed characterization of liposomes obtained by this method see Düzgüneş et al.[16]

*Small Unilamellar Vesicles*

Small unilamellar vesicles (SUVs) are prepared by treatment with ultrasonic sound.[17,18] A total of 4 $\mu$mol of phospholipids is suspended in 0.5 ml of the appropriate buffer and sonicated on ice 10 times for 2 min. To obtain a liposome population with a reasonable size distribution the sonicated suspension is subjected to gel filtration with a Sepharose 4B column. The second elution peak usually appears as two halves, the second of which is collected and used further. As reported by Litman,[19] only this vesicle fraction exhibits a linear correlation between turbidity and lipid concentration. The liposome suspension is concentrated by ultrafiltration with an Amicon Ultrafilter (YM-100 membrane; Millipore, Bedford, MA) to a final concentration of approximately 10 m$M$.

*Interdigitation Fusion Vesicles*

When SUVs composed of saturated symmetrical chain phospholipids are exposed to high ethanol concentrations, $c$(EtOH) $\geq$ 1.5 $M$, they coalesce to form a viscous suspension of interdigitated sheets. Incubation of these sheets at temperatures above the main phase transition temperature $T_m$ of the phospholipids leads to spontaneous formation of large, predominantly unilamellar vesicles called interdigitation fusion vesicles (IFVs).[20,21]

[15] N. Düzgüneş, J. Wilschut, R. Fraley, and D. Papahadjopoulos, *Biochim. Biophys. Acta* **642,** 182 (1981).

[16] N. Düzgüneş, J. Wilschut, K. Hong, R. Fraley, C. Perry, D. S. Friend, T. L. James, and D. Papahadjopoulos, *Biochim. Biophys. Acta* **732,** 289 (1983).

[17] Y. Lu, M. D. Bazzi, and G. L. Nelsestuen, *Biochemistry* **34,** 10777 (1995).

[18] F. M. Megli, M. Selvaggi, A. DeLisi, and E. Quagliariello, *Chem. Phys. Lipids* **22,** 89 (1993).

[19] B. J. Litman, *Biochemistry* **12,** 2545 (1973).

[20] P. L. Ahl, L. Chen, W. R. Perkins, S. R. Minchey, L. T. Boni, T. F. Taraschi, and A. S. Janoff, *Biochim. Biophys. Acta* **1195,** 237 (1994).

[21] P. L. Ahl and W. R. Perkins, *Methods Enzymol.* **367,** 80 (2003).

If the SUVs are composed of a phospholipid mixture, care must be taken that the ethanol addition is performed at temperatures below the $T_m$ of the mixture. Ethanol is removed after formation of the IFVs by additional incubation above the $T_m$ and a gentle stream of $N_2$. IFVs are washed three times with 150 m$M$ NaCl, 10 m$M$ Tris (pH 7.4) by centrifugation (12,000 $g$, 15 min, room temperature).

### Reverse Micelles

Reverse micelles are used as an optically transparent model of membrane–water interfaces when the availability of water molecules for hydration is limited.[22] A typical preparation protocol[23] uses sodium bis(2-ethylhexyl) sulfosuccinate (Aerosol OT) and water in isooctane. Dry Aerosol OT is dissolved in the desired volume of isooctane and $H_2O$ is added with a Hamilton syringe; possible molar ratios are in the range of $x = 2.0–30$, with $x = n(H_2O)/n$(Aerosol OT). Preweighed quantities of dry annexin are added, and the sample is shaken and sonicated for 2 min. After incubation for several hours at room temperature the sample is centrifuged (2000 $g$) to remove undissolved protein.

### Binding Assays

#### Centrifugation Assay with Direct Determination of Protein

Calcium-dependent binding of annexins to liposomes can be elucidated by a centrifugation assay. The lipid vesicles and membrane-bound annexin are sedimented by centrifugation at ultrahigh speed. One can either determine the amount of unbound annexin in the supernatant or directly quantify the amount of protein copelleted with the liposomes.

For ultraviolet (UV) spectroscopic determination, using LUVs by the Reeves–Dowben method,[24] a sample with a total volume of 500 $\mu$l consists of a 20-$\mu$l aliquot of PS/PE liposome suspension (molar ratio, 3:1), which is equivalent to 0.1 $\mu$mol of total phospholipid content, as well as the desired amount of $CaCl_2$ and buffer containing 180 m$M$ saccharose and 10 m$M$ HEPES (pH 7.4). After addition of 2 nmol of annexin the sample is incubated at room temperature for 15 min and subjected to ultracentrifugation with a Beckman Coulter (Fullerton, CA) Airfuge (130,000 $g$, 30 min).

[22] P. L. Luisi and L. J. Magid, *Crit. Rev. Biochem.* **20,** 409 (1986).
[23] J. Gallay, M. Vincent, C. Nicot, and M. Waks, *Biochemistry* **26,** 5738 (1987).
[24] A. Hofmann, J. Proust, A. Dorowski, R. Schantz, and R. Huber, *J. Biol. Chem.* **275,** 8072 (2000).

The supernatant is removed carefully, and the UV absorption at 280 nm is determined. A protein-free sample, which is subjected to the same procedure, is used as a control for baseline correction. A second control sample, containing annexin but no calcium, is used to normalize the absorption data. The difference between the total amount of annexin and the spectroscopically determined unbound annexin yields the portion of membrane-bound protein.

Densitometric determination of membrane-bound protein is a popular method[25–32] because it measures directly the amount of protein copelleted with the liposomes and, therefore, yields more accurate results than the UV assay. The procedure described in Refs. 25 and 26 uses 100-$\mu$l aliquots of LUVs as prepared by the Reeves–Dowben method. The liposomes are suspended in 10 m$M$ piperazine-$N,N'$-bis(2-ethanesulfonic acid) (PIPES, pH 6.8) and the desired amount of $CaCl_2$ and 100 $\mu$g of annexin are added to yield a final volume of 200 $\mu$l. After incubation for 20 min the liposomes are collected by centrifugation (47,000 $g$, 15 min) and the pellet is washed with 200 $\mu$l of calcium-containing buffer. After a second centrifugation step the pellet is resuspended in 10 m$M$ PIPES (pH 6.8), 5 m$M$ EDTA, and 0.1% Triton X-100. Twenty microliters of the first supernatant and the resuspended final pellet are subjected to sodium dodecyl sulfate–polyacrylamide gel electrophoresis (SDS–PAGE) (12%), and the gel is stained with silver[25] or Coomassie blue.[27,32] Densitometric analysis can be carried out with ImageQuant 1.2[25] or QuickQuant III.[27] Whereas the result from the first supernatant shows the amount of free annexin, the quantification of the resuspended pellet yields the amount of annexin bound to the membrane. A modification of this assay uses immunostained gels based on anti-annexin antibodies to detect the membrane-bound protein.[31]

[25] T. Dubois, J. P. Mira, D. Feliers, E. Solito, F. Russo-Marie, and J. P. Oudinet, *Biochem. J.* **330**, 1277 (1998).

[26] A. Hofmann, C. Raguénès-Nicol, B. Favier-Perron, J. Mesonero, R. Huber, F. Russo-Marie, and A. Lewit-Bentley, *Biochemistry* **39**, 7712 (2000).

[27] R. A. Blackwood and J. D. Ernst, *Biochem. J.* **266**, 195 (1990).

[28] M. A. Kaetzel, Y. D. Mo, T. R. Mealy, B. Campos, W. Bergsma-Schutter, A. Brisson, J. R. Dedman, and B. A. Seaton, *Biochemistry* **40**, 4192 (2001).

[29] D. R. Patel, C. C. Jao, W. S. Mailliard, J. M. Isas, R. Langen, and H. T. Haigler, *Biochemistry* **40**, 7054 (2001).

[30] F. Porte, P. de Santa Barbara, S. Phalipou, J. P. Liautard, and J. Sri Widada, *Biochim. Biophys. Acta* **1293**, 177 (1996).

[31] T. Kirsch, H. D. Nah, D. R. Demuth, G. Harrison, E. E. Golub, S. L. Adams, and M. Pacifici, *Biochemistry* **36**, 3359 (1997).

[32] D. D. Schlaepfer and H. T. Haigler, *J. Biol. Chem.* **14**, 6931 (1987).

Centrifugation assays using radioactive detection present an accurate, although labor-intense, method. Annexin is labeled with [125]I by incubation with Na[125]I (0.25 mCi, i.e., about $1 \times 10^7$ Bq) in IODO-GEN-coated tubes (Pierce Chemical, Rockford, IL) and then dialyzed extensively against a KI solution. The membrane binding of annexin is investigated by a protocol described above, except that determination of the labeled protein is carried out by scintillation counting.[32]

Because modifications in these assays are easy to carry out, effects of third components on calcium-mediated membrane binding of annexins have been studied using the above-described procedures. The influence of divalent metal ions[32] or benzodiazepine derivatives[33] on annexin–membrane interactions has been investigated employing this assay architecture. However, every component introduced into the annexin–membrane system must be checked carefully with respect to its effect on the liposomes. Metal ions might alter membrane fluidity or induce membrane aggregation/fusion, and organic molecules, such as benzodiazepines, can bind to or penetrate the liposomes. Appropriate control measurements are necessary to ensure correct interpretation of the data.

*Determination of Membrane-Bound Annexin by Light Scattering*

The binding of annexin to liposomes yields particles with an increased molecular weight as compared with the liposomes themselves. This can be used in a light-scattering experiment in which the scattered light perpendicular to the incident beam is monitored at a wavelength of $\lambda = 320$ nm.[6,34] SUVs and protein are added to a buffer containing 100 m$M$ NaCl, 50 m$M$ Tris (pH 7.5), and 0.01% NaN$_3$. On binding of protein to the liposomes the weight average molecular weight of the scattering species in solution increases. If $I_2$ and $I_1$ are the scattering intensities in the presence and absence of the protein and $(\delta n_2/\delta c_2)/(\delta n_1/\delta c_1)$ is the ratio of the changes of refractive indices with concentration for the two species,[6] the ratio of the weight average molecular weights, $M_2/M_1$, is obtained by

$$\left(\frac{M_2}{M_1}\right)^2 = \frac{I_2/I_1}{[(\delta n_2/\delta c_2)/(\delta n_1/\delta c_1)]^2} \tag{1}$$

Nelsestuen and Lim[6] presented a detailed description of this method and reported refractive indices $(\delta n/\delta c)$ for protein and protein–phospholipid mixtures to be in the range of 0.172–0.192 for prothrombin. Errors of up

[33] A. Hofmann, A. Escherich, A. Lewit-Bentley, J. Benz, C. Raguénès-Nicol, F. Russo-Marie, V. Gerke, L. Moroder, and R. Huber, *J. Biol. Chem.* **273**, 2885 (1998).
[34] T. C. Evans and G. L. Nelsestuen, *Biochemistry* **33**, 13231 (1994).

to 15% might be attached to the determination of $M_2/M_1$ because of uncertainties in refractive index and concentration measurements. In addition, the value of $I_2$ must be corrected for light scattering due to unbound protein. This is done by assuming that $(\delta n_2/\delta c_2) = (\delta n_1/\delta c_1)$ and an estimation of the percentage of free and bound protein from the value of $I_2/I_1$. Because the major contribution to the final error is systematic, results are presented ideally in a comparative manner. Using the above-stated assumption for the ratio of changes of the refractive indices, $M_2/M_1$ can be estimated. This value is used to recalculate $(\delta n_2/\delta c_2)/(\delta n_1/\delta c_1)$, and a second value for $M_2/M_1$ is obtained. The iteration is continued until a convergence for the molecular weight ratio is achieved.

Several annexins are known to cause aggregation or fusion of phospholipid vesicles, a process that might interfere with interpretation of light-scattering results in terms of the pure binding process. According to Evans and Nelsestuen,[34] aggregation can be distinguished from initial protein–membrane association, because the latter appears as a rapid increase in light-scattering intensity to a final stable value. Aggregation and fusion of liposomes, in contrast, exhibit a continuous, nonsaturable increase of intensity. These processes require several minutes and thus are on a different time scale than the binding process. In addition, because aggregation and fusion phenomena are sensitive to concentration, their occurrence can be minimized by the use of dilute solutions and large vesicles rather than small ones.

## Direct Determination of Membrane-Bound Annexin by Fluorescence

Fluorescence spectroscopy is one of the most versatile techniques to investigate protein behavior in solution, and it has been employed frequently for monitoring annexin–membrane binding. Protocols for binding assays take advantage of either fluorescence resonance energy transfer (FRET) or fluorescence quenching.

If an appropriate fluorescence label is present within the liposome population, the fluorescence emission of protein tryptophan residues (donor) is transferred to the liposome-anchored label (acceptor) provided that the distance criterion of FRET is fulfilled ($r < 100$ Å), which is the case on binding of protein to the vesicle surface. Because an overlap between donor emission and acceptor absorption spectra is a requirement for FRET, appropriate labels for these assays are pyrene[35] or dansyl groups,[36] such as 1-hexadecanoyl-2-(1-pyrenedecanoyl)-sn-glycero-3-phosphoglycerol

[35] S. Maezawa, T. Yoshimura, K. Hong, N. Düzgüneş, and D. Papahadjopoulos, *Biochemistry* **28**, 1422 (1989).
[36] M. D. Bazzi and G. L. Nelsestuen, *Biochemistry* **26**, 115 (1987).

ammonium salt (pyrene-PG; Molecular Probes, Eugene, OR). For pyrene labels, the recording of emission spectra is done from $\lambda_{em} = 300-560$ nm. Tryptophan fluorescence is excited at $\lambda_{exc} = 290$ nm in order to minimize pyrene fluorescence emission. The FRET is calculated from the recorded fluorescence emission intensity, using the following normalization[37]:

$$\text{FRET} = \frac{I^{\lambda_{int}} - I_0^{\lambda_{int}}}{I_0^{\lambda_{int}}} \tag{2}$$

where $I_0$ denotes the intensity in the absence of protein. The intensities are integrated over the wavelength range of $\lambda_{int} = 370-560$ nm. With dansyl as the acceptor group, protocols have been reported in which fluorescence emission is recorded at a single wavelength only ($\lambda_{em} = 520$ nm) and the excitation wavelength is $\lambda_{exc} = 284$ nm throughout the experiment.[17,34,38]

The use of pyrene as a membrane-residing fluorescence probe allows also for further studies of the binding process, because the pyrene monomer is visible at $\lambda_{em} = 377$ nm, 395 nm and the excimer at $\lambda_{em} = 480$ nm.[39] The excimer-to-monomer ratio is dependent on the concentration of pyrene within the membrane, the lateral diffusion coefficient, and the lifetime of the excimer.[39] Meers et al.[40] carefully determined the fluorescence lifetimes of pyrene-labeled vesicles in the presence and absence of annexin A5 and polylysine and found only small changes, while at the same time binding of annexin causes a decrease in the excimer-to-monomer ratio. Accordingly, they conclude that annexin binding causes changes in the lateral diffusion coefficient of the fluorescence probe (decrease in "membrane fluidity") because the effects of annexin on the lifetime of the probe are insignificant and a drastic change in the two-dimensional concentration of the probe within the membrane cannot be reasonably explained. Furthermore, these authors found the polycation spermine to displace membrane-bound annexin, which emphasizes the prevailing ionic character of annexin–membrane interactions in this state.

Finally, the reverse design of a fluorescence-monitored binding assay is also used: annexin protein is labeled with fluorescein 5-isothiocyanate (FITC) and the quenching of its fluorescence emission on binding to SUVs is monitored. The labeling procedure for annexin A5 as described by Tait et al.[41] requires incubation of 50 $\mu M$ annexin with 50 $\mu M$ FITC

[37] G. Köhler, U. Hering, O. Zschörnig, and K. Arnold, *Biochemistry* **36,** 8189 (1997).
[38] M. D. Bazzi and G. L. Nelsestuen, *Biochemistry* **30,** 971 (1991).
[39] H. J. Galla and E. Sackmann, *Biochim. Biophys. Acta* **339,** 103 (1974).
[40] P. Meers, D. Daleke, K. Hong, and D. Papahadjopoulos, *Biochemistry* **30,** 2903 (1991).
[41] J. F. Tait, D. Gibson, and K. Fujikawa, *J. Biol. Chem.* **264,** 7944 (1989).

for 1 h at 37° in a buffer containing 150 m$M$ NaCl, 1 m$M$ EDTA, and 50 m$M$ Na$_2$B$_4$O$_7$ (pH 9.0). The reaction mixture is quenched with 10 m$M$ glycine and dialyzed against 80 m$M$ NaCl, 1 m$M$ EDTA, and 20 m$M$ HEPES (pH 8.0). Further purification via anion exchange, using a MonoQ column, yields several fractions of labeled annexin A5, with the monola-beled protein eluting at 310–350 m$M$ NaCl. A content of 1.03–1.08 mol of fluorescein per mole of protein is reported for this fraction, and the molar extinction coefficient is determined to be $\varepsilon_{494 \text{ nm}}$ = 78,000 liters/(mol · cm).

Measurement is carried out by setting the excitation wavelength at $\lambda_{exc}$ = 495 nm and monitoring the emission at $\lambda_{em}$ = 520 nm. A careful fine-tuning of the setup including, for example, beam attenuation to mini-mize the nonspecific fluorescence signal, allows for protein concentrations as low as 1 p$M$. Although photobleaching is not an immediate problem, the shutter should be closed between measurements. The liposome suspension is titrated into the cuvette holding FITC-annexin in 100 m$M$ NaCl, 1.2 m$M$ CaCl$_2$, ovalbumin (5 $\mu$m/ml), 50 m$M$ HEPES (pH 7.4), and 3 m$M$ NaN$_3$. Fresh cuvettes are used for each sample; 5 m$M$ EDTA is added to the sample after completion of the measurement, and the fluorescence emis-sion is recorded again (desorption of the protein). The fluorescence emission intensity in the presence of EDTA is used as an internal correction to compensate for (apparent) minor differences in protein con-centration between individual samples. Annexin binding can then be calculated simply from

$$x_{\text{Anx (bound)}} = 1 - \frac{I}{I_0} \tag{3}$$

where $I$ and $I_0$ are the emission intensities in the presence and absence of EDTA, respectively.

These protocols have been used to investigate the effect of PS content and the calcium dependence of annexin A5 binding to the membrane, as well as to determine the dissociation constant for the ternary complex (annexin–calcium–vesicle); however, the phenomena have been treated as simple noncooperative binding to homogeneous sites.[41,42]

### Indirect Determination of Membrane-Bound Annexin Using Phospholipase A$_2$ Activity Assay

Pyrene-PG-labeled vesicles are prepared at a total lipid concentration of 2 m$M$ in 1 ml of buffer containing 0.5 $M$ NaCl, 1 m$M$ EGTA, and

[42] J. F. Tait and D. Gibson, *Arch. Biochem. Biophys.* **298,** 187 (1992).

50 mM Tris (pH 7.5). Fatty acid-free bovine serum albumin (BSA, 0.1%), 10 $\mu$l of the protein to be tested, 1 $\mu$g of bee venom phospholipase $A_2$ ($PLA_2$) (i.e., a final concentration of 70 nM), and 10 mM $CaCl_2$ are subsequently added to the sample. $PLA_2$ activity is monitored by recording the monomer emission of pyrene.[43] An annexin binding curve is obtained by performing a dose-dependent experiment[26]; the $PLA_2$ activity is indirectly proportional to the fraction of annexin bound.

### Binding Kinetics Studied by FRET and Light Scattering

Dansyl-labeled liposomes have also been used to study kinetic aspects of annexin-membrane binding.[34,44] Annexins are added to phospholipid vesicles containing N-dansyl-L-$\alpha$-phosphatidylethanolamine (dansyl-PE). The addition of a 40-fold excess of vesicles without fluorescent label introduces a competition for binding sites for the protein and would capture all of the unbound protein, as well as any protein that dissociates from the labeled vesicles. Dissociation of the initial annexin–calcium–vesicle complex can therefore be monitored by a decrease in FRET. Whereas virtually no dissocation is found at calcium concentrations of 1 and 0.1 mM (after several hours), dissociation occurs rapidly after addition of EGTA.[34,44]

The fast kinetics of annexin A6 binding to liposomes of various sizes was investigated by combining the FRET and light-scattering protocols with a stopped-flow setup.[17] Pseudo-first-order kinetics can be applied to the first half of the binding reaction and yield rate constants on the order of $1.5 \times 10^{10}$ $M^{-1}$ $s^{-1}$ (second-order rate constants at 1 mM $Ca^{2+}$, normalized with respect to vesicle concentration). Deviation from linearity in the second half of the binding reaction suggests that the membrane sites are crowded with protein, thus not allowing for a simple isothermic model. The authors also report that the collisional phase of annexin–membrane interactions lacks selectivity with respect to membrane composition; a possible preference for certain phospholipids occurs in the equilibrium and dissociation phase of the binding reaction. These protocols also allow for investigation of the calcium dependence of annexin binding to liposomes, and the authors report variations of the half-maximal calcium concentrations with variation of membrane composition as well as the actual protein-to-vesicle ratio. Different equilibrium behavior seems to be ruled mainly by the calcium dissociation rates.

[43] F. Radvanyi, L. Jordan, F. Russo-Marie, and C. Bon, *Anal. Biochem.* **177,** 103 (1989).
[44] M. D. Bazzi and G. L. Nelsestuen, *Biochemistry* **30,** 7970 (1991).

### Determination of Calcium Stoichiometry in Ternary Annexin–Calcium–Vesicle Complex

Three protocols for investigation of calcium stoichiometry for the annexin–calcium–vesicle complex have been reported so far: detection of radioactive calcium in a gel-filtration procedure[6,44,45] and in a copelleting assay,[29] as well as studies using isothermal titration calorimetry (ITC).[29]

The group of Nelsestuen[6,34,44] has reported an application of the Hummel–Dryer technique[45] to monitor the dissociation of calcium from the protein–phospholipid complex. Sephacryl S-300 columns are equilibrated with 100 m$M$ NaCl, 10% glycerol, 0.5 m$M$ dithiothreitol (DTT), 20 m$M$ Tris (pH 7.9), and 50 $\mu M$ CaCl$_2$ (Ca$^{2+}$ or $^{45}$Ca$^{2+}$). First, 180 $\mu$g of PS/PC SUVs (molar ratio, 1:3) is mixed with 2–20 nmol of annexin and/or buffer in the presence of 50 $\mu M$ $^{45}$Ca$^{2+}$ with a final sample volume of 400 $\mu$l. After incubation for 20 min, the samples are applied to the equilibrated columns and eluted with the same buffer, thereby achieving a separation of free and membrane-bound protein. The $^{45}$Ca$^{2+}$ content and protein content of each fraction are determined. The two fractions with the highest amounts of membrane-bound protein are combined and subjected to a second gel-filtration column equilibrated with unlabeled Ca$^{2+}$ buffer, allowing for an exchange of unlabeled against labeled calcium. Fractions obtained from elution of this column are analyzed with respect to protein and $^{45}$Ca$^{2+}$ contents. The effect of nonspecific binding of $^{45}$Ca$^{2+}$ to liposomes and the contribution of free $^{45}$Ca$^{2+}$ in the elution buffer are checked by running protein-free samples. On the time scale of these experiments (several hours), the exchange rate for $^{45}$Ca$^{2+}$ is rapid. Because the authors reasonably argue that the dissociation of membrane-bound protein does not occur during this time span, these experiments prove the exchange of calcium ions of the ternary annexin–calcium–vesicle complex. A variant of this technique uses immobilized annexin on a phospholipid affinity gel.[34]

The copelleting assay mentioned earlier can be combined with a radioassay technique as reported by Patel et al.,[29] to determine the amount of calcium bound within the annexin B12–calcium–vesicle complex. Fifty micrograms of protein and LUVs (final phospholipid mass per sample, 250 $\mu$g) are added to 100 m$M$ Tris (pH 7.5), yielding samples with a total volume of 1 ml; $^{45}$Ca$^{2+}$ is present at an activity concentration of 41–52 Bq/nmol (1.1–1.4 nCi/nmol). Pellets and supermatants after centrifugation are subjected to liquid scintillation counting to determine the total amount of $^{45}$Ca$^{2+}$ bound within the ternary complex. Thus, the level of

---

[45] J. P. Hummel and W. J. Dryer, *Biochim. Biophys. Acta* **63,** 530 (1962).

nonspecific binding of $^{45}Ca^{2+}$ must be determined in protein-free samples and substracted from the total counts of the protein-containing samples.

Yet another approach, also described by Patel et al.,[29] uses ITC, and the authors report excellent agreement between the results obtained from either the copelleting radioassay or calorimetry. The 1.37-ml cell of a MicroCal (Northampton, MA) MCS-ITC instrument is loaded with 25 $\mu M$ annexin B12 and 10 $\mu g$ of PS/PC LUVs (molar ratio, 2:1). A 5 m$M$ CaCl$_2$ solution is titrated in 400-s increments in portions of 5 $\mu l$ into the sample cell, and the rate of heat released on injection is measured and integrated over 32 injections. Two control titrations, one into a vesicle-only sample and a second one into a sample with protein only, are used to correct the data from experiments with the ternary complex.

*Membrane Determinants for Annexin Binding to Liposomes*

ITC is also used to study thermodynamic effects elicited by variations in the membrane component. Plager and Nelsestuen[46] presented a study dealing with liposomes of different sizes, namely SUVs and LUVs, and different membrane compositions (head group types) and the effects on the binding enthalpies of annexins A5 and A6. A detailed comparison reveals that enthalpies observed from annexin–membrane interactions obviously involve contributions from the membrane as well as from the protein. Thus, annexin binding to membranes exerts large effects on the membrane, which cannot be neglected.

Stability Assay

Destabilizing effects of proteins or ions on liposomes can be elucidated by an efflux assay as reported by Wilschut et al.[47] The inclusion of the water-soluble fluorescence dye 5(6)-carboxyfluorescein (CF) into liposomes keeps the dye in self-quenching concentrations as long as the vesicles stay intact. Once disruption of the lipid layer occurs, the dye dilutes into the extravesicular solution, which causes an increase in fluorescence emission intensity. This assay has been employed frequently to ensure stability of (FURA-2 - loaded) liposomes in the presence of different annexin proteins.[24,33,48]

[46] D. A. Plager and G. L. Nelsestuen, *Biochemistry* **33,** 13239 (1994).

[47] J. Wilschut, N. Düzgüneş, R. Fraley, and D. Papahadjopoulos, *Biochemistry* **19,** 6011 (1980).

[48] M. Garbuglia, M. Verzini, A. Hofmann, R. Huber, and R. Donato, *Biochim. Biophys. Acta* **1498,** 192 (2000).

To prepare CF liposomes, a standard LUV preparation (Reeves–Dowben method) is used, with 2 ml of incubation buffer containing the fluorescence dye: 20 mg of CF, 180 $\mu M$ EDTA, 162 m$M$ saccharose, and 5 m$M$ HEPES (pH 7.4). After collection of the liposomes by centrifugation, they are resuspended in 200 $\mu$l of wash buffer [200 $\mu M$ EDTA, 180 m$M$ saccharose, 10 m$M$ HEPES (pH 7.4)]. An S-200 MicroSpin column (Pharmacia, Uppsala, Sweden) is equilibrated by spinning first 150 $\mu$l and then 50 $\mu$l of wash buffer through the column (1300 $g$, 30 s). The resuspended vesicles are purified via spin gel filtration with the pre-equilibrated S-200 column. The eluting vesicle suspension is mixed with 300 $\mu$l of wash buffer and again subjected to centrifugation at high speed (12,000 $g$, 10 min, 4°), and the supernatant is removed carefully. The pellet is resuspended in 200 $\mu$l of wash buffer, and the liposome preparation is used preferably within the same day. For better stability, all buffers in this protocol should be saturated with argon or $N_2$ before use.

A typical sample for the stability assay consists of 20 $\mu$l (0.1 $\mu$mol of lipids) of CF liposome suspension, 5 $\mu$l (0.5 m$M$) of $CaCl_2$, and 1 $\mu M$ protein in 500 $\mu$l of wash buffer. The sample is excited at $\lambda_{exc} = 480$ nm, and fluorescence emission is monitored at $\lambda_{em} = 520$ nm; data are collected in time increments of 1 min, and the shutter is closed between the measurements. After recording the baseline for 4 min, the protein is added to the liposome suspension and fluorescence is monitored for 20 min. Triton X-100 is then added to a final concentration of 0.1%, and fluorescence is monitored again for 5 min (disruption). The detergent disrupts the vesicular structure and leads to a dilution of the fluorescence dye into the buffer. This yields the maximal possible fluorescence emission intensity of the particular sample. A normalized CF leakage curve can be calculated as

$$\delta = \frac{I - \bar{I}(0-4\,\text{min})}{\bar{I}(20-25\,\text{min}) - \bar{I}(0-4\,\text{min})} \tag{4}$$

with $I$ being the emission intensity and using the averaged emission intensities of the baseline and disruption phase, respectively, to normalize the data. $\delta_0$ is determined as a control with a protein-free sample. Values of $\delta/\delta_0 < 1$ therefore indicate stabilization with respect to a protein-free, calcium-containing sample.

Because results from these assays are to be compared for different proteins or for the same protein under different conditions, the choice of the parameter to analyze and compare is somewhat arbitrary. Because of the rather small slope of the CF leakage curves with annexins, we chose to focus on a certain time point within these curves. For comparison, $\delta/\delta_0$ values at $t = 10$ min are plotted for different annexins (Fig. 1A).

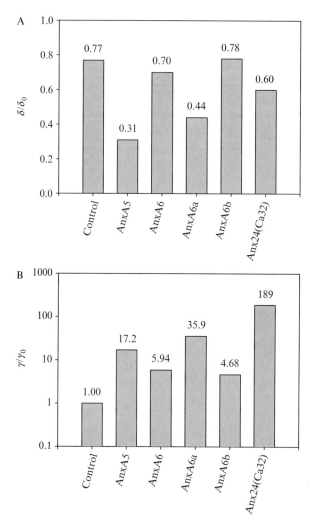

FIG. 1. (A) Results from stability assays (CF leakage assay) with PS/PE LUVs (molar ratio, 3:1) with different annexins ($c = 1\ \mu M$) in the presence of 500 $\mu M$ CaCl$_2$. The control was conducted with a liposome-only sample. Because the data are normalized with respect to a protein-free, calcium-containing sample ($\delta_0$), calcium *per se* destabilizes the liposomes. (B) Activity parameters $\gamma/\gamma_0$ for the same annexin proteins as in (A) obtained by the calcium influx assay with PS/PE LUVs (molar ratio, 3:1). The control was conducted with a protein-free, calcium-containing sample. Data shown are the average of at least two independent measurements and obtained from A. Hofmann, "Strukturelle und funktionelle Untersuchungen an Annexinen." Ph.D. thesis, Technische Universität München, Munich, Germany, 1998.

Calcium Influx Assay

To characterize annexin-induced membrane permeabilization, various groups are using a calcium influx assay[49–52] based on the calcium-sensitive fluorescent dye FURA-2, which is enclosed in phospholipid vesicles. In the presence of annexin, liposomes might be permeable to $Ca^{2+}$ ions, which migrate from the extravesicular solution into the vesicles. In the presence of calcium the dye changes its fluorescence emission characteristics and therefore provides a convenient observable for the calcium flux.[53]

A criticism of this assay concerns whether one is observing the efflux of FURA-2 out of the liposomes rather than the influx of calcium. Studies with the CF leakage assay (see above) prove that fluorescent dye-loaded liposomes are stable under the conditions used in these assays and therefore the influx assay can produce valid results.

A standard LUV liposome preparation for FURA liposomes follows the protocol described above for the stability assay, in which the rehydration buffer is exchanged against 100 $\mu M$ FURA-2, 180 $\mu M$ EDTA, 162 m$M$ saccharose, and 5 m$M$ HEPES (pH 7.4).

*Continuous Monitoring*

The calcium influx assay is used to characterize the time-dependent annexin-induced calcium flux through the phospholipid membrane of liposomes. As the time course of annexin-induced calcium influx into liposomes is rather slow (on the order of 1 h), the assay is usually run in a "continuous monitoring" mode.[24,31,33,48–51]

Fluorescence emission of FURA-2 is monitored by excitation of the sample at $\lambda_{exc1} = 340$ nm and $\lambda_{exc2} = 380$ nm, respectively, and recording the emission intensity at $\lambda_{em} = 510$ nm. For further analysis the intensity ratio $F$ is used:

$$F = \frac{I_{340\,nm}(510\,nm)}{I_{380\,nm}(510\,nm)} \tag{5}$$

Data are collected in time increments of 1 min, and the shutter is closed between the measurements. After recording the baseline for 4 min the

[49] R. Berendes, A. Burger, D. Voges, P. Demange, and R. Huber, *FEBS Lett.* **317,** 131 (1993).
[50] E. L. Goossens, C. P. Reutelingsperger, F. H. Jongsma, R. Kraayenhof, and W. T. Hermens, *FEBS Lett.* **359,** 155 (1995).
[51] J. Benz, Ph.D. thesis, Kristallisation und Strukturaufklärung des humanen Annexin VI-Molekular biologische, biochemische und Strukturelle Charakterisierung humaner Annexin V Mutanten, Technische Universität München, Munich, Germany, 1996.
[52] N. Kaneko, R. Matsuda, M. Toda, and K. Shimamoto, *Biochim. Biophys. Acta* **1330,** 1 (1997).
[53] G. Grynkiewcz, M. Poenie, and R. Y. Tsien, *J. Biol. Chem.* **260,** 3440 (1985).

protein (and/or substrates to be tested) is added. With annexins, calcium influx is monitored for 36 min, then the maximal possible intensity ratio $F$ is determined by adding 3 $\mu$l of Bromo-A23187 to stimulate a rapid $Ca^{2+}$ exchange between extra- and intravesicular space.[54] This effect is recorded for 5 min and is sufficient to reach equilibrium. Alternatively, one can use a detergent such as Tween 20 or Triton X-100 to disrupt the vesicle structure and expose the total amount of the fluorescence dye to $Ca^{2+}$. However, in our hands, the fluorescence characteristics of FURA-2 were affected frequently by the presence of these detergents. Therefore, usage of this method is not recommended, although it must be noted that we did not investigate this phenomenon in great detail. Samples are always mixed in an Eppendorf tube and retransferred into the cuvette after addition of newly introduced compounds. A homogeneous distribution within the sample is achieved in this way. To compare results from different experiments, it is necessary to normalize the data with respect to the maximal possible value of $F$, which is obtained at the end of the experiment. The influx curve can be analyzed with respect to three different parameters:

1. *Initial activity:* The curve is fitted within the section from 4 to 12 min, and the slope at $t = 4$ min is calculated. This value is defined as initial activity $\beta$.
2. *Steady state activity:* The influx curve can be fitted with a linear function in the section from 15 to 35 min. The slope of this function is defined as steady state activity $\alpha$.
3. *Total influx:* The difference in the intensity ratios at $t = 35$ min and $t = 4$ min corresponds to the total amount of calcium that transfers from the extra- to the intravesicular space under the influence of annexin. The maximal possible difference of $F$, $\Delta F_{max}$, is given by the observed intensity ratios at the end and the beginning of the experiment. Therefore, the total annexin-induced calcium influx can be expressed as a number smaller than 1, with $\Delta F(35 \text{ min})$ reflecting the total amount of calcium transferred:

$$\frac{\Delta F(35 \text{ min})}{\Delta F_{max}} = \frac{F(35 \text{ min}) - \overline{F}(0-4 \text{ min})}{F(36-40 \text{ min}) - \overline{F}(0-4 \text{ min})} \tag{6}$$

The protocol assumes that the system is equilibrated at $t = 35$ min, which seems to be reasonable on the grounds of the observation that the intensity ratio $F$ does not change significantly after that time. For a more convenient comparison, $\alpha$ and $\beta$ are merged into one parameter, $\gamma$, which represents the length of a vector in an $\alpha-\beta$ plot[24]:

[54] C. M. Deber, J. Tom-Kun, E. Mack, and S. Grinstein, *Anal. Biochem.* **146,** 349 (1985).

$$\gamma = \sqrt{\left(\frac{\alpha}{1\,\mathrm{min}^{-1}}\right)^2 + \left(\frac{\beta}{1\,\mathrm{min}^{-1}}\right)^2} \qquad (7)$$

This allows comparison of a single number each for different annexins (see Fig. 1B).

*Burst Method*

Kaneko *et al.*[52] reported a variation of the FURA-2 assay, which uses a different observation technique. LUVs are added to a buffer solution containing calcium and annexin (and/or additional substrates) and incubated for 10 min. After adding 400 $\mu M$ EGTA to capture free calcium, the liposomes are collected by centrifugation and resuspended in 480 $\mu$l of a buffer containing 1 $\mu M$ FURA-2, 180 $\mu M$ EGTA, 160 m$M$ saccharose, and 10 m$M$ HEPES (pH 7.4). After incubation for 1 min, the liposomes are burst by addition of 0.2% Triton X-100 and the fluorescence is determined at the above-mentioned wavelength settings.

Liposome Aggregation and Fusion Assays

The most popular technique to investigate liposome aggregation is by monitoring turbidity changes. However, this assay might not be able to distinguish between aggregation and fusion phenomena. For a clear insight, liposome aggregation can be verified by phase-contrast microscopy.[27] Alternatively, to test for fusion effects one can employ an intermixing assay. Intermixing assays can monitor the mixing of either the phospholipids or the contents of two different liposome populations and might show contradicting results: whereas the phospholipid intermixing might indicate fusion, the contents-intermixing assay might not support this finding. This is probably due to a phenomenon called semifusion, in which two liposome surfaces are fused but their contents are still separated by a diaphragm-like lipid layer.[55]

*Annexin Effects on Liposome Aggregation as Monitored by Turbidity*

Samples containing liposome suspensions show considerable light scattering because of the size of the phospholipid particles. This phenomenon can be used to monitor liposome aggregation, as the degree of scattering is correlated to the size of the scattering particle. The absorbance recorded from these turbid samples with a spectrophotometer is an "apparent

[55] N. Düzgüneş, T. M. Allen, J. Fedor, and D. Papahadjopoulos, *Biochemistry* **26,** 8435 (1987).

absorbance" or "optical density." Therefore, this method is generally termed turbidimetry. Although turbidimeters (nephelometers, Klett meters) are not common in every laboratory, one can also use a conventional UV/Vis spectrophotometer for this purpose; some careful inspection for suitability of the instrument should be done before the experiment, however, because the effectiveness is dependent on the distance between the sample and the detector.[56] As to the choice of wavelength, lower wavelengths provide greater sensitivity of the turbidity measurement; however, a longer wavelength will give greater linearity.[56]

The background absorbance of samples containing LUVs (final phospholipid concentration, 0.2 $\mu M$) and the appropriate amount of calcium is monitored at a wavelength of $\lambda = 450$ nm[27,57] (or 350 nm[28,58]) for 500 s. After addition of the protein (up to 0.3 nmol) the turbidity changes are recorded continuously for 8 min. Bitto and Cho[57] report changes in turbidity within the first 2–3 min to be almost at the maximal level. To construct a graph of the calcium dependency of annexin-mediated liposome aggregation, the maximal turbidity signals are plotted versus the calcium concentrations of the various samples, thus enabling the calculation of a half-maximal aggregation concentration, $c_{1/2}(Ca^{2+})$.[27,57,59] Other cations[28,58] and various protein constructs, including point and truncation mutants, have been investigated as well.[27,28,30]

PS LUVs spontaneously fuse in the presence of 2–3 m$M$ Ca$^{2+}$.[47,60] For annexins that do not possess intrinsic aggregation activity, the assay can be used to measure their ability to inhibit spontaneous aggregation.[27,58]

## Monitoring of Fusion by Phospholipid Intermixing

Fluorescence quenching between two dyes, N-(7-nitro-2,1,3-benzoxadiazol-4-yl) (NBD) and N-(Lissamine rhodamine B sulfonyl) (Rh), is the basis for an intermixing assay monitoring the fusion of phospholipid layers of two liposome populations (Struck et al.[61]; and see [15] in this volume[62]).

Two populations of PS/PE LUVs are prepared: one unlabeled and the other containing 1% NBD-PE and 1% Rh-PE. The high local concentration

[56] R. K. Poole and U. Kalnenieks, in "Spectrophotometry and Spectrofluorimetry: A Practical Approach" (M. G. Gore, ed.). Oxford University Press, New York, 2000.
[57] E. Bitto and W. Cho, Biochemistry 37, 10231 (1998).
[58] G. Lee and H. B. Pollard, Anal. Biochem. 252, 160 (1997).
[59] E. Bitto and W. Cho, Biochemistry 38, 14094 (1999).
[60] R. Fraley, J. Wilschut, N. Düzgüneş, C. Smith, and D. Papahadjopoulos, Biochemistry 19, 6021 (1980).
[61] D. K. Struck, D. Hoekstra, and R. E. Pagano, Biochim. Biophys. Acta 649, 751 (1981).
[62] N. Düzgüneş, Methods Enzymol. 372, [15], 2003 (this volume).

of both fluorescent dyes results in considerable quenching of fluorescence emission. On fusion of a labeled vesicle with an unlabeled vesicle, the dyes dilute into additional membrane area and the quenching effect becomes less pronounced, leading to an increase in fluorescence emission intensity. For a sample of 2 ml, 70 n$M$ protein and a final phospholipid concentration of 50 $\mu M$ in a ratio of 9:1 (unlabeled:labeled vesicles) are used. Fluorescence is monitored by exciting at $\lambda_{exc} = 450$ nm and monitoring emission intensity at $\lambda_{em} = 530$ nm. The assay is started by addition of calcium.[27,55]

Köhler et al.[37] presented a data analysis that uses the emission intensity ratios of NBD ($\lambda_{em} = 520$ nm) and Rh ($\lambda_{em} = 588$ nm) rather than the more unspecific overall intensity increase as mentioned above. Normalization is achieved by dissolving the liposomes at the end of the experiment. With $F_0$ being the intensity ratio before addition of calcium and $F_{final}$ being the ratio after dissolution of the liposomes in 0.1% Triton X-100 the authors calculate the extent of phospholipid mixing, $M$, as

$$M = \frac{F - F_0}{F_{final} - F_0} \quad \text{with} \quad F = \frac{I_{520\,nm}}{I_{588\,nm}} \tag{8a,b}$$

### Monitoring of Fusion by Contents Intermixing

A contents intermixing assay, described by Wilschut and co-workers,[47,63] takes advantage of the opposite fluorescence phenomenon. Two LUV populations are prepared, one encapsulating $Tb^{3+}$ ions and the other the dipicolinate anion (DPA; 2,6-pyridinedicarboxylic acid). On fusion of vesicles belonging to the two different populations the complex $[Tb(DPA)_3]^{3-}$ is formed,[64] the fluorescence of which is enhanced by a factor of $10^4$ compared with the $Tb^{3+}$ ion.[65] Because $Tb^{3+}$ ions have been shown to destabilize liposomes, they are masked during vesicle preparation with the citrate anion.[47] Buffers for preparation of the two liposome populations might be[51] 2.5 m$M$ $TbCl_3$, 50 m$M$ sodium citrate, 160 m$M$ saccharose, 20 m$M$ HEPES (pH 7.4), and 50 m$M$ dipicolinic acid, 20 m$M$ NaCl, 160 m$M$ saccharose, 20 m$M$ HEPES (pH 7.4). The fluorescence experiment is carried out by exciting at $\lambda_{exc} = 276$ nm and monitoring the emission intensity at $\lambda_{em} = 545$ nm.

The opposite effect, quenching on fusion, is used in the ANTS–DPX assay, originally introduced by Ellens et al.[66] 1-Aminonaphthalene-3,6,8-trisulfonic acid (ANTS) and its quencher $N,N'$-$p$-xylylenebis (pyridinium

[63] J. Wilschut and D. Papahadjopoulos, Nature **281**, 690 (1979).
[64] I. Grenthe, J. Am. Chem. Soc. **83**, 360 (1961).
[65] T. D. Barela and A. D. Sherry, Anal. Biochem. **71**, 351 (1976).
[66] H. Ellens, J. Bentz, and F. C. Szoka, Biochemistry **24**, 3099 (1985).

bromide) (DPX) are encapsulated in two different populations of liposomes. In the event of fusion of liposomes from different populations, the fluorescence emission of ANTS is quenched.

## Assays Probing Biophysical Membrane Properties

### Annexin Effects on Membrane Surface Hydrophobicity

A decrease in the membrane surface dielectric constant is the result of dehydration of the phospholipid head groups either by binding of protein to the membrane surface or by the creation of water-free interfaces between two vesicles, which occurs during aggregation.[67] For annexin A5, which does not induce vesicle aggregation,[68] a strong decrease in the surface dielectric constant is observed in the presence of 200 $\mu M$ CaCl$_2$, which can be understood as binding of the protein to the membrane.[37]

The emission wavelength of $N$-(6-dimethylaminonaphthalene-2-sulfonyl)-1,2-dihexadecanoyl-$sn$-glycero-3-PE (6-dansyl-PE) has been shown to be proportional to the dielectric constant of the probe environment.[69] By using a set of organic solvents[67,70] an empiric law is derived allowing the calculation of dielectric constants from the emission wavelength of the fluorescence probe. LUVs with 0.3–0.5% 6-dansyl-PE are used for elucidation of annexin effects on membrane surface hydrophobicity; the samples are excited at $\lambda_{exc} = 340$ nm, whereas fluorescence emission is recorded in the range $\lambda_{em} = 400$–600 nm.[37]

### Lateral Diffusion Coefficients

As already mentioned above, the excimer-to-monomer fluorescence emission intensity ratio of pyrene within liposomes is dependent on, among other parameters, pyrene concentration and lateral diffusion within the membrane layer.[39] Köhler et al.[37] describe a protocol for determination of the lateral diffusion coefficient in PC/PS LUVs (molar ratio, 1:9) labeled with 10% pyrene-PC. The emission intensities at $\lambda_{exc} = 470$ nm and $\lambda_{exc} = 394$ nm are recorded for the excimer (E) and monomer (M), respectively. With $\Phi$ being the quantum yield, $I$ the emission intensity, $k_f$ the fluorescence rate constant, $\tau_0$ the lifetime, $k_a$ the

[67] S. Ohki and K. Arnold, J. Membr. Biol. **114**, 195 (1990).
[68] D. Hoekstra, K. Buist-Arkema, K. Klappe, and C. P. M. Reutelingsperger, Biochemistry **32**, 14194 (1993).
[69] E. von Lippert, Z. Elektrochem. **61**, 962 (1957).
[70] Y. Kimura and A. Ikegami, J. Membr. Biol. **85**, 225 (1985).

second-order rate constant for excimer formation, and $c$ the concentration of pyrene in the membrane, the excimer formation can be assessed according to

$$\frac{\Phi^E}{\Phi_M} = \kappa \frac{I_{470\,nm}}{I_{394\,nm}} = \frac{k_f^E}{k_f^M} \tau_0^E k_a c \qquad (9)$$

$\kappa$ is a special conversion factor that is introduced to use fluorescence intensities rather than quantum yields; its value was determined to be $\kappa = 1.64$.[37] The diffusion coefficent $D$ can be calculated by assuming a two-dimensional random walk model of phospholipids in which the molecules randomly jump between neighboring lattice sites. $D$ is therefore determined by the jump length and frequency.[71]

Membrane fluidity can also be investigated by electron spin resonance (ESR) with the help of spin labels present in different locations within the membrane bilayer of liposomes. ESR requires either the presence of paramagnetic metal centers or the covalent attachment of groups with unpaired electrons to the biomolecule to be investigated. The latter method is preferred with liposomes and carried out by introduction of stable organic nitroxide radicals, such as oxazolidine, piperidine, or pyrrolidine derivatives. Once bound to the alkyl chains of the phospholipid molecules, these probes have a given orientation with respect to the molecule and the applied magnetic field.[72] Neglecting the hyperfine coupling of the probe with naturally occurring $^{13}C$, these radicals typically exhibit a three-line spectrum in isotropic solvents. Their rigid orientation within the membrane bilayer, however, leads to a spectral anisotropy that affects the hyperfine splitting constant $A$ and the Landé factor $g$. Maximal values of anisotropy $(A_V - A_H)$ are observed when the probe is totally immobilized in a direction parallel to the magnetic field $(A_{ZZ}; A_V)$. For isotropic motion of the probe, anisotropy reaches the minimal value $(A_{XX}, A_{YY}$ if axial symmetric; $A_H)$. Determination of the maximal anisotropy requires recording of an ESR spectrum with a completely immobilized spin probe $(A_{par}$ and $A_{per})$. The mobility of the probe for each sample can then be described by the order parameter $S$ according to

$$S = \frac{A_{V\,obs} - A_{H\,obs}}{A_{par} - A_{per}} \qquad (10)$$

[71] H. J. Galla, W. Hartmann, U. Theilen, and E. Sackmann, *J. Membr. Biol.* **48**, 215 (1979).
[72] L. J. Berliner, "Spin Labeling: Theory and Applications." Molecular Biology: An International Series of Monographs and Textbooks. Academic Press, New York, 1976.

Accordingly, total immobilization is characterized by an order parameter of $S = 1$, whereas an isotropic molecular movement yields a value of $S = 0$. Because liposomes are spherical vesicles in which the probes are distributed randomly, one must consider signal overlaps leading to a significant broadening of the signal within the spectra.[72,73]

Megli et al.[74] describe detailed protocols for preparation of spin-labeled liposomes, as well as a careful procedure for spectral analysis of ESR spectra in the presence of annexin and calcium by spectral fitration. These authors marked different positions along the fatty acid acyl chain, as well as the phospholipid head group with a spin label, and thereby obtained a flexibility profile of a phospholipid molecule within the liposome membrane layer. In the absence of annexin the order parameter $S$ decreases with increasing position of the spin label, which is in accordance with an increase in mobility along the acyl chain.[72] Annexin A5 in the presence of calcium does not change this behavior, but generally shifts the order parameter to higher values without changing the shape of the profile.

*Lipid Phase Transition*

Polarization of fluorescence emission is dependent on the orientation of the transition dipole momentum and therefore tied to the movement of the fluorescent molecule, which includes rotation and flexibility in diverse segments of the molecule. Monitoring fluorescence polarization of probes within the membrane layers of liposomes is therefore a technique to investigate lipid phase transitions, because changes in viscosity and orientation are observed on these transitions.

In a combined NMR and fluorescence study, Saurel et al.[75] used, among other lipid systems, labeled LUV preparations (1,6-diphenyl-1,3,5-hexatriene; label concentration in the membrane, 0.5%) to measure temperature-dependent fluorescence polarization (4–40°, 0.7°/min).[76] Polarization $p$ is given as

$$p = \frac{I_V - I_H}{I_V + I_H} \tag{11}$$

[73] H. M. Swartz, J. R. Bolton, and D. C. Borg, "Biological Applications of Electron Spin Resonance." John Wiley & Sons, New York, 1972.

[74] F. M. Megli, M. Selvaggi, S. Liemann, E. Quagliariello, and R. Huber, *Biochemistry* **37,** 10540 (1998).

[75] O. Saurel, L. Cézanne, A. Milon, J. F. Tocanne, and P. Demange, *Biochemistry* **37,** 1403 (1998).

[76] N. Leborgne, L. Dupou-Cézanne, C. Teulières, H. Canut, J. F. Tocanne, and A. M. Boudet, *Plant Physiol.* **100,** 246 (1992).

where $I_V$ and $I_H$ are the fluorescence emission intensity components parallel and perpendicular to the incident beam, respectively.

In the absence of annexin or calcium, a continuous depolarization of the probe is found, consistent with a liposome model in which the lipids are in the fluid phase and no transition from the gel to the liquid crystalline phase occurs. In general agreement with the ESR results mentioned above, these authors find that addition of annexin A5 and calcium does not change this behavior, indicating that even at saturating concentrations of annexin the protein has no influence on bilayer fluidity. Because the polarization of a fluorophore is somewhat dependent on the fluorescence lifetime $\tau$, these authors also determined lifetimes for various protein concentrations and found a constant value throughout, which supports the conclusion that the protein performs a simple adsorption reaction without disturbing the membrane bilayer properties.

*Thermal Effects as Monitored by Differential Scanning Calorimetry*

Microcalorimetry allows the measurement of heat capacities enabling determination of thermal parameters. If used in a denaturation experiment, transition temperatures $T_{1/2}$, which characterize the denaturation process, can be obtained as well as van't Hoff enthalpies $\Delta H_{vH}$ (for monomolecular processes).[77] Differential scanning calorimetry (DSC) has been used to elucidate annexin A5 unfolding in the presence of PC/PS LUVs (molar ratio, 4:1) and various calcium concentrations.[14] Annexin A5 at concentrations of 10–15 $\mu M$ is filled into the sample cell (0.47 ml) of a DASM-4 microcalorimeter (MicroCal) together with the liposomes (0.68 m$M$, final lipid concentration) in 100 m$M$ KCl, 50 m$M$ Tris (pH 8.0). Calcium concentrations are either 0, 0.5 m$M$, or 50 m$M$, and the instrument is run at a heating rate of 1.98 K/min (thermal denaturation behavior of annexin A5 has been shown to be dependent on the heating rate[13]). For calorimetric details see Refs. 14 and 43. Without protein, only the lipid phase transition is visible in the thermograms and vesicle aggregation can be detected at high calcium concentrations. Thermograms recorded with annexin A5 in the absence of calcium reveal two transitions: the lipid phase transition and the protein-unfolding process. In the presence of calcium, in the first heating process from 15 to 90° three transitions are visible in the thermogram. The first transition corresponds to the lipid phase transition ($T_{1/2}$ of ~45°), whereas the second one can be attributed to the denaturation of the protein ($T_{1/2}$ of ~60°). A third, broad transition occurs with a $T_{1/2}$ of approximately 80° and belongs to denaturation of

---

[77] P. L. Privalov and S. A. Potekhin, *Methods Enzymol.* **131,** 4 (1986).

membrane-bound protein. In a second heating process, only the lipid transition is visible because all the protein is irreversibly unfolded. Notably, vesicle aggregation is also visible at high calcium concentrations at this stage, because annexin A5 is no longer protecting the vesicle surfaces.

### Bilayer Structure

The bilayer structure of interdigitated fusion vesicles (IFVs) in the presence of annexin A5 and calcium was investigated by Saurel et al. by NMR spectroscopy.[75] IFVs possess reduced lateral diffusion of lipids and thus produce fewer artifacts in NMR experiments. A 50-$\mu$mol quantity of PC/PS IFVs (molar ratio, 9:1) is prepared in 100 m$M$ NaCl and 10 m$M$ HEPES (pH 7.5) and incubated overnight with 1 m$M$ CaCl$_2$ for stabilization.

Vesicles are sedimented by centrifugation (13,400 $g$, 30 min), and 0.32 $\mu$mol of annexin A5 is added to the resuspended pellet. The vesicles are sedimented again and resuspended, and the final concentration of lipids and annexin is determined, yielding a ratio of 43 phospholipid molecules per annexin.

As reported by the authors, the presence of annexin A5 does not modify the shape of the $^{31}$P NMR spectrum, which exhibits a typical powder pattern consistent with spectra well known from LUVs (width, $\Delta\sigma = 44$ ppm). Their spectrum also shows a sharp peak at $\delta = 0$ ppm, which is attributed to a small fraction of SUVs left in the preparation undergoing isotropic motions. The presence of annexin A5 slightly increases the shoulder of the IFV peak at $\delta = 23$ ppm, which might be due to subtle modifications of the vesicle shape, thus deviating from a strictly spherical geometry.[78]

### Folding Studies in Presence of Liposomes

### Annexin Secondary Structure in Membrane-Bound State as Monitored by Circular Dichroism

The simplest approach to elucidate effects of membrane binding on the annexin fold is certainly to measure the circular dichroism (CD) of annexin in the presence of liposomes and calcium. As a precaution in these studies one should check for the lack of spectral dependence on the cuvette position with respect to the detector in order to ensure the absence of light-scattering effects.[46] All reports agree that only minor changes, if any, occur

---

[78] X. Qiu, P. A. Mirau, and C. Pidgeon, *Biochim. Biophys. Acta* **1147,** 59 (1993).

for annexins (AnxA5, AnxA6) in the presence of liposomes.[14,46,79] Various kinds of liposomes have been used for these experiments, and even reverse micelles are reported to work well for CD spectroscopy.[79]

## Monitoring Annexin Fold by ESR

ESR spectroscopy has already been mentioned earlier with respect to the elucidation of membrane fluidity in the presence and absence of annexin. Although this requires spin labeling of the liposomes, one can also take the opposite experimental approach and attach the spin label to the protein, a technique called site-directed spin labeling.[80] The paramagnetic nitroxide side chain is coupled to the protein via a thioether linkage with a substituted cysteine residue. To scan a certain amino acid sequence within a protein, all residues within that sequence must be substituted independently to cysteine and labeled subsequently with the spin probe.[80] The mobility and/or accessibility of the probe(s) might be sufficient in certain cases to identify a secondary structure element.[81] Langen *et al.*[82] applied this technique to PS/PC LUVs (molar ratio, 2:1) and annexin B12 mutants to characterize the membrane-bound form of AnxB12 in a low-pH environment. These authors claim to have identified an amino acid region ranging from helix IID to the linker between domains II and III, which forms a continuous transmembrane helix in the membrane-bound state.

## Monitoring Conformational Effects by Fluorescence

The position of the unique tryptophan residue in annexin A5, Trp-187, on membrane binding of the protein was a question addressed early in annexin research. Fluorescence emission spectra (excitation at $\lambda_{exc} = 295$ nm and $\lambda_{exc} = 280$ nm) of annexin A5 in the presence of PS/PC LUVs (molar ratio, 1:1) with or without 100 $\mu M$ CaCl$_2$ revealed a large increase in emission intensity, which is reversible on addition of EDTA.[83] Doxyl spin labels in three different positions (positions 5, 12, and 16) on phospholipids of another LUV preparation were used to quench the fluorescence emission by Trp-187.[83] Because quenching is observed with a label at position 5 only

[79] J. Sopkova, M. Vincent, M. Takahashi, A. Lewit-Bentley, and J. Gallay, *Biochemistry* **38,** 5447 (1999).

[80] W. L. Hubbell, H. S. Mchaourab, C. Altenbach, and M. A. Lietzow, *Structure* **4,** 779 (1996).

[81] H. S. Mchaourab, M. A. Lietzow, K. Hideg, and W. L. Hubbell, *Biochemistry* **273,** 810 (1996).

[82] R. Langen, J. M. Isas, W. L. Hubbell, and H. T. Haigler, *Proc. Natl. Acad. Sci. USA* **95,** 14060 (1998).

[83] P. Meers, *Biochemistry* **29,** 3325 (1990).

and the fluorescence emission of annexin A5 does not show a blue shift on membrane binding, but rather a small red shift, Meers[83] concluded that Trp-187 is positioned in the interfacial region of the membrane surface. These data and the rapid reversibility of binding strongly argue for a peripheral membrane binding of annexin A5 without considerable membrane penetration under these conditions.

The matter was investigated further by Follenius-Wund *et al.*,[84] using time-resolved fluorescence spectroscopy with Trp-187. Because steady state fluorescence superimposes the emission of different fluorophore species, it might be possible to distinguish between fluorophores with different empiric formulas (and therefore different fluorescence properties) but it practically never allows for identification of conformers. Because annexin A5 possesses only one unique tryptophan residue, which can be selectively excited, the recording of the fluorescence decay can reveal insights into different populations of this residue, that is, the presence of different conformers. For experimental details see Follenius-Wund *et al.*,[84] which contains a detailed description of the experimental setup and analysis procedures. Basically, the deconvolution of fluorescence decays yields the lifetime $\tau_i$ and the preexponential factor $\alpha_i$ for $N$ species. Knowing $N$, the fractional intensity $f_i$ can be calculated for each species from the recorded intensities $I(t)$:

$$I(t) = I_0 \sum_i^N \alpha_i e^{-t/\tau_i} \qquad f_i = \frac{\alpha_i \tau_i}{\sum_i^N \alpha_i \tau_i} \qquad (12a,b)$$

with $I(t)$ and $I_0$ being the intensities at time $t$ and $t = 0$, respectively. The fluorescence decay curves can be recorded as a function of the emission wavelength $\lambda_{em}$ (310–380 nm), and steady state emission spectra associated with each component of the decay (i.e., the different conformers) can be back-calculated.[85] It was indeed found that on membrane binding Trp-187 has (at least) two conformers: One conformer with a lifetime of $\tau_1 = 6.9$–$7.2$ ns is located in a partially polar environment, the interfacial region. A second conformer with a lifetime of $\tau_2 = 2.0$–$2.2$ ns is hydrophobically embedded and most likely penetrates into the lipid membrane. This latter species agrees well with results from the quenching experiments with the doxyl spin label (see above).

A similar study[79] used SUVs and reverse micelles for fluorescence investigation of Trp-187. Whereas with liposomes, annexin A5 binds to the

---

[84] A. Follenius-Wund, E. Piémont, J. M. Freyssinet, D. Gérard, and C. Pigault, *Biochem. Biophys. Res. Commun.* **234,** 111 (1997).

[85] B. Donzel, P. Gauduchon, and P. Wahl, *J. Am. Chem. Soc.* **96,** 801 (1974).

outer surface of the vesicles, in the case of reverse micelles the protein is encapsulated. Lifetime measurements provided by Sopkova et al.[79] reveal three populations in which the long-lived species was attributed to the membrane-bound state of the protein. Analysis of the coupling between the rotational correlation times and the lifetimes[86] indicated the existence of local conformers with different subnanosecond mobilities. This means that even the membrane-bound state in this study allows for restricted motions of the tryptophan residue, thereby raising the question of how tightly the third domain is attached to the lipid surface in the membrane-bound state.

In a unique study, Silvestro and Axelsen[87] applied transmission infrared spectroscopy (IR) to annexin A5 samples in the presence and absence of PC/PS LUVs (molar ratio, 4:1) and 2 m$M$ CaCl$_2$. However, they could not obtain high-quality transmission spectra of the membrane-bound protein, and interpretation of any sort is possible only by analyzing the shape of the amide I/II bands. As they have shown, IR yields much more informative results in another variant, the polarized attenuated total internal reflectance-Fourier transform IR (PATIR-FTIR) which, however, uses supported lipid monolayers instead of liposomes.

## Concluding Remarks

As a huge family of proteins, annexins show an impressive structural conservation: they share a characteristic topology, calcium-binding sites/membrane loops, and the ability to bind to membrane surfaces. These well-defined properties are contrasted by an enormous variety of possible functions and involvements in physiological pathways. The landmark feature of these proteins is the peripheral membrane binding, which is calcium-mediated in the case of the vertebrate members of this protein family. Evidence collected suggests that this is also the place where annexin biology takes place. Despite being cytosolic proteins, annexins seem to fulfill their physiological functions in membrane-bound states or by the timed attachment to or desorption from membranes.

It is thus not surprising that annexin–membrane interactions have been investigated by using liposomes as an object mimicking physiological membranes. Experiments like those mentioned in this chapter have proven that liposomes are a versatile and powerful tool to characterize protein–membrane interactions. Whereas a cellular environment presents far too

---

[86] N. Rouvière, M. Vincent, C. T. Craescu, and J. Gallay, *Biochemistry* **36,** 7339 (1997).
[87] L. Silvestro and P. H. Axelsen, *Biochemistry* **38,** 113 (1999).

many different influences in order to keep track of basic molecular events, liposomes enable a much needed scientific simplification to gain insights into generic mechanisms at the protein–membrane interface. At the same time, one must be aware of the simplification and great care is to be applied when extrapolating the findings into the physiological environment.

Annexin research has profited enormously from the availability of liposomes. As this field enters a new era with investigation of annexin–protein complexes at membrane surfaces, liposomes will definitely play a significant role in the process of understanding.

### Acknowledgments

We thank Claudia Jatzke for helpful discussions. A. H. thanks Alex Wlodawer for support.

## [12] Anchoring of Glycosylphosphatidylinositol–Proteins to Liposomes

*By* Olivier Nosjean and Bernard Roux

### Introduction

#### Glycosylphosphatidylinositol–Proteins and Interest in Membrane Reconstitutions

Membrane proteins associate with the hydrophobic region of phospholipid bilayers by either exposing a favorable peptidic segment or by inserting a posttranslational lipidic moiety into the leaflet. Among these so-called membrane anchors is glycosylphosphatidylinositol (GPI), a complex C-terminal glycolipid that is transferred to proteins bearing the appropriate C-terminal signal sequence (see the comprehensive review by McConville and Ferguson[1]). The GPI structure was discovered as such in 1985[2] and a consensus core structure quickly appeared: phosphoethanolamine–(mannose)$_3$–glucosamine–phosphatidylinositol. The GPI

---

[1] M. J. McConville and M. A. J. Ferguson, *Biochem. J.* **294,** 305 (1993).
[2] A. H. Futerman, M. G. Low, K. E. Ackerman, W. R. Sherman, and I. Silman, *Biochem. Biophys. Res. Commun.* **129,** 312 (1985).

moiety is widely distributed from yeast to mammals and is associated with miscellaneous proteins of various sizes and functions (for a review see Nosjean *et al.*[3]). Despite intensive research, there is still no functional model able to explain the conservation of such a complex lipid anchor during evolution. Nevertheless, in the mid-1990s GPI–proteins were proposed to associate with cholesterol, glycosphingolipids, and signal transduction proteins to form functional membrane microdomains,[3] described at the nanoscale by atomic force microscopy.[4] These membrane components can be isolated by their common insolubility in detergent at low temperature, and are therefore often referred to as detergent-resistant membrane complexes. The concept has been the subject of intensive work and debates (see comments in Nosjean *et al.*[3]), and the topological organization of GPI–proteins on membranes still represents an active field of investigation. Therefore, the membrane reconstitution of GPI–proteins represents a constructive approach to evaluate the formation of these functional microdomains. In addition, the method of membrane reconstitution described hereafter can be relevant for building oriented membranes bearing any lipid-anchored molecules (e.g., for cellular targeting of liposomes).

## Necessity to Control Membrane Orientation

*In vivo*, GPI-anchored proteins concentrate exclusively into the exoleaflet of eukaryotic cell plasma membranes, together with cholesterol and most of the glycolipids. In addition, surface antigens, receptors, and transporters also adopt an orientation relevant to their involvement in communication/exchange functions with the external milieu. As a consequence, the physiological membrane environment of GPI–proteins is highly oriented, a characteristic that should be taken into account as much as possible for reconstitution of biomimetic membranes.

Assembling lipids and proteins or lipid-anchored proteins into proteolipidic complexes is feasible with the simple use of sonication or, more favorably, nonionic detergent. But the resulting structures might be different from biomimetic membranes, with most probably high variability in liposome size. The following evaluation shows a possibility for random distribution of GPI–proteins in the inner and outer leaflets of the liposomes. Alkaline phosphatase is a globular protein of about 5 nm in diameter,[5] and the hydrophilic part of the GPI glycan is not expected to

[3] O. Nosjean, A. Briolay, and B. Roux, *Biochim. Biophys. Acta* **1331,** 153 (1997).
[4] M. C. Giocondi, V. Vié, E. Lesniewska, J. P. Goudonnet, and C. Le Grimellec, *J. Struct. Biol.* **131,** 38 (2000).

exceed 5 nm if stretched to its maximum (personal evaluations). Thus, the resulting overall size of GPI-anchored alkaline phosphatase can be estimated at 10 nm. Hence, there is no steric hindrance in a liposome of 100 nm, and several GPI–proteins could be anchored in its inner leaflet. As a consequence, accurate control of reconstitution conditions is crucial for preparing liposomes with oriented GPI-anchored proteins. In this respect, we suggest the application of the procedure described below, based on detergent-mediated incorporation of GPI–proteins into preformed liposomes. This approach was inspired by a method originally described for membrane insertion of $Ca^{2+}$-ATPase.[6–8] We adapted the protocol to the handling of lipid-anchored proteins, and the technique was validated with various preparations of GPI–proteins.[9,10] In the following sections, we first describe the prerequisites in terms of liposome preparation and GPI–protein purification, then we detail the procedure for ectoplasmic incorporation of GPI–proteins into liposomes, and last we present some technical approaches for characterizing the correct preparation of liposomes containing ectoplasmic GPI-anchored proteins. Detailed conditions are given where necessary. Alternatively, whenever the procedure can be freely adapted to a particular sample, the details are given in parentheses as examples only. Furthermore, most techniques require the assay of the membrane components. In this respect, radiolabeling of the lipids is recommended, as well as the use of an easily assayed GPI–protein.

Description of Methods

The procedure we describe for the controlled ectoplasmic anchoring of a GPI–protein on a liposome necessitates the appropriate preparation of biological samples. Indeed, in this method the liposomes are not considered a bulk phospholipidic support for the protein, but, rather, as a starting material as homogeneous and well defined as possible. In addition, the purified GPI–protein must contain only minute amounts of contaminating lipidic compounds if clean and reproducible results are to be obtained. Briefly, the procedure of reconstitution consists of destabilization

[5] M. H. Le Du, T. Stigbrand, M. J. Taussig, A. Menez, and E. A. Stura, *J. Biol. Chem.* **276,** 9158 (2001).

[6] D. Lévy, A. Gulik, A. Bluzat, and J.-L. Rigaud, *Biochim. Biophys. Acta* **1107,** 283 (1992).

[7] U. Kragh-Hansen, M. le Maire, J. P. Noel, T. Gulik-Krzywicki, and J. V. Møller, *Biochemistry* **32,** 1648 (1993).

[8] J. L. Rigaud, B. Pitard, and D. Levy, *Biochim. Biophys. Acta* **1231,** 223 (1995).

[9] M. Angrand, A. Briolay, F. Ronzon, and B. Roux, *Eur. J. Biochem.* **250,** 168 (1997).

[10] O. Nosjean and B. Roux, *Eur. J. Biochem.* **263,** 865 (1999).

of liposomes by the appropriate use of detergent, incubation with the GPI–protein, and elimination of excess detergent by dialysis. These steps are described herein, together with additional comments concerning assessment of correct reconstitution.

*Preparation of Liposomes*

In our studies, we tend to consider 100-nm, 10-mg/ml egg yolk phosphatidylcholine (PC), dipalmitoyl phosphatidylcholine (DPPC), or dimyristoyl phosphatidylcholine (DMPC) unilamellar liposomes as the material of reference. Although less "biomimetic," DPPC and DMPC have the advantage over PC of being homogeneous in structure and, therefore, of having more precise phase transition curves. These liposomes can be conveniently prepared by the detergent method, which is summarized herein, but is also described in more detail by Schubert.[10a] Briefly, egg-yolk L-$\alpha$-phosphatidylcholine (PC) and *n*-octyl-$\beta$-D-glucoside (OG) are dissolved in CHCl$_3$–methanol (2:1, v/v) at the exact PC:OG molar ratio of 0.2. The solution is evaporated under nitrogen flow to form a thin film in a test tube, and then resuspended at a PC concentration of 10 mg/ml in the buffer chosen for the reconstitution, for example, 10 m$M$ Tris or phosphate-buffered saline, pH 7.4. The resulting micellar suspension is dialyzed in order to eliminate the detergent progressively, leading to the conversion of the mixed micelles into liposomes. The diameter of the liposomes obtained depends highly on the conditions of dialysis (temperature, speed). The utilization of a Liposomat apparatus (Harvard Apparatus, Holliston, MA), although much more expensive than simple cellulose tubing dialysis, affords fast, comfortable, and reliable support for the preparation of liposomes. Under our conditions [10,000 molecular weight cutoff (MWCO) membranes, sample at 0.5 ml/min, buffer at 2.5 ml/min, room temperature], we obtain 100-nm liposomes after 90 min of dialysis, as assessed by quasi-elastic light scattering (QELS). If radiolabeling of the liposomes is desired, tritiated PC (e.g., 0.1 $\mu$Ci/mg total PC) can be added to the PC before formation of the lipidic film. Alternatively, fluorescent lipid analogs can be used, but we have no experience in this field and the reader is invited to consult the appropriate reports.

*Preparation of GPI–Protein*

For convenience of detection, and whenever the nature of the GPI-anchored protein has little importance for the model membrane, the best protein to choose is an enzyme. Alkaline phosphatase is the most

---

[10a] R. Schubert, *Methods Enzymol.* **367,** 46 (2003).

convenient GPI–protein, as it is broadly expressed with a GPI anchor in mammalian tissues. It is easy to find substrates for alkaline phosphatase, the most common being $p$-nitrophenylphosphate, which, after hydrolysis, yields $p$-nitrophenol with significant absorption at 420 nm ($E_{420} = 18,500 \text{ cm}^{-1} \cdot M^{-1}$). 5′-Ectonucleotidase is another common GPI-anchored enzyme, but its assay is not as inexpensive and easy as that of alkaline phosphatase. Other GPI-anchored proteins exist [e.g., Thy-1, decay-accelerating factor (DAF), Fcγ receptor III (FCγRIII)] for which an enzyme-linked immunosorbent assay (ELISA) will be the standard assay procedure. It is noteworthy that commercial preparations of proteins such as alkaline phosphatase contain only minute amounts of the GPI-tethered protein, even when prepared from mammalian sources. Hence, custom purification of the GPI–protein is often mandatory. GPI–proteins are highly amphipathic, with a water-soluble polypeptide chain and a necessarily hydrophobic C-terminal phosphatidylinositol. Hence, purification of GPI–proteins must involve the use of detergents for solubilization from biological samples and, thereafter, for stabilization during chromatographic steps. The choice of a detergent with high critical micellar concentration (CMC) is a must, as it will allow elimination of excess detergent by dialysis. In this respect, OG represents one of the best choices (CMC = 19–25 m$M$, depending on experimental conditions[11]), as it is now commercially available at low price from several suppliers. Furthermore, OG has proved to solubilize GPI–proteins efficiently as compared with other non-ionic detergents.[12–14] The use of detergent throughout the chromatographic procedures is mandatory, but final steps should contribute to lower the detergent concentration. Furthermore, for optimal results it is highly recommended that the purification procedure be designed to minimize copurification of membrane components (glycolipids, membrane proteins, and other lipid-tethered proteins). This can be achieved through the use of density gradients in the early steps of purification, solvent-mediated delipidation whenever the GPI–protein can stand it, or repeated chromatographic steps under various conditions. The use of cultured cells as a primary source of GPI–proteins can alleviate some of the copurification problems, but tissue samples have the advantage of providing rich sources of GPI–proteins (e.g., liver, kidney cortex, and intestinal mucosa). A purification procedure for kidney alkaline phosphatase is presented below.

[11] M. le Maire, P. Champeil, and J. V. Møller, *Biochim. Biophys. Acta* **1508,** 86 (2000).

[12] P. Banerjee, J. B. Joo, J. T. Buse, and G. Dawson, *Chem. Phys. Lipids* **77,** 65 (1995).

[13] T. J. Cain, Y. Liu, T. Takizawa, and J. M. Robinson, *Biochim. Biophys. Acta* **1235,** 69 (1995).

[14] N. M. Hooper and A. J. Turner, *Biochem. J.* **250,** 865 (1988).

This procedure has been applied successfully with bovine or porcine kidney cortex and intestinal mucosa, but may require slight adaptations for intertissue or interspecies differences. The first steps are described in detail and are critical for obtaining a well-solubilized protein with minimal contaminating membrane components or detergent. The nature of the buffer must be chosen in agreement with the protein to be purified, and supplemented with the classic protease inhibitors until the first delipidation step [e.g., 0.5 m$M$ phenylmethylsulfonyl fluoride (PMSF), 1 $\mu M$ antipain, aprotinin (0.5 $\mu$g/ml), 1 $\mu M$ leupeptin, and 1 $\mu M$ pepstatin A, or any suitable simpler mixture]. Cellular membranes are prepared from the tissue, washed with the buffer of preparation (4 ml/g initial tissue), and further washed three times with the buffer supplemented with Triton X-100 (0.1%, v/v) (4 ml/g initial tissue) in order to eliminate loosely bound membrane material. The membranes are resuspended in a minimal volume of buffer (e.g., 0.1 ml/g initial tissue) and partially delipidated by treatment with CHCl$_3$–methanol (1:2, v/v; 3 ml/g initial tissue). After a 10-min centrifugation at 2000 $g$, the pelleted material is dried under a flow of nitrogen and resuspended in the buffer (3 ml/g initial tissue), using a tissue grinder apparatus. The sample is adjusted to 60 m$M$ OG and is incubated for 60 min at 4° under agitation. The solubilized GPI–protein is isolated as a supernatant after ultracentrifugation (100,000 $g$, 30 min). The solubilisate is then delipidated with CHCl$_3$–methanol (1:2, v/v), at 3 ml/ml supernatant, and recovered as a 2000 $g$ pellet. The delipidated material is dried under nitrogen as described above and resolubilized in the buffer supplemented with Triton X-100 (0.1%, v/v) (1 ml/g initial tissue), with brief sonication if necessary. The sample is then ready for further purification by chromatographic techniques, ideally affinity methods. During preparation of the sample, delipidations are key steps and must be performed with care. The incubations are performed at room temperature for 15–20 min under gentle agitation. The drying under nitrogen must be sufficient to obtain material with solid consistence, but must not be prolonged after the appearance of overdried zones of a typical dark brown color. After chromatographic purification, the GPI–protein must be in a buffer supplemented with Triton X-100 (0.1% v/v).

### Determination of Liposome Stability

For an amphipathic compound to insert into a lipidic bilayer, the hydrophilic barrier of phopholipid heads must first be disrupted in order to give access to the hydrophobic core of the membrane. Detergents can readily be used for this purpose. Le Maire and collaborators[7,11,15] and other laboratories[16–20] have characterized well the detergent-mediated solubilization

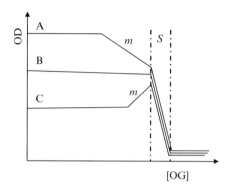

FIG. 1. Typical profiles of solubilization of phosphatidylcholine (PC) liposomes by octylglucoside (OG). These solubilization profiles are based on Refs. 6, 16, 18, and 19, and our own experience. On addition of OG, liposomes first incorporate detergent molecules in their membrane without any change in turbidity [stable optical density (OD), e.g., at 450 nm]. Then 100-nm liposomes (A) undergo morphological changes traced by a decrease in OD ($m$ phase). Liposomes < 70 nm (C) also show structural changes ($m$ phase), but that lead here to an increase in size and turbidity. Finally, for a given concentration of PC, all liposomes, including intermediate-sized liposomes (B), start to release mixed PC/OG micelles at the same detergent concentration, until complete solubilization of the phospholipid bilayers ($S$ phase).

process of liposomes. Briefly, during increasing detergent concentration the membrane incorporates detergent molecules in a noncooperative manner. After saturation, the detergent incorporation is cooperative and the bilayer starts to dismantle and progressively releases detergent/phospholipid binary micelles, leading to complete solubilization of the phospholipids. The process is monitored easily by turbidimetry, that is, measurement of light scattering at 450–600 nm with a spectrophotometer—the lower the wavelength, the better the signal. Visually, turbidity means a "milky" appearance of the solution, which becomes clearer as the liposomes are being solubilized. OG solubilization of PC liposomes can give either a monophasic or a biphasic solubilization profile, depending on the size of the liposomes (Fig. 1). Polydispersion of liposomes must be minimal in order to give a satisfactory cooperative solubilization profile. Size determination by QELS gives information about liposome diameter,

[15] U. Kragh-Hansen, M. le Maire, and J. V. Møller, *Biophys. J.* **75**, 2932 (1998).
[16] M. Paternostre, O. Meyer, C. Grabielle-Madelmont, S. Lesieur, M. Ghanam, and M. Ollivon, *Biophys. J.* **69**, 2476 (1995).
[17] A. de La Maza and J. L. Parra, *Eur. J. Biochem.* **226**, 1029 (1994).
[18] M. Keller, A. Kerth, and A. Blume, *Biochim. Biophys. Acta* **1326**, 178 (1997).
[19] P. K. Vinson, Y. Talmon, and A. Walter, *Biophys. J.* **56**, 669 (1989).
[20] M. R. Wenk, T. Alt, A. Seelig, and J. Seelig, *Biophys. J.* **72**, 1719 (1997).

as well as about population polydispersion around the mean size. Hence, again, the initial liposome preparation must be conducted under highly controlled and reproducible conditions leading to constant size with low dispersity.

Each novel liposome preparation must be characterized thoroughly in terms of solubilization in order to determine the detergent saturation point to be used for insertion of GPI–proteins. The characterization of the solubilization profile of liposomes must be performed under exactly the same conditions as are planned for the reconstitution: identical detergent, buffer composition, liposome concentration, and sample temperature. In practice, the optical diffusion of the liposome suspension (e.g., 500 $\mu$l, PC at 1 mg/ml) is measured continuously in a spectrophotometer (e.g., at 450 nm), with intermittent addition of aliquots of concentrated detergent (e.g., 1 $\mu$l of 500 m$M$ OG). There can be either continuous or intermittent gentle stirring of the solution, and the optical density must be recorded only after stabilization. Concentrations of OG along the procedure are calculated after correction for dilution. Typical solubilization profiles are displayed in Fig. 2, illustrating the impact of liposome concentration on the solubilization profile. In these examples, the starting points of phospholipid solubilization are 18 m$M$ OG (PC at 0.25 mg/ml), 19 m$M$ OG (PC at 0.5 mg/ml), and 21 m$M$ OG (PC at 1 mg/ml). These concentrations are to be considered as starting points for determining the optimal detergent concentration for GPI–protein insertion, which is expected to be within a 1 m$M$ range around these values.

*Insertion of Lipoconjugates*

After the detergent saturation/phospholipid solubilization point is determined, the corresponding detergent concentration can be used for the insertion of GPI–proteins and other membrane components. The reconstitution in itself is a simple step, consisting of mixing the membrane components together with the detergent. The detergent is removed afterward by dialysis. Table I gives an example of a complex membrane reconstitution involving GPI–proteins, cholesterol, and lactocerebroside, a simple glycosphingolipid. Because lipids are not water soluble, it can be convenient to stabilize lipid compounds in aqueous stock solution, using detergent. In our example, cholesterol and lactocerebroside stock solutions contain 50 m$M$ OG. The detergent brought by these samples is taken into account when determining the total detergent concentration, which is adjusted by addition of OG stock solution. The sample is thereafter dialyzed in order to eliminate the detergent, which is why it is important to work with high-CMC detergents such as alkylglucosides. To facilitate dialysis, it is recommended that a high

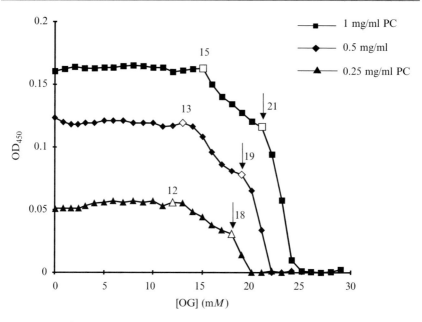

Fig. 2. Solubilization profiles of PC liposomes at various concentrations. PC liposomes (100 nm) are prepared as a 10-mg/ml suspension as described in text. Suspensions of PC (1, 0.5, and 0.25 mg/ml) are prepared, and the solubilization is performed by addition of 1-$\mu$l aliquots of 500 m$M$ OG, with gentle pipette stirring at each step. The turbidity is monitored by optical density at 450 nm, using 0.5-cm pathlength microcuvettes. We observe that all profiles are similar in shape, and that phase breaks vary with the different liposome concentrations. The arrows represent the phase breaks around which to investigate for the determination of the optimal OG concentration for GPI–protein incorporation.

size-exclusion material (e.g., MWCO 10,000) be used for a constant period of time (e.g., 12 h). The method of detergent-assisted membrane reconstitution as initially described did not involve lipidic membrane compounds other than phospholipids,[7,8] and the detergent was removed by incubation of the preparation with porous hydrophobic beads. We determined that the pores of beads such as the Amberlite XAD-2 are capable of binding some cholesterol but, moreover, a large amount of GPI–protein (Table II), thereby withdrawing some of the material from the liposomes. In addition, the mechanical action of the beads on the liposomes seems to be responsible for structural changes such as formation of multilamellar liposomes[21] and inside-out bilayer inversion.[9,10] As a consequence, hydrophobic beads or chromatographic steps must be avoided for removing detergent when oriented membranes are constructed with GPI–proteins.

[21] O. Lambert, D. Levy, J.-L. Ranck, G. Leblanc, and J.-L. Rigaud, *Biophys. J.* **74,** 918 (1998).

TABLE I

CONDITIONS OF INCUBATION OF COMPONENTS FOR MEMBRANE RECONSTITUTION[a]

|  | Stock solution | Volume | Quantity | Final concentration | Molar ratio | MW |
|---|---|---|---|---|---|---|
| Liposomes | 10 mg/ml | 100 $\mu$l | 1 mg | 1.3 m$M$ | 100 | 766 |
| Buffer |  | qsp 1 ml |  |  |  |  |
| Cholesterol | 0.52 mg/ml | 100 $\mu$l | 52 $\mu$g | 0.13 m$M$ | 10 | 400 |
| LacCer | 1.17 mg/ml | 50 $\mu$l | 58.5 $\mu$g | 0.065 m$M$ | 5 | 900 |
| GPI–protein | 50 U/ml | 100 $\mu$l | 5 U | 65 nM | 0.05 | 65,000 |
| OG | 100 m$M$ | qsp 20 m$M$ | 5.85 mg | 20 m$M$ | 1,540 | 292.4 |

*Abbreviations:* GPI, glycosylphosphatidylinositol; LacCer, lactocerebroside; OG, octyl-glucoside; gnf, quasi-static pressure.

[a] These conditions are designed for 100-nm, 1-mg/ml PC liposomes, in agreement with the solubilization profile previously determined (see Fig. 2), where optimal OG concentration is 20 m$M$. Cholesterol and lactocerebroside (LacCer) stock solutions are stabilized by 50 m$M$ OG, which is taken into account for total OG in the incubation medium. Sample components are added in the order given and mixed after each step by gentle pipette stirring. Under our conditions, the buffer is TBS (pH 8.5)–1 m$M$ MgCl$_2$ and the GPI–protein is bovine kidney alkaline phosphatase purified as described in text, with additional antibody affinity chromatography.[10]

TABLE II

YIELD OF RECOVERY OF MEMBRANE COMPONENTS AFTER
ELIMINATION OF DETERGENT BY TWO DIFFERENT METHODS[a]

|  | [$^3$H]PC (%) | GPI–protein (%) | [$^{14}$C]cholesterol |
|---|---|---|---|
| XAD-2 | 94 | 40 | 86 |
| Dialysis | 100 | 95 | 99 |

[a] Methods include incubation with Amberlite XAD-2 hydrophobic porous beads, and dialysis using 10,000 MWCO acetate cellulose dialysis membranes. The GPI–protein used in this experiment is purified bovine alkaline phosphatase and is enzymatically assayed as described in text.

## Characterization of Reconstituted Membranes

Only proper characterization of reconstituted membranes allows conclusions to be drawn about the experimental observations made with this material. In this respect, we can distinguish two sets of methods for characterizing reconstituted GPI–proteins: first, those determining the formation of proteoliposomes and, second, those giving information about the orientation of the bilayer.

*Did we obtain GPI–proteins inserted into liposome membranes?* The formation of stable lipoprotein complexes is monitored easily by centrifugation on density gradients. The principle of this technique is to separate low-density material, for example, lipid components that will "float" in a sucrose gradient, from high-density components, for example, proteins, which will remain at the bottom of the centrifugation tube. This method is used here to assess the association of GPI–protein with liposomes. Briefly, the sample is diluted 2-fold with the same buffer supplemented with 60% sucrose and Triton X-100 (0.1%, v/v), and is placed at the bottom of a centrifugation tube. A 20–0% discontinuous sucrose gradient is then added, with the most concentrated fractions close to the sample. The various solutions can be added one onto the other, if sufficient care is taken to avoid mixing the different sucrose concentrations. Alternatively, the solutions can be added in reverse order, using a device consisting of a long needle going directly to the bottom of the tube. During a centrifugation of 3 h at 160,000 $g$, the gradient linearizes and the liposomes migrate to a position usually corresponding to 10–15% sucrose. Fractionation of the contents of the tube is performed with a needle connected to a peristaltic pump and a fraction collector. The liposomes can be detected by turbidimetry at 650 nm, or by radioactivity detection if the PC is radiolabeled. Enzymatic assays or ELISAs allow quantification of the protein in the different fractions collected after density gradient centrifugation. The example shown in Fig. 3 illustrates the importance of temperature on membrane reconstitution. When using lipid components such as cholesterol, known to change the properties of lipid bilayers in a temperature-dependent manner,[22] the temperature of reconstitution must be adequate to reach stable association of membrane components. In our example (Fig. 3), ternary membrane complexes are formed with PC, cholesterol, and a GPI–protein, alkaline phosphatase, at 20° but not at 4°. Therefore, although the protein is more stable *in vitro* at 4°, the reconstitution with cholesterol necessitates operating at a higher temperature.

Alternatively, QELS is a nondestructive technique giving information about liposome diameter and polydispersity. In addition to being important for characterizing the liposomes used as the starting material for GPI–protein insertion, this method can also be used to check the final size of the vesicles. Moreover, low polydispersity of the liposome diameter is a tracer of spherical and homogeneous vesicles. In the example in Fig. 4, cholesterol slightly increases the apparent diameter of the liposomes and dramatically raises the polydispersity. In contrast, the GPI–protein slightly

---

[22] D. C. Lee and D. Chapman, *Symp. Soc. Exp. Biol.* **41,** 35 (1987).

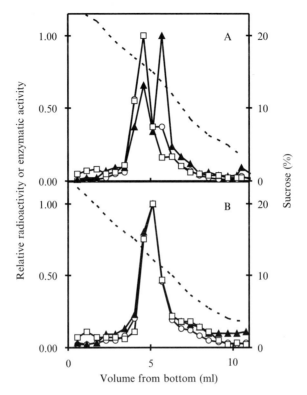

Fig. 3. Density gradient of unsuccessful and successful membrane reconstitutions. Ternary membrane complexes were prepared with PC, cholesterol, and a GPI–protein, bovine kidney alkaline phosphatase, according to the procedure described in text. The dialysis is performed at 4 or $20°$. The reconstituted membranes are analyzed by centrifugation on a density gradient. After centrifugation, the contents of the tube are fractionated, and aliquots are used for enzymatic activity and radioactivity determinations. The sucrose concentration is determined with a hand-held polarimeter. The values are ratioed to the maximal value in the gradient: [$^3$H]PC, open circles; [$^{14}$C]cholesterol, solid triangles; GPI–protein, open squares. For protein stability, $4°$ is more appropriate than $20°$, but the results show incomplete membrane incorporation at $4°$ (A), whereas incubation at $20°$ gives satisfactory cohesion between membrane components (B).

reduces the liposome mean diameter. The important increase in polydispersity most probably reflects a change in vesicle shape, which may indicate structure more complex than simple spheres.

*Are GPI–proteins preferentially inserted into the exoleaflet?* After focusing on liposome preparation (defining conditions for GPI–protein insertion, performing membrane reconstitution, and controlling macromolecular assembly of membrane components) comes the moment for

A                                          B

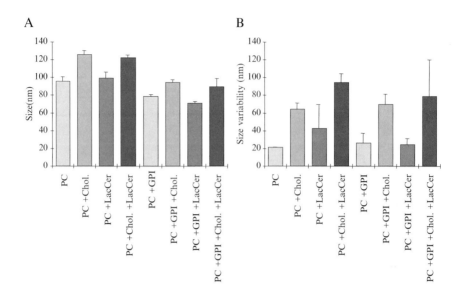

FIG. 4. Liposome size (A) and size variability (B) after reconstitution. Liposomes of various compositions are prepared according to the procedures described in text. The samples are analyzed in a QELS apparatus, and the resulting apparent particle diameter (A) and liposome population diameter variability (B) are presented. PC liposomes have the expected size (96 ± 22 nm) but, interestingly, incorporation of other membrane components leads to significant changes in apparent vesicle size, suggesting important morphological changes. Most dramatic changes are observed after the incorporation of cholesterol (Chol.), which increases liposome size and size polydispersity. On the other hand, the GPI–protein alkaline phosphatase (GPI) tends to slightly reduce the liposome size.

determining the orientation of the biomimetic membrane. With a classic transmembrane protein, membrane orientation can be determined by peptide profiling after proteolysis, or by antibody labeling followed by size-exclusion chromatography. The proteolysis of a GPI–protein is likely to be complete, because the peptidic structure is entirely accessible to an exogenous protease and this property can be used as a criterion for determining GPI–protein orientation. However, the above-described techniques are somewhat difficult to use for quantification purposes. In contrast, the GPI anchor has the interesting property of being hydrolyzed by phosphatidylinositol-specific phospholipase C (PIPLC). This enzyme, commonly prepared from *Bacillus thuringiensis* or *Bacillus cereus,* is available from reagent suppliers. PI-PLC cuts the anchor between the phosphate moiety and the diacylglycerol of the phosphatidylinositol, leading to liberation of the protein from the membrane, to which the diacylglycerol remains attached. Of course, the GPI is hydrolyzed only when it is accessible to the PI-PLC, that is, when the protein is inserted into the exoleaflet of the liposome membrane. Incubation of a GPI–protein with PI-PLC is usually performed at 30° for 1–2 h. The PI-PLC:GPI–protein ratio must be determined empirically, although for purified material it is reasonable to expect efficient GPI hydrolysis when using 0.5–1 U of PI-PLC per microgram of GPI–protein. The optimal pH for PI-PLC is pH 6–7, and cationic ions are neither activators nor inhibitors of the enzyme, although commercial preparations often include EDTA. By using centrifugation on a sucrose density gradient, it is easy to separate intact liposome-bound GPI–protein from water-soluble high-density protein without the GPI moiety. Figure 5 shows an example of liposomes containing a GPI–protein, alkaline phosphatase, incorporated with cholesterol into PC liposomes. Incubation of the ternary liposomes at 30° did not release any protein, whereas PI-PLC treatment released 90% of the protein, demonstrating its ectoplasmic location on liposomes. It is noteworthy, however, that not all GPI moieties are good PI-PLC substrates. Indeed, acylation of the inositol occasionally occurs *in vivo,* resulting in PI-PLC resistance,[23,24] and, in addition, glycosylations in the C-terminal region of the GPI–protein can sterically hinder the action of PI-PLC.[25] Therefore, it is highly recommended that the hydrolysis of GPI–protein in the absence of liposomes be controlled. GPI hydrolysis can be evidenced by anti-cross-reactive determinant (CRD) antibodies, directed against the cyclophosphate group that spontaneously forms on the inositol ring after PLC action.[25,26] Alternatively, phase partition experiments allow the quantification of GPI-anchored or free protein after PI-PLC incubation.[27] Interestingly, alkaline phosphatase is robust enough to remain active after SDS–PAGE under nonreducing conditions. Indeed, if the sample is not heated before electrophoresis, protein activity can be detected in gels by the use of 0.24 m$M$ 5-bromo-4-chloro-3-indolyl phosphate (BCIP) and 0.25 m$M$ nitroblue tetrazolium (NBT) in 100 m$M$ Tris-HCl (pH 9.6), 100 m$M$ NaCl, and 5 m$M$ MgCl$_2$.[10,28] The GPI anchor confers high hydrophobicity to the protein, which, hence, bears an excessive

[23] W. L. Roberts, J. J. Myher, A. Kuksis, M. G. Low, and T. L. Rosenberry, *J. Biol. Chem.* **263,** 18766 (1988).

[24] A. Stieger, M. L. Cardoso de Almeira, M.-C. Blatter, U. Brodbeck, and C. Bordier, *FEBS Lett.* **199,** 182 (1986).

[25] M. L. Guther, M. L. Cardoso de Almeida, T. L. Rosenberry, and M. A. J. Ferguson, *Anal. Biochem.* **219,** 249 (1994).

[26] S. E. Zamze, M. A. J. Ferguson, R. Collins, R. A. Dwek, and T. W. Rademacher, *Eur. J. Biochem.* **176,** 527 (1988).

[27] B. Seetharam, C. Tiruppathi, and D. H. Alpers, *Arch. Biochem. Biophys.* **253,** 189 (1987).

[28] I. Koyama, Y. Arai, M. Miura., H. Matsuzaki, Y. Sakagishi, and T. Komoda, *Clin. Chim. Acta* **139,** 139 (1988).

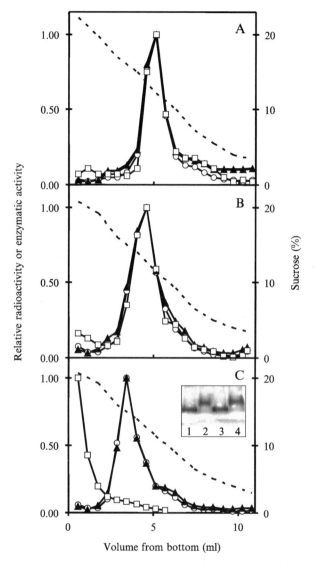

Fig. 5. Demonstration of ectoplasmic anchoring of a reconstituted GPI–protein. The sample is prepared and evaluated as described in the legend to Fig. 3B. To demonstrate the insertion of the GPI–protein in the outer leaflet of the liposomes, the vesicles are nontreated (A) or incubated for 60 min at $30°$ without (B) or with (C) PI-PLC (0.25 U/ml) before centrifugation on a sucrose density gradient. After treatment with PI-PLC, but not under control conditions, we observe the release of the GPI–protein, which concentrates at the bottom of the tube, corresponding to the highest densities. In addition, nondenaturing SDS–PAGE analysis is performed as described in text [C, inset: control protein in solution (lane 1);

amount of SDS on the acyl chains, leading to an increase in negative charge and a decreased apparent molecular weight. This property is convenient to control the PI-PLC hydrolysis of alkaline phosphatase GPI, either in solution or after membrane recontitution (Fig. 5C, inset). Another phospholipase can hydrolyze the GPI moiety, the GPI–phospholipase D, but its efficiency depends on membrane composition[29,30] and the enzyme is not commercialized to the same extent as PI-PLC.

Furthermore, it can be satisfying to demonstrate the conservation of coherent vesicles with an intravesicular milieu separated from the bulk solution by the liposome membrane. The most evident way to proceed is by the use of a small cation (e.g., 100 $\mu M$ Ca$^{2+}$), included in all buffers used to prepare liposomes and GPI–protein-containing liposomes. The final liposome suspension is then dialyzed in order to eliminate free cations, the entrapped population remaining inside the liposomes. The solubilization profile of the resulting liposome suspension is determined in the presence of a fluorescent calcium probe (e.g., 100 $\mu M$ Fura-2 or Quin-2) and monitored by fluorescence (e.g., $\lambda_{exc} = 340$ nm, $\lambda_{em} = 510$ nm for Fura-2; $\lambda_{exc} = 340$ nm, $\lambda_{em} = 490$ nm for Quin-2). If the insertion of GPI–proteins in liposomes results in intact vesicles with Ca$^{2+}$ encapsulated in the lumenal space, then the solubilization of the vesicles must lead to a progressive leakage of Ca$^{2+}$ from the liposomes, easily detected by the Ca$^{2+}$ probe. However, divalent calcium ions are known to induce fusion of phosphoethanolamine-containing liposomes, and for this reason calcium ions must be avoided in these studies. On the basis of a similar principle, we have developed an alternative method specific for alkaline phosphatase,[10] but that can be adapted to any GPI-anchored two-substrate enzyme. In this method, Mg$^{2+}$ ions are trapped into the liposomes during the preparation of the reconstituted membranes in Tris-buffered saline (TBS) supplemented with 1 m$M$ MgCl$_2$. Because alkaline phosphatase is a metalloenzyme depending on Mg$^{2+}$ for its activity, dialysis of the final preparation results in a loss of enzymatic activity, as measured under isotonic conditions. The progressive solubilization of these liposomes results in a release

---

protein in solution treated with PI-PLC (lane 2); control protein after liposome reconstitution and incubation at 30$^\circ$ (lane 3); reconstituted protein treated with PI-PLC (lane 4)]. The change in electrophoretic mobility after GPI hydrolysis is due to the removal of acyl chains, which bear several molecules of SDS. This results in an increase in apparent molecular weight. [$^3$H]PC, open circles; [$^{14}$C]cholesterol, solid triangles; protein, open squares.

---

[29] M. G. Low and K. O.-S. Huang, *Biochem. J.* **279,** 483 (1991).
[30] A. S. Bergman and S. R. Carlsson, *Biochem. J.* **298,** 661 (1994).

of magnesium ions, and a subsequent reactivation of the enzyme.[10] The same principle can be applied with any enzyme, using, for instance, its cosubstrate as the encapsulated material for tracing liposome integrity. Obviously, this method applies only to ectoplasmic reconstitutions of GPI-linked enzymes.

## Concluding Remarks

Membrane reconstitutions have the advantage of providing a biomimetic membrane environment with precise and defined composition. The reconstitution of GPI–proteins in lipid bilayers may seem to be an easy process: simple mixing of components with, or even without, detergent has a reasonable probability of resulting in some incorporation of the GPI anchors into the membranes. But we believe that this somewhat naive approach is by no means appropriate for the investigation of the behavior of membrane components, especially when considering ternary, quaternary, or more complex models. Therefore, for biochemical and biophysical approaches, extreme attention must be paid to the preparation of such membranes, in order to be able to draw pertinent conclusions. We propose here a set of methods for accurately building GPI–protein-containing liposomes. This methodology, although initially requiring some time, opens the way to preparing more complex membrane reconstitutions, with the inclusion of natural compounds of increasing complexity. These models can therefore become increasingly relevant to biological membranes, which, in the end, is the context in which we wish to understand the mechanisms of membrane dynamics.

## Acknowledgments

We are grateful to Marc le Maire for helpful comments during the preparation of this manuscript.

# Section II

Liposomes in Molecular Cell Biology

# [13]  Uptake of Liposomes by Cells: Experimental Procedures and Modeling

*By* SHLOMO NIR and JOSÉ L. NIEVA

## Introduction

Quantification of the kinetics of liposome uptake by cells is a first step toward understanding the mechanisms of uptake. It is also an essential prerequisite in the design of liposomes as drug carriers. The uptake of liposomes by cells is viewed as a sequence of two major steps: binding to sites on the plasma membrane, followed by entry into the cell.[1] The method of analysis, which is described in the second section, is applicable both for plain lipid vesicles and for targeted liposomes.[1a-3] The membranes of targeted liposomes include ligands that can bind with high affinity to particular receptors on the cell surface, for example, liposomes that include phosphatidylethanolamine (PE) lipid derivatives containing hyalurononan (HA) oligosaccharides.[3] These liposomes (HA-PE) were shown to target selectively to murine melanoma cells that overexpress the CD44 receptor, which binds to HA.[3] The mathematical model can be applied to a mode of liposome endocytosis via coated pits or via caveolae. We attempt to outline a search for experimental procedures for discriminating between these two modes of entry, which can coexist.

Clearly, understanding and quantitating the effect of certain drugs on cells require as accurate an estimate as possible of the kinetics of drug entry into the cells, rather than total uptake. Furthermore, it is anticipated that the effect of certain drugs on the cells may depend critically on their mode of entry, that is, via coated pits or via caveolae. Hence, it is advantageous to develop procedures that can estimate the fraction of cellular entry of liposomes via either mode. One of the points emphasized by the kinetic model is that the number of liposomes bound to the plasma membrane in the presence of inhibitors of cell entry can be severalfold larger than the number of liposomes bound when endocytosis operates. Consequently, estimating cellular entry of liposomes by the difference between their total uptake and their uptake in the presence of inhibitors of endocytosis is expected to yield an underestimate.

---

[1] N. Düzgüneş and S. Nir, *Adv. Drug Deliv. Rev.* **40**, 3 (1999).
[1a] R. J. Lee and P. S. Low, *J. Biol. Chem.* **269**, 3198 (1994).
[2] R. J. Lee and P. S. Low, *Biochim. Biophys. Acta* **1233**, 134 (1995).
[3] R. E. Eliaz and F. C. Szoka, Jr., *Cancer Res.* **61**, 2592 (2001).

Kinetic Analysis of Binding and Total Uptake of Liposomes by Cells

A kinetic analysis of liposome uptake by cells consists of the following sequence of steps.

1. Experimental determination of the kinetics and final extent of liposome binding to cells in the absence of liposome entry into the cells: This can be done at $4°$, or at higher temperatures in the presence of inhibitors of liposomal entry. The liposomes can be radiolabeled or fluorescently labeled with rhodamine-PE, or with other probes.
2. Application of a binding model to determine the number of "receptor sites" per cell and the rate constants of binding and dissociation: In Lee et al.,[4] the calculations considered one type of receptor site, whereas in Nir et al.[5] the treatment was extended to two types of sites. A receptor site on the cell can consist of one or several receptor molecules.
3. Experimental determination of the kinetics of total liposome uptake in the absence of inhibitors of liposomal entry: In this context, it is also suggested to apply inhibitors of endocytosis or entry via caveolae, and vice versa.
4. Application of model calculations that determine the rate constant(s) of liposome entry and simulate the experimental results: Model calculations were shown to yield predictions for liposome binding and uptake by cells.[4,5]

After studying liposome binding and uptake by cells of type A and then types B, C, and so on, the model calculations can estimate liposome uptake by each type of cell in a suspension containing several cell types.

Model Equations

*Equations for Competitive Binding of Particles to Several Types of Cells or Receptors in Absence of Endocytosis*

Let the molar concentration of cells of type $i$ ($i = 1, \ldots, n$) be $G_{0i}$. The molar concentration of total and free (unbound) liposomes will be denoted by $L_0$ and $L(t)$, respectively, where $t$ is the time of incubation. Initially, at time $t = 0$, $L(t) = L_0$. The molar concentration of cells of type $i$ with $I$ surface-bound particles is denoted $A_i(I)$. Mass conservation of liposomes is expressed as[6]

[4] K.-D. Lee, S. Nir, and D. Papahadjopoulos, *Biochemistry* **32,** 889 (1993).
[5] S. Nir, R. Peled, and K.-D. Lee, *Colloids Surf. A Physicochem. Eng. Aspects* **89,** 45 (1994).
[6] S. Nir, K. Klappe, and D. Hoekstra, *Biochemistry* **25,** 2155 (1986).

$$L_0 = L + \sum_{i=1}^{n} \sum_{L=1}^{N_i} A_i(I)I \tag{1}$$

in which $N_i$ is the largest number of particles that can bind to a cell of type $i$. Mass conservation equations for the cells are

$$G_{0i} = \sum_{I=1}^{N_i} A_i(I) \tag{2}$$

The binding of particles to each cell type is described by two parameters: $C_i(M^{-1}\ s^{-1})$, the rate constant of association or adhesion, and $D_i(s^{-1})$, the dissociation rate constant. The kinetics of adhesion are described by Eqs. (3) and (4):

$$dA_i(I)/dt = C_i L A_i(I-1)(N_i+1-I)/N_i + D_i A_i(I+1)(I+1)$$
$$-C_i L A_i(I)(N_i-I)/N_i - D_i A_i(I)I \tag{3}$$

$$dL/dt = -L \sum_{i=1} C_i \sum_{I=0} A_i(I)(N_i-I)/N_i + \sum_{i=1} D_i \sum_{I-1} A_i(I)I \tag{4}$$

The solution of these nonlinear differential equations is based on Taylor series expansion. A Fortran program is available[5] for the case of binding to $n = 2$, but the extension to any order is straightforward.

The application to the case of binding to $n$ types of independent receptors on the same cell requires that $G_{0i} = G_0$. An extension of these equations to the case in which liposome binding is followed by endocytosis is given in Refs. 4 and 5. When the details of the distribution of cells with respect to number of liposomes bound and endocytosed are not needed, a significant simplification can be introduced. Equations (1)–(4) are equivalent to a bimolecular reaction with the following substitutions[5]:

$$R_0 = G_0 N \tag{5}$$

and

$$C = cN \tag{6}$$

in which $R_0$ is the total molar concentration of receptor sites available for the binding of a liposome and $c$ is the association rate constant between a liposome and a hypothetical cell that includes a single binding site. We have used this simplification in solving the equations for binding to several types of receptors.

At equilibrium, Eqs. (1)–(6) yield

$$r = (L_0 - L)/G_0 = KL/(1 + KL/N) \tag{7}$$

where $r$ is the number of liposomes bound per cell. This equation may be rearranged to the form of a Scatchard plot,

$$r/L = K - rK/N \tag{8}$$

whose intercept and slope are $K$ and $K/N$, respectively, or

$$r/L = kN - rk \tag{9}$$

where $kN = K$, $k = c/D$, and $K = C/D$. In practice, Scatchard plots are drawn by implicitly assuming binding to a single type of receptor, whereas more types of receptor may exist. An initial guess can be made by splitting the Scatchard plots into two linear segments.[5] This provides initial values for $k_1$, $N_1$, $k_2$, and $N_2$. Then refinements are obtained by choosing the best fits. For small liposome concentrations, the important parameters are $k_1$ and $N_1$, which pertain to the high-affinity sites. Their values are close to the initial guess obtained from the linear segment of Scatchard plots corresponding to lower $r$ values.

The initial time points of the kinetics of binding provide the values of $C$. At this stage the results are less sensitive to $D$ values because the fraction bound is relatively small. The parameters $D_{1,2}$ are already fixed by

$$D_1 = C_1/K_1 \quad \text{and} \quad D_2 = C_2/K_2 \tag{10}$$

*Binding and Cell Entry*

In the presence of two types of binding sites, the simplified equations employ the relation

$$R_{0i} = G_0 N_i \tag{11}$$

Let $E_i$ denote the molar concentration of liposomes that have entered the cells via receptors of type $i$. In the following, we consider two types of receptors and correspondingly $\varepsilon_{1,2}$ will denote the rate constants of liposome entry, which can imply endocytosis and entry via caveolae, or two types of sites for endocytosis.[7] The total concentration of ingested liposomes is

$$E = E_1 + E_2 \tag{12}$$

$$dL/dt = -c_1 L R_1 - c_2 L R_2 + D_1(R_{01} - R_1) + D_2(R_{02} - R_2) \tag{13}$$

$$dE_1/dt = \varepsilon_1(R_{01} - R_1) \tag{14}$$

[7] I. Nunes-Correia, J. Ramalho-Santos, S. Nir, and M. C. Pedroso de Lima, *Biochemistry* **38,** 1095 (1999).

$$dE_2/dt = \varepsilon_2(R_{02} - R_2) \tag{15}$$

$$dR_1/dt = -c_1LR_1 + D_1(R_{01} - R_1) + \varepsilon_1(R_{01} - R_1) \tag{16}$$

$$dR_2/dt = -c_2LR_2 + D_2(R_{02} - R_2) + \varepsilon_2(R_{02} - R_2) \tag{17}$$

The mass conservation equation for liposomes is

$$(R_{01} - R_1) + (R_{02} - R_2) + E + L = L_0 \tag{18}$$

At $t = 0$, $L(0) = L_0$, $R_i(0) = R_{0i}$, and $E_i(0) = 0$. Hence, the first derivatives $dL/dt$, $dR_i/dt$, and $dE/dt$ in Eqs. (13)–(17) are known. The second and third derivatives of these equations are taken, followed by a Taylor expansion. When two (or more) types of cells are considered, the rate constants, $\varepsilon_i$, are characteristic of cells of type $i$.

*Kinetics of Binding and Endocytosis: Examples*

In Table I, we illustrate calculated results of total uptake of liposomes by cells. Two sets of cases are presented: (1) plain liposomes composed of phosphatidylcholine–cholesterol (PC–Chol, 2:1) and J774 (murine macrophage) cells[4] and (2) targeted liposomes (HA-PE) and B16F10 (murine melanoma) cells.[3] In both cases, the calculations are capable of yielding simulations and predictions of experimental results of binding and total uptake of liposomes. In the second case (Eliaz and Szoka[3]; R. E. Eliaz, S. Nir, and F. C. Szoka, Jr., unpublished results, 2002), work is still in progress. The emphasis here is on the analysis of a wide range of hypothetical conditions. In both sets of calculations, the number of cells and liposomes and the parameters describing liposome binding to the corresponding cells are fixed, but the values of $\varepsilon$, the rate constant of endocytosis, are varied from 0 (binding only; in the presence of inhibitors of endocytosis) to a relatively large value, $\varepsilon = 0.005$ s$^{-1}$, which somewhat exceeds the largest reported values.[4,7,8]

For $\varepsilon = 0.001$ s$^{-1}$, and times of incubation of 3 h, the total uptake is about three times more than the amount bound. The contribution of endocytosis to the total uptake becomes noticeable after 30 min. The calculations indicate that the uptake of the targeted liposomes by B16F10 cells is about 16 times larger than that of the PC–Chol (2:1) liposomes by J774 cells. However, the actual difference in the rate of uptake is about three orders of magnitude, because the number of J774 cells was $10^7$/ml, whereas the number of B16F10 cells was $2 \times 10^5$/ml.

[8] C. M. Waters, K. C. Oberg, G. Carpenter, and K. A. Overholser, *Biochemistry* **29**, 3563 (1990).

TABLE I
CALCULATED PERCENTAGE OF BINDING AND TOTAL UPTAKE FOR
TARGETED AND NONTARGETED LIPOSOMES

| Time (min) | Nontargeted liposomes[a] | | Targeted liposomes[b] | |
|---|---|---|---|---|
| | % total uptake | % bound | % total uptake | % bound |
| | $\varepsilon = 0$ | | | |
| 10 | 0.14 | 0.14 | 2.7 | 2.7 |
| 30 | 0.35 | 0.35 | 6.5 | 6.5 |
| 60 | 0.51 | 0.51 | 9.7 | 9.7 |
| 120 | 0.72 | 0.72 | 12.2 | 12.2 |
| 180 | 0.78 | 0.78 | 12.9 | 12.9 |
| | $\varepsilon = 0.0001\ \mathrm{s}^{-1}$ | | | |
| 10 | 0.14 | 0.13 | 2.7 | 2.6 |
| 30 | 0.35 | 0.32 | 6.6 | 6.0 |
| 60 | 0.58 | 0.47 | 10.3 | 8.4 |
| 120 | 0.87 | 0.58 | 15.0 | 9.7 |
| 180 | 1.10 | 0.60 | 18.5 | 9.7 |
| | $\varepsilon = 0.001\ \mathrm{s}^{-1}$ | | | |
| 10 | 0.14 | 0.10 | 2.7 | 2.0 |
| 30 | 0.38 | 0.17 | 7.3 | 3.2 |
| 60 | 0.73 | 0.19 | 13.3 | 3.3 |
| 120 | 1.40 | 0.19 | 24.3 | 2.9 |
| 180 | 2.1 | 0.19 | 33.9 | 2.6 |
| | $\varepsilon = 0.005\ \mathrm{s}^{-1}$ | | | |
| 10 | 0.144 | 0.05 | 2.8 | 0.9 |
| 30 | 0.430 | 0.05 | 8.1 | 0.9 |
| 60 | 0.850 | 0.05 | 15.5 | 0.8 |
| 120 | 1.700 | 0.05 | 28.7 | 0.7 |
| 180 | 2.500 | 0.05 | 39.8 | 0.6 |

[a] The lipid concentration was 10 $\mu M$, corresponding to a liposome concentration of $1.6 \times 10^{-10}M$. The number of cells was $10^7$/ml, which amounts to $G_0 = 1.67 \times 10^{-14}M$. The parameters used in the calculations were $N = 306$, $c = 4.9 \times 10^5 M^{-1}\ \mathrm{s}^{-1}$, $D = 2.3 \times 10^{-4}\ \mathrm{s}^{-1}$. The cells were J774 murine macrophages (parameters from Nir et al.[5]).

[b] The concentration of targeted liposomes was $2.08 \times 10^{-10}M$. The number of cells was $2 \times 10^5$/ml, or $G_0 = 3.33 \times 10^{-16}M$. The parameters used were $N = 3 \times 10^5$, $c = 5 \times 10^5 M^{-1}\ \mathrm{s}^{-1}$, and $D = 2.4 \times 10^{-4}\ \mathrm{s}^{-1}$. The cells were B16F10 murine melanoma cells.

The calculations indicate that after a certain period of incubation, whereas the percentage of total uptake has increased, the number (percentage) of liposomes bound to the plasma membrane has decreased severalfold below the value in the absence of endocytosis, 4- to 5-fold for

$\varepsilon = 0.001$ s$^{-1}$ and 16-to 21-fold for $\varepsilon = 0.005$ s$^{-1}$. This effect follows from Eqs. (13)–(17). In the absence of endocytosis, the equilibrium concentration of bound liposomes is reached (for one type of binding sites) when $CLR = D(R_0 - R)$, that is, $R_0 - R$, which is the concentration of bound liposomes equaling $CLR/D$, or $kLR$. However, in the presence of endocytosis, the concentration of bound liposomes is reduced at a rate of $\varepsilon(R_0 - R)$. Initially, the rate of endocytosis of liposomes is slower, because the number of bound liposomes is small. As the number of bound liposomes increases, the rate of endocytosis also increases, but the decrease in $R$, the number of unoccupied receptor sites, results in a decrease in the rate of liposome binding, which is given by $CL(t)R(t)$, and consequently the overall rate of uptake is reduced. The importance of this effect is minimal in the case of PC–Chol (2:1) liposomes (10 $\mu M$ lipid) and J774 cells in Table I, but it is more significant in the case of targeted liposomes.

In the case of nontargeted liposomes, the amount of bound liposomes reaches its maximal values after shorter periods of incubation than in the absence of endocytosis. In the case of targeted liposomes, the same effect is observed for $\varepsilon = 0.001$ s$^{-1}$, but for $\varepsilon = 0.001$ and 0.005 s$^{-1}$, the bound amount reaches a maximal value, which decreases after 60 or 10 min, respectively.

Lee et al.[4] deduced somewhat higher $\varepsilon$ values from (1) an analysis based on total liposome uptake according to Eqs. (12)–(18) than those obtained from (2) measurements of spectral pH-dependent changes of a fluorescent dye preencapsulated in the liposomes. The average 50% difference was within the estimated uncertainty, but the origin of this difference could stem from an implicit assumption in case 2 that the number of bound liposomes stayed relatively constant from 10 to 30 min. Table I indicates clearly that the deviations between deduced $\varepsilon$ values could be larger if the simulations cover larger incubation times.

In a study monitoring the uptake of albumin–Au (A–Au) particles by bovine lung microvascular cells,[9] it was found that the particles interact preferentially with caveolae. After a 30-min incubation, fewer A–Au particles were observed at the cell surface than after a 10-min incubation. As we pointed out (Table I), the model equations do predict this phenomenon without assuming slow recycling of the vesicles and/or A–Au receptors back to the cell surface. However, we did encounter with targeted liposomes (Eliaz and Szoka[3]; R. E. Eliaz, S. Nir, and F. C. Szoka, Jr., unpublished results, 2002), a plateau in their uptake after 3 h, which may reflect this effect. If further tests confirm this effect (which was also pointed out previously with targeted liposomes[1]), then the model equations will

[9] J. E. Schnitzer, P. Oh, E. Pinney, and J. Allard, J. Cell Biol. **127**, 1217 (1994).

need to be somewhat extended, albeit at the expense of adding one or two parameters whose introduction was not needed at times of incubation of up to 3 h.

### Experimental Procedures for Monitoring Liposome Uptake

We do not attempt to give here an exhaustive description of methods, but rather to present selected useful procedures devised to characterize quantitatively cell surface binding and endocytosis of liposomes.

*Endocytosis via Coated Pits.* Total uptake, or association of liposomes with cells as a consequence of surface binding and subsequent endocytosis via a coated pit/coated vesicle system, may be evaluated by several means. Invariably, all the procedures involve the incubation of cells with labeled liposomes under defined experimental conditions. For example, macrophage-like endocytotic cell lines have been incubated with radiolabeled liposomes in monolayers.[10] Cells are previously depleted of serum by rinsing with fetal bovine serum-free medium. Total uptake is measured in serum-free medium after incubating cells and radiolabeled liposomes at 37°. Monolayers are washed subsequently several times with cold phosphate-buffered saline (PBS) to remove unbound liposomes. Cells are finally digested in 0.5 N NaOH (2 h, 20°)[11–13] before determination of cell-associated radioactivity in a scintillation spectrometer. Simultaneous determination of cellular protein content allows the normalized expression of uptake as nanomoles of lipid per milligram of protein.[10,13]

Equivalently, fluorescently labeled liposomes can be used.[4,13–15] In Lee *et al.*[4] incubations are additionally carried out in cell suspensions ($10^6$ cells in 1 ml), and under continuous shaking. In this case, at defined incubation times unbound liposomes are removed by centrifugation. Mild centrifugation for 5 min at 1000 rpm (200 g) followed by resuspension in cold PBS (repeated twice) removes unbound liposomes from the cell suspension incubation medium. Before fluorescence determinations, cells in the monolayers are detached from the plastic substrate by trypsinization[13] or by incubation with EDTA[4,15] and subsequently counted with a hemacytometer. The amount of bound liposomes is determined routinely by measuring the amount of fluorescence associated with a defined number of

[10] C. Chu, J. Dijkstra, M. Lai, K. Hong, and F. C. Szoka, *Pharm. Res.* **7,** 824 (1990).
[11] J. Dijkstra, M. van Galen, and G. L. Scherphof, *Biochim. Biophys. Acta* **804,** 58 (1984).
[12] J. Dijkstra, M. van Galen, D. Regts, and G. L. Scherphof, *Eur. J. Biochem.* **148,** 391 (1985).
[13] I. Wrobel and D. Collins, *Biochim. Biophys. Acta* **1235,** 296 (1995).
[14] K.-D. Lee, K. Hong, and D. Papahadjopoulos, *Biochim. Biophys. Acta* **1103,** 185 (1992).
[15] C. R. Miller, B. Bondurant, S. D. McLean, K. A. McGovern, and D. F. O'Brien, *Biochemistry* **37,** 12875 (1998).

cells. Fluorescence intensity may be translated into lipid concentration, using a calibration curve, and cell–liposome association may be expressed as nanomoles of lipid per cell.[4,15]

Fluorometric determinations may also be carried out directly on cells grown on glass coverslips.[16] In this case, after the incubation of cells with liposomes and subsequent washouts, the coverslip must be inserted diagonally into the fluorometer cuvette oriented to minimize direct reflection of excitation light. These experiments usually monitor an intensive property of the fluorescent probe, which depends on the changing environmental conditions along the endocytotic pathway (see below).

Several procedures for liposome labeling with stable markers have been described. Radiolabeled liposomes have been produced that include in the lipid composition radioactive phospholipids such as [125]I-labeled p-hydroxybenzamidine dihexadecylphosphatidylethanolamine,[10] or phosphatidyl[[14]C]choline,[11,12,17] or cholesterol derivatives such as [[3]H]cholesterylhexadecyl ether.[13] Alternatively, radioactive aqueous markers such as metabolically inert [[3]H]inulin may be encapsulated inside vesicles.[10–12]

Fluorescently labeled vesicles may be obtained with lipidic probes such as N-(Lissamine rhodamine B sulfonyl)phosphatidylethanolamine (Rho-PE),[4,14,15] N-(7-nitro-benz-2-oxa-1,3-diazol-4-yl) phosphatidylethanolamine (NBD-PE),[18] or octadecylrhodamine B (R18).[19] A complementary procedure involves the use of aqueous markers such as fluorescein isothiocyanate (FITC)-dextrans[19] or pyranine-based 8-hydroxy-1,3, 6-pyrene-trisulfonate (HPTS).[17]

At the physiological temperature of $37°$ both surface-associated and endocytosed liposome fractions are quantitated. To estimate specifically the kinetics and final extent of binding to cell surfaces, the endocytosis process can be blocked by incubating labeled liposomes and cells at $4°$.[1,4,13] It should be emphasized that liposome binding to cells at $4°$ should not be the same as binding at $37°$ (with inhibitors), although it can be a useful indicator, in particular for examining whether a particular pretreatment of cells, for example, with inhibitors of endocytosis, could affect the number and function of receptors. In the previous section, we criticized the procedure of taking the difference of the two values for estimating liposome uptake at $37°$. An alternative procedure for the inhibition of endocytotic uptake involves incubations at physiological temperatures, in the presence of metabolic inhibitors[4,13,20] (see below).

[16] R. M. Straubinger, D. Papahadjopoulos, and K. Hong, *Biochemistry* **29**, 4929 (1990).

[17] B. Lundberg, K. Hong, and D. Papahadjopoulos, *Biochim. Biophys. Acta* **1149**, 305 (1993).

[18] K. Kono, T. Igawa, and T. Takagishi, *Biochim. Biophys. Acta* **1325**, 143 (1997).

[19] N. Higashi and J. Sunamoto, *Biochim. Biophys. Acta* **1243**, 386 (1995).

[20] S. Kessner, A. Krause, U. Rothe, and G. Bendas, *Biochim. Biophys. Acta* **1514**, 177 (2001).

The effectiveness of internalization inhibitors has been confirmed with HPTS-loaded liposomes.[4] HPTS was entrapped inside vesicles according to the procedure described by Straubinger et al.[16] Endocytosis may be monitored specifically from spectral changes occurring because of HPTS excitation dependence on changes in the pH.[4,14–16] HPTS maximal fluorescence emission at 510 nm increases with pH when excitation is at 450 nm and decreases when excitation is at 350–405 nm, whereas an isosbestic wavelength is observed to occur at 413 nm. Thus, measurements at invariant excitation of 413 nm allow direct quantitative estimation of total liposome uptake in a way similar to that described for the lipidic Rho-PE probe. The pH dependence of HPTS excitation is not affected by encapsulation of the probe into liposomes. The ratio of fluorescence measured at excitation wavelengths of 450 and 413 nm ($\gamma$ parameter) increases linearly within the pH 4.5–7.5 range. Using this parameter, it has been shown that 1 h after liposome addition most cell-associated liposomes are internalized and reside at a mean pH of ~6.5.

Thus, $\gamma$ may in principle be used as an indicator of the pH sensed by the probe while being endocytosed. The fraction of liposomes inside cells, out of the total cell-associated population, may then be calculated from the following expression:

$$\text{Fraction inside} = (\gamma_{\text{pH7.4}} - \gamma_{\text{measured}})/(\gamma_{\text{pH7.4}} - \gamma_{\text{pH6.5}})$$

where $\gamma_{\text{pH7.4}}$ is the fluorescence ratio at pH 7.4 (outside cells), $\gamma_{\text{measured}}$ is the fluorescence ratio measured at a defined liposome–cell incubation time (time = 1 h), and $\gamma_{\text{pH6.5}}$ is the ratio measured at pH 6.5 (inside cells at time = 1 h). Using this approach in combination with total uptake determinations, it was shown that incubation at low temperature, or incubation at 37° in the presence of ATP-depleting compounds, arrested endocytosis completely.[4]

In a complementary approach the internalized liposome fraction is estimated after stripping cells of their surface-bound liposomes through acidic saline washes.[1,20] The cells are washed several times with cold buffer as previously described (for removal of unbound liposomes) and then incubated with acid buffer (pH 3.0) for 3 min (acid wash to remove receptor-bound liposomes) before washing again with buffer twice. In this way the amount of fluorescence associated with cells reflects only endocytosed liposomes.

Kessner et al.[20] also describe a relatively new procedure to estimate directly liposome internalization based on the use of liposomes asymmetrically labeled in the outer monolayer with N-(7-nitro-benz-2-oxa-1,3-diazol-4-yl)phosphatidylethanolamine (NBD-PE). Quenching of liposomal NBD fluorescence may be achieved by chemically reducing NBD into the corresponding nonfluorescent derivative, using dithionite. Because dithionite cannot permeate the cell membrane, liposome quenching occurs only outside the cells. Therefore for internalization studies, cell monolayers incubated with liposomes are treated for 3 min with dithionite. After several washings, the NBD fluorescence associated with the cells is considered to represent the internalized fraction.

*Endocytosis via Caveolae.* An alternative route of internalization into cells that is likely to contribute to total liposome uptake is the caveolar system. Caveolae are omega-shaped invaginations of the plasma membrane with a diameter of 50–100 nm. These invaginations can detach from the plasma membrane to form noncoated plasmalemmal vesicles within the cell cytoplasm. Caveolae found in a variety of cells have been implicated in a number of important cellular functions including signal transduction, potocytosis, and the endocytotic and transcytotic movements of macromolecules.

The caveolar system has not been explored systematically as yet as a route of liposome internalization. In this regard it is worth mentioning here the results reported by Low and co-workers.[1a,2,21] These authors targeted liposomes to cells expressing folate receptors. Folate vitamin receptors must be clustered in caveolae for potocytosis of this compound.[22,23] Folate was covalently linked to lipid-conjugated PEG on the liposome surface.[1a] The obtained liposomes (mean diameter of 66 nm) were subsequently incubated with folate receptor-expressing cells. Quantitation of surface-bound and internalized liposomes by these authors yielded a total of $2.5 \times 10^5$ liposomes per cell after 4 h, without further increase, whereas $1.3 \times 10^5$ liposomes bound per cell at 4°. These values are comparable to, or even higher than, values recorded for endocytosis via coated pits.

The main advantages found for the use of this route is an increased affinity due to receptor clustering within caveolae and endocytosis into nonlysosomal compartments. However, a route from caveolae to endosomes has also been suggested to exist.[9,24,25]

[21] S. Wang, R. J. Lee, G. Cauchon, D. G. Gorenstein, and P. S. Low, *Proc. Natl. Acad. Sci. USA* **92**, 3318 (1995).

[22] E. J. Smart, D. C. Foster, Y. S. Ying, B. A. Kamen, and R. G. W. Anderson, *J. Cell Biol.* **124**, 307 (1994).

[23] E. J. Smart, C. Mineo, and R. G. W. Anderson, *J. Cell Biol.* **134**, 1169 (1996).

[24] B. F. Roettger, R. U. Rentsch, D. Pinon, E. Holicky, E. Hadac, J. M. Larkin, and L. J. Miller, *J. Cell Biol.* **128**, 1029 (1995).

[25] G. Raposo, I. Dunia, C. Delavier-Klutchko, S. Kaveri, A. D. Strosberg, and E. L. Benedetti, *Eur. J. Cell Biol.* **50**, 340 (1989).

*Inhibitors of Endocytosis via Coated Pits.* Inhibition of endocytotic uptake of liposomes in coated vesicles has been demonstrated to be promoted by ATP depletion.[26] A combination of inhibitors of oxidative metabolism (antimycin A or NaN$_3$) and glycolysis (NaF or 2-deoxy-D-glucose) inhibits liposome internalization[4,11,20] but not surface binding.[13] Restriction of the metabolic activity of cells inhibits both phagocytosis and coated pit-mediated endocytosis. Specific inhibition of the latter routes may be attained under defined incubation conditions. Hypertonicity and cell depletion of K$^+$ inhibit receptor-mediated endocytosis of liposomes but not actin-dependent phagocytosis.[20,27] Both treatments cause coated pits to disappear, and both induce abnormal clathrin polymerization into empty microcages.[28] Hypertonic incubation solutions are prepared by adding ~0.45 $M$ sucrose.[27,28] To deplete cells of intracellular K$^+$, cultures are incubated in K$^+$-free buffer that contains nigericin.[29] In contrast, microfilament-disrupting agents, such as cytochalasin D, which inhibits actin polymerization, may be used to selectively block liposome phagocytosis, but not endocytosis through the clathrin-coated pit/clathrin-coated vesicle system.[20,27] It should be noted, however, that actin microfilaments were reported to play a critical role in receptor-mediated endocytosis at the apical surface of polarized epithelial cells.[30]

Widely used inhibitors of endosomal acidification such as ionophores, weak bases, and inhibitors of v-ATPase do not interfere with internalization of liposomes from the cell surface. An illustrative example of the previous statement is the work by Wrobel and Collins.[13] These authors compared the effect of different inhibitors of endocytosis on binding and internalization of cationic liposomes by HepG2 cells. Binding at 4° was not affected by the presence of inhibitors. Uptake was increased several times at 37°, consistent with activation of endocytosis at physiological temperature. In the presence of the metabolic inhibitors azide plus 2-deoxy-D-glucose or under hypertonic conditions, the total uptake at 37° decreased to the levels of pure binding observed at 4°. By contrast, the use of monensin, an inhibitor of endosomal acidification, did not affect uptake under similar experimental conditions. These data indicate that inhibitors of endocytosis allow binding but not cellular entry of liposomes, and that uptake may proceed in the presence of inhibitors acting downstream in the endocytotic route.

[26] R. M. Straubinger, K. Hong, D. S. Friend, and D. Papahadjopoulos, *Cell* **32,** 1069 (1983).
[27] W. J. Muller, K. Zen, A. B. Fisher, and H. Shuman, *Am. J. Physiol.* **269,** L11 (1995).
[28] J. E. Heuser and R. G. W. Anderson, *J. Cell Biol.* **108,** 389 (1989).
[29] J. M. Larkin, M. S. Brown, J. L. Goldstein, and R. G. W. Anderson, *Cell* **33,** 273 (1983).
[30] T. A. Gottlieb, I. E. Ivanov, M. Adesnik, and D. D. Sabatini, *J. Cell Biol.* **120,** 695 (1993).

Another widely used inhibitor is $N$-ethylmaleimide (NEM),[31–33] which also inhibits endocytosis by caveolae and transcytosis. According to Anderson,[33] it is not clear whether tubular or vesicular caveolae ever fuse with endosomes from coated pits. It also remains to be tested whether the metabolic inhibitors that prevent endocytosis via coated pits cannot inhibit part of the uptake via caveolae, by which material is transported to endosomes.

*Inhibitors of Caveolae.* As mentioned previously, liposomes may also be internalized into cells in uncoated vesicles. Functioning of caveolae as an operative route of entry into cells was initially less established, at least in part until a method was found to inhibit selectively this putative pathway. Opportunities exist now to evaluate the specific contribution of this route to the total uptake of liposomes by cells. It has been shown that sterol-binding agents such as filipin, nystatin, or digitonin selectively inhibit caveolae-mediated intracellular and transcellular transport of selected molecules, while the clathrin-dependent pathway is not affected.[9] Also, activators of protein kinase C such as phorbol 12-myristate 13-acetate (PMA) are potent inhibitors of potocytosis,[22] suggesting that this compound inhibits uptake by inactivating caveolae internalization. Such selective inhibition may be useful for distinguishing uptake mediated by caveolae versus clathrin-coated vesicles.

For instance, using specific inhibitors it was concluded that caveolae constituted the portal for simian virus 40 (SV40) entry into cells.[34] SV40 infectious entry was not sensitive to treatments blocking coated pit-mediated endocytosis, but was blocked by treating cells with the phorbol ester PMA or nystatin. In one report[35] λ phage particles displayed an HIV-1. Tat peptide on their surfaces have been used to deliver genes into cells. Gene delivery is not affected by endosomotropic agents such as monensin or chloroquine but is partially impaired by inhibitors of caveolae formation such as filipin or nystatin, suggesting that Tat-phages may penetrate the plasma membrane through caveolae.

More systematic studies on the effect of caveolae inhibiting agents on liposome uptake are needed in order to evaluate the relevance and specific traits of the caveolar route of entry. Papadimitriou and Antimisiaris[36] have

[31] J. E. Schnitzer, J. Allard, and P. Oh, *Am. J. Physiol.* **268,** H48 (1995).

[32] D. Predescu, R. Horvart, S. Predescu, and G. E. Palade, *Proc. Natl. Acad. Sci. USA* **91,** 3014 (1994).

[33] R. G. W. Anderson, *Annu. Rev. Biochem.* **67,** 199 (1998).

[34] H. A. Anderson, Y. Chen, and L. C Norkin, *Mol. Biol. Cell* **7,** 1825 (1996).

[35] A. Eguchi, T. Akuta, H. Okuyama, T. Senda, H. Yokoi, H. Inokuchi, S. Fujita, T. Hayakawa, K. Takeda, M. Hasegawa, and M. Nakanishi, *J. Biol. Chem.* **276,** 26204 (2001).

[36] E. Papadimitriou and S. G. Antimisiaris, *J. Drug Target.* **8,** 335 (2000).

reported on the uptake of small unilamellar liposomes (mean diameter of about 80 nm) by human umbilical vein endothelial cells (HUVECs) or human promyelocytic leukemia cells (HL60). These authors found that internalization was clathrin independent, because it was not inhibited by sodium azide and deoxyglucose, but was affected by filipin. Consistent with these data, Kessner et al.[20] concluded that liposomes depleted of specific ligands may be internalized into HUVECs through caveolae.

In the previous section we reviewed results that indicated that widely used inhibitors of endosomal acidification do not interfere with internalization of liposomes via coated pits. Mineo and Anderson[37] reported that bafilomycin $A_1$ inhibited the uptake of 5-methyltetrahydrofolate into MA104 cells. These authors deduced that acidification can occur in plasmalemmal vesicles. Consequently, a measurement of intracellular pH by means of HPTS preencapsulated in liposomes may not necessarily constitute a proof for the residence of the liposomes in endosomes.

### Acknowledgments

J.L.N. was supported by DGCYT (Grant PB96-0171), the Basque Government (EX-1998-28; PI-1998-32), and the University of the Basque Country (UPV 042.310-EA085/97; UPV 042.310-G03/98).

[37] E. Mineo and R. G. W. Anderson, *Exp. Cell Res.* **224,** 237 (1996).

## [14]  Assembly of Endocytosis-Associated Proteins on Liposomes

*By* Markus R. Wenk and Pietro De Camilli

### Introduction

Synapses are intercellular junctions of a neuron with its target cells, and they are the sites of neurotransmission. At chemical synapses, depolarization of the synaptic plasma membrane triggers $Ca^{2+}$ entry into the presynaptic nerve terminal, which leads to fusion of synaptic vesicles with the plasma membrane (exocytosis) and release of their contents into the synaptic cleft. After exocytosis, synaptic vesicle membranes are retrieved and recycled to generate new neurotransmitter-filled vesicles for a subsequent round of release. A major route of synaptic vesicle membrane retrieval involves clathrin-mediated endocytosis.[1] The protein machinery underlying this endocytotic reaction has been characterized in detail. In addition to the major integral components of the clathrin coat, such as clathrin and the "classic" clathrin adaptor (AP-2), accessory factors (e.g., AP-180, amphiphysin, endophilin, eps-15, epsin, dynamin, Hip1R, intersectin, and synaptojanin) participate in coat assembly, pit formation, and vesicle fission.[2–4]

Cell-free reconstitution assays utilizing chemically defined liposomes (rather than biological membranes) as donor membranes, and cytosolic fractions or purified proteins as ligands, have been developed in order to identify and/or characterize the molecular requirements for vesicle biogenesis, including the biogenesis of clathrin-coated vesicles. Cell-free complete or partial reconstitutions have been reported for COPI-, COPII-, and clathrin-mediated vesicle budding,[5–10] as well as for membrane tubulation reactions often observed *in vivo* at sites of vesicle budding.[10–12]

All these studies have emphasized the critical role of protein–lipid interfaces in the formation of tubulovesicular elements within the cell. Protein binding to lipids often involves both ionic interactions with the head groups of phospholipids, as well as hydrophobic interactions with the nonpolar domain of the membrane, and with specific lipids such as cholesterol and fatty acids.[13] In addition, growing evidence indicates that lipids that can undergo rapid and reversible changes within the membrane play an important role in regulating the recruitment of cytosolic factors to

[1] P. De Camilli, V. I. Slepnev, O. Shupliakov, and L. Brodin, *in* "Synapses" (M. W. Cowan, T. C. Sudhof, and C. F. Stevens, eds.), p. 217. Johns Hopkins University Press, Baltimore, MD, 2000.

[2] M. Marsh and H. T. McMahon, *Science* **285,** 215 (1999).

[3] P. De Camilli, H. Chen, J. Hyman, E. Panepucci, A. Bateman, and A. T. Brunger, *FEBS Lett.* **513,** 11 (2002).

[4] V. I. Slepnev and P. De Camilli, *Nat. Rev.* **1,** 161 (2000).

[5] A. Spang, K. Matsuoka, S. Hamamoto, R. Schekman, and L. Orci, *Proc. Natl. Acad. Sci. USA* **95,** 11199 (1998).

[6] K. Matsuoka, L. Orci, M. Amherdt, S. Y. Bednarek, S. Hamamoto, R. Schekman, and T. Yeung, *Cell* **93,** 263 (1998).

[7] M. Bremser, W. Nickel, M. Schweikert, M. Ravazzola, M. Amherdt, C. A. Hughes, T. H. Sollner, J. E. Rothman, and F. T. Wieland, *Cell* **96,** 495 (1999).

[8] M. Kinuta, H. Yamada, T. Abe, M. Watanabe, S. A. Li, A. Kamitani, T. Yasuda, T. Matsukawa, H. Kumon, and K. Takei, *Proc. Natl. Acad. Sci. USA* **99,** 2842 (2002).

[9] M. G. Ford, B. M. Pearse, M. K. Higgins, Y. Vallis, D. J. Owen, A. Gibson, C. R. Hopkins, P. R. Evans, and H. T. McMahon, *Science* **291,** 1051 (2001).

[10] K. Takei, V. Haucke, V. Slepnev, K. Farsad, M. Salazar, H. Chen, and P. De Camilli, *Cell* **94,** 131 (1998).

[11] S. M. Sweitzer and J. E. Hinshaw, *Cell* **93,** 1021 (1998).

[12] K. Takei, V. I. Slepnev, V. Haucke, and P. De Camilli, *Nat. Cell Biol.* **1,** 33 (1999).

[13] W. B. Huttner and A. Schmidt, *Curr. Opin. Neurobiol.* **10,** 543 (2000).

the bilayer.[14] Among such lipids, phosphoinositides (phosphorylated derivatives of phosphatidylinositol) are of critical importance and are best characterized. Rapid phosphorylation–dephosphorylation at different positions of the inositol ring by a multiplicity of enzymes generates a variety of stereoisomers, and phosphoinositides are now believed to be key players in the regulation of clathrin coat recruitment and dynamics.[15,16]

Although cytosolic proteins alone can generate from liposomes, membrane buds, and tubules that are similar to the different types of coated vesicle buds and tubules observed in the cell, integral membrane proteins often act as receptors for coat proteins and their accessory factors. In doing so, they contribute to the regulation of coating in time and space in the living cytoplasm. Thus, a synergistic effect of both membrane components, lipids and proteins, is central to the regulation of assembly reactions on membrane surfaces.

Here we describe the methodology of a cell-free system utilizing liposomes and cytosol to study assembly reactions of proteins involved in clathrin-mediated endocytosis. Liposomes, when incubated with brain cytosol and nucleotides, undergo morphological changes and fragmentation into smaller structures that resemble closely intermediates of clathrin-mediated endocytosis[8,10] (Fig. 1). Biochemical and biophysical analysis of these reactions show that elevated levels of phosphatidylinositol 4,5-bisphosphate [$PI(4,5)P_2$, or $PIP_2$] correlate with increased recruitment of clathrin and clathrin adaptors to the liposomal surface[17,18] (Fig. 2). Furthermore, inclusion into the liposomes of a surface-exposed recombinant fragment of synaptotagmin enhances the recruitment of clathrin coat proteins, demonstrating the synergistic actions of proteins and lipids[19] (Fig. 3). Three modes of analysis are applied: (1) observation of liposomal structure by electron microscopy, (2) protein biochemistry to assess amounts of endocytotic proteins recruited to the liposomal surface, and (3) lipid biochemistry to measure the metabolism of phosphoinositides during the incubation reaction.

[14] J. H. Hurley and T. Meyer, *Curr. Opin. Cell Biol.* **13**, 146 (2001).

[15] O. Cremona and P. De Camilli, *J. Cell Sci.* **114**, 1041 (2001).

[16] T. F. Martin, *Curr. Opin. Cell Biol.* **13**, 493 (2001).

[17] O. Cremona, G. Di Paolo, M. R. Wenk, A. Luthi, W. T. Kim, K. Takei, L. Daniell, Y. Nemoto, S. B. Shears, R. A. Flavell, D. A. McCormick, and P. De Camilli, *Cell* **99**, 179 (1999).

[18] M. R. Wenk, L. Pellegrini, V. A. Klenchin, G. Di Paolo, S. Chang, L. Daniell, M. Arioka, T. F. Martin, and P. De Camilli, *Neuron* **32**, 79 (2001).

[19] V. Haucke, M. R. Wenk, E. R. Chapman, K. Farsad, and P. De Camilli, *EMBO J.* **19**, 6011 (2000).

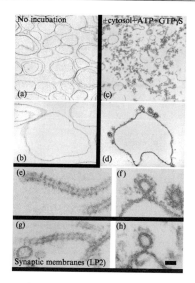

FIG. 1. Electron micrographs demonstrating the effect of brain cytosol on liposomes composed of brain lipids. Preparations were analyzed after plastic embedding and thin sectioning. (a and b) Control liposomes not incubated with cytosol. (c–f) Liposomes incubated with rat brain cytosol, ATP, and GTPγS. High-power observation reveals the presence of dynamin-like rings and clathrin-coated profiles (d–f). Fields (e)–(h) demonstrate the similarity of dynamin-coated tubules and clathrin-coated pits observed on liposomes (e and f) and synaptic membranes (g and h) incubated with brain cytosol, ATP, and GTPγS. Calibration bar: 500 nm in (a) and (c); 150 nm in (b) and (d); and 50 nm in (e)–(h). Reprinted from *Cell,* Vol. 94: K. Takei, V. Haucke, V. Slepnev, K. Farsad, M. Salazar, H. Chen, and P. De Camilli, Generation of coated intermediates of clathrin-mediated endocytosis on protein-free liposomes, pp. 131–141. Copyright (1998), with permission from Elsevier Inc.[10]

Reconstitution assays such as this one are a powerful tool with which to study the molecular events that are required for the formation as well as the dynamics of protein coats on membrane surfaces. We have used this integrated approach to detect alterations in phosphoinositide metabolism and clathrin coat dynamics in cytosolic fractions prepared from knockout mice.[17,18,20] Furthermore, purified proteins, such as amphiphysin and dynamin,[12,21] endophilin,[22] or phosphatidylinositol-4-phosphate 5-kinase (PIP

[20] G. Di Paolo, S. Sankaranarayanan, M. R. Wenk, L. Daniell, E. Perucco, B. J. Caldarone, R. Flavell, M. Picciotto, T. A. Ryan, O. Cremona, and P. De Camilli, *Neuron* **33,** 789 (2002).
[21] K. Takei, V. I. Slepnev, and P. De Camilli, *Methods Enzymol.* **329,** 478 (2001).
[22] K. Farsad, N. Ringstad, K. Takei, S. R. Floyd, K. Rose, and P. De Camilli, *J. Cell Biol.* **155,** 193 (2001).

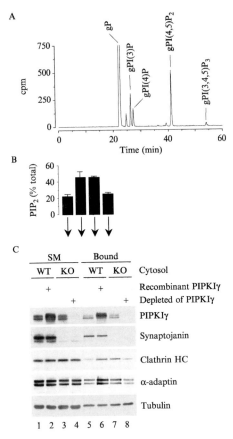

Fig. 2. Phosphoinositide metabolism and clathrin coat recruitment to liposomes. Liposomes composed of brain lipids are incubated with brain cytosol, $[\gamma\text{-}^{32}P]ATP$, and GTP and analyzed for incorporation of phosphate into phosphoinositides (A and B) and for binding of clathrin coat proteins to the liposomal surface (C). After the reaction, phosphoinositides are extracted and deacylated, and glycerophosphoinositols are separated by HPLC. A typical chromatogram is shown in (A). (B) Quantification of $PI(4,5)P_2$ levels in an experiment utilizing different sources of cytosol as indicated in (C) [knockout (KO) and wild-type (WT) denote cytosolic fractions from synaptojanin 1 knockout mice and control litter mates, respectively]. These cytosolic fractions were used in lanes 5 and 7 and were supplemented (lane 6) or depleted (lane 8) of PIP kinase I$\gamma$, the major $PIP_2$-synthesizing enzyme in brain cytosol. Levels of $PIP_2$ generated during the incubation of liposomes with cytosol (B) correlate nicely with the amount of clathrin coat that is recruited to the liposomes [elevated levels of clathrin heavy chain (HC) and $\alpha$-adaptin in lanes 6 and 7 compared with control incubations (lanes 5 and 8)]. A control protein (tubulin) does not show any significant variations between the different incubation conditions. Reprinted from *Neuron,* Vol. 32: M. R. Wenk, L. Pellegrini, V. A. Klenchin, G. Di Paolo, S. Chang, L. Daniell, M. Arioka, T. F. Martin, and P. De Camilli, PIP kinase I$\gamma$ is the major $PI(4,5)P_2$ synthesizing enzyme at the synapse, pp. 79–88. Copyright (2001), with permission from Elsevier Inc.[18]

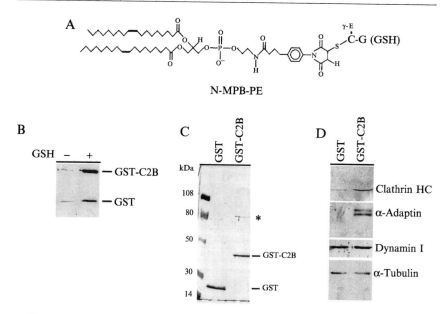

FIG. 3. Liposomes containing glutathione-linked GST fusion proteins. (A) Chemical structure of glutathione (GSH)-linked N-MPB-PE. (B) High-affinity binding of GST fusion proteins to liposomes containing GSH-PE (+) compared with control liposomes devoid of GSH-PE (−). (C) Liposomes with GSH-PE in their membranes were incubated with GST or GST-synaptotagmin I C2B (GST-C2B) proteins to allow coupling of GST fusion proteins to the liposomal membrane via GSH-PE. An aliquot (50 $\mu$g of total lipid) was then analyzed by SDS–PAGE and Coomassie blue staining. (D) Liposomes containing glutathione-PE-linked GST or GST-C2B (C) were incubated with rat brain cytosol, reisolated, and washed, and bound proteins were analyzed by SDS–PAGE and immunoblotting. The presence of GST-C2B on the liposomal surface greatly facilitates clathrin/AP-2 recruitment, but has no effect on the binding of tubulin or dynamin I. Reprinted from *EMBO J.*, Vol. 19: V. Haucke, M. R. Wenk, E. R. Chapman, K. Farsad, and P. De Camilli, Dual interaction of synaptotagmin with $\mu$2- and $\alpha$-adaptin facilitates clathrin-coated pit nucleation, pp. 6011–6019. By permission of Oxford University Press.[19]

kinase)[18] can be used in this system in addition to, or instead of, cytosolic fractions. Functionalized liposomes, such as the proteoliposomes described here, are ideally suited to investigate the synergistic effect of membrane lipids and membrane proteins on coat structure and function. They are also a useful tool for future applications in the field of liposome research, such as targeting for drug delivery or the development of liposome-based affinity probes for proteins and lipids.

Methods

## Preparation of Liposomes

*Pure Lipid Vesicles.* Large unilamellar vesicles are prepared as described[23] with some modifications. A defined amount of lipid (4 mg) [bovine brain extract, type I Folch fraction I; Sigma, St. Louis, MO; stored as a 20-mg/ml stock solution in chloroform–methanol (1:2) at $-20°$] is transferred to a 10-ml borosilicate glass tube and spiked with 0.05% (w/w) NBD-PC {1-palmitoyl-2-[12-[(7-nitro-2-1,3-benzoxadiazol-4-yl) amino]dodecanoyl]-*sn*-glycero-3-phosphocholine; Avanti Polar Lipids, Alabaster, AL} or, alternatively, [³H]PC (L-α-dipalmitoyl-[2-palmitoyl-9,10-³H($N$)]-phosphatidylcholine) (PerkinElmer Life Sciences, Boston, MA) at 0.05 $\mu$Ci/mg lipid. The low amounts of fluorescent or radioactive lipids are used for quantification of recovery after reisolation of the liposomes from the incubation reaction (see below). The lipid solution is then dried under a gentle stream of nitrogen. A thin and even film of lipid is formed on the glass wall by slowly (manually) rotating the glass tube during evaporation of the organic solvent. The lipid is further dried in a desiccator over phosphorous pentoxide (Sigma) and high vacuum for 2 h. The dried lipid film is then hydrated with a gentle stream of water-saturated nitrogen until it loses some of its opalescent appearance. A defined amount of 0.3 $M$ sucrose (in water) (2 ml) is gently added to the test tube, which is flushed with nitrogen and sealed. Liposomes are allowed to form spontaneously for 2 h at 37°. The suspension is then agitated gently to resuspend the liposomes, and large aggregates and debris are removed by brief centrifugation. Thus obtained liposomes (final lipid concentration, 2 mg/ml) are mostly unilamellar spheres with diameters of >500 nm (Fig. 1a and b), which can be stored at 4° for several days. Instead of a lipid extract from a biological tissue, such as the total brain lipids described above, chemically defined components (i.e., synthetic lipids) can be used to generate these liposomes. For that aim, individual lipids are mixed in organic solvent at a defined ratio and then dried and processed as described above. The effect of the individual lipid species on protein recruitment can hence be studied.[10]

*Lipoprotein-Containing Liposomes.* Recombinant glutathione S-transferase (GST) fusion proteins can be chemically linked to a glutathione (GSH)-derivatized phosphatidylethanolamine (GSH-PE)[7,19] (Fig. 3A). The latter is synthesized by mixing equal volumes (2 ml each) of [4-(*p*-maleimidophenyl)butyryl] phosphatidylethanolamine (MPB-PE, 5 mg/ml in

[23] J. P. Reeves and R. M. Dowben, *J. Cell. Physiol.* **73**, 49 (1969).

CHCl$_3$; Avanti; Polar Lipids) and glutathione [5 m$M$ in dimethylforma-mide (DMF); Sigma] supplemented with 10 m$M$ morpholinepropanesul Fonic acid (MOPS), pH 7.5. After incubation for 2 h at room temperature, free maleimido residues are blocked with 10 m$M$ 2-mercaptoethanol, and the mixture is further incubated for 30 min at room temperature. The GSH-PE lipid is extracted by adding 4 ml of chloroform and 2 ml of methanol to the mixture, followed by vortexing and brief centrifugation to separate phases. The lower organic phase is then carefully removed, the aqueous phase is reextracted twice with 2 ml of chloroform, and the organic phases are pooled, dried, and resuspended in chloroform–methanol (2:1, v/v). Lipids (bovine brain extract; see above) and GSH-PE are mixed in chloroform at a ratio of 8:2 (w/w) and 0.05% (w/w) NBD-PE {1-palmitoyl-2-[12-[(7-nitro-2-1,3-benzoxadiazol-4-yl)amino]dodeca-noyl]-$sn$-glycero-3-phosphatidylethanolamine; Avanti Polar Lipids} is added as a fluorescent marker for quantification. The lipid film (5 mg) is dried under a stream of nitrogen and resuspended in 0.5 ml of 0.6 $M$ sucrose [in phosphate-buffered saline (PBS)] by vigorously vortexing the suspension. The resulting multilamellar liposomes are then filtered five times through polycarbonate filters (pore size, 400 nm), using an extruder system (Liposofast; Avestin, Ottawa, ON, Canada) to form unilamellar liposomes. The nonencapsulated sucrose is removed by washing the lipo-somes in 3 ml of PBS followed by centrifugation for 20 min at 43,600 $g$ [35,000 rpm in a Beckman Coulter (Fusterton, CA) TLA 100.2 rotor] and resuspension in 0.5 ml of PBS (final lipid concentration, 10 mg/ml).

GSH-PE-containing liposomes are then coupled to GST or a GST fusion protein (GST-C2B domain of synaptotagmin) by incubation (200 $\mu$g/ml of lipid) with protein (40 $\mu$g/ml) in PBS for 4 h at 4°. An aliquot (50 $\mu$g of total lipid) is analyzed by sodium dodecyl sulfate–polyacrylamide gel electrophoresis (SDS–PAGE) (Fig. 3B). As expected, the presence of glutathione greatly enhances the binding of GST or GST-C2B to liposomes (Fig. 3B). Liposomes containing similar amounts of either GST or GST-C2B (Fig. 3C) are then incubated (0.5 mg/ml; 50 $\mu$g of total lipid) with brain cytosol (2.5 mg/ml) for 10 min at 37° to assess the ability of the synaptotagmin C2B domain to recruit clathrin and AP-2. After the reac-tion, liposomes are reisolated by centrifugation (20 min at 35,000 rpm in a Beckman Coulter TLA 100.2 rotor) and washed (once in 1 ml of PBS), and bound proteins are analyzed by SDS–PAGE and Western blotting (Fig. 3D). Liposomes carrying the C2B domain are much more efficient in the recruitment of AP-2 and clathrin from cytosol to the liposome surface (Fig. 3D).

*Preparation of Cytosol*

Cytosol is prepared from fresh or frozen rat brains by high-speed centrifugation followed by a desalting step and ammonium sulfate precipitation to concentrate the sample.[10] The brainstem is removed from 40 rat brains, and the brains are rinsed in breaking buffer (BB) [500 m$M$ KCl, 10 m$M$ MgCl$_2$, 250 m$M$ sucrose, 25 m$M$ Tris-HCl (pH 8.0), 1 m$M$ dithiothreitol (DTT), 2 m$M$ EGTA, protease inhibitors (complete EDTA-free tablets; Roche Diagnostics, Mannheim, Germany)] and minced coarsely into small pieces with a razor blade. The tissue is then homogenized in 100 ml of breaking buffer with a Polytron blender (Ultra Turrax; Janke & Kunkel, Staufen, Germany) at medium speed for 1 min, and centrifuged for 60 min at 9000 $g$ (8700 rpm in a Beckman Coulter SS34 rotor). The supernatant is recovered and centrifuged at 184,000 $g$ (50,000 rpm in a Beckman Coulter 70Ti rotor) for 2 h. The supernatant is then dialyzed two times for 2 h at 4° against 4 liters of dialysis buffer (DB) [50 m$M$ KCl, 25 m$M$ Tris-HCl (pH 8.0), 1 m$M$ DTT], using Spectra/Por 3 membranes (Spectrum, Gardena, CA). Ammonium sulfate [(NH$_4$)$_2$SO$_4$] is added slowly to 60% saturation (over a duration of approximately 30 min), and the solution is stirred for an additional 30 min. The precipitate is recovered by centrifugation at 9000 $g$ for 30 min, resuspended in 13 ml of buffer from the second dialysis step, and dialyzed for another 2 h against 4 liters of the same buffer, followed by 2 h against DB (4 liters) without DTT. Finally, the cytosol is centrifuged at 100,000 $g$ (47,000 rpm in a 70Ti rotor) for 2 h and recovered in 15 ml of DB (final protein concentration is approximately 20 mg/ml), aliquoted (1.5-ml Eppendorf tubes), and snap-frozen by dropping the tubes into liquid nitrogen. Samples can be stored for several months at −70°.

*Incubation of Liposomes with Cytosol and Nucleotides*

A common incubation protocol that allows for morphological and biochemical analysis is presented here: liposomes, brain cytosol, nucleotides, and a stock (10×) of cytosolic buffer (CB) {final concentrations after dilution: 25 m$M$ HEPES-KOH (pH 7.4), 25 m$M$ KCl, 2.5 m$M$ magnesium acetate [Mg(CH$_3$COO)$_2$], 150 m$M$ potassium glutamate, 10 $\mu M$ Ca$^{2+}$} are mixed in 1.5-ml Eppendorf tubes to obtain the following final concentrations: liposomes (0.1 mg/ml), cytosol (4 mg/ml), GTP (200 $\mu M$), ATP-regenerating system [ATP (2 m$M$), phosphocreatine (17 m$M$; Sigma), creatine phosphokinase (17 units/ml; Sigma)]. For lipid and protein biochemistry (Figs. 2 and 3), total reaction volumes of 100 and 400 $\mu l$, respectively, are sufficient, whereas larger amounts (1.5 ml) are needed for plastic embedding and thin sectioning (Fig. 1). The samples are incubated for 15 min at 37° and processed as described below.

## Analysis of Bound Protein by SDS–PAGE and Western Blotting

At the end of incubation, liposomes are loaded onto a step gradient (600 $\mu$l of 0.5 $M$ sucrose over a cushion of 180 $\mu$l of 2 $M$ sucrose) and centrifuged for 30 min at 150,000 $g$ (in a Beckman Coulter TLA 100.2 rotor) at 4° to separate bound from unbound protein. Liposomes are recovered from the interface of the step gradient, washed once with 400 $\mu$l of CB, and resuspended in 200 $\mu$l of CB. The recovery of liposomes is assessed by measuring NBD fluorescence ($\lambda_{ex}$ 360 nm, $\lambda_{em}$ 430 nm) of an aliquot (10 $\mu$l), using a spectrofluorometer. If a radiolabeled lipid tracer ([$^3$H]PC) is used, a small aliquot (10 $\mu$l) is measured by liquid scintillation counting. On the basis of these measurements, equal amounts of lipid are then loaded for SDS–PAGE and Western blot analysis. Endocytosis-associated proteins are detected with the following primary antibodies: clathrin HC [American Type Culture Collection (ATCC), Manassas, VA], $\alpha$-adaptin (AP-2) (Affinity BioReagents, Golden, CO), dynamin 1 (Transduction Laboratories, Lexington, KY), and tubulin 1 (Sigma), which serves as a control (Figs. 2 and 3). For semiquantitative purposes, secondary antibodies conjugated to horseradish peroxidase in combination with a chemiluminescence kit (SuperSignal West Pico peroxide; Pierce, Rockford, IL) can be used (Fig. 3D). A more reliable and more quantitative analysis, however, requires detection by secondary antibodies conjugated to [125]I, followed by quantitative autoradiography with a Storm PhosphoImager (Molecular Dynamics-Amersham Biosciences, Piscataway, NJ).[24]

## Electron Microscopy

Incubation mixtures are fixed in suspension by addition of an equal volume of 2× fixative [final concentration, 3% paraformaldehyde (Electron Microscopy Sciences, Fort Washington, PA) (freshly prepared from powder), 2% glutaraldehyde in 50 m$M$ HEPES-KOH (pH 7.4)] for 30 min, and liposomes are pelleted by centrifugation in a Beckman Coulter TLA 100.3 rotor at 50,000 rpm for 10 min. The pellets are washed in 0.1 $M$ sodium cacodylate (Electron Microscopy Sciences) buffer, pH 7.3, postfixed with 1% OsO$_4$ in the same buffer for 1 h, and washed three times in buffer. They are then *en bloc* stained with 2% aqueous uranyl acetate (Electron Microscopy Sciences) for 1 h in the dark and dehydrated with increasing concentrations of ethanol (once each with 10, 50, 70, and 95%, and three times with 100%, for 10 min each), followed by substitution with

[24] H. Gad, N. Ringstad, P. Low, O. Kjaerulff, J. Gustafsson, M. Wenk, G. Di Paolo, Y. Nemoto, J. Crun, M. H. Ellisman, P. De Camilli, O. Shupliakov, and L. Brodin, *Neuron* **27**, 301 (2000).

propylene oxide (three times for 10 min each). The samples are embedded in Epon (Electron Microscopy Sciences) by incubation in a mixture of Epon–propylene oxide (1:1, v/v) in a rotating mixer at approximately 8 rpm (Pelco R2; Ted Palla, Redding, CA), uncapped and overnight. This Epon is then replaced with fresh Epon, and incubation is continued for several hours. The pellets are embedded in fresh Epon and baked for 48 h at 60° in Eppendorf tubes. Blocks are then trimmed and thin-sectioned at 40 nm, using a microtome (Leica Ultracut; Leica, Wetzlar, Germany). Sections are stained on a grid with 2% aqueous uranyl acetate followed by lead citrate (0.4% in water) and observed at 80 kV in a transmission electron microscope. For some samples, $OsO_4$ postfixation is followed by impregnation with 1% tannic acid in distilled water to enhance visualization of coat proteins.[25]

## Lipid Biochemistry: Measurement of Phosphoinositide Levels

For lipid analysis, incubations are performed in the presence of 5 $\mu$Ci of $[\gamma\text{-}^{32}P]ATP$, at a final ATP concentration of 50 $\mu M$, and typically in a reaction volume of 50–100 $\mu$l in 1.5-ml screw-cap plastic Eppendorf tubes. Reactions are stopped by addition of 400 $\mu$l of crude brain phosphoinositides (20 $\mu$g/ml; Sigma) in chloroform–methanol (2:1, v/v), followed by 400 $\mu$l of chloroform and 300 $\mu$l of 1 $M$ HCl. After vigorous vortexing, phases are separated by brief centrifugation (2 min at high speed in a table-top centrifuge) and the lower organic phase is removed with a Pasteur pipette or a transfer pipette. Care should be taken not to remove any of the upper aqueous phase and protein–aqueous interphase. The aqueous phases are reextracted with an additional 400 $\mu$l of chloroform, and the organic phases are pooled and dried under a stream of nitrogen. A sample concentrator, which can hold and dry multiple samples in parallel (Dri-Block; Techne, Princeton, NJ), is used for this purpose. Work should be carried out under a fume hood. Phosphoinositides can be separated without further processing by thin-layer chromatography (TLC). Lipids are resuspended in 40 $\mu$l of chloroform–methanol (2:1 v/v), and a 10-$\mu$l aliquot is spotted onto a TLC plate [Merck Silica gel 60 TLC plate that has been preimpregnated with potassium oxalate: TLC plates are dipped into a solution of 1% potassium oxalate (Sigma) in methanol–water (1:1, v/v), air dried, and activated for 30 min at 120° before use]. Plates are developed in a solvent system of chloroform–acetone–methanol–acetic acid–water (64:30:24:32:14, by volume) and autoradiographed at $-70°$ for several hours or overnight. Major radioactive spots [$PIP_2$, phosphatidylinositol

[25] L. Orci, B. S. Glick, and J. E. Rothman, *Cell* **46,** 171 (1986).

4-phosphate (PIP), and phosphatidic acid (PA)] are identified by comigration with purified lipid standards (Sigma and Avanti Polar Lipids) and visualized by placing the TLC plate into a tank saturated with iodine vapor [crystals of iodine (Sigma) are added to a TLC tank, the tank is covered with the lid, and iodine vapors are allowed to form by sublimation for 2 h]. Associated radioactivity is measured by quantitative autoradiography with a PhospholImager (Molecular Dynamics-Amersham Biosciences) or by scraping the area of silica gel on the TLC plate that harbors the spot into a scintillation vial, followed by liquid scintillation counting.

Alternatively, lipids are deacylated for analysis by strong anion-exchange (SAX) high-performance liquid chromatography (HPLC).[26] Lipids are deacylated with methylamine as described.[27] A mixture of 40% aqueous methylamine (Fluka, Milwaukee, WI)–water–$n$-butanol–methanol (36:8:9:47, by volume) (2 ml) is added to the dried lipid extract and incubated at 50° for 45 min in a tightly stoppered tube. The mixture is then cooled on ice and evaporated to dryness under vacuum in a Savant SpeedVac (Thermo Savant, Holbrook, NY). A mixture of $n$-butanol–petroleum ether–ethyl formate (20:40:1, by volume) (2 ml) is added, followed by water (2 ml). After thorough mixing and gentle centrifugation to separate phases, the lower aqueous phase, containing the deacylated lipids, is carefully removed. The organic phase is reextracted with 2 ml of water, and the combined aqueous phases are dried in a SpeedVac. Samples are taken up in water, loaded on a Partisphere SAX ion-exchange column (4.6 × 125 mm; Whatman, Clifton, NJ), and eluted at a flow rate of 1 ml/min, using the following gradient profile[28]: 0–10 min, 0% B; 10–55 min, 0–35% B; and 55–70 min, 35–100% B [buffer B is 1.4 $M$ $(NH_4)_2HPO_4$, pH 3.7]. Fractions (1 ml) are collected and counted for radioactivity by scintillation counting. Alternatively, radioactivity is measured with an in-line liquid scintillation counter (Packard, Meriden, CT).[29] Peaks are identified by coelution with standards, most of which must be generated by enzymatic reactions using phosphoinositide kinases and phosphatases.[26,30] Phophatidylinositol-[inositol-2-$^3$H($N$)] ([$^3$H]PI) and phosphatidylinositol-[inositol-2-$^3$H($N$)]-4,5-bisphosphate ([$^3$H]PI(4,5)P$_2$) are available from Perkin Elmer Life Sciences.

[26] L. A. Serunian, K. R. Auger, and L. C. Cantley, *Methods Enzymol.* **198**, 78 (1991).
[27] C. J. Kirk, A. J. Morris, and S. B. Shears, *in* "Peptide Hormone Action: A Practical Approach" (K. Siddle and J. C. Hutton, eds.), p. 151. IRL Press, Oxford, 1990.
[28] I. M. Bird, *Methods Mol. Biol.* **105**, 25 (1998).
[29] L. Zhang and I. L. Buxton, *Methods Mol. Biol.* **105**, 47 (1998).
[30] X. Zhang, J. C. Loijens, I. V. Boronenkov, G. J. Parker, F. A. Norris, J. Chen, O. Thum, G. D. Prestwich, P. W. Majerus, and R. A. Anderson, *J. Biol. Chem.* **272**, 17756 (1997).

Concluding Remarks

The methodology described above allows for the investigation of key aspects of lipids and their metabolism in coat dynamics. Among the advantages offered is the possibility to study the role of individual proteins or of cytosol and cytosolic fractions under similar assay conditions. The power of this methodology is further enhanced by the use of proteoliposomes, rather than pure lipid vesicles, to study the cooperativity of proteins and lipids in membrane dynamics. Some limitations of *in vitro* systems such as this one include the relatively low efficiency, slow kinetics, and lack of physiological "compartmentalization" of the components (e.g., the lack of segregation of proteins and lipids into distinct subcompartments). Clearly, the validity of results obtained by these techniques must be ultimately assessed by studies *in situ* and by observations from genetic approaches.

# [15] Fluorescence Assays for Liposome Fusion

*By* Nejat Düzgüneş

Introduction

Membrane fusion is a biophysical reaction that is of fundamental importance in biological systems and takes place in such diverse processes as fertilization, viral infection, exocytosis, and intracellular membrane traffic.[1] Because their membrane composition can be manipulated readily, liposomes have provided a convenient model system to study the molecular determinants and mechanisms of membrane fusion.[2,3] The roles of membrane composition, the ionic environment, and membrane phase state have been elucidated by the use of liposome fusion assays. The role of cytosolic or viral membrane proteins in fusion, as well as the role of fusion activity of peptides derived from viral proteins, have been studied in detail with liposomes.[2,3] Liposomes containing fluorescent lipids have been used to study the fusion characteristics of intracellular membranes

---

[1] N. Düzgüneş and F. Bronner (eds.), "Membrane Fusion in Fertilization, Cellular Transport and Viral Infection," p. xviii and 384. Academic Press, New York, 1988.

[2] N. Düzgüneş and S. Nir, *in* "Liposomes as Tools in Basic Research and Industry" (J. R. Philippot and F. Schuber, eds.), p. 103. CRC Press, Boca Raton, FL, 1995.

[3] N. Düzgüneş, *in* "Trafficking of Intracellular Membranes: From Molecular Sorting to Membrane Fusion" (M. C. Pedroso de Lima, N. Düzgüneş, and D. Hoekstra, eds.), p. 97. Springer-Verlag, Berlin, 1995.

such as secretory granules, microsomes, and the Golgi apparatus.[4–8] In this chapter we present details of fluorescence assays used to study the kinetics of membrane fusion. Membrane fusion involves the simultaneous inter-mixing of the aqueous compartments bounded by two membranes and of the membrane components. Two widely used assays for the intermixing of internal aqueous contents, the Tb–DPA and the ANTS–DPX assays, and several assays for the intermixing of lipids based on fluorescence resonance energy transfer, are described.

## Intermixing of Internal Aqueous Contents

### Terbium–Dipicolinic Acid Assay

The assay involves the detection of fluorescence generated on the interaction of terbium (Tb) and dipicolinic acid (DPA) initially encapsulated in two separate populations of liposomes[9] (Fig. 1). Fluorescence is generated via internal energy transfer from DPA to Tb within the $[Tb(DPA)_3]^{3-}$ chelation complex, enhancing the fluorescence intensity of Tb by four orders of magnitude. The assay has been used to investigate the effect on membrane fusion of various factors including phospholipid composition[2,3,10–14]; monovalent, divalent, and trivalent cations[14–20];

[4] M. Bental, P. I. Lelkes, J. Scholma, D. Hoekstra, and J. Wilschut, *Biochim. Biophys. Acta* **774**, 296 (1984).

[5] P. Meers, J. D. Ernst, N. Düzgüneş, K. Hong, J. Fedor, I. M. Goldstein, and D. Papahadjopoulos, *J. Biol. Chem.* **262**, 7850 (1987).

[6] C. Martinez-Bazenet, C. Audigier-Petit, J. Frot-Coutaz, R. Got, C. Nicolau, and R. Létoublon, *Biochim. Biophys. Acta* **943**, 35 (1988).

[7] T. Kobayashi and R. E. Pagano, *Cell* **55**, 797 (1988).

[8] S. Kagiwada, M. Murata, R. Hishida, M. Tagaya, S. Yamashina, and S.-I. Ohnishi, *J. Biol. Chem.* **268**, 1430 (1993).

[9] J. Wilschut, N. Düzgüneş, R. Fraley, and D. Papahadjopoulos, *Biochemistry* **19**, 6011 (1980).

[10] N. Düzgüneş, J. Wilschut, R. Fraley, and D. Papahadjopoulos, *Biochim. Biophys. Acta* **642**, 182 (1981).

[11] R. Sundler, N. Düzgüneş, and D. Papahadjopoulos, *Biochim. Biophys. Acta* **649**, 751 (1981).

[12] R. Sundler and D. Papahadjopoulos, *Biochim. Biophys. Acta* **649**, 743 (1981).

[13] J. Wilschut, N. Düzgüneş, K. Hong, D. Hoekstra, and D. Papahadjopoulos, *Biochim. Biophys. Acta* **734**, 309 (1983).

[14] N. Düzgüneş, S. Nir, J. Wilschut, J. Bentz, C. Newton, A. Portis, and D. Papahadjopoulos, *J. Membr. Biol.* **59**, 115 (1981).

[15] J. Wilschut, N. Düzgüneş, and D. Papahadjopoulos, *Biochemistry* **20**, 3126 (1981).

[16] J. Bentz, N. Düzgüneş, and S. Nir, *Biochemistry* **22**, 3320 (1983).

[17] S. Nir, N. Düzgüneş, and J. Bentz, *Biochim. Biophys. Acta* **735**, 160 (1983).

[18] N. Düzgüneş, J. Paiement, K. B. Freeman, N. G. Lopez, J. Wilschut, and D. Papahadjopoulos, *Biochemistry* **23**, 3486 (1984).

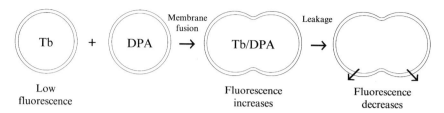

FIG. 1. The terbium–dipicolinic acid (DPA) assay for monitoring the intermixing of aqueous contents during liposome fusion. Terbium citrate and DPA are encapsulated in different populations of liposomes. Fusion results in increased fluorescence intensity.

temperature[18,21–23]; osmotic pressure[21]; cholesterol[24]; α-tocopherol[25]; polyamines[26]; lectins[27,28]; the cytoplasmic $Ca^{2+}$-binding protein, synexin (annexin VII)[5,29–31]; sulfatides[32]; β-bungarotoxin[33]; and poly(ethylene glycol).[34]

Three sets of large unilamellar liposomes are prepared, encapsulating different solutions: (1) terbium liposomes: 2.5 mM $TbCl_3$ (Alfa, Danvers, MA), 50 mM sodium citrate (Sigma, St. Louis, MO), 10 mM N-tris (hydroxymethyl) methyl-2-aminoethane sulfonic acid (TES; Sigma), pH 7.4; (2) DPA liposomes: 50 mM sodium dipicolinate (Sigma), 20 mM NaCl (Sigma), 10 mM TES, pH 7.4; and (3) terbium–DPA liposomes: a 1:1 mixture of sets 1 and 2. The sodium dipicolinate solution takes time to prepare, because solvation and titration to adjust the pH to neutral are lengthy processes. The encapsulated solutions are slightly hypoosmotic compared

[19] J. Bentz and N. Düzgüneş, *Biochemistry* **24,** 5436 (1985).
[20] J. Bentz, D. Alford, J. Cohen, and N. Düzgüneş, *Biophys. J.* **53,** 593 (1988).
[21] S. Ohki, *J. Membr. Biol.* **77,** 265 (1984).
[22] J. Wilschut, N. Düzgüneş, D. Hoekstra, and D. Papahadjopoulos, *Biochemistry* **24,** 8 (1985).
[23] J. Bentz, N. Düzgüneş, and S. Nir, *Biochemistry,* **24,** 1064 (1985).
[24] S. A. Shavnin, M. C. Pedroso de Lima, J. Fedor, P. Wood, J. Bentz, and N. Düzgüneş, *Biochim. Biophys. Acta* **946,** 405 (1988).
[25] F. J. Aranda, M. P. Sanchez-Migallon, and J. C. Gomez-Fernandez, *Arch. Biochem. Biophys.* **333,** 394 (1996).
[26] F. Schuber, K. Hong, N. Düzgüneş, and D. Papahadjopoulos, *Biochemistry* **22,** 6134 (1983).
[27] N. Düzgüneş, D. Hoekstra, K. Hong, and D. Papahadjopoulos, *FEBS Lett.* **173,** 80 (1984).
[28] D. Hoekstra and N. Düzgüneş, *Biochemistry* **25,** 1321 (1986).
[29] K. Hong, N. Düzgüneş, and D. Papahadjopoulos, *J. Biol. Chem.* **256,** 3641 (1981).
[30] K. Hong, N. Düzgüneş, R. Ekerdt, and D. Papahadjopoulos, *Proc. Natl. Acad. Sci. USA* **79,** 4642 (1982).
[31] K. Hong, N. Düzgüneş, and D. Papahadjopoulos, *Biophys. J.* **37,** 297 (1982).
[32] X. Wu and Q. T. Li, *J. Lipid Res.* **40,** 1254 (1999).
[33] S. Rufini, J. Z. Pedersen, A. Desideri, and P. Luly, *Biochemistry* **29,** 9644 (1990).
[34] W. A. Talbot, L. X. Zheng, and B. R. Lentz, *Biochemistry* **36,** 5827 (1997).

with 100 m$M$ NaCl–10 m$M$ TES, the medium in which the liposomes are suspended during the assay. If experiments will be carried out in media of higher salt concentrations (e.g., 150 m$M$ NaCl), the osmolality of the encapsulation solutions should be increased by adding more NaCl. The osmolality can be checked readily by the use of a vapor pressure os-mometer (Wescor, Logan, UT). The presence of citrate is essential to chelate the $Tb^{3+}$, which would otherwise interact with and precipitate negatively charged phospholipids. When Tb is to be encapsulated in phosphatidic acid-containing liposomes, nitrilotriacetic acid is used instead of citrate, because Tb appears to interact with this acidic lipid even in the presence of citrate.[11,12] The preparation of large unilamellar liposomes is described elsewhere.[35]

Small unilamellar liposomes may also be utilized, but the low encapsu-lated volume may present problems in the detection of fluorescence distinct from light-scattering artifacts. Thus, the use of a fluorometer with a high-intensity light source is recommended. Also, higher concentrations of the reactants are used to achieve a favorable signal-to-noise ratio. Three sets of small unilamellar liposomes are prepared,[35] encapsulating different solu-tions: (1) small terbium liposomes: 15 m$M$ TbCl$_3$, 150 m$M$ sodium citrate, 10 m$M$ TES, pH 7.4; (2) small DPA liposomes: 150 m$M$ sodium dipicoli-nate, 10 m$M$ TES, pH 7.4; and (3) small terbium–DPA liposomes: a 1:1 mixture of sets 1 and 2. These solutions are hyperosmotic compared with the eventual medium in which the fusion reaction is normally carried out (100 m$M$ NaCl–10 m$M$ TES). Nevertheless, the contents do not leak out during storage, because small unilamellar liposomes are not osmotically active.

During the preparation of the liposomes, three Sephadex G-75 or G-50 (Pharmacia, Uppsala, Sweden) columns (1 × 20 cm; Bio-Rad, Hercules, CA) are equilibrated with 100 m$M$ NaCl, 1 m$M$ EDTA, 10 m$M$ TES, pH 7.4 (buffer A). The reason for the inclusion of EDTA in the buffer is to pre-vent $Tb^{3+}$ binding to the vesicle membrane as the citrate is diluted during gel filtration. Each vesicle preparation is chromatographed on a separate column to remove the unencapsulated material. After letting the liposome suspension permeate the gel at the top of the column, buffer A is placed carefully on top of the gel and the reservoir is connected via plastic tubing to the top of the column. The eluted solution is measured with a 15-ml graduated polystyrene culture tube. The first 4 ml is discarded, and the next 3 ml is collected in a similar culture tube. Argon gas is flushed over the liposome suspensions for 10 s, and the tubes are capped tightly, sealed with Parafilm, and stored in ice.

[35] N. Düzgüneş, *Methods Enzymol.* **367**, 23 (2003).

The terbium–DPA liposomes can be used for calibration of the assay (see below). Alternatively, 1 ml of the terbium liposomes is chromatographed on a Sephadex column equilibrated with 100 m$M$ NaCl, 10 m$M$ TES, pH 7.4 (buffer B) to eliminate the EDTA in the medium because it interferes with the formation of the Tb–DPA complex when the liposomes are lysed with 0.5% (w/v) sodium cholate (Sigma) to set the fluorescence to 100%. The resulting liposomes are called "terbium-minus-EDTA" liposomes.

The lipid concentration of the liposomes is measured in each of the culture tubes by inorganic phosphate analysis.[35] To calibrate the assay, an aliquot of the terbium-minus-EDTA liposomes equivalent to the amount of terbium liposomes used in the actual assay is placed in a fluorometer cuvette. In routine experiments, this amount is equivalent to 25 $\mu M$ lipid. An appropriate volume of buffer B is added to the cuvette to bring the final volume to 1 ml. From a 2 m$M$ DPA–10 m$M$ TES (pH 7.4) stock, 10 $\mu$l is added to the cuvette (final DPA concentration, 20 $\mu M$). This concentration of free DPA is sufficient to chelate all the Tb$^{3+}$ when the liposomes are lysed subsequently; higher concentrations of DPA should be avoided to prevent inner filter effects. This step is followed by the addition of 50 $\mu$l of either 10% (w/v) sodium cholate (final concentration, 0.5%) or 16 m$M$ C$_{12}$E$_8$ (final concentration, 0.8 m$M$; Calbiochem, La Jolla, CA) to release the liposome contents. As an alternative method to calibrate the assay, terbium–DPA liposomes (50 $\mu M$) are diluted from a 10× stock (i.e., 500 $\mu M$ lipid) into buffer B and their fluorescence is set to 100%.[16,17] These vesicles correspond to the fusion product of all the terbium liposomes and DPA liposomes in the fusion assay, assuming that they have the same size distribution and are used at the same lipid concentration as the combination of the terbium liposome and DPA liposome populations (25 $\mu M$ each). It is necessary to wait about 0.5 h after transferring the terbium–DPA liposomes from 0° (at which temperature the liposomes are stored) to room temperature for the fluorescence to equilibrate.

The excitation wavelength of the fluorometer is set at 276–278 nm, and the emission wavelength is set at 545 nm, using intermediate slit widths. A high-pass cutoff filter (>530 nm, e.g., a 3-68 filter; Corning, Corning, NY) is placed before the emission monochromator to minimize any contributions from light scattering. Crossed polarizers can also be used to minimize light-scattering contributions. The fluorescence of the terbium-minus-EDTA liposomes lysed in the presence of DPA is set to 100% fluorescence. SLM 4000, SLM 8000, Spex Fluorolog 2, and PerkinElmer MPF43 fluorometers have been used for our experiments. Obviously, other instruments with similar light intensity and sensitivity, and with a stirred cell compartment, can be used to perform these assays.

To perform the assay, a 1:1 mixture of the terbium liposomes and the DPA liposomes is prepared as a 10× stock solution in a culture tube (e.g., at a lipid concentration of 500 $\mu M$ for an assay utilizing 50 $\mu M$ lipid in the cuvette). The stock solution (0.1 ml) is diluted into 0.9 ml of buffer B in a cuvette containing a small "flea" stir bar. For fluorometers requiring larger amounts of buffer (2–3 ml) because of the location of the light path, a "castle" stir bar is recommended. The EDTA concentration in the medium is diluted in the process to 0.1 m$M$. The EDTA is necessary to prevent the interaction of any released $Tb^{3+}$ ions with leaked DPA in the external medium. (Divalent cations, which may be used to induce fusion, further inhibit the interaction.) For a fluorometer with a strip chart recorder, the residual fluorescence of this suspension is set to 0% $F_{max}$, using the offset function of the fluorometer. The gain function is used to make necessary adjustments to the 100% level with the calibration liposomes in place. The calibration procedure should be repeated after a set of measurements to ensure that the lamp intensity has not changed in the meantime. For a fluorometer with computerized data acquisition, the initial level of fluorescence [$I(0)$] is subtracted from the data set and the resulting data set is divided by the numerical (fluorescence intensity) difference between the fluorescence intensity of the calibration vesicles [$I(\infty)$] and the initial fluorescence. To obtain percentage values, the result is multiplied by 100. Thus, the extent of fusion, $F(t)$, as a percentage of maximal fluorescence, is given by $F(t) = 100 \times [I(t) - I(0)]/[I(\infty) - I(0)]$, where the fluorescence intensity at time $t$ is $I(t)$. Fusion is initiated by the injection of the fusogenic agent to be studied, using a Hamilton syringe (Fisher Scientific, Pittsburgh, PA; www.fishersci.com), preferably via a pinhole at the top of the chamber to avoid opening the chamber lid.

*ANTS–DPX Assay*

The ANTS–DPX assay is based on the collisional quenching of aminonaphthalene trisulfonic acid (ANTS) fluorescence by *p*-xylene bis(pyridinium) bromide (DPX), each initially encapsulated in different populations of liposomes[36] (Fig. 2). Intermixing of aqueous contents during membrane fusion results in the quenching of ANTS fluorescence. Leakage of liposome contents into the medium does not result in quenching, because the low concentration of DPX that may be have leaked into the medium is not effective in quenching ANTS. The assay was utilized initially in studies of low pH-or $Ca^{2+}$-induced fusion of liposomes[36,37] and later was

[36] H. Ellens, J. Bentz, and F. C. Szoka, *Biochemistry* **24**, 3099 (1985).

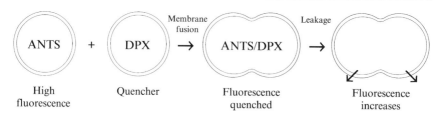

FIG. 2. The ANTS–DPX assay for measuring the intermixing of aqueous contents during liposome fusion. Aminonaphthalene trisulfonic acid (ANTS) and *p*-xylene bis(pyridinium) bromide (DPX) are encapsulated in different populations of liposomes. Fusion results in decreased fluorescence intensity.

used to study the induction of membrane fusion by poly(ethylene glycol),[38] phospholipase C,[39,40] and a peptide derived from the rod outer segment protein Peripherin/RDS.[41]

Three sets of large unilamellar liposomes are prepared,[35] containing the following solutions: (1) ANTS liposomes: 25 m$M$ aminonaphthalene trisulfonic acid (Molecular Probes, Eugene, OR), 40 m$M$ NaCl, 10 m$M$ TES, pH 7.4; (2) DPX liposomes: 90 m$M$ DPX (Molecular Probes), 10 m$M$ TES, pH 7.4; and (3) ANTS–DPX liposomes: a 1:1 mixture of sets 1 and ·2. The liposomes are chromatographed on Sephadex columns as described above for the Tb–DPA assay, except that the columns are equilibrated with 100 m$M$ NaCl, 0.1 m$M$ EDTA, 10 m$M$ TES, pH 7.4 (buffer C). If the assay is to be performed in the presence of physiological concentrations of NaCl (~150 m$M$), the NaCl concentration in the ANTS solution should be 90 m$M$. Likewise, the DPX solution should also contain 50 m$M$ NaCl. Because adjusting the pH of solutions can change their osmolality, the pH should be adjusted first, and then the osmolality should be adjusted by adding an appropriate amount of NaCl. As with all fluorophores, solutions containing ANTS should be kept in the dark (simply by wrapping aluminum foil around the container), at 4°. The lipid concentrations of the chromatographed liposome preparations are determined by assaying for inorganic phosphate as in the Tb–DPA assay.[35]

[37] N. Düzgüneş, R. M. Straubinger, P. A. Baldwin, D. S. Friend, and D. Papahadjopoulos, *Biochemistry* **24,** 3091 (1985).

[38] R. A. Parente and B. R. Lentz, *Biochemistry* **25,** 6678 (1986).

[39] J. L. Nieva, F. M. Goñi, and A. Alonso, *Biochemistry* **28,** 7364 (1989).

[40] A. S. Luk, E. W. Kaler, and S. P. Lee, *Biochemistry* **32,** 6965 (1993).

[41] K. Boesze-Battaglia, O. P. Lamba, A. A. Napoli, Jr., S. Sinha, and Y. Guo, *Biochemistry* **37,** 9477 (1998).

For the assay, the excitation monochromator is set to 360 nm, and the emission monochromator is set to 530 nm, with relatively wide slit widths (10–20 nm). A high-pass filter (e.g., Corning 3-68) should be placed in the emission channel to eliminate any potential effects of light scattering. In a typical assay with a total lipid concentration of 50 $\mu M$, the fluorescence of 25 $\mu M$ ANTS liposomes plus 25 $\mu M$ DPX liposomes in the appropriate amount of buffer C (usually 1–2 ml, depending on the fluorometer) is set to 100%. The fluorescence of 50 $\mu M$ ANTS–DPX liposomes is set to 0%. The fluorescence of 50 $\mu M$ ANTS–DPX liposomes is set to 0%. These liposomes represent the theoretical fusion product of all the component liposomes in the assay. Flea or castle stir bars should be placed in the cuvette to facilitate rapid mixing of the "fusogen" with the liposomes. Fusion is monitored as the decrease in fluorescence and can be quantified using the same formalism as for the Tb–DPA assay. Thus, the extent of fusion, $F(t)$, as a percentage of maximal fluorescence, is given by $F(t) = 100 \times [I(0) - I(t)]/[I(0)-I(\infty)]$, where the fluorescence intensity at time $t$ is $I(t)$, and the fluorescence intensities of ANTS liposomes and ANTS–DPX liposomes are $I(0)$ and $I(\infty)$, respectively.

In our laboratory, attempts to use the ANTS–DPX assay to monitor the fusion of small unilamellar liposomes have not produced reliable results, most likely because ANTS binds excessively to such vesicles, although it does not bind appreciably to large unilamellar liposomes. Some laboratories have, however, published their experiments with small liposomes.[40,42]

## Intermixing of Membrane Lipids

### NBD–Rhodamine Resonance Energy Transfer Assay

The concept of using resonance energy transfer (RET) between two fatty acid-bound fluorophores to detect membrane fusion was introduced by Keller et al.[43] The method was extended to detect liposome fusion by several laboratories.[44–49] In RET, the emission band of the energy

[42] L. Oshry, P. Meers, T. Mealy, and A. I. Tauber, Biochim. Biophys. Acta 1066, 239 (1991).
[43] P. M. Keller, S. Person, and W. Snipes, J. Cell Sci. 28, 167 (1977).
[44] G. A. Gibson and L. M. Loew, Biochem. Biophys. Res. Commun. 88, 135 (1979).
[45] P. Vanderwerf and E. F. Ullman, Biochim. Biophys. Acta 596, 302 (1980).
[46] P. S. Uster and D. W. Deamer, Arch. Biochem. Biophys. 209, 385 (1981).
[47] D. K. Struck, D. Hoekstra, and R. E. Pagano, Biochemistry 20, 4093 (1981).
[48] N. Düzgüneş and J. Bentz, in "Spectroscopic Membrane Probes" (L. Loew, ed.), p. 117. CRC Press, Boca Raton, FL, 1988.
[49] D. Hoekstra and N. Düzgüneş, Methods Enzymol. 220, 15 (1993).

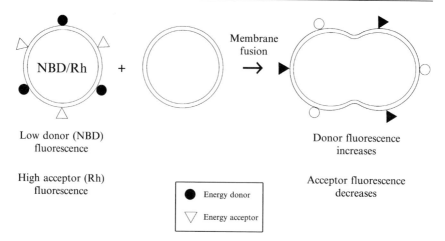

FIG. 3. The NBD–rhodamine (Rh) assay for the intermixing of lipids during liposome fusion. The donor, $N$-(7-nitrobenz-2-oxa-1,3-diazol-4-yl) phosphatidylethanolamine, and the acceptor, $N$-(Lissamine rhodamine B sulfonyl) phosphatidylethanolamine, are incorporated in one population of liposomes. Dilution of the probes into unlabeled liposomes results in an increase in donor fluorescence or a decrease in acceptor fluorescence, if the donor fluorescence is excited. The solid symbols indicate low fluorescence; the open symbols denote high fluorescence.

donor fluorophore overlaps with the excitation band of the energy acceptor fluorophore, and the excited state energy of the donor generated by an absorbed photon is transferred nonradiatively to the acceptor, which in turn fluoresces. The overlap of the emission spectrum of the donor and the absorption spectrum of the acceptor, and the inverse sixth power of the distance between the two fluorophores, determine the rate and efficiency of energy transfer. The assays depend on the dilution of a donor–acceptor pair from "labeled" liposomes to "unlabeled" liposomes as a result of lipid mixing during membrane fusion (Fig. 3). The efficiency of energy transfer is then decreased. Low concentrations of probes, usually less than 1 mol% of total lipid, are used for RET assays, minimizing the extent of membrane perturbation. The NBD–rhodamine RET assay was first described by Struck *et al.*[47] It involves the use of the probes $N$-(7-nitro-benz-2-oxa-1,3-diazol-4-yl) phosphatidylethanolamine ($N$-NBD-PE) and $N$-(Lissamine rhodamine B sulfonyl) PE ($N$-Rh-PE), which do not exchange between membranes to any significant extent and do not cause appreciable perturbation of bilayer packing.

The assay has been employed to study proton- and divalent cation-induced fusion of liposomes,[50–54] the role of cholesterol in fusion,[24] fusion of liposomes induced by clathrin[55] and myelin basic protein,[56] anion-induced fusion of cationic liposomes,[57] and fusion of cationic liposomes with anionic liposomes[57] and of nonphospholipid vesicles with phospholipid liposomes.[58] The assay has been useful in studying the fusion of liposomes with biological membranes, such as vesicles derived from bacteria,[59] neutrophil plasma membranes in the presence of annexin I,[60] and cells expressing influenza hemagglutinin.[61]

Three sets of liposomes are prepared for the assay: (1) labeled liposomes containing 0.8 mol% of each of the fluorophores (both from Avanti Polar Lipids, Alabaster, AL) in the initial chloroform mixture, (2) unlabeled liposomes of the same lipid composition, but without the fluorophores, and (3) calibration liposomes containing 0.08 mol% of each fluorophore. Large or small unilamellar liposomes are prepared as described,[35] utilizing 10 or 20 $\mu$mol of total lipid in a final buffer C volume of 1 ml. Other physiological buffer systems may be substituted, avoiding the use of $CaCl_2$ or $MgCl_2$. The lipid concentration of each liposome preparation is assessed by assaying for inorganic phosphate as outlined in detail in Düzgüneş.[35] Aliquots from the labeled and unlabeled liposomes are mixed at a ratio of 1:9. For a lipid-mixing assay utilizing 50 $\mu M$ lipid, this would correspond to 5 $\mu M$ labeled liposomes and 45 $\mu M$ unlabeled liposomes in 1 or 2 ml of buffer C, depending on the requirements for the fluorometer. The excitation monochromator is adjusted to 460 nm, the excitation monochromator is adjusted to 530 nm, and 100% (or "maximal") fluorescence is set by using 50 $\mu M$ calibration liposomes. The residual fluorescence of the mixture of labeled and unlabeled liposomes is designated as 0% fluorescence.

[50] N. Düzgüneş, R. M. Straubinger, P. A. Baldwin, D. S. Friend, and D. Papahadjopoulos, *Biochemistry* **24**, 3091 (1985).

[51] N. Düzgüneş, T. M. Allen, J. Fedor, and D. Papahadjopoulos, *Biochemistry* **26**, 8435 (1987).

[52] J. Rosenberg, N. Düzgüneş, and C. Kayalar, *Biochim. Biophys. Acta* **735**, 173 (1983).

[53] J. Wilschut, J. Scholma, M. Bental, D. Hoekstra, and S. Nir, *Biochim. Biophys. Acta* **821**, 45 (1985).

[54] A. Walter and D. P. Siegel, *Biochemistry* **32**, 3271 (1993).

[55] S. Maezawa, T. Yoshimura, K. Hong, N. Düzgüneş, and D. Papahadjopoulos, *Biochemistry* **28**, 1422 (1989).

[56] Y. Cajal, J. M. Boggs, and M. K. Jain, *Biochemistry* **36**, 2566 (1997).

[57] N. Düzgüneş, J. A. Goldstein, D. S. Friend, and P. L. Felgner, *Biochemistry* **28**, 9179 (1989).

[58] M. El Baraka, E. I. Pecheur, D. F. Wallach, and J. R. Philippot, *Biochim. Biophys. Acta* **1280**, 107 (1996).

[59] A. J. Driessen, D. Hoekstra, G. Scherphof, R. D. Kalicharan, and J. Wilschut, *J. Biol. Chem.* **260**, 10880 (1985).

[60] L. Oshry, P. Meers, T. Mealy, and A. I. Tauber, *Biochim. Biophys. Acta* **1066**, 239 (1991).

[61] G. van Meer, J. Davoust, and K. Simons, *Biochemistry* **24**, 3593 (1985).

When lipid mixing occurs as a result of the introduction of a fusogen, NBD fluorescence increases as $N$-NBD-PE and $N$-Rh-PE are diluted into the unlabeled liposomes, the average distance between the probes increases, and RET decreases. The percentage of lipid mixing as a function of time is given by $M(t) = 100 \times [I(t) - I(0)]/[I(\infty) - I(0)]$, where $I(t)$ is the fluorescence intensity at time $t$, $I(0)$ is the residual fluorescence, and $I(\infty)$ is the fluorescence of the calibration liposomes.

When utilizing RET pairs, it is important to establish that no lipid mixing takes place under conditions in which no fusion is expected, or in which the liposomes are expected to aggregate but do not fuse.[51] Under some conditions, lipid mixing can be observed in the absence of contents mixing. This observation is interpreted as arising from "semifusion," the intermixing of the outer monolayers of interacting liposomes, in the absence of any direct communication of the aqueous interiors.[51,52]

*Inner Monolayer Mixing Assay*

In the inner monolayer mixing assay, the fluorophores exposed on the outer monolayer of liposomes are reduced by dithionate, eliminating the fluorescence emanating from outer monolayer fluorophores.[62] Thus, the potential alteration of the fluorescence intensity of the probes by ions, proteins, or other molecules, for example, by lateral phase separation, can be avoided.

Three sets of liposomes are prepared for the assay: (1) labeled liposomes containing 0.75 mol% of $N$-NBD-phosphatidylserine ($N$-NBD-PS; Avanti Polar Lipids) (this probe being less prone to transbilayer movement than $N$-NBD-PE after reduction), and either $N$-Rh-PE or 1,1'-dioctadecyl-3,3,3',3'-tetramethylindodicarbocyanine perchlorate [diI (5)C18; Molecular Probes], (2) unlabeled liposomes of the same lipid composition, but without the fluorophores, and (3) calibration liposomes containing 0.075 mol% of each fluorophore. The liposome suspension and the sodium dithionate (stock solution of 200 m$M$ sodium dithionate, 100 m$M$ Tris, pH 10, made shortly before use) are mixed, so that the final dithionate concentration is 80–100 m$M$, and the phospholipid concentration is 10 m$M$. The mixture is incubated for 30–45 min on ice. The dithionate is removed by centrifuging 100-$\mu$l aliquots through Bio-Gel A-30 spin columns (0.8 ml; Bio-Rad). Inner monolayer mixing is monitored as for the NBD–rhodamine assay described above.

---

[62] P. Meers, S. Ali, R. Erukulla, and A. S. Janoff, *Biochim. Biophys. Acta* **1467,** 227 (2000).

## BODIPY Assay

The BODIPY assay is based on the dilution of phospholipids containing fluorescently labeled acyl chains from labeled liposomes to unlabeled liposomes. The labeled lipids, 2-(4,4-difluoro-5,7-diphenyl-4-bora-3a, 4a-diaza-s-indacene-3-dodecanoyl)-1-hexadecanoyl-sn-glycero-3-phosphocholine (BODIPY 500-PC; donor) and 2-(4,4-difluoro-5,7-diphenyl-4-bora-3a,4a-diaza-s-indacene-3-dodecanoyl)-1-hexadecanoyl-sn-glycero-3-phosphoethanolamine (BODIPY 530-PE; acceptor), are incorporated into one population of liposomes at 0.5 mol% each.[63] The donor is excited at 500 nm and the fluorescence is measured at 520 nm, at which wavelength there is no fluorescence due to the acceptor. The labeled liposomes are mixed with unlabeled liposomes at a ratio of 1:4. The percentage of lipid mixing can be determined by two different methods: (1) setting the fluorescence of liposomes containing 0.1 mol% of each probe to 100% and determining the percentage of lipid mixing as in the case of the NBD–Rh assay described above and (2) calculating the fraction of lipid mixing from the ratio of the donor fluorescence ($F_d$) to the acceptor fluorescence ($F_a$), and the exponential constant $k$ describing the relationship of this ratio to the lipid-to-probe ratio,[63] given by $M(t) = (1/n)\{\exp k[F_d/F_a] - (F_d/F_a)_0] - 1\}$, where $n$ is the ratio of unlabeled to labeled liposomes, and $(F_d/F_a)_0$ is the donor-to-acceptor fluorescence ratio in the labeled liposomes before fusion.

Results obtained with this assay using dioleoylphosphatidylcholine (DOPC) liposomes fusing in the presence of poly(ethylene glycol) (PEG 8000) indicated that the rate of lipid mixing is faster than that obtained with the NBD–Rh assay, but similar to that measured with diphenylhexatriene substituted for one of the acyl chains of PC.[63] It was proposed that the bulky head groups of NBD-PE and Rh-PE may inhibit their movement through the fusion stalk between fusing membranes, in contrast to lipids labeled in the acyl chain region.

## Leakage of Contents

Fusion of liposomes may be accompanied by leakage of the internal aqueous contents. This process may be slow compared with fusion in certain types of liposomes, whereas in others it may be rapid and extensive. Certain bacterial toxins such as colicins, or peptides including mellitin, may cause the leakage of internal contents via disruption of bilayer integrity or channel formation. Fluorescence assays for leakage can conveniently monitor these processes.

[63] V. S. Malinin, M. E. Haque, and B. R. Lentz, *Biochemistry* **40**, 8292 (2001).

*Carboxyfluorescein or Calcein Leakage*

The release of aqueous contents that may accompany divalent cation-induced liposome fusion has been monitored by using liposomes containing carboxyfluorescein.[9,10] The intracellular fate of liposomes and the intracytoplasmic delivery of liposome contents was assessed with either carboxyfluorescein[64] (which is permeable through liposome and endosome membranes at low pH),[65] or calcein[66] (which is normally retained in liposomes at low pH).[67,68] These fluorophores have also been used in monitoring the channel-forming properties of bacterial and other toxins,[69,70] membrane destabilization by viral proteins,[71] immune complex-mediated lysis of liposomes,[72] leakage of contents in immobilized liposome chromatography,[73] the stability of liposomes containing archaeal bolaform lipids,[74] and the pH sensitivity of liposomes.[67,68,75]

To monitor the leakage of aqueous contents of liposomes, carboxyfluorescein[64] or calcein[65] (Molecular Probes) is encapsulated at self-quenching concentrations inside liposomes. Their leakage into the external medium, and hence dilution, results in the relief of self-quenching and in an increase in fluorescence. In the case of large unilamellar liposomes, 50 m$M$ carboxyfluorescein, or calcein (sodium salt), with 10 m$M$ TES, pH 7.4, is encapsulated. The unencapsulated material is separated by gel permeation on Sephadex G-50 or G-75, using buffer C as the elution buffer, as described above. Higher concentrations (100 m$M$) of carboxyfluorescein or calcein can be used for small unilamellar liposomes, as these liposomes are not osmotically active. This concentration of carboxyfluorescein encapsulated in osmotically active large unilamellar liposomes would result in leakage during column chromatography.

[64] J. N. Weinstein, S. Yoshikami, P. Henkart, R. Blumenthal, and W. A. Hagins, *Science* **195**, 489 (1977).

[65] R. M. Straubinger, K. Hong, D. S. Friend, and D. Papahadjopoulos, *Cell* **32**, 1069 (1983).

[66] T. M. Allen and L. G. Cleland, *Biochim. Biophys. Acta* **597**, 418 (1980).

[67] R. M. Straubinger, N. Düzgüneş, and D. Papahadjopoulos, *FEBS Lett.* **179**, 148 (1985).

[68] V. A. Slepushkin, S. Simoes, P. Dazin, M. S. Newman, L. S. Guo, M. C. Pedroso de Lima, and N. Düzgüneş, *J. Biol. Chem.* **272**, 2382 (1997).

[69] C. Kayalar and N. Düzgüneş, *Biochim. Biophys. Acta* **860**, 51 (1986).

[70] D. Rapaport, R. Peled, S. Nir, and Y. Shai, *Biophys. J.* **70**, 2502 (1996).

[71] P. Tian, J. M. Ball, C. Q. Zeng, and M. K. Estes, *J. Virol.* **70**, 6973 (1996).

[72] B. Babbitt, L. Burtis, P. Dentinger, P. Constantinides, L. Hillis, B. McGirl, and L. Huang, *Bioconjug. Chem.* **4**, 199 (1993).

[73] X. Y. Liu, C. Nakamura, Q. Yang, and J. Miyake, *Anal. Biochem.* **293**, 251 (2001).

[74] Q. Fan, A. Relini, D. Cassinadri, A. Gambacorta, and A. Gliozzi, *Biochim. Biophys. Acta* **1240**, 83 (1995).

[75] K. Kono, K. Zenitani, and T. Takagishi, *Biochim. Biophys. Acta* **1193**, 1 (1994).

The maximal fluorescence (100%) is established by lysing the vesicles with 0.1% (w/v) Triton X-100 or 0.8 m$M$ $C_{12}E_8$. The extent of leakage, $L(t)$, as a percentage of maximal fluorescence, may also be calculated from $L(t) = 100 \times [I(t) - I(0)]/[I(\infty) - I(0)]$, where the fluorescence intensity at time $t$ is $I(t)$, the initial residual fluorescence of the liposomes is $I(0)$, and the fluorescence achieved by lysing the liposomes with detergent is $I(\infty)$. Fluorescence is excited at 490–493 nm and monitored at 520–530 nm. It is preferable to use a high-pass filter (e.g., Corning 3-68 filter, > 530 nm) in the emission channel to eliminate scattering artifacts, because the excitation and emission wavelengths are close to one another.

## Tb–DPA Leakage

In liposome systems that undergo leaky membrane fusion, the Tb–DPA fusion signal can be corrected for released contents or dissociated Tb–DPA complex due to the entry of $Ca^{2+}$ and EDTA into the liposome.[16,17] The dissociation of Tb–DPA has also been used to assess the destabilization of glycolipid-containing liposomes in the presence of lectins and fusogens.[28] Terbium–DPA liposomes are prepared as described in the section on the Tb–DPA assay (see above). Because the release of Tb–DPA into a medium containing EDTA and $Ca^{2+}$ results in dissociation of the complex and reduction in Tb fluorescence, the assay is initiated at a fluorescence value of $I(0)$. Maximum release is obtained by lysing the liposomes with 0.5% cholate, and the fluorescence is reduced to a residual level, here designated as $I(\infty)$. Thus, the extent of leakage as a function of time can be expressed as $L(t) = 100 \times [I(0) - I(t)]/[I(0) - I(\infty)]$, where the fluorescence intensity at time $t$ is $I(t)$.

## ANTS–DPX Leakage

Leakage of ANTS–DPX from liposomes results in an increase in fluorescence and was used initially to examine the destabilization of liposomes at low pH.[76] The assay has been employed to assess the interaction with liposomal membranes of surfactant-associated proteins,[77,78] erythrocyte protein 4.1,[79] peptides derived from viral fusion proteins,[80,81] the synthetic amphipathic peptide GALA,[82] antimicrobial peptides (defensins),[83] and

[76] H. Ellens, J. Bentz, and F. C. Szoka, *Biochemistry* **23**, 1532 (1984).
[77] K. Shiffer, S. Hawgood, N. Düzgüneş, and J. Goerke, *Biochemistry* **27**, 2689 (1988).
[78] R. Chang, S. Nir, and F. R. Poulain, *Biochim. Biophys. Acta* **1371**, 254 (1998).
[79] K. Shiffer, J. Goerke, N. Düzgüneş, J. Fedor, and S. B. Shohet, *Biochim. Biophys. Acta* **937**, 269 (1988).
[80] N. Düzgüneş and F. Gambale, *FEBS Lett.* **227**, 110 (1988).
[81] N. Düzgüneş and S. A. Shavnin, *J. Membr. Biol.* **128**, 71 (1992).

the $\kappa$-opioid receptor-selective heptadecapeptide dynorphin.[84] ANTS–DPX liposomes are prepared as described in the section on the ANTS–DPX fusion assay (see above). The assay is calibrated to maximal (100%) fluorescence by lysing the vesicles with 0.1% (w/v) Triton X-100 or 0.8 m$M$ $C_{12}E_8$. The extent of leakage as a function of time can be calculated according to $L(t) = 100 \times [I(t) - I(0)]/[I(\infty) - I(0)]$, where the fluorescence intensity at time $t$ is $I(t)$, the fluorescence intensity of the initial ANTS–DPX liposomes is $I(0)$, and that of the lysed liposomes is $I(\infty)$.

## Concluding Remarks

The simplicity and sensitivity of fluorescence assays for membrane fusion and contents leakage have rendered these assays highly useful. Some of these assays, such as the NBD–Rh assay, have also found use in monitoring the fusion of cationic liposomes or their DNA complexes with cells,[85] the fusion of secretory granules with liposomes,[5] and virus fusion with liposomes.[86,87] Improvements of the first-generation assays, such as the BODIPY and inner monolayer mixing assays, will help in the elucidation of the molecular mechanisms of membrane fusion in pure liposome systems as well as in biological membranes.

[82] F. Nicol, S. Nir, and F. C. Szoka, Jr., *Biophys. J.* **78,** 818 (2000).

[83] K. Hristova, M. E. Selsted, and S. H. White, *J. Biol. Chem.* **272,** 24224 (1997).

[84] D. Alford, V. Renugopalakrishnan, and N. Düzgüneş, *Int. J. Peptide Protein Res.* **47,** 84 (1996).

[85] P. Pires, S. Simões, S. Nir, R. Gaspar, N. Düzgüneş, and M. C. Pedroso de Lima, *Biochim. Biophys. Acta* **1418,** 71 (1999).

[86] T. Stegmann, S. Nir, and J. Wilschut, *Biochemistry* **28,** 1698 (1989).

[87] J. Ramalho-Santos, M. C. Pedroso de Lima, and S. Nir, *J. Biol. Chem.* **271,** 23902 (1996).

## [16] Liposome Fusion Assay to Monitor Intracellular Membrane Fusion Machines

By BRENTON L. SCOTT, JEFFREY S. VAN KOMEN, SONG LIU, THOMAS WEBER, THOMAS J. MELIA, and JAMES A. McNEW

## Introduction

Cellular compartmentalization and function rely on the faithful delivery of protein and lipid cargo in vesicle carriers between membrane-bound compartments.[1] The intricate process of vesicle production and delivery ultimately concludes with the specific membrane merger between the transport vesicle and the target membrane. Work has shown that specific membrane proteins, collectively known as SNAREs, are utilized to provide energy to promote the fusion reaction.[2–4] At the same time, the specific pairing of vesicle SNAREs (v-SNAREs) and target membrane SNAREs (t-SNAREs) not only provides the driving force for bilayer merger, but also provides the final layer of specificity in the process of transport vesicle docking and fusion.[5,6]

Fluorescence resonance energy transfer (FRET)-based lipid-mixing assays have long been used to study a variety of membrane fusion events[7–9] (Fig. 1). Among these are fusion events between pure artificial lipid membranes catalyzed by divalent cations or protons,[7,10,11] viral membrane fusion,[12–14] as well as experiments to study intracellular fusion events.[15–17] Many different assays have been used to demonstrate lipid mixing and thus fusion.[9] In the early 1980s Struck *et al.* developed a lipid-mixing assay based on FRET between two fluorophores, NBD and rhodamine, linked to phospholipids.[7] Because of its versatility and robustness, this assay is currently the most widely used assay.

[1] J. E. Rothman, *Nature* **372,** 55 (1994).

[2] W. Nickel, T. Weber, J. A. McNew, F. Parlati, T. H. Söllner, and J. E. Rothman, *Proc. Natl. Acad. Sci. USA* **96,** 12571 (1999).

[3] F. Parlati, T. Weber, J. A. McNew, B. Westermann, T. H. Söllner, and J. E. Rothman, *Proc. Natl. Acad. Sci. USA* **96,** 12565 (1999).

[4] T. Weber, B. V. Zemelman, J. A. McNew, B. Westermann, M. Gmachl, F. Parlati, T. H. Söllner, and J. E. Rothman, *Cell* **92,** 759 (1998).

[5] J. A. McNew, F. Parlati, R. Fukuda, R. J. Johnston, K. Paz, F. Paumet, T. H. Sollner, and J. E. Rothman, *Nature* **407,** 153 (2000).

[6] T. Söllner, S. W. Whiteheart, M. Brunner, H. Erdjument-Bromage, S. Geromanos, P. Tempst, and J. E. Rothman, *Nature* **362,** 318 (1993).

[7] D. K. Struck, D. Hoekstra, and R. E. Pagano, *Biochemistry* **20,** 4093 (1981).

[8] N. Düzgüneş and J. Bentz, in "Spectroscopic Membrane Probes" (L. M. Loew, ed.), p. 117. CRC Press, Boca Raton, FL, 1988.

[9] D. Hoekstra and N. Düzgüneş, *Methods Enzymol.* **220,** 15 (1993).

[10] N. Düzgüneş, J. Wilschut, and D. Papahadjopolous, in "Physical Methods on Biological Membranes and Their Model Systems" (F. Conti, W. E. Blumberg, J. de Gier, and F. Pocchiari, eds.), p. 193. Plenum Press, New York, 1985.

[11] N. Düzgüneş, K. Hong, P. A. Baldwin, J. Bentz, S. Nir, and D. Papahadjopolous, in "Cell Fusion" (A. E. Sowers, ed.), p. 241. Plenum Press, New York, 1987.

[12] S. Gibson, C. Y. Jung, M. Takahashi, and J. Lenard, *Biochemistry* **25,** 6264 (1986).

[13] T. Stegmann, D. Hoekstra, G. Scherphof, and J. Wilschut, *Biochemistry* **24,** 3107 (1985).

[14] G. van Meer, J. Davoust, and K. Simons, *Biochemistry* **24,** 3593 (1985).

[15] T. Kobayashi and R. E. Pagano, *Cell* **55,** 797 (1988).

[16] J. G. Orsel, I. Bartoldus, and T. Stegmann, *J. Biol. Chem.* **272,** 3369 (1997).

[17] K. Hu, J. Carroll, S. Fedorovich, C. Rickman, A. Sukhodub, and B. Davletov, *Nature* **415,** 646 (2002).

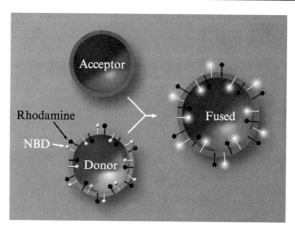

FIG. 1. Lipid-mixing assay model. A schematic version of the lipid-mixing assay utilizing the FRET pair NBD (light spheres) and rhodamine (black spheres) is depicted. A quenched mixture of head group-labeled phospholipids is reconstituted in the same donor vesicle bilayer, and the emission of NBD is muted by fluorescence resonance energy transfer. When fusion occurs between fluorescent donor and unlabeled acceptor vesicles, the fluorescent probes are diluted in the new membrane and the emission of NBD is enhanced.

Lipid-mixing assays were used to characterize the fusogenic properties of enveloped viruses by fusing viruses with liposomes.[12,13] Similar experiments have been performed with virosomes/proteoliposomes that contain the viral membrane proteins thought to be responsible for viral fusion.[18,19] From these studies considerable insight has been gained into biological fusion mechanisms in general and viral fusion in particular.

The successful application of lipid-mixing assays to viral fusion and the identification of many of the proteins involved in intracellular transport[1] led to the choice of a fluorescence-dequenching assay to analyze the eukaryotic machinery of intracellular membrane fusion. The results of these efforts ultimately led to the identification of SNARE proteins[6] as the minimal machinery for membrane fusion in eukaryotes.[4]

Similar to the reconstitution of virosomes, our application of this technique required the technical challenge of incorporating proteins into liposomes. In our case, however, both liposome populations contain proteins along with the fluorescent labels used to monitor lipid mixing. This requirement put specific constraints on our method of liposome preparation and utilization. To make sure that the fusion observed was indeed

[18] O. Eidelman, R. Schlegel, T. S. Tralka, and R. Blumenthal, *J. Biol. Chem.* **259,** 4622 (1984).
[19] M. C. Harmsen, J. Wilschut, G. Scherphof, C. Hulstaert, and D. Hoekstra, *Eur. J. Biochem.* **149,** 591 (1985).

protein mediated, we reduced the potential fusion-promoting effects of certain lipids. Importantly, we avoided the inclusion of hexagonal ($H_{II}$) phase-promoting lipids such as phosphatidylethanolamine that can enhance fusion.[10,14,20–22] We chose specifically the relatively inert lipid phosphatidylcholine as the primary lipid in our assay, based on a substantial body of work in protein-free fusion systems.[10] In addition, we purposefully increased the concentration of the FRET donor and acceptor pair up to 1.5 mol% of each fluorophore. This increase makes the system less sensitive and provides a negligible background. In most experiments, we included small amounts (15 mol%) of phosphatidylserine to aid in the reconstitution of some SNARE proteins. This addition is not required, and SNARE-mediated fusion will occur in its absence as well as in other lipid mixtures.[4,20]

## Protein Production

### General Comments

Because our main goal is to determine the fusogenic capacity of reconstituted proteins, the method of generating these proteins is crucial. To date, 21 of the 23 *Saccharomyces cerevisiae* SNARE proteins have been made as recombinant proteins in suitable quantity for reconstitution, including the exceedingly insoluble Spo20p. In addition, various mammalian SNARE proteins, as well as some invertebrate SNAREs, have been expressed and reconstituted. These purification efforts have relied entirely on affinity purification to isolate SNARE proteins expressed in bacteria. Although other expression systems may be beneficial, the quantities required for reconstitution suggest that bacterial expression is currently the best choice. We have successfully employed a number of tags in various arrangements. The primary choice for affinity tags has been a polyhistidine tag or a glutathione-*S*-transferase (GST) tag. His$_6$ tags have been located on either the N or C terminus, whereas GST is exclusively N terminal. The GST moiety often includes a protease cleavage site for tag removal. GST can be removed from the pure protein before complex formation and/or reconstitution; alternatively, GST may be cleaved once successful reconstitution has been achieved. This approach was necessary to properly

[20] B. Brugger, W. Nickel, T. Weber, F. Parlati, J. A. McNew, J. E. Rothman, and T. Sollner, *EMBO J.* **19,** 1272 (2000).

[21] H. Ellens, D. P. Siegel, D. Alford, P. L. Yeagle, L. Boni, L. J. Lis, P. J. Quinn, and J. Bentz, *Biochemistry* **28,** 3692 (1989).

[22] B. Kachar, N. Fuller, and R. P. Rand, *Biophys. J.* **50,** 779 (1986).

reconstitute the endoplasmic reticulum (ER)–Golgi t-SNARE Sed5p.[3] In some cases, removal of the GST moiety is not necessary, as seen for GST–Sec9c (Fig. 2).

A functional SNARE complex consists of a four-helix bundle.[3,23,24] Each of these helices may be derived from separate proteins, as is the case for most intracellular membranes, or one protein may provide two helical segments, as is the case for the plasma membrane.[4,5] Most SNARE proteins possess a transmembrane domain (TMD), and additional steps must be taken to preserve the solubility of these proteins. Depending on the particular trafficking step being studied, between two and four of the SNARE proteins will contain transmembrane domains. We begin with some general strategies applicable to any SNARE and then focus our attention on yeast exocytosis and the production of the SNAREs Sso1p, Sec9p, and Snc2p involved in this process.

Although every new protein is different and conditions must be determined empirically, we have found some general principles. Most membrane-integrated SNAREs have their TMD at the extreme carboxy terminus. The TMD is flanked by only two to five amino acids located inside the vesicle lumen, with a few notable exceptions. This topology likely allows the SNAREs to exist as protein rosettes in bacteria and not be integrated into the *Escherichia coli* membrane. This has not been tested directly *in vivo* in *E. coli;* however, rosettes may occur *in vitro* in the absence of detergent. For instance, the neuronal t-SNARE complex Syntaxin1A–SNAP25 is exceedingly resistant to precipitation when the purified protein is diluted with water below the critical micellar concentration (CMC) of octyl-$\beta$-D-glucoside [OG, 0.56–0.73% (w/v), 19–25 m$M$[25]]. In fact, a 10-fold dilution [0.1% (w/v), 3.3 m$M$ OG] followed by centrifugation at $\sim$100,000 $g$ resulted in virtually complete recovery of protein in the soluble fraction.[26]

Typically, we express the v-SNARE components with a C-terminal His$_6$ tag. Although this location puts the tag in the lumen of the vesicle and immediately adjacent to the membrane surface, we have not encountered substantial reconstitution difficulties. We have also successfully employed N-terminal tags for v-SNAREs. The most significant factor in determining which tag location is used is the expression level in bacteria, because tag location can affect adversely expression or solubility. The tags used to purify

[23] M. A. Poirier, W. Xiao, J. C. Macosko, C. Chan, Y. K. Shin, and M. K. Bennett, *Nat. Struct. Biol.* **5,** 765 (1998).

[24] R. B. Sutton, D. Fasshauer, R. Jahn, and A. T. Brunger, *Nature* **395,** 347 (1998).

[25] M. le Maire, P. Champeil, and J. V. Moller, *Biochim. Biophys. Acta* **1508,** 86 (2000).

[26] J. A. McNew, unpublished data (2002).

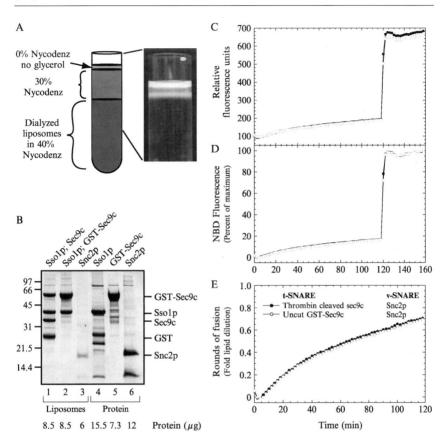

FIG. 2. SNARE-mediated membrane fusion. (A) Density gradient purification of t-SNARE liposomes. A schematic diagram of a 5-ml (13 × 41 mm) SW55 Ti tube is shown with liposomes and gradient medium additions drawn to scale. Shown next to this is a representative image of the top half of the t-SNARE gradient. (B) Coomassie-stained gel of proteoliposomes and input protein resolved on a 10% Bis-Tris NuPAGE Novex gel (Invitrogen) with MES running buffer. The proteoliposomes were prepared in the presence of 200 m$M$ KCl instead of the previously reported 400 m$M$ KCl.[5] Lanes 1 and 2 are Sso1; Sec9c t-SNARE liposomes. The proteoliposomes shown in lane 1 have been treated with human thrombin (0.05 U/$\mu$l; Sigma) for 2 h at room temperature after reconstitution, whereas those in lane 2 are untreated. Some GST-Sec9c remains uncut because of protease inaccessibility to the lumenal side of the liposome. Lane 3 shows Snc2p-containing v-SNARE liposomes. Lanes 4–6 show affinity-purified His$_8$–Sso1p, GST–Sec9c, and Snc2p–His$_6$, respectively. (C) Raw fluorescence data directly from a Fluoroskan II is plotted versus time. Fusion assays were performed, in which 5 $\mu$l of fluorescently labeled donor Snc2p liposomes (B, lane 3) were mixed with 45 $\mu$l of unlabeled acceptor t-SNARE liposomes before (open circles) or after (solid circles) thrombin cleavage. (D) The raw fluorescence was normalized to maximal detergent signal after the addition of $n$-dodecylmaltoside (DM) [0.5% (w/v), final concentration]. (E) The percentage of maximal DM fluorescence was converted to "rounds of fusion" or fold lipid dilution by an empirically derived calibration curve (see text).

t-SNARE components have much more variability. In the case of the mammalian plasma membrane, the t-SNARE complex was formed *in vivo* in bacteria by coexpressing Syntaxin1A (contributes one helix) and SNAP25 (contributes two helices). Several combinations of tags and locations were used for these two proteins. Three versions of Syntaxin1A have been made with His$_6$ tags at either the N or C terminus or with no tag at all. SNAP25 has been successfully produced with an N-terminal GST or an N-terminal His$_6$ tag. The combination that proved to be most successful was untagged Syntaxin1A coexpressed with His$_6$–SNAP25.[27]

Similar conditions were attempted for the compartmentally analogous yeast proteins Sso1p (syntaxin), Sec9p (SNAP25), and Snc2p (VAMP). However, in this case, coexpression was not required to produce a functional t-SNARE and, in fact, coexpression was severely detrimental. Coexpression of the yeast t-SNARE complex produced a conformation that was much more susceptible to *in vivo* proteolysis in bacteria. This example dramatically illustrates the point that all proteins are different and should be approached empirically.

We routinely use 25 m$M$ HEPES-KOH at a final pH of 7.4 in all our buffers for protein production and reconstitution. When buffers are made from stock solutions of 1 $M$ HEPES, this stock must be at pH 7.7 for the the 40-fold dilution (25 m$M$) to be pH 7.4. When large volumes are needed, such as for reconstitution and dialysis, buffers are made from powder stocks and the final pH of the complete buffer (without reducing agents) is adjusted to pH 7.4 with concentrated potassium hydroxide.

*Production of Snc2p–His$_6$*

The majority of our His$_6$-tagged vectors is derived from pET vectors (Novagen, Madison, WI) and are expressed in *E. coli* BL21(DE3) (Novagen). The construction of bacterial expression vectors by standard molecular biological techniques has been described for several SNAREs.[3,4,28–31] The yeast Golgi–plasma membrane v-SNARE Snc2p (or its functionally

[27] T. Weber, F. Paralti, J. A. McNew, R. J. Johnston, B. Westermann, T. H. Söllner, and J. E. Rothman, *J. Cell Biol.* **149,** 1063 (2000).

[28] R. Fukuda, J. A. McNew, T. Weber, F. Parlari, T. Engel, W. Nickel, J. E. Rothman, and T. H. Söllner, *Nature* **407,** 198 (2000).

[29] J. A. McNew, T. Weber, D. M. Engelman, T. H. Söllner, and J. E. Rothman, *Mol. Cell* **4,** 415 (1999).

[30] J. A. McNew, T. Weber, F. Parlati, R. J. Johnston, T. H. Söllner, and J. E. Rothman, *J. Cell Biol.* **150,** 105 (2000).

[31] F. Parlati, J. A. McNew, R. Fukuda, R. Miller, T. H. Söllner, and J. E. Rothman, *Nature* **407,** 194 (2000).

redundant homolog Snc1p) is expressed from the pET-derived vector pJM81 (or pJM91 for Snc1p). pJM81 (Snc2p–His$_6$) is produced by introducing a 366-bp polymerase chain reaction (PCR) fragment containing the *SNC2* open reading frame into the *NcoI–XhoI* sites of pET28a, resulting in an in-frame fusion to produce the C-terminal His$_6$ tag.[5]

*Cell Growth.* The following protocol describes the growth of 8 liters of cell culture. It may be scaled up or down accordingly. Each 8-liter preparation requires ~5 ml of medium for a "preculture," 400 ml of medium in a 2-liter baffled flask (Belco glass) for a "preculture," and 8 liters of medium for the growth. These 8 liters are separated into two 4-liter volumes, each in a 6-liter baffled flask (VWR Scientific, San Francisco, CA). If baffled flasks are not available, the bacteria can be grown in Erlenmeyer flasks lacking the baffles if the bacterial culture does not exceed one -third of the total volume of the flask.

A single colony of *E. coli* that has been transformed freshly with pJM81 is used to inoculate a 3- to 5-ml preculture of rich medium containing kanamycin (50 μg/ml) in a 14-ml round-bottom Falcon tube (BD Biosciences Discovery Labware, Bedford, MA). The preculture is grown at 37° for 8–10 h in a shaking incubator (C25KC; New Brunswick Scientific, Edison, NJ) at ~200 rpm. We routinely use super broth (SB: 32 g of Bacto-Tryptone, 20 g of yeast extract, 5 g of sodium chloride, pH 7.5) (Qbiogene, Carlsbad, CA) and have found that expression levels for Snc2p are increased when SB is used rather than 2×YT or Luria broth (LB). In addition, if a single plasmid is to be expressed, as is the case for all v-SNAREs, we prefer to use a vector that contains the kanamycin resistance marker. We have had problems in the past with β-lactamase secretion and plasmid loss for some genes (specifically with the mammalian v-SNARE VAMP2). Historically, we prevented this by carefully removing the conditioned medium from precultures and routinely adding new antibiotic throughout the growth. All this added cost and effort can be avoided by using the intracellular kanamycin phosphotransferase marker. The preculture is used to inoculate a preculture, usually 400 ml or 5% of the final culture volume, of SB containing kanamycin (50 μg/ml) in a 2-liter baffled flask. This preculture is grown overnight (12–15 h) at 37° with shaking at ~200 rpm. The cells are harvested by centrifugation in two 200-ml aliquots in sterile 250-ml conical-bottom tubes (Corning, Corning, NY) in an Allegra 6R tabletop centrifuge (Beckman Coulter, Fullerton, CA) at 3000 rpm for 10 min. The medium is removed, and the cells are resuspended in 20–30 ml of SB containing kanamycin (50 μg/ml). Each 250-ml bottle is resuspended from one of the 4-liter flasks containing medium. In this way, a similar inoculum is used for each of the separate 4-liter growths. The optical density (OD$_{600}$) of the starting culture is measured, and the cells are grown at 37° with shaking at ~200 rpm. The initial $OD_{600}$ ranges from 0.05 to 0.2, depending on the length of the preculture growth. The cells are grown until the $OD_{600}$ reaches 0.6–1.0. The cells are then induced by the addition of isopropyl-$\beta$-D-thiogalactopyranoside (IPTG, 0.2 m$M$). We have induced cells with up to 1 m$M$ IPTG, but we have not systematically varied this parameter. Induction proceeds for 4 h at 37°. The cells are harvested by two sequential centrifugations in four 1-liter bottles in a Beckman Coulter J6-MC equipped with a JS 4.2 swinging-bucket rotor at 4000 rpm for 20 min. The pelleted cells are resuspended in buffer [25 m$M$ HEPES-KOH, pH 7.4 (prepared by a 40-fold dilution of 1 $M$ HEPES-KOH, pH 7.7 stock), and 400 m$M$ KCl] and pooled into one 500-ml bottle. The washed cells are pelleted by centrifugation in a Sorvall RC-5B with a GS-3 rotor at 7000 rpm for 10 min. The wash solution is removed, and the cell pellet is frozen at −20°. These pellets may be stored at −20° for days to months.

*Preparation of Cell Extract.* Cell lysis may be achieved by any number of methods. We prefer a pneumatic homogenizer manufactured by Avestin (Ottawa, ON, Canada; www.avestin.com) called the Emulsiflex C5. This machine is reliable and capable of handling large amounts of cells. It is useful for relatively small-scale growths of 4–8 liters and capable of handling greater than 1 liter of cell paste (~300 g of wet cell paste) derived from ~100 liters of fermenter growth.

The frozen cell pellet derived from 8 liters of culture volume is resuspended in 80 ml of lysis buffer [25 m$M$ HEPES-KOH (pH 7.4), 400 m$M$ KCl, 2 m$M$ 2-mercaptoethanol, and EDTA-free protease inhibitor tablets (Roche Molecular Biochemicals, Indianapolis, IN)] on ice with a 10-ml plastic pipette. Sometimes lysis due to freeze–thawing is noticed as an increase in viscosity. If this is seen, detergent is added immediately, regardless of the degree of resuspension. Detergent [20 ml of 20% (w/v) Triton X-100, 4% (w/v) final concentration] is added to the resuspended cell paste for protein solubilization. The cell paste must be thoroughly resuspended before passage through the homogenizer. Residual cell aggregates are either disrupted by passing the extract through an 18-gauge needle or removed by a low-speed (~500-rpm) spin in a centrifuge. Lysis is achieved by two sequential passages though the Emulsiflex C5 at greater than 10,000 lb/in². The lysate is now less viscous and darker in color. Cell debris and residual *E. coli* membranes are removed by centrifugation at ~125,000 $g_{ave}$ (40,000 rpm in a Beckman Coulter–type 45 Ti rotor) for 60 min. The cleared extract is recovered into two 50-ml conical tubes, using a disposable 25-ml pipette, and the pellet is discarded.

*Affinity Purification.* The His$_6$-tagged protein is purified from the whole cell extract by affinity chromatography. We use an ÄKTAprime low-pressure chromatography system (Amersham Biosciences, Piscataway,

NJ) equipped with a 50-ml superloop and a 1-ml HiTrap HP metal-chelating column loaded with nickel according to the manufacturer's instructions. However, any convenient chromatography system may be used. We do suggest that a chromatography system be used as opposed to batch purification because detergent exchange, as well as gradient elution, is required. The HiTrap HP metal-chelating column is equilibrated in the lysis buffer containing 1% Triton X-100, and the cleared extract is passed over the column at a flow rate of 1.0 ml/min. The column is washed with 20 column volumes of lysis buffer containing 1% Triton X-100. An excess of this wash buffer is used to ensure that unbound proteins are passed through the system because bulk protein elution, seen by monitoring $A_{280}$, is obscured by the absorbance of Triton X-100. The low-CMC detergent Triton X-100 is exchanged for the high-CMC detergent $n$-octyl-$\beta$-D-glucopyranoside (OG) by washing the column in 25 column volumes of OG wash buffer [25 m$M$ HEPES-KOH (pH 7.4), 100 m$M$ KCl, 10% (w/v) glycerol, 2 m$M$ 2-mercaptoethanol, 50 m$M$ imidazole acetate (pH 7.0), 1% (w/v) OG] or until a stable, low $A_{280}$ is achieved. In this case, $A_{280}$ monitors the removal of Triton X-100 as well as contaminating proteins that are loosely bound to the affinity matrix and are eluted with 50 m$M$ imidazole. Specific protein elution is achieved by a 10-column volume linear imidazole gradient from 50 to 500 m$M$ at 1.0 ml/min, collecting 1.0-ml fractions. The final salt concentration in the purified protein is usually adjusted during the OG wash and imidazole elution steps. We often purify Snc2p in 400 m$M$ KCl to better match the salt condition in cognate t-SNARE preparations. Salt concentration is an important consideration when attempting to identify regulatory interacting proteins that impinge on SNARE complex formation or function.

*Characterization and Storage.* Gradient fractions are analyzed by sodium dodecyl sulfate–polyacrylamide gel electrophoresis (SDS–PAGE), and the protein peak is pooled. A representative 8-liter preparation will be 60–80% pure protein, with the major contaminant being an ~10-kDa N-terminal truncated form of Snc2p, and yield ~6 ml of protein at a concentration of ~1.2 mg/ml (Fig. 2B, lane 6). Aliquots of 100 $\mu$l are frozen in liquid nitrogen and stored at $-80°$.

## Production of His$_8$–Sso1p

The His$_8$-tagged t-SNARE component Sso1p is produced in much the same way as Snc2p–His$_6$. We usually produce this protein in 8-liter preparations as well and have found little difference in expression levels in various media. Cells are grown and induced using conditions similar to those described above. We have also attempted changing the growth and induction temperatures to improve expression levels. Slightly more protein may be produced when the cells are induced overnight at 16°; however, we routinely use 37° growth. A typical 8-liter preparation yields ~18 mg of protein. The major contaminants appear to be C-terminal proteolytic fragments (Fig. 2B, lane 4) that do not reconstitute (Fig. 2B, compare lanes 2 and 4).

*Production of GST–Sec9c*

GST–Sec9c is an N-terminally truncated version of Sec9p.[32] Sec9p is 651 amino acids in length, and the C-terminal ~250 amino acids are the region of clear homology with the neuronal t-SNARE component SNAP25.[33] GST–Sec9c is expressed from BB442[32] in an Epicurian Coli BL21-CodonPlus (DE3)-RIL (Stratagene, La Jolla, CA) *E. coli* strain. GST–Sec9c is clearly one of the best expressed SNAREs, and this is likely attributable to the fact that the BB442 plasmid was constructed as a synthetic gene, removing significant codon bias in this portion of the *SEC9* coding region. The CodonPlus strain of *E. coli* is used routinely when codon bias is an issue; however, in this case it is not required because of plasmid engineering. Four-liter cultures are used routinely but can be scaled easily. Culture conditions are the same as those for Snc2p–His$_6$ with the following exceptions: the cells are grown at 25° at all times and the induction time with 0.2 m*M* IPTG is reduced to 2 h. A short (2-h) induction is imperative to reduce the amount of *in vivo* proteolysis of GST–Sec9c. The cells are harvested by centrifugation and resuspended in 50 ml of lysis buffer containing 25 m*M* HEPES-KOH (pH 7.4), 400 m*M* KCl, 10% (w/v) glycerol, 2 m*M* 2-mercaptoethanol, and one complete protease inhibitor tablet. Cells are lysed by passage through the Emulsiflex C5 as described above, and cell debris is removed by centrifugation. Affinity purification of GST–Sec9c is performed in batch. Approximately 200 mg of dry glutathione–agarose beads (Sigma) is suspended in lysis buffer and swollen in a 50-ml conical tube for ~30 min at 4° on a rotating wheel. The clarified extract is bound to the equilibrated glutathione–agarose beads overnight at 4° on a rotating wheel. Unbound protein is removed by isolating the beads by centrifugation at ~2000 rpm in a Beckman Coulter Allegra 6R tabletop centrifuge and removing the supernatant with a 10-ml plastic pipette. The ~0.5 ml of packed beads is resuspended in ~10 ml of lysis buffer and transferred to a 10-ml disposable Poly-Prep column (Bio-Rad,

---

[32] L. Katz, P. I. Hanson, J. E. Heuser, and P. Brennwald, *EMBO J.* **17**, 6200 (1998).

[33] P. Brennwald, B. Kearns, K. Champion, S. Keranen, V. Bankaitis, and P. Novick, *Cell* **79**, 245 (1994).

Hercules, CA). The initial wash is removed by spinning the column at ~2000 rpm for 3–5 min in the Allegra 6R tabletop centrifuge to near dryness and collecting the wash in a 14-ml conical bottom tube. The wash is repeated three or four times with 10 ml of lysis buffer. GST–Sec9c is eluted in batch by adding 2 ml of lysis buffer containing 10 m$M$ reduced glutathione, followed by a 1-ml elution with the same elution buffer. The resin is incubated for 20–30 min at room temperature for each elution, followed by an ~2000-rpm spin for 3–5 min in the Allegra 6R tabletop centrifuge to recover the eluate. Sec9c can also be removed from the resin by thrombin cleavage. In this case, 3 ml of lysis buffer containing 80–100 units of human thrombin (T-1063; Sigma) is added and the resin is turned on a rotating wheel for 2 h at room temperature. Proteolysis is terminated by the addition of 2 m$M$ 4-(2-aminoethyl) benzenesulfonyl fluoride (Calbiochem AEBSF; EMD Biosciences, San Diego, CA). For most applications, glutathione elution is preferable because some precipitation may occur during resin-bound thrombin cleavage. A typical 4-liter preparation yields ~25–30 mg of GST–Sec9c.

## Liposome Preparation

### Lipid Stocks

Vesicle reconstitution begins with preparation of lipid stocks in chloroform. All lipid stocks are obtained from Avanti Polar Lipids (Alabaster, AL; www.avantilipids.com). A lipid mixture consisting of 85 mol% 1-palmitoyl-2-oleoyl phosphatidylcholine (POPC) and 15 mol% 1,2-dioleoyl phosphatidylserine (DOPS) is prepared at a final lipid concentration of 15 m$M$. These lipids will be used to generate unlabeled acceptor t-SNARE vesicles. A second lipid mixture containing these lipids in addition to the head group-labeled fluorescent lipids ($N$-(7-nitro-2,1,3-benzoxadiazole-4-yl)-1,2-dipalmitoyl phosphatidylethanolamine (NBD-DPPE) and $N$-(Lissamine rhodamine B sulfonyl)-1,2-dipalmitoyl phosphatidylethanolamine (rhodamine-DPPE) at an 82:15:1.5:1.5 mole ratio is prepared at a final lipid concentration of 3 m$M$. The second lipid mixture will be used to prepare the labeled donor v-SNARE vesicles. The initial choice of which protein to reconstitute into donor or acceptor liposomes is somewhat arbitrary. Fusion will also occur if the fluorescent label is located in the t-SNARE liposome population.[4] The seemingly reduced fusion efficiency with fluorescently labeled t-SNAREs can now be readily explained when the number of v- and t-SNAREs per liposome is considered (see below). These lipid stocks are usually prepared in large quantities (100–200 ml) and aliquoted into glass ampoules. The ampoules are heat-sealed under nitrogen and stored at $-80°$. This prevents variations in lipid concentration that could occur as a result of evaporation of the volatile solvent chloroform.

Working stocks of lipids are prepared by transferring an ampoule of lipid ($\sim$3–4 ml) to a 10-ml Reacti-vial (Pierce Biochemicals, Rockford, IL) followed by the addition of trace amounts ($\sim$1 $\mu$l of a 250-$\mu$Ci/ml solution per 1 ml of lipid) of tritiated 1,2-dipalmitoyl phosphatidylcholine ([$^3$H]DPPC; Amersham Biosciences). The resulting lipids yield $\sim$1000 cpm/$\mu$l of working stock lipids. For each reconstitution, 100 $\mu$l of the lipids (1500 nmol of nonfluorescent acceptor lipid for t-SNAREs and 300 nmol of fluorescent lipids for v-SNAREs) is dried in 10 $\times$ 75 mm glass test tubes by a gentle stream of nitrogen gas and any remaining traces of chloroform are then removed under vacuum for 30 min. Multiple samples can be dried, using an Evap-O-Rac apparatus (Cole Parmer, Vernon Hills, IL). The resulting lipid films are used for reconstitution.

*Lipid Resuspension*

The lipid films are first dissolved by the addition of SNARE proteins in detergent (1% OG). For v-SNARE reconstitutions, this is accomplished by adding 100 $\mu$l of the v-SNARE protein to the lipid. If less than 100 $\mu$l of v-SNARE protein is used, the remaining volume should include the "reconstitution buffer" [25 m$M$ HEPES-KOH, 400 m$M$ KCl, 10% (w/v) glycerol, 1 m$M$ dithiothoeitol (DTT)] containing 1% OG. For the yeast v-SNARE Snc2p, 50 $\mu$l of protein is usually a sufficient quantity. For t-SNARE gradients 500 $\mu$l of t-SNARE is used. In the case of the yeast t-SNAREs, the t-SNARE complex of Sso1p and Sec9c is preformed by incubating the proteins overnight at $4°$ before reconstitution. This reaction typically contains 250 $\mu$l of His$_8$–Sso1p ($\sim$400 $\mu$g, $\sim$12 $\mu$mol, $\sim$24 $\mu M$), 250 $\mu$l of GST–Sec9c ($\sim$1.8 mg, $\sim$33 $\mu$mol, $\sim$66 $\mu M$), and 13 $\mu$l of 20% (w/w) OG. Additional OG is required to maintain a 1% OG concentration, because GST–Sec9c is prepared without detergent. The $\sim$3-fold molar excess of GST–Sec9c is essential to drive the binary reaction to completion in an 18- to 24-h time period.[34] The lipid films are resuspended by vortexing vigorously for 15 min at room temperature. This time is usually sufficient to completely resuspend the lipid film; however, if visual inspection suggests that part of the film remains, vortexing should be continued until it is fully dissolved.

---

[34] K. L. Nicholson, M. Munson, R. B. Miller, T. J. Filip, R. Fairman, and F. M. Hughson, *Nat. Struct. Biol.* **5,** 793 (1998).

*Detergent Dilution*

The protein–detergent–lipid micelles are converted to liposomes by detergent dilution followed by dialysis. The 100-$\mu$l v-SNARE mixture is diluted with 200 $\mu$l of room temperature reconstitution buffer (1 m$M$ final lipid concentration, 0.33% OG), and the 500-$\mu$l t-SNARE mixture is diluted with 1.0 ml of room temperature reconstitution buffer (1 m$M$ final lipid concentration, 0.33% OG). The dilution below the CMC of OG [0.56–0.73% (w/v), 19–25 m$M$[25]] is performed slowly by adding buffer dropwise to a vigorously vortexed sample. The clear solution of t-SNARE protein–detergent–lipid micelles will cloud briefly, then appear opalescent. The micelle–vesicle transition likely occurs at this step.[35] The v-SNAREs will behave similarly, but it is more difficult to notice because of the red coloration of the fluorescent lipids.

*Dialysis*

After detergent dilution, detergent monomers are removed by dialysis against reconstitution buffer [25 m$M$ HEPES-KOH, 400 m$M$ KCl, 10% (w/v) glycerol, 1 m$M$ DTT]. This buffer is made in a 4-liter volume, and the pH of the final solution is adjusted to pH 7.4 with concentrated KOH. Dialysis may be achieved by simple bulk dialysis, using a dialysis bag in a large (4- to 5-liter) reservoir of buffer or, more conveniently, in a controlled flow-rate microdialysis chamber. The latter allows a greater number of samples to be dialyzed in an equivalent volume of buffer. We utilize a GIBCO-BRL (Gaithersburg, MD) microdialysis chamber. The apparatus is available in 8-well and 24-well sizes. Both work equally well, and sample preparation number determines which apparatus is used. Flat dialysis membranes (20-kDa cutoff) may be purchased from GIBCO-BRL; however, we have found that traditional dialysis tubing (Spectrapor, 6–8 kDa; Spectrum Laboratories, Rancho Dominguez, CA) may be used by carefully cutting the flat tubing to make a flat sheet. The dialysis tubing is prepared by hydrating it in water for 5–10 min before use. The microdialysis chamber is assembled and moved to a 4° cold room, along with the samples and dialysis buffer. The microdialysis chamber is attached to a peristaltic pump (Minipuls 3; Gilson, Middleton, WI), the samples are added to individual chambers, and the empty chambers are filled with reconstitution buffer. Dialysis buffer is pumped through the apparatus at a flow rate of ~3–4 ml/min for ~12–15 h (~3 liters of buffer).

[35] M. Ollivon, S. Lesieur, C. Grabielle-Madelmont, and M. Paternostre, *Biochim. Biophys. Acta* **1508**, 34 (2000).

Although dialysis is efficient, it should be noted that small quantities of residual detergent might remain because of detergent trapped in the inner leaflet of the newly formed vesicles that exchanges slowly with the detergent-free dialysis buffer.

### Proteoliposome Recovery

A key difference between traditional liposome preparation and proteoliposome preparation is the requirement to separate proteoliposomes from unincorporated protein. We achieve this separation by flotation in a density gradient. The proteoliposomes will float because of the presence of lipid and the lower buoyant density that the lipids provide, while free protein or protein aggregates remain at the bottom of the gradient. Several density gradient materials have been tested, and by far the best results are obtained with Nycodenz (5-[N-(2,3-dihydroxypropyl)acetamido]-2,4,6-triiodo-N,N'-bis(2,3-dihydroxypropyl) isophthalamide). We currently purchase Nycodenz from Accurate Chemicals (Westbury, NY), who have renamed the product Accudenz. Proteoliposomes containing neuronal SNAREs can also be isolated by sucrose density gradient centrifugation; however, recoveries are substantially lower, likely because of the much higher osmotic activity of sucrose. Nevertheless, the liposomes recovered by this method retain their fusion activity.

*Preparation of Gradient Material.* We currently use a three-step density gradient to isolate proteoliposomes (Fig. 2A). For these gradients, solutions of Nycodenz at 80% (w/v) and 30% (w/v) are needed. These are prepared by weighing the appropriate mass of Nycodenz powder in a 50-ml conical tube. One-fifth volume of a 5-fold-concentrated reconstitution buffer [125 m$M$ HEPES-KOH, 2 $M$ KCl, 50% (w/v) glycerol, final pH 7.7] (without reducing agent) is added to the powder. Additional water is added to dissolve the Nycodenz. This solution is turned on a rotating wheel overnight to fully resuspend the Nycodenz, and the volume is finally adjusted with water after complete resuspension. Kept at 4°, these solutions are stable for weeks to months; alternatively, they may be frozen at −20°. The 80% (w/v) Nycodenz solution will be extremely viscous when kept at 4°. Concentrated reducing agent (from a 1 $M$ DTT stock) is added to a final concentration of 1 m$M$ to all gradient solutions immediately before use.

*Gradient Preparation.* The dialyzed liposomes are recovered from the dialysis chamber into a 1.5-ml microcentrifuge tube for the ∼300 $\mu$l of v-SNARE liposomes and into a 15-ml conical-bottom tube for the ∼1.5 ml of t-SNARE liposomes and kept on ice. Alternatively, the t-SNARE liposomes may be recovered directly into a 5-ml SW 55 Ti

Ultra-Clear tube (see below). An equal volume (300 μl for v-SNAREs and
1.5 ml for t-SNAREs) of 80% Nycodenz in reconstitution buffer is added
to the appropriate proteoliposomes and mixed, resulting in a 40% (w/v)
Nycodenz solution containing the dialyzed proteoliposomes and
unincorporated proteins. This solution is transferred to ultracentrifuge
tubes. The original description[4] of liposome recovery by gradient flotation
called for two gradient systems: an SW 55 Ti rotor for v-SNARE
flotation and an SW 60 Ti rotor for t-SNARE flotation. We have subsequently adapted this protocol such that a single rotor, an SW 55 Ti, can
be used. The v-SNARE gradients are produced in a manner identical to
that previously reported.[3–5,27–31] The ~600 μl of 40% (w/v) Nycodenz solution containing the dialyzed v-SNARE proteoliposomes and unincorporated protein is split equally into two 900-μl (5 × 41 mm) SW 55 Ti
Ultra-Clear tubes (344090; Beckman Coulter). The next layer is 250 μl of
30% (w/v) Nycodenz in reconstitution buffer, which is layered carefully
on top of the 40% Nycodenz and proteoliposome mixture. The final layer
is 50 μl of reconstitution buffer without glycerol (25 mM HEPES-KOH,
400 mM KCl, and 1 mM DTT adjusted to pH 7.4). This buffer contains
no Nycodenz and no glycerol.

The t-SNARE gradients have been modified to conform to a standard
5-ml (13 × 51 mm) SW 55 Ti Ultra-Clear tube (344057; Beckman Coulter).
The ~3.0 ml of 40% (w/v) Nycodenz solution containing the dialyzed
t-SNARE proteoliposomes and unincorporated protein is transferred to
the SW 55 Ti tube or simply mixed if the liposomes were recovered directly
in the tube. The second layer of this gradient is also 30% Nycodenz. For the
modified SW 55 gradient, 1.5 ml of the 30% solution is layered onto the
40% solution, as opposed to 750 μl for the smaller SW 60 tube. The final
layer is 250 μl of reconstitution buffer without glycerol.

The gradients are loaded into the SW 55 Ti rotor (with split adapters
for the v-SNARE gradients) and centrifuged at 48,000 rpm (maximal
allowable velocity with split adapters) for 4 h at 4°. Because each v-
SNARE reconstitution requires two small gradients, a maximum of three
v-SNARE reconstitutions can be performed with one SW 55 rotor. An
additional benefit of the modified t-SNARE gradient is the ability to
centrifuge v-SNARE and t-SNARE gradients in the same rotor at the
same time.

It is also possible to produce v-SNARE liposomes on a larger scale
using 5-ml (13 × 51 mm) SW 55 Ti Ultra-Clear tubes and t-SNARE
liposomes using a 13.2-ml (14 × 89 mm) SW 41 tube when large volumes
of a single proteoliposome type are desired.

*Gradient Harvesting.* Both the t-SNARE and v-SNARE proteoliposomes float to the 0%/30% Nycodenz interface. The t-SNAREs appear

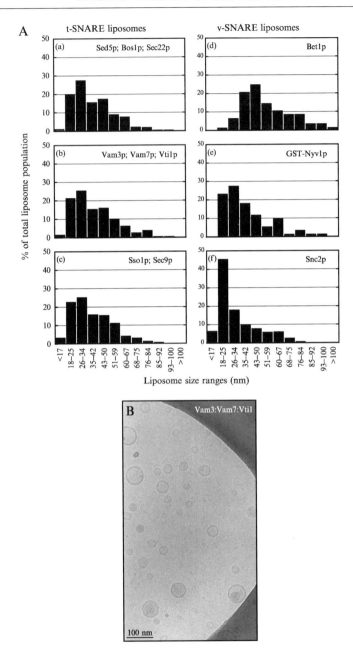

FIG. 3. Size distributions of reconstituted liposomes with various yeast v-SNAREs or t-SNARE complexes. All t-SNARE and v-SNARE liposomes have approximately equal concentrations of SNARE complexes. (A) Liposome diameters were measured directly from

as an opaque band against a black background (Fig. 2). For each t-SNARE gradient, 400 $\mu$l of proteoliposomes is recovered from the density gradient with a pipette. For each v-SNARE gradient, 75 $\mu$l of the red fluorescent proteoliposomes is recovered and the two gradients from a single v-SNARE reconstitution are pooled, resulting in 150 $\mu$l of v-SNAREs. Proteoliposomes may be stored for days to weeks at 4°. When the proteoliposomes are flash-frozen in liquid nitrogen (which does not change their size appreciably), they can be stored for months to years at $-80°$ without apparent loss of activity.

### Proteoliposome Characterization

The recovered proteoliposomes are characterized routinely by determining lipid recovery and protein content. Lipid recovery is determined by scintillation counting [³H]DPPC in the recovered material compared with the input lipid stocks. Lipid recoveries range from 40% to 70% of input lipid.

Protein content is analyzed by SDS–PAGE (Fig. 2) and quantitated by amido black staining.[36] This protein assay is not affected significantly by lipid, detergent, or reducing agent present in the various samples of pure protein or proteoliposomes. SNARE reconstitution efficiency varies from ~10% to 50%, depending on which particular SNARE protein or complex is used. These estimates may vary depending on the methods used to estimate recovery (SDS–polyacrylamide gel comparisons versus total protein measured by amido black).

*Electron Microscopy.* We have utilized cryoelectron microscopy to examine the appearance of proteoliposomes. This method has certain advantages over our previous negative stain transmission electron microscopy (TEM) analysis, which suffered from sample dehydration and particle flattening. As a result, cryoelectron microscopy provides a more reliable method to determine the size distribution of various liposome populations. Figure 3A shows the quantitative results of such analysis.

---

[36] W. Schaffner and C. Weissmann, *Anal. Biochem.* **56,** 502 (1973).

---

cryoelectron micrographs under a reticulated magnifying glass. The histograms represent measurements from two or three separate cryoelectron grids and four or more individual electron micrographs. The statistics for the liposomes included in the histograms are as follows: $n$ = number of liposomes, mean ± SD: Sed5p; Bos1p; Sec22p, $n$ = 439, 42 ± 15 nm; Vam3p; Vam7p; Vti1p, $n$ = 319, 43 ± 16 nm; Sso1p; Sec9p, $n$ = 285, 41 ± 15 nm; Bet1p, $n$ = 98, 59 ± 18 nm; GST–Nyv1p, $n$ = 96, 43 ± 17 nm; Snc2p, $n$ = 275, 35 ± 15 nm. (B) A representative cryoelectron micrograph is shown for liposomes containing Vam3p; Vam7p; Vti1p.

Gradient-purified proteoliposome samples containing various yeast SNARE proteins are applied to perforated carbon–Formvar electron microscope grids (400 mesh copper; E. F. Fullam, Latham, NY), formed by evaporating a thin layer of carbon onto a Formvar meshwork.[37] Initially, only liposomes suspended in the vitreous ice within a hole in the carbon mesh were imaged, as this provides the greatest degree of contrast. However, we observed that liposomes of some protein compositions appeared to stick almost exclusively to the carbon mesh and were absent from the holes. Therefore, we routinely cover the entire preformed carbon–Formvar meshwork with an additional, thin layer of carbon formed by evaporation onto the clean surface of mica (E. F. Fullam). Immediately before application of sample (usually 5–10 $\mu$l of a 150–300 $\mu M$ lipid sample), the grids are exposed to a short glow discharge. The proteoliposome samples are left on the grid for $\sim$2 min, and then excess solution is wicked off and the grids are rapidly plunged into liquid ethane and stored in liquid nitrogen. The frozen samples are imaged on a Philips CM12 electron microscope (FEI, Hillsboro, OR) operating at 120 kV. Typical magnifications are between $\times$30,000 and $\times$65,000. The images are taken at a nominal defocus of 2 $\mu$m. Figure 3B illustrates a representative image of the yeast vacuolar t-SNARE complex (Vam3p, Vti1p, and Vam7p[5,28]) used in the quantitation shown in Fig. 3A. All liposome diameters are measured directly off film to the nearest 0.5 mm under a magnifying glass with associated reticle and placed into bins accordingly. The recorded diameters are the longest measurable distances parallel to the long axis of the electron micrograph.

Fusion Assay

*Equipment*

Lipid mixing is monitored at 37° with a Fluoroskan II (MTX Labsystems, Vienna, VA; www.mtxlsi.com) fluorescence plate reader connected to a PC running Spectrosoft software. We have also utilized an older version of this software for the Macintosh, called DeltaSoft. Although DeltaSoft is no longer supported by MTX Labsystems, it has several features that make it preferable to Spectrosoft, such as the ability to define specific wells to be read rather than reading the entire 96-well plate at every time point. NBD fluorescence is monitored with filters set at 460 nm (excitation, half-bandwidth, 25 nm) and 538 nm (emission, half-bandwidth, 25 nm) at 2-min intervals. Fusion reactions are performed and analyzed in

---

[37] A. Fukami and K. Adachi, *J. Electron Microsc.* **14,** 112 (1965).

FluoroNunc PolySorp white 96-well microtiter plates (Nalge Nunc, Rochester, NY). Similar results have been obtained with a SPECTRAmax Gemini plate reader (Molecular Devices, Sunnyvale, CA; www.molecular-devices.com/) and, when scaled up, with a cuvette-based PerkinElmer (Norwalk, CT) LS 50B spectrophotometer.[4]

## Fusion Assays

Typically, 5 μl of fluorescently labeled v-SNARE donor vesicles (5.7 nmol of phospholipid, 2 μg of Snc2p, corresponding to 145 pmol of Snc2p) is mixed with 45 μl of unlabeled t-SNARE acceptor vesicles (96 nmol of phospholipid, 78 μg of t-SNARE complex corresponding to 861 pmol) directly in a microtiter plate on ice. These volumes were originally based on the reconstitution efficiency of the neuronal SNARE complex.[4] The 9:1 volume ratio of t-SNARE liposomes to v-SNARE liposomes is used to offset the difference in protein content.[3,4] An approximately equal molar ratio of t-SNARE protein to v-SNARE protein is produced with these volumes. As reconstitution efficiencies have improved, these values are somewhat different, although we still maintain the 9:1 volume ratio in all experiments. For yeast SNAREs, the plate is usually placed directly in the 37° fluorescence plate reader without any prior preincubation. If the reactions require preincubation, the plate wells are sealed with clear tape and wrapped in aluminum foil, and the plate is agitated gently in the dark at 4° overnight. NBD fluorescence is monitored for 2 h at 2-min intervals. At the end of 120 min, 10 μl of 2.5% (w/v) n-dodecylmaltoside (DM) (Roche Molecular Biochemicals) is added to determine NBD fluorescence at infinite dilution and the reaction is continued for 40 min. Several detergents have been analyzed to avoid the reduction in NDB quantum yield seen with Triton X-100. Triton X-100 typically results in 30–40% reduction in total NBD fluorescence at infinite dilution compared with n-dodecylmaltoside.[38] Our observed reduction of NBD fluorescence yield with Triton X-100 is in good agreement with the correction factor derived by measuring NDB quantum yield directly with low quantities of fluorophore in the presence or absence of Triton X-100.[9] We have chosen to substitute n-dodecylmaltoside for Triton X-100 to minimize this effect.

## Data Analysis

The raw fluorescence kinetic fusion data (Fig. 2C) are exported from Spectrosoft as a Microsoft Excel file and imported into KaleidaGraph

---

[38] B. L. Scott and J. A. McNew, unpublished data (2002).

(Synergy Software, Reading, PA) graphing software. Each time point is converted to percent detergent signal (Fig. 2D) by first subtracting the lowest fluorescence value from all time points. This value for each time point is then divided by the highest value after detergent addition and multiplied by 100. This is followed by the conversion to "rounds of fusion" (Fig. 2E) or, more precisely, fold lipid dilution, using an empirically derived calibration curve.

*"Rounds of Fusion" Calculations.* Lipid-mixing assay data have traditionally been represented as a percentage of NBD fluorescence at "infinite dilution," that is, in the presence of a large excess of detergent (Fig. 2D). However, when this method is used, a working knowledge of the lipid-mixing assay itself is needed to easily glean quantitative information from these results. To allow kinetic analysis and more easily present quantitative fusion data, we sought to represent our fusion data in a more intuitive unit, specifically "rounds of fusion." We decided to calibrate the detergent-normalized NBD fluorescence to fold lipid dilution in much the same way that protein assay standard curves are generated. In this way, a given percentage of NBD fluorescence can be correlated mathematically to a physical parameter, fold donor lipid dilution. With a few stated assumptions, fold donor lipid dilution is identical to rounds of donor vesicle fusion.[3]

The calibration curve used to convert percent of maximal detergent signal into fold lipid dilution is generated by preparing a series of proteoliposomes that contain different ratios of donor and acceptor lipids.[3] Fluorescent donor lipids are premixed with unlabeled acceptor lipids in chloroform to yield a specific mixture that would be produced by a specified number of fusion events of labeled donor vesicles with unlabeled acceptor vesicles. For example, if every donor liposome in the population participated in one round of fusion with acceptor liposomes, the resulting lipids would be a 1:1 mixture of fluorescent donor lipids and unlabeled acceptor lipid. Lipid mixtures that mimicked between 0 and 8 rounds of fusion are used to prepare proteoliposomes. v-SNARE protein is included in all these preparations to provide a protein-to-lipid ratio comparable to the standard liposome preparation. These proteoliposomes are made and isolated as described above. The samples are standardized by the degree of total lipid recovery (determined by tracer [³H]DPPC).

The NBD fluorescence of a fixed amount of total fluorescent lipid, equivalent to 5 μl of undiluted liposomes, is measured for all the liposome samples in the dilution series. For example, 5 μl of undiluted donor liposomes contains ~5 nmol of total lipid. To analyze a similar amount of total fluorescent lipid for the 8-fold diluted liposomes, ~45 nmol of total lipid must be measured. In this case, 40 nmol of bulk lipid must be added to the undiluted liposome sample to maintain a constant total phospholipid concentration.

The NBD fluorescence of these lipid mixtures is measured, and the absolute amount of NBD fluorescence is determined before and after detergent addition. The difference in fluorescence before and after detergent dilution at each fold lipid dilution is normalized relative to the undiluted sample. The fold lipid dilution ($y$ axis) is then plotted versus percentage of maximum NBD fluorescence ($x$ axis), and a double exponential fitting procedure yields the following equation: $Y = 0.49666 \times e^{(0.036031X)} - 0.50597 \times e^{-0.053946X}$, where $Y$ is the fold lipid dilution and $X$ is the percent dodecylmaltoside signal at a given time interval. Interestingly, the curve obtained in this manner is similar to a curve based on theoretical calculations[3,39] (using a Förster radius of 65 Å for the FRET pair NBD and rhodamine[8]). A different, but quantitatively similar, equation is obtained when different absolute amounts of fluorescent lipids are used (1.5 mol% versus 2.0 mol%) as well. In addition, the relationship does not change when different sources of v-SNAREs are used (VAMP2 versus Snc2p versus Nyv1p), illustrating that the protein content of these liposomes makes a negligible contribution to the resulting curve.

As mentioned above, the fold lipid dilution calculation may be interpreted as rounds of vesicle fusion by making a few assumptions: (1) the v- and t-SNARE liposomes are equal in size, such that a fusion event results in a doubling of surface area, and (2) all liposomes in the population have an equal probability of fusion. Slight deviations from these assumptions have predictable outcomes, and their effects can be corrected in data analysis. For example, the mean size of Snc2p v-SNAREs (~35 nm) is somewhat smaller than those of the cognate t-SNARE population Ssolp; Sec9c (~41 nm) (Fig. 3A). This means that for a given fusion event, the fluorescent lipid in the Snc2p liposomes will be diluted more in the new vesicle membrane than would be predicted if the vesicles were equal in size. This fact will overestimate slightly the rounds of fusion calculation by the same factor as the difference in surface area, in this example 1:1.4 instead of 1:1 for vesicles of equal size.

It should also be noted that the proteoliposomes in our (or others) assays are not uniform in size, but are a population centered on a mean (Fig. 3A). Consequently, it is possible that proteoliposomes of different sizes have different propensities to fuse. In pure lipid systems, small liposomes have a comparatively higher probability to fuse.[10] It is unclear

[39] B. K. Fung and L. Stryer, *Biochemistry* **17,** 5241 (1978).

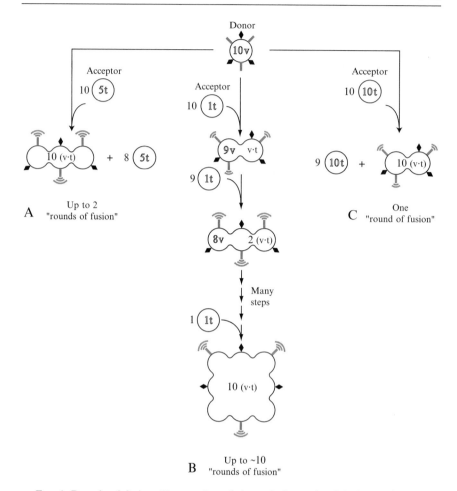

FIG. 4. Rounds of fusion. The number of theoretical rounds of fusion is determined primarily by the relative surface density of the SNARE proteins. A 10:1 volume ratio is maintained throughout. Scheme B illustrates an approximately 10-fold greater v-SNARE surface density (**10v** in the fluorescently labeled donor liposome compared with unlabeled t-SNARE vesicles (**1t**). After one round of vesicle fusion, the lipid probes rhodamine (black diamond) and NBD (gray bar) are diluted in the 2-fold surface area increase. This results in a measured increase in NDB fluorescence. The new vesicle now contains one equivalent of an inert *cis*-SNARE complex (v·t) containing both v- and t-SNARE in the same membrane as well as nine equivalents of unreacted, functional v-SNAREs. This new, larger donor vesicle can participate in further fusion events until all of the v-SNARE resides in *cis*-SNARE complexes. Each new fusion event dilutes further the lipid probes and increases the NDB fluorescent signal; however, each subsequent fusion event provides a smaller increase than the last, because of the strong distance relationship of fluorescence resonance energy transfer. Scheme A shows an increase in t-SNARE surface density (**5t**) such that

whether SNARE-containing liposomes of different sizes are also subject to these fusion differences because smaller liposomes will inevitably also have a reduced number of SNAREs per liposomes. This question awaits experimental resolution.

The major determinant of the ability of SNARE-containing proteoliposomes to fuse is the surface density of the incorporated SNARE proteins. Various SNARE proteins have defined affinities for a specific complex or subcomplexes. For this reason, the overall concentration of the SNARE proteins is an important parameter and one of the major reasons that the total amount of lipid in our fusion assays is at least an order of magnitude greater than the lipid-only or viral fusion systems (1–2 m$M$ versus 50–100 $\mu M$). This amount of material is necessary to mimic conditions seen *in vivo* and provides the SNARE proteins sufficient mass to interact and fuse on a reasonable time scale.

The extent of SNARE incorporation also determines the potential number of rounds of fusion a given vesicle may undergo (Fig. 4). The extent of SNARE reconstitution is quantified by measuring the lipid-to-protein ratio of a purified proteoliposome population. Determination of lipid recovery is easy and reliable because it is measured directly from co-reconstituted tracer [$^3$H]DPPC. The amido black protein assay of purified proteoliposomes also provides a quantifiable measure of protein recovery. In the context of potential rounds of fusion, it is useful to approximate the average number of SNAREs per liposome. This calculation takes into consideration the mean diameter of the liposome population (measured by EM) and makes the following assumptions: a bilayer thickness of 4 nm and a phospholipid head group area of 65 Å$^2$. The number of phospholipid molecules in both the inner and outer leaflets is calculated from the surface area of a sphere of a fixed diameter and the assumed head group area of lipid. Once the total number of lipid molecules is determined, this number is divided by the lipid-to-protein ratio, yielding the total number of SNAREs per vesicle.

Although this calculation is useful to evaluate the average number of SNAREs per liposome, it does not take into account variability within the population. For example, it is possible that reconstitution results in a variable number of SNAREs per liposome with some liposomes containing few, if any, reconstituted proteins. This interpretation is supported by the

---

the ratio of v-SNARE to t-SNARE is 2:1. In this case, five equivalents of *cis*-SNARE complexes are produced in the first round of vesicle fusion, leaving five functional v-SNAREs remaining to be spent in the second and final fusion event. Scheme C depicts equivalent surface densities of t- and v-SNAREs, in which one fusion event utilizes all of the SNAREs involved.

appearance of the diffuse band of recovered proteoliposomes in the density gradient (Fig. 2A), although detailed analysis of specific regions within the diffuse band has not been performed. It is technically difficult, if not impossible, to quantitatively analyze these populations.

In the original incarnation of the rounds of fusion calculation,[3] the number of v-SNAREs reconstituted was different from the number of t-SNAREs. In fact, the ratio of v-SNAREs to t-SNAREs was approximately 10:1. This was the primary reason we chose a 9:1 (45 $\mu$l:5 $\mu$l) volume ratio for input liposomes in the fusion assay. In this way, the total amount of input v- and t-SNAREs would be equivalent. In this scenario, the theoretical maximum number of rounds of fusion would be approximately 10 (Fig. 4, scheme B). If the number of v-SNARE proteins per liposome were more similar to the number of t-SNARE complexes in a merging population, then the theoretical fusion maximum for these populations would be decreased (Fig. 4, scheme A). For equivalent densities of v- and t-SNAREs, the result of a single fusion event would yield a new vesicle with double the surface area (assuming equal size) and the new membrane would contain only inert *cis*-SNARE complexes (Fig. 4, Scheme C).

Although the rounds of fusion (fold donor lipid dilution) calculation is a preferable way in which to express the rate and extent of fusion, care must be taken, because this parameter is also somewhat dependent on the relative protein amount incorporated into the different vesicle populations.

Conclusions

We have adapted a well-established FRET-based lipid-mixing assay to measure fusion driven by reconstituted SNARE proteins. Our application of this assay requires the functional reconstitution of a few to several membrane proteins along with diagnostic lipids. This modified assay has been used successfully to reconstitute fusion driven by seven different sets of SNARE proteins.[4,5,28,31,40,41]

One of the remaining challenges is the development of a routine liposome content-mixing assay to complement the present lipid-mixing assay. Content mixing driven by SNAREs has been demonstrated unambiguously, utilizing a novel oligonucleotide-based content-mixing assay.[2]

[40] F. Parlati, O. Varlamov, K. Paz, J. A. McNew, D. Hurtado, T. H. Sollner, and J. E. Rothman, *Proc. Natl. Acad. Sci. USA* **99**, 5424 (2002).

[41] F. Paumet, B. Brugger, F. Parlati, J. A. McNew, T. H. Sollner, and J. E. Rothman, *J. Cell Biol.* **155**, 961 (2001).

Unfortunately, this assay requires large amounts of modified oligonucleotide and is expensive. The prohibitive cost makes this assay untenable as a routine content-mixing assay. We have made several attempts to adapt traditional fluorophore-based content-mixing assays,[8,42] to the SNARE-mediated fusion system with little success. The ANTS–DPX assay was reported to be problematic with small unilamellar liposomes,[8,42] and we have since confirmed this observation. The other major assay, generation of the fluorescent terbium–dipicolinic acid chelate complex, has significant drawbacks for our application as well. The major obstacle for this assay is the presence of significant quantities [$\sim$10% (w/v)] of Nycodenz in our gradient-purified proteoliposomes. Nycodenz has an absorption maximum at 245 nm and a large, experimentally determined extinction coefficient ($\varepsilon_{245} = \sim$28,000 $M^{-1}$ $cm^{-1}$),[43] and the primary excitation wavelength for terbium is at 276 nm. We made several attempts to remove Nycodenz by extensive dialysis and gel filtration; unfortunately, we could not remove sufficient quantities to allow practical application of this assay. We also attempted gradient purification with other density media such as sucrose, glycerol, and Ficoll, as well as various combinations of the above to eliminate the need for Nycodenz. Although these other gradients worked to some degree, poor recovery compared with Nycodenz makes their application impractical.

*Future Directions*

The availability of a relatively simple and extremely robust assay to study SNARE-mediated membrane fusion permits *in vitro* analysis of many events that precede and regulate the fusion event. This assay, which provides a fully reconstituted system, is philosophically different from other assays that use broken cells[44] or intact organelles.[45,46] Each of these assays has its own advantages and disadvantages, and they are, in many ways, complementary.

We are currently utilizing reconstituted yeast post-Golgi transport reactions to analyze the role of regulatory proteins such as Sec1p and Sec4p in the assembly and function of the SNARE complex. In addition to focusing on particular proteins with biochemically or genetically established roles, we are beginning to develop assays to uncover potentially unanticipated function. These efforts are directed at bridging the gap between the

[42] N. Düzgüneş and J. Wilschut, *Methods Enzymol.* **220,** 3 (1993).
[43] S. Liu and J. A. McNew, unpublished data (2002).
[44] Y. A. Chen, S. J. Scales, S. M. Patel, Y. C. Doung, and R. H. Scheller, *Cell* **97,** 165 (1999).
[45] A. Haas, B. Conradt, and W. Wickner, *J. Cell Biol.* **126,** 87 (1994).
[46] W. Wickner and A. Haas, *Annu. Rev. Biochem.* **69,** 247 (2000).

completely synthetic system and the broken cell or whole organelle assays. To this end, we are attempting to reconstitute fusion between secretory vesicles purified from yeast with t-SNARE-containing proteoliposomes. An analogous effort is being made to generate inside-out plasma membrane vesicles from yeast[47] to serve as a fusion partner for synthetic v-SNARE proteoliposomes. These assays can address requirements for membrane proteins present in their native membranes, as well as the potential for specific lipid effects.

A step in this direction has been taken with functionally reconstituted t-SNARE vesicles and purified synaptic vesicles.[17] This group showed that synthetic t-SNARE-containing liposomes were capable of fusing with purified synaptic vesicles containing native membrane and all the additional proteins present in the synaptic vesicle membrane. This work suggests that additional proteins in the vesicle associate with the v-SNARE VAMP to regulate the availability of this SNARE. These types of experiments are now possible and should provide important insights into the regulation of the process of protein-mediated membrane fusion.

### Acknowledgments

The authors thank Jim Rothman, in whose laboratory this work was initiated, and Craig Foster (Foster Medical Communication) for Fig. 1. Grant support was provided by the Robert A. Welch Foundation (J.A.M. and S.L.), and by NSF-IGERT training grant DGE-0114264 (J.V.K.).

[47] A. Menendez, C. Larsson, and U. Ugalde, *Anal. Biochem.* **230,** 308 (1995).

## [17] Liposomes in the Study of Membrane Fusion in Neutrophils

*By* James E. Smolen

### Introduction

Exposure of human neutrophils to a variety of particulate and soluble stimuli evokes a series of responses, including chemotaxis, phagocytosis, degranulation, hexose monophosphate shunt stimulation, generation of reactive derivatives of oxygen, release of membrane-bound calcium, and reorganization of the cytoskeleton. All these responses and the signal transduction processes mediating them are of considerable scientific and clinical interest. However, the least understood of these responses is degranulation, particularly the fusion of granule membranes with phagosomes or plasma membrane (with the accompanying discharge of granule contents and expression of granule membrane components on the cell surface).

Degranulation is rapid[1] and apparently highly efficient, because intracellular liberation of granule contents has not been reported. This efficiency and rapidity have not yet been demonstrated *in vitro,* hindering our comprehension of degranulation. The importance of elevated free $Ca^{2+}$ levels in degranulation has been demonstrated directly in permeabilized neutrophils.[1-5] Our subsequent studies on *in vitro* fusion systems, discussed herein, emphasize the importance of this divalent cation.[6-11] These studies have also demonstrated the importance of lipid composition[11] together with enzymes that can modify these compositions, such as phospholipase $A_2$[7] and phospholipase D.[8]

A number of experimental and conceptual factors have hindered studies of *in vitro* fusion in neutrophils. The first is the lack of a suitable assay. Neutrophils are terminal cells whose protein contents are relatively fixed. Because the proteins cannot be readily manipulated, almost all fusion work uses a lipid-mixing assay, such as relief of octadecyl rhodamine (R18) self-quenching.[12] These assays are easy to employ and, theoretically, can be used on any membrane. But whereas fusion always involves lipid mixing, the converse is not true.[13,14] Lipid exchange could occur across apposed

[1] J. E. Smolen, S. J. Stoehr, A. E. Traynor, and L. A. Sklar, *J. Leukoc. Biol.* **41,** 8 (1987).

[2] M. M. Barrowman, S. Cockcroft, and B. D. Gomperts, *Nature* **319,** 504 (1986).

[3] J. E. Smolen and S. J. Stoehr, *J. Immunol.* **134,** 1859 (1985).

[4] J. E. Smolen, R. F. Todd. III, and L. A. Boxer, *Am. J. Pathol.* **124,** 281 (1986).

[5] J. E. Smolen and R. R. Sandborg, *Biochim. Biophys. Acta* **1052,** 133 (1990).

[6] R. A. Blackwood, J. E. Smolen, R. J. Hessler, D. M. Harsh, and A. Transue, *Biochem. J.* **314,** 469 (1996).

[7] R. A. Blackwood, A. Transue, D. M. Harsh, R. C. Brower, S. J. Zacharek, J. E. Smolen, and R. J. Hessler, *J. Leukoc. Biol.* **59,** 663 (1996).

[8] R. A. Blackwood, J. E. Smolen, A. T. Transue, D. M. Harsh, R. J. Hessler, and R. C. Brower, *Am. J. Physiol.* **272,** C1279 (1997).

[9] J. W. Francis, J. E. Smolen, K. J. Balazovich, R. R. Sandborg, and L. A. Boxer, *Biochim. Biophys. Acta* **1025,** 1 (1990).

[10] J. W. Francis, K. J. Balazovich, J. E. Smolen, D. I. Margolis, and L. A. Boxer, *J. Clin. Invest.* **90,** 537 (1992).

[11] T. G. Brock, K. Nagaprakash, D. I. Margolis, and J. E. Smolen, *J. Membr. Biol.* **141,** 139 (1994).

[12] D. Hoekstra, T. de Boer, K. Klappe, and J. Wilschut, *Biochemistry* **23,** 5675 (1984).

[13] N. Düzgüneş, T. M. Allen, J. Fedor, and D. Papahadjopoulos, *Biochemistry* **26,** 8435 (1987).

[14] L. Song, Q. F. Ahkong, D. Georgescauld, and J. A. Lucy, *Biochim. Biophys. Acta* **1065,** 54 (1991).

membrane surfaces. Attempts to label biological membranes with highly hydrophobic compounds (such as R18) could leave adherent micelles of the probe that, under suitable conditions, might then dissolve or partition into available membranes, giving a false-positive signal. Membrane disruption due to lytic agents, added exogenously or generated endogenously (e.g., lipids and phospholipases), also gives a false-positive signal. Thus, the action of a detergent might be interpreted as fusion, if viewed uncritically.

The key to conducting unambiguous fusion studies is to use a content-mixing assay. In these assays, membrane vesicles (either natural or artificial) containing two different materials are allowed to fuse, such that the contents react within a single resulting vesicle. Devising a content-mixing assay is greatly simplified when endogenous vesicle contents can be manipulated. For example, in some cell types endocytosis can be used to create populations of vesicles containing desired reactants.[15–17] Alternatively, cultured cells can be manipulated by molecular biological techniques to obtain granules with desired contents.[16,18] Neither of these tactics will work with neutrophils, as (1) these are terminal, short-lived cells that are not amenable to molecular biological techniques, and (2) granules are created at earlier stages of granulocytopoiesis such that their contents are fixed in mature cells. Consequently, content-mixing systems for neutrophils must use the materials "at hand" and already contained in the granules. Because of these difficulties, it is convenient to use liposomes as one of the fusion partners, as the contents of these vesicles can be manipulated readily.

Of vital importance in content-mixing assays is a means of accounting for leakage of encapsulated materials or vesicle lysis, either of which would allow the product reactions to occur in the surrounding medium. The best way to deal with the leakage/lysis problem is to make sure that extravesicular reactants are quenched and cannot give rise to a signal. Trapping or quenching of released reactants is the rationale employed by those studying endosome and Golgi fusion and has allowed great progress.[15–17] Such systems, which automatically cancel out artifacts, allow the introduction of lipids and other membrane-disruptive agents that could play important roles in biological fusion. The disadvantages of the content-mixing assay are as follows.

[15] M. Wessling-Resnick and W. A. Braell, *J. Biol. Chem.* **265**, 16751 (1990).
[16] J. E. Rothman and L. Orci, *FASEB J.* **4**, 1460 (1990).
[17] L. S. Mayorga, R. Diaz, and P. D. Stahl, *J. Biol. Chem.* **263**, 17213 (1988).
[18] P. F. Wick, R. A. Senter, L. A. Parsels, M. D. Uhler, and R. W. Holz, *J. Biol. Chem.* **268**, 10983 (1993).

1. *High false-negative rate:* If fusion did occur, but the vesicles were leaky, one or more of the reactants might be quenched, leading to a reduced signal. However, this eventuality is not as bad as false positives obtained with lipid-mixing assays.

2. *Specificity:* The reactants and traps must be carefully chosen for the vesicles and contents to be studied. This is much more laborious than attempting simple lipid labeling.

3. *Sealed vesicles:* Content mixing requires closed membrane vesicles and cannot be used with plasma membrane sheets.

Both lipid-mixing and content-mixing assays are useful for studying fusion in neutrophils, as long as the advantages and disadvantages of each are kept in mind. In the following pages, both systems are described.

General Methods

*Buffer System*

Membrane fusion in most biological systems involves the interactions of both cytoplasmic faces of the membranes. Consequently, models of such fusion must use buffer systems that mimic the intracellular space, with a relatively high concentration of $K^+$. The buffer system employed must also contain EGTA, in order to buffer $Ca^{2+}$ in a systematic and reproducible manner, as this divalent cation is vital in membrane fusion.[1-5] The medium must also have sufficient pH-buffering capacity to control $H^+$ exchanges accompanying alterations in the $Ca^{2+}$-to-EGTA ratio or by the addition of low millimolar concentrations of other potential molecules of interest (e.g., nucleotides and fatty acids). Our buffer K consists of the following: 100 m$M$ KCl, 20 m$M$ NaCl, 1.00 m$M$ EGTA, 30 m$M$ HEPES, (pH 7.0). When preparing this buffer, it is vital that the pH be set at 7.0, in order that the $Ca^{2+}$-buffering characteristics be reproducible. Furthermore, the EGTA concentration should be as precise as possible, for the same reason. However, because the actual content of each batch of EGTA varies slightly, it is essential that each preparation of buffer be calibrated with a calcium electrode.[19]

*Calibration of Buffer K*

1. Standards of $10^{-5}$, $10^{-4}$, $10^{-3}$, $10^{-2}$, and $10^{-1}$ $M$ CaCl$_2$ (Ca$_{tot}$) are prepared in buffer K without EGTA.

---

[19] D. M. Bers, *Am. J. Physiol.* **242**, C404 (1982).

2. Using the instructions provided by the manufacturer of the calcium electrode, the millivolts obtained with each of the standards is measured, and a standard curve is constructed. The plot of $\log(Ca_{tot})$ versus millivolts should be linear.

3. A stock of 100 m$M$ CaCl$_2$ is prepared with volumetric precision.

4. A sample of buffer K is carefully titrated with increasing amounts of total calcium over a range of 50 $\mu M$ to 3 m$M$, recording the millivolts from the electrode at each point. Careful attention should be paid to the region of EGTA equivalence (1 m$M$) where the relationship of $[Ca_{fr}^{2+}]$ to $[Ca_{tot}^{2+}]$ is particularly sensitive.

5. Using the standard curve generated in step 2 above, a plot of $[Ca_{fr}^{2+}]$ to $[Ca_{tot}^{2+}]$ is prepared. It is often convenient to prepare a table for experimentally useful conditions, giving the $[Ca_{tot}^{2+}]$ required to obtain a desired $[Ca_{fr}^{2+}]$.

### Preparation of Large Unilamellar Vesicles

Virtually all our experiments are done with large unilamellar vesicles (LUVs). This is important because (1) they resemble the biological targets that are enclosed by single membranes; (2) by presenting their entire membrane surface and enclosed volume for fusion (unlike multilamellar vesicles), the maximal experimental signal can be obtained; (3) their large size reduces strain in the membranes that might be introduced into small vesicles; and (4) their large size resembles that of the biological targets. Liposomes of a uniform size are prepared with a Lipex biomembrane extruder (Northern Lipids, Vancouver, BC, Canada) with polycarbonate filters. Variously sized filters can be employed to change the diameters of the liposomes. When solutes are trapped within the liposomes, unencapsulated materials are removed by exclusion chromatography.

The following generic description employs our standard fusogenic lipid mixture prepared in buffer K and is a modification of the reversed phase evaporation method of Düzgüneş et al.,[13] and published by Blackwood and Ernst.[20] Lipids and buffers are varied, depending on the fusion system being employed. These other details are provided in discussions of the individual fusion systems.

### Evaporation

1. All phospholipids utilized in our laboratory are from Avanti Polar Lipids (Alabaster, AL). Other sources are suitable. The lipids are prepared at 10 mg/ml in chloroform. Once opened, they are stored under argon gas at $-20°$ to reduce oxidation.

[20] R. A. Blackwood and J. D. Ernst, *Biochem. J.* **266,** 195 (1990).

2. The standard fusion system consists of phosphatidic acid–phosphatidylethanolamine (PA–PE; 1:3).[21] From their respective frozen stocks, PE (6 mg) and PA (2 mg) are combined in a test tube.

3. The lipids are dried under a flow of nitrogen or argon gas in a water bath at 40–50°.

4. Ether (1 ml) is added and then mixed by vortexing.

5. Buffer K (1 ml) is added and then mixed by vortexing.

6. A dispersion is prepared by sonication. We sonicate twice for 15 s with a small tip at 35% power (model 300i; Fisher Scientific, Pittsburgh, PA).

7. Ether is removed from the dispersion by rotating the tube under reduced pressure in a rotary evaporator (Büchi Rotovapor; Brinkmann, Westbury, NY). The tube mouth is first covered with Teflon tape to prevent the contents from being lost if evaporation becomes overly vigorous. The tube is then placed in a round-bottom Rotovapor flask. The entire apparatus is tilted at an angle of about 30° from horizontal, which allows the dispersion to coat the sides of the test tube. Rotation at approximately 50 rpm is established first and then a gentle vacuum is applied. It is essential that this vacuum be carefully monitored. If the vacuum is too strong, violent evaporation of the ether will blow the dispersion from the tube.

8. After 5–15 min of controlled evaporation, the phospholipids form an amorphous dispersion in buffer K.

9. An additional 1 ml of buffer K is added and mixed again by vortexing.

10. Evaporation in the Rotovapor is reestablished and continued until there is no smell of ether or chloroform in the tube (20–70 min).

### Extrusion

The extrusion process forces the liposomes through filters of a constant size, preparing uniform vesicles. The process described below is used in our laboratories with a Lipex biomembrane extruder. It is important to follow the manufacturer's instructions that accompany the extrusion device.

1. The lipid dispersion is subjected to three cycles of freeze–thawing.

2. The extruder device is prepared. Two 0.1-$\mu$m pore size polycarbonate filters are stacked between metal filters, according to manufacturer's instructions. Filters of various sizes can be employed.

---

[21] K. Hong, N. Düzgüneş, R. Ekerdt, and D. Papahadjopoulos, *Proc. Natl. Acad. Sci. USA* **79,** 4642 (1982).

3. The stack is placed into the extruder and wetted with buffer K. Excess buffer is removed with a glass pipette held vertically at the edge of the filters. It is essential that the filter stack contain no air bubbles. Assembly of the extruder is completed, and the entire apparatus is equilibrated at 40°.

4. As an initial wash, some buffer K is added with a long plastic pipette and pumped through with low-pressure argon.

5. The liposome suspension is then placed in the extruder and pushed through the filters with higher pressure argon gas (not exceeding 750 lb/in$^2$). This process should be repeated, for a total of five passages through the apparatus. The filters do not need to be replaced until a different liposome preparation is used.

### Gel-Exclusion Chromatography to Remove Untrapped Solutes

1. Sephadex G-75 is swollen in buffer K.

2. The top of a 10-ml plastic pipette is removed, and the tip is gently plugged with glass wool. The pipette is then attached to ring stand.

3. The column is packed with 10 ml of the gel, avoiding air bubbles and leakage of the gel past the glass wool. Buffer K is added continuously so that the column does not run dry.

4. When ready, the column is permitted to run down to the top of the gel and then the liposome suspension (up to 1 ml) is layered at the top of the column.

5. A test tube with a mark at 3 ml is placed under the column to collect effluent. The liposomes are run through the column, adding buffer K as necessary.

6. When 3 ml has been collected, liposomes with trapped solutes emerge in the void volume. A fresh tube is used to collect 2 ml of the turbid liposomes. Untrapped solutes will be retained in the gel and can be discarded.

### Lipid-Mixing Assays

#### Octadecyl Rhodamine

One of the simplest means of measuring fusion is by use of the probe octadecyl rhodamine (R18). R18 consists of a rhodamine fluorophore coupled to a single long-chain aliphatic group that anchors it in the membrane. Liposomes are prepared with a low concentration ($\leq 2\%$) of R18 that is nonetheless sufficient for rhodamine to become self-quenched. When the R18-labeled liposomes fuse with an unlabeled membrane, the R18 is diluted, self-quenching is relieved, and rhodamine fluorescence increases (Fig. 1). Advantages of this system include (1) use of only one probe compound; (2) flexibility of the unlabeled target membrane, which can be a liposome, plasma membrane, granule membrane, unsealed vesicle, planar membrane, and so on; and (3) easy, real-time readout. Disadvantages include (1) sensitivity to removal of the probe from the membrane, by cleavage, extraction, or hemifusion; and (2) sensitivity to swelling of the liposomes. Disadvantage 1 is particularly important as it can lead to false-positive results. It is essential to demonstrate that none of the observed signal changes attributable to fusion occur in mixtures containing only labeled liposomes (at the same total lipid concentration as the original mixture). This control will cover most of the potential artifacts.

Lipid mixing is assayed by quantitating increased rhodamine fluorescence due to the relief of self-quenching of R18 in fusing membranes, using a recording spectrofluorimeter. R18-labeled liposomes, granules, or plasma membrane-enriched fractions are stirred with unlabeled liposomes or granules at 37°. Fusion is induced by adding various combinations of divalent cations and/or fusogens. All fluorescence signals are expressed as percentages of relative maximum fluorescence, obtained by adding 0.1% Triton X-100 to lipid mixtures. The phospholipid ratio of R18-labeled to unlabeled vesicles or liposomes is 1:4 for all assays.

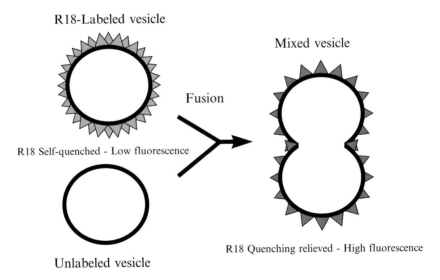

FIG. 1. Use of R18 to measure fusion: Concept of fusing liposomes containing self-quenched R18 with unlabeled vesicles. The resulting lipid dilution causes R18 fluorescence to increase.

SPECIFIC METHODS

1. For standard PA–PE liposomes, the following stocks of lipids are prepared in chloroform and stored under argon at $-20°$: PE (10 mg/ml), PA (10 mg/ml), and R18 (1 mg/ml).

2. Appropriate aliquots (such as 150 $\mu$l of PE, 50 $\mu$l of PA, and 4 $\mu$l of R18) of the stocks are combined and evaporation performed as detailed on pp. 304–305. Unlabeled target membranes are prepared separately. Phospholipid concentrations are determined by the phosphate assay of Morrison.[22]

3. The lipid mixtures are combined in a fluorescence cuvette containing prewarmed buffer K. A labeled liposome-to-unlabeled target ratio of 1:4 is optimal for obtaining dilution of R18 during fusion. A final lipid concentration of 100–200 $\mu$g/ml in buffer K is appropriate for most fusion measurements.

4. The rhodamine fluorescence of the mixture is monitored continuously (with a chart printout) at an excitation wavelength of 560 nm and an emission wavelength of 610 nm. Ideally, the spectrofluorimeter should be equipped with a temperature-regulated cell holder and magnetic stirring device.

5. After establishment of a baseline, $Ca^{2+}$ or some other fusogen is added and the resultant signal is continuously measured.[9,11] At the end of the experiment, the liposomes are disrupted with 0.1% Triton X-100, completely relieving self-quenching. This provides a total signal for R18 that can be used as a denominator for final calculations. We customarily calculate the amount of fusion as the percentage of total R18 signal obtained over a fixed period of time (such as 10 min).

### NBD-Rhodamine-PE

Another lipid-mixing assay employs two labeled lipids, *N*-(Lissamine rhodamine B sulfonyl)-phosphatidylethanolamine (Rho-PE) and *N*-(7-nitro-2,1,3-benzoxadiazol-4-yl)-phosphatidylethanolamine (NBD-PE). Each of these lipids is incorporated into liposomes at a 1% concentration. At these concentrations of NBD-PE and Rho-PE, the distance between NBD and Rho moieties is sufficiently close that Rho quenches NBD fluorescence (Fig. 2). Fusion between a labeled membrane and an unlabeled target membrane results in an increase in the distance between NBD and Rh and a decrease in quenching. The resulting increase in NBD fluorescence is detected by a spectrofluorimeter.

---

[22] W. R. Morrison, *Anal. Biochem.* **7,** 218 (1964).

NBD and Rho-labeled vesicle

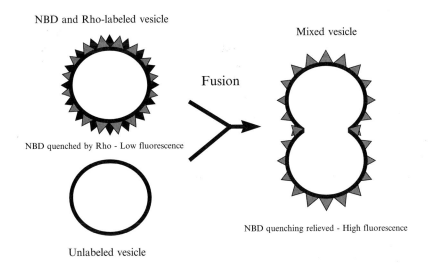

Mixed vesicle

Fusion

NBD quenched by Rho - Low fluorescence

NBD quenching relieved - High fluorescence

Unlabeled vesicle

FIG. 2. Use of Rho-PE and NBD-PE to measure fusion: Concept of fusing liposomes containing NBD that is quenched by Rho with unlabeled vesicles. The resulting lipid dilution causes NBD fluorescence to increase.

As a lipid-mixing assay, the NBD/Rho-PE system shares most of the advantages and disadvantages of the R18 system. By using two components, it is slightly more complex than the R18 assay. However, it has the distinct advantage that each fluorochrome is attached to a phospholipid with two fatty acids. Thus, the membrane anchors are much more secure.

SPECIFIC METHODS

1. For standard PA–PE liposomes, the following stocks of lipids are prepared in chloroform and stored under argon at $-20°$: PE (10 mg/ml), PA (10 mg/ml), Rho-PE (1 mg/ml), and NBD-PE (1 mg/ml).

2. Appropriate aliquots (such as 600 $\mu$l of PE, 200 $\mu$l of PA, 80 $\mu$l Rho-PE, and 80 $\mu$l of NBD-PE) of the stocks are combined, and the liposomes are prepared as detailed on pp. 304–305. Unlabeled target membranes are prepared separately. Phospholipid concentrations are determined by the phosphate assay of Morrison.[22]

3. The lipid mixtures are combined in a fluorescence cuvette containing prewarmed buffer K. A labeled liposome-to-unlabeled target ratio of 1:9 is optimal for obtaining dequenching of NBD during fusion. A final lipid concentration of 100–200 $\mu$g/ml in buffer K is appropriate for most fusion measurements.

4. The NBD fluorescence of the mixture is monitored continuously (with a chart printout) at an excitation wavelength of 450 nm and an emission wavelength of 530 nm. Ideally, the spectrofluorimeter should be equipped with a temperature-regulated cell holder and magnetic stirring device.

5. After establishment of a baseline, $Ca^{2+}$ or some other fusogen is added and the resultant fusion is measured continuously.[6,23] At the end of the experiment, the liposomes are disrupted with 0.1% Triton X-100, completely relieving quenching of NBD by rhodamine. This provides a total signal for NBD that can be used as a denominator for final calculations. We customarily calculate the amount of fusion as the percentage of total NBD signal obtained over a fixed period of time (such as 10 min).

## Content-Mixing Assays

From the point of view of scientific rigor in assessing the presence or absence of membrane fusion, content-mixing assays are superior to lipid-mixing assays. These issues are discussed in the Introduction. We currently use two different fusion systems to determine mixing of contents, both based on the fluorophore 8-aminonaphthalene-1,3,6-trisulfonic acid (ANTS) and its quencher p-xylene-bis(pyridinium) bromide (DPX). In system I, ANTS and DPX are encapsulated in separate liposomes. In system II, they are both encapsulated in the same liposome. In each system, steps are taken to prevent leakage or lysis or the vesicles from creating any artifactual fusion signals.

### ANTS–DPX System I: For Two Liposomes

The ANTS–DPX assay for content mixing between liposomes[24] is well established.[25] In this scheme, one liposome contains an aqueous fluorophore (ANTS) while the other liposome contains a quencher (DPX). If the liposomes fuse, the fluorescence is quenched (Fig. 3). If lysis/leakage occurs, then (1) leaking ANTS creates no signal change, and (2) leaking DPX is too dilute to cause any signal change. Thus, dilution in the extravesicular space serves as an effective trap for extravesicular reactions. Furthermore, one can directly assay leakage of ANTS by including a high concentration of DPX (4.5 m$M$) on the outside.

---

[23] J. W. Francis, S. J. Stoehr, D. I. Margolis, L. A. Boxer, and J. E. Smolen, *J. Cell Biol.* **111**, 77a (1990). (Abstract)

[24] H. Ellens, J. Bentz, and F. C. Szoka, *Biochemistry* **24**, 3099 (1985).

[25] P. Meers, J. Bentz, D. Alford, S. Nir, D. Papahadjopoulos, and K. Hong, *Biochemistry* **27**, 4430 (1988).

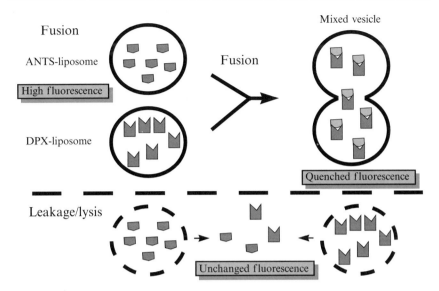

Fig. 3. Use of ANTS and DPX (system I) to measure fusion: Concept of allowing liposomes containing ANTS and DPX to fuse with each other. The resulting mixing of internal contents causes ANTS fluorescence to decrease.

Advantages of this system are as follows: (1) it is an unambiguous test of fusion, and (2) it is relatively simple for a content-mixing assay. Disadvantages include the following: (1) the need to trap reagents into two different vesicles effectively precludes use of biological membranes, (2) it is substantially more time-consuming than a lipid-mixing assay, and (3) it provides a negative signal (decreasing fluorescence with increasing fusion).

SPECIFIC METHODS

1. For standard PA–PE liposomes, the following stocks of lipids are prepared in chloroform and stored under argon at $-20°$: PE (10 mg/ml) and PA (10 mg/ml). To maintain osmolarity at the high concentration of ANTS to be employed, buffer K is replaced with 25 m$M$ ANTS, 19.5 m$M$ KCl, 30 m$M$ HEPES (pH 7.0) for preparing one batch of liposomes. A high concentration of DPX is required to quench ANTS. Hence, all KCl and NaCl are removed from buffer K and replaced with DPX in the other batch of liposomes. The composition of this solution is 117 m$M$ DPX, 1 m$M$ EGTA, 20 m$M$ HEPES (pH 7.0).

2. Appropriate aliquots (such as 600 $\mu$l of PE and 200 $\mu$l of PA) of the stocks are combined, and the liposomes are prepared as detailed on

pp. 304–305. A set of DPX-containing liposomes is prepared separately by the same method. Untrapped reagents are removed by gel-exclusion chromatography as detailed on p. 306. These liposomes remain stable for several days at 4°. Phospholipid concentrations are determined by the phosphate assay of Morrison.[22]

3. The lipid mixtures are combined in a fluorescence cuvette containing prewarmed buffer K. A labeled liposome-to-unlabeled target ratio of 1:1 is optimal for obtaining quenching of ANTS during fusion. A final lipid concentration of 100–200 μg/ml in buffer K is appropriate for most fusion measurements.

4. The ANTS fluorescence of the mixture is monitored continuously (with a chart printout) at an excitation wavelength of 370 nm and an emission wavelength of 530 nm. Ideally, the spectrofluorimeter should be equipped with a temperature-regulated cell holder and magnetic stirring device.

5. After establishment of a baseline, $Ca^{2+}$ and/or some other fusogen is added and the resultant fusion is measured continuously as a decrease in fluorescence.[10,25] At the end of the experiment, the liposomes are disrupted with 0.1% Triton X-100, completely relieving quenching of ANTS by DPX. This provides a total signal for ANTS that can be used as a measure of the encapsulated fluorochrome. DPX (4.5 m$M$) is then added to quench the ANTS signal fully. The difference between the free and quenched ANTS signals represents the total potential fusion signal. The amount of fusion is calculated generally as the percentage of total ANTS signal obtained over a fixed period of time (such as 10 min).

### ANTS–DPX System II: For Two Generic Vesicles

A second ANTS–DPX system was conceived by the author's collaborator, A. Blackwood (University of Michigan Medical Center, Ann Arbor, MI), for use with any two sealed vesicles.[6] This system takes advantage of fluorescence properties of ANTS and the quenching properties of DPX. ANTS fluorescence is based on the total concentration of ANTS in the solution, whereas DPX quenching is determined by its local concentration. When fusion occurs between a liposome containing both ANTS and DPX and an empty target vesicle, DPX is diluted, ANTS is unquenched, and increased fluorescence is detected (Fig. 4). This target vesicle is usually a liposome, but can in principle be any sealed membrane vesicle (plasma membrane, phagosome, granule membrane, etc.). The key to obtaining an unambiguous fusion signal is the extravesicular trap, in this case external DPX. The amount of DPX outside the vesicle is carefully calibrated such that lysis gives no signal, either positive or negative. This is an extremely

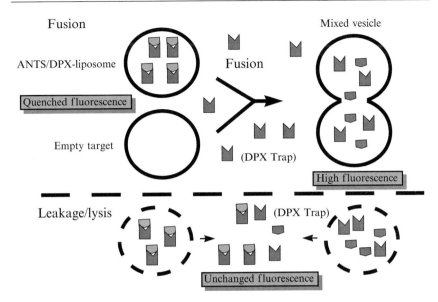

Fig. 4. Use of ANTS and DPX (system II) to measure fusion: Concept of fusing liposomes containing both ANTS and DPX to empty vesicles. The resulting mixing of internal contents causes ANTS fluorescence to increase.

useful method, particularly for examining the fusion of different types of granules with liposomes.

Advantages of this system include the following: (1) it provides an unambiguous test of fusion; (2) trapping the reagents in a single vesicle, such as a liposome, permits other sealed vesicles of biological origin to be used as fusion partners; and (3) it provides a positive signal (increasing with fusion). Disadvantages include the following: (1) careful calibration is required for controlling leakage/lysis signal, and (2) it is substantially more time-consuming than a lipid-mixing assay.

Specific Methods

1. For standard PA–PE liposomes, the following stocks of lipids are prepared in chloroform and stored under argon at $-20°$: PE (10 mg/ml) and PA (10 mg/ml). High concentrations of ANTS and DPX are required to be trapped within the liposomes. To maintain isomolarity, a modified buffer system is employed. This consists of 55 mM ANTS, 50 mM DPX, 10 mM HEPES (pH 7.0). Alternatively, the 50 mM DPX can be replaced with 20 mM DPX and 30 mM KCl to increase the sensitivity of the assay.

2. Appropriate aliquots (such as 600 μl of PE and 200 μl of PA) of the stocks are combined, and the liposomes are prepared as detailed on

pp. 304–305. Unlabeled target membranes are prepared separately. Untrapped reagents are removed by gel-exclusion chromatography as detailed on p. 306. These liposomes remain stable for several days at 4°. Phospholipid concentrations are determined by the phosphate assay of Morrison.[22]

3. The lipid mixtures are combined in a fluorescence cuvette containing prewarmed buffer K. LUVs containing both ANTS and DPX are combined in a stirred cuvette with unlabeled target vesicles at a ratio of 1:9 (180 $\mu$g of total lipid).

4. The ANTS fluorescence of the mixture is monitored continuously (with a chart printout) at an excitation wavelength of 370 nm and an emission wavelength of 530 nm. Ideally, the spectrofluorimeter should be equipped with a temperature-regulated cell holder and magnetic stirring device.

5. It is essential to determine the amount of external DPX required to control any leakage/lysis signal. This must be done on a daily basis. An aliquot of the fusion mixture is placed in the fluorometer, and the fluorescence is noted. Triton X-100 (0.1%) is added, producing an unquenched signal (Fig. 5). Concentrated DPX is then added back until the original fluorescence is restored. Another reaction mixture is treated with this concentration of DPX. If there is any untrapped ANTS due to leakage or incomplete removal by gel filtration, then there will be a slight drop in fluorescence (Fig. 5). This reaction mixture is now lysed (in the presence of DPX). If there is no further change in signal after the lysis step, then this concentration of DPX will be employed for the remainder of the experiment. If some change in fluorescence is observed during lysis, then the concentration of DPX used for trapping is modified slightly until no change occurs. This concentration is generally 25–30 m$M$.

6. After establishment of a baseline, Ca$^{2+}$ and/or some other fusogen is added and the resultant fusion is measured continuously by increased fluorescence of ANTS. Fusion is measured as the initial rate of fusion (during the first 12 s) and as total fusion in 90 s. Each rate is calculated as the percentage of maximal change in fluorescence per minute. For liposomes containing both ANTS and DPX, the maximal change in fluorescence is measured as the difference between the quenched (intact liposomes) and unquenched ANTS signal (lysed with 0.1% Triton X-100 in the absence of external DPX). This change in fluorescence signal represents a theoretical maximal unquenching within the system and is related only indirectly to individual fusion events. However, this scale permits a comparison between individual experiments and different fusion systems.

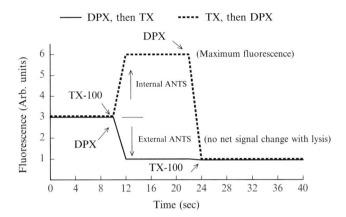

FIG. 5. Calibration of external trap for ANTS and DPX (system II): Rationale for preparing and calibrating the external trap necessary to neutralize any ANTS liberated by leakage/lysis. When properly established, leakage of ANTS will produce no fluorescence signal.

## Target Membranes and Vesicles for Neutrophil Fusion Studies

The tools described can be used in a number of different studies related to membrane fusion in neutrophils. One way is to use vesicles as mere targets to assess the fusogenic potential of proteins isolated from neutrophils (fusogens). We have used this rationale for studying annexins, glyceraldehyde-3-phosphate dehydrogenase (GAPDH), and phospholipases in fusion.[8,10,26] However, such studies do not model fusion in neutrophils and are not discussed further. A second approach is to employ liposomes that are "designed" to resemble neutrophil membranes. These "complex" liposomes are far less prone to fusion than are "standard" PA–PE liposomes.[11] The third method is to employ actual membranes isolated from neutrophils as one of the fusion partners. Plasma membranes, granules, or phagosomes are isolated from neutrophils and fused with an artificial partner, generally a standard or complex liposome containing probes appropriate for measuring lipid mixing or content mixing.[6,7,9,10] This approach is a convenient compromise in that it does not attempt to incorporate label into the neutrophil membranes, which is a highly problematic venture. The final approach is to fuse two endogenous membranes, such as granules and plasma membranes, *in vitro*. Methods for this

[26] R. J. Hessler, R. A. Blackwood, T. G. Brock, J. W. Francis, D. M. Harsh, and J. E. Smolen, *J. Leukoc. Biol.* **63,** 331 (1998).

approach have not yet been devised, as labeling the participant vesicles has not been established satisfactorily.

## Use of Artificial Membranes to Study Fusion in Neutrophils

A wide range of liposome compositions can be employed. This was epitomized by our study of fusion using many different lipid compositions that were designed to mimic either plasma membranes or granule membranes.[11] To simulate plasma membranes, we use several phospholipids, namely phosphatidylcholine (PC), phosphatidylethanolamine (PE), sphingomyelin (SM), phosphatidylinositol (PI), and phosphatidylserine (PS) at molar ratios for PC:PE:SM:PI:PS of 24:27:20:16:13. Natural membranes also have significant concentrations of cholesterol, which we added to our standard complex liposomes at a concentration of 50 mol%.[7]

Specifically, each of the lipids to be employed is prepared as a concentrated stock of 10 mg/ml in chloroform. The desired contents of each are combined as described on pp. 304–305. If desired, labels for lipid-mixing assays can be added at this time. Alternatively, labels for content mixing can be added to the medium for preparing the liposomes.

## Use of Isolated Membranes to Study Fusion in Neutrophils

Endogenous targets of neutrophil fusion studies are obtained by subcellular fractionation. We have employed these membranes to determine how well they fuse with liposomes of "standard" composition, such as PA–PE.[6,8–10] In general, we have found that plasma membranes are easily fused with $Ca^{2+}$. Specific granules are distinctly more resistant to fusion, whereas azurophil granules are highly refractory to fusion. Methods for isolating the subcellular fractions from neutrophils are described below.

### SPECIFIC METHODS

1. Neutrophils are isolated from whole blood by Hypaque/Ficoll gradients, dextran sedimentation, and hypotonic lysis of erythrocytes.[27,28] The cells (>95% neutrophils) are washed and suspended in phosphate-buffered saline (PBS) buffer (138 m$M$ NaCl, 2.7 m$M$ KCl, 8.1 m$M$ $Na_2HPO_4$, and 1.5 m$M$ $KH_2PO_4$) before use.

2. Endogenous proteases are inactivated with diisopropyl fluorophosphate (DFP). Cell suspensions are treated with DFP (1 $\mu$l/ml) for 5 min at room temperature. The cells are then washed twice with PBS to remove excess DFP. *Caution:* DFP is extremely toxic.

[27] A. Boyum, *Scand. J. Clin. Lab. Invest.* **21,** 77 (1968).
[28] R. B. Zurier, S. Hoffstein, and G. Weissmann, *Proc. Natl. Acad. Sci. USA* **70,** 844 (1973).

3. Cells ($1 \times 10^8$/ml) are suspended in relaxation buffer: 100 m$M$ KCl, 3 m$M$ NaCl, 3.5 m$M$ MgCl$_2$, 10 m$M$ piperazine-$N,N'$-bis(2-ethanesulfonic acid) (PIPES), and 1 m$M$ ATP (pH 7.3) in preparation for nitrogen cavitation.

4. Nitrogen cavitation and subcellular fractionation are performed according to Jesaitis et al.[29] as modified in our laboratory. We employ a cavitation "bomb" manufactured by Parr Instruments (Moline, IL), whose instructions should be followed.

5. The cell sample, along with a stirring bar, is placed in a 50-ml beaker. This beaker is placed in the "bomb" on top of a stirring plate in a cold room. Stirring is started, and the sample is equilibrated with N$_2$ (375 lb/in$^2$) for 20 min at 4°.

6. The sample should be released slowly into a capped test tube containing a hole in the cap. This technique prevents spraying of the cell homogenate as it is released from the high-pressure N$_2$.

7. The cavitate is centrifuged for 10 min at 1500 rpm at 4°, and the pelleted nuclei are discarded.

8. A discontinuous sucrose density gradient is prepared to isolate subcellular fractions, using 15%, 43%, 53%, and 60% layers.[6] The postnuclear cavitate is layered on this gradient and centrifuged for 1 h at 93,000 g in a Beckman Coulter (Fullerton, CA) L5-50B ultracentrifuge.

9. Fractions are removed from the interfaces between each layer and are (from top to bottom) plasma membrane, specific granules, and azurophil granules.

10. Sucrose is removed from the plasma membrane and specific granule fractions by gel filtration on a Sephadex G-25 column equilibrated in buffer K. We found that $71 \pm 1\%$ (SEM, $n = 5$) of the plasma membrane fraction was both sealed and oriented inside-out.[6]

11. Because azurophil granules bind to the Sephadex G-25 column, they are first concentrated by recentrifugation onto a 60% sucrose cushion. This band is harvested with minimal excess sucrose and then resuspended in buffer K before use. (Other studies in our laboratory show that up to 20% sucrose has no effect on the fusion assay.) Granules are used at an OD$_{450}$ of 0.17 to ensure comparable amounts of phospholipids between preparations.

12. These subcellular fractions can be used as targets for both lipid-mixing and content-mixing assays. As an example, we describe a typical experiment using the NBD-Rho system.

[29] A. J. Jesaitis, J. R. Naemura, R. G. Painter, L. A. Sklar, and C. G. Cochrane, Biochim. Biophys. Acta 719, 556 (1982).

13. For standard PA–PE liposomes, the following stocks of lipids are prepared in chloroform and stored under argon at $-20°$: PE (10 mg/ml), PA (10 mg/ml), Rho-PE (1 mg/ml), and NBD-PE (1 mg/ml).

14. Appropriate aliquots (such as 600 $\mu$l of PE, 200 $\mu$l of PA, 80 $\mu$l of Rho-PE, and 80 $\mu$l of NBD-PE) of the stocks are combined, and the liposomes are prepared as detailed on pp. 304–305.

15. The NBD-Rho fusion system is used as described on pp. 308–310. A labeled liposome-to-unlabeled target ratio of 1:9 is optimal for obtaining dequenching of NBD during fusion. A final lipid concentration of 100–200 $\mu$g/ml in buffer K is appropriate for most fusion measurements.

16. Plasma membranes or granules (90 $\mu$g of lipid) are combined with 10 $\mu$g of NBD-Rho-labeled liposomes. The total volume of buffer K is made up to 1 ml. If sucrose solution is present (as in the azurophil granule preparations), these volumes must be noted so that the correct final concentrations of $Ca^{2+}$ can be obtained with buffer K.

17. Mixtures are preincubated, $Ca^{2+}$ is added, and fusion is measured as described on pp. 308–310. We have used this system to show that phospholipase $A_2$ can greatly reduce the $Ca^{2+}$ requirements for fusion of specific granules.[7]

Conclusions

A number of tools are available for studying membrane fusion and degranulation in neutrophils. Both lipid-mixing and content-mixing assays have been successfully employed. The lipid-mixing assays are easy to perform and are reliable from day to day. These assays also do not require closed membrane vesicles, which can be an advantage when an uncharacterized membrane preparation is being used as a target. On the other hand, lipid-mixing assays are prone to artifacts producing false-positive results. False-positive results can be eliminated by using content-mixing assays. However, these assays are more difficult to perform and require careful calibration. Our approach has been to use lipid-mixing assays for most routine studies and then to follow up with content-mixing assays to ensure that fusion, but not lysis, is the phenomenon being observed.

A limitation of this work is that neutrophil granules and plasma membranes cannot be directly labeled with fusion probes in an unambiguous manner. Although labeling itself can be accomplished, it is difficult to ensure that the probes are incorporated into the membranes in the manner supposed. In fact, this assumption is usually unacknowledged, leading to considerable uncertainty about the phenomenon being observed. We have avoided these uncertainties by using liposomes, within which homogeneous incorporation of the probes can be readily accomplished, as targets.

Unfortunately, this strategy means that liposomes will always be one partner of the fusion pair, and this limits the biological relevance of the system. Further progress in this field awaits the development of fusion assay systems using the endogenous components of neutrophil granule membranes and plasma membranes. This is a daunting task given the difficulty of manipulating these cells genetically or biochemically. Until such time, the malleability of liposome compositions can be used to advantage. There are still many questions to be answered, using liposomes as fusion partners.

# [18] Delivery of Macromolecules into Cytosol using Liposomes Containing Hemolysin

By Manas Mandal, Elizabeth Mathew, Chester Provoda, and Kyung Dall-Lee

## Introduction

The cytosol and nucleus of mammalian cells, two topologically connected subcellular compartments, are important delivery target sites for various macromolecular therapeutic agents. Proteins including antigens and toxins, as well as oligonucleotides, ribozymes, and plasmid DNA, all belong to a class of therapeutic macromolecules that has seen rapid expansion in recent decades but faces critical delivery challenges, as these macromolecules can exert their pharmacological effects only when delivered into the cytosol or nucleus. Among the many delivery systems utilized for this purpose, liposomes have tremendous potential in formulations designed for macromolecules because of their superb pharmaceutical characteristics such as their intrinsically noncytotoxic nature, their ability to protect molecules from degradation in plasma, and their potential for modifiable pharmacokinetic properties and targetability.

In this chapter, we focus on liposomal drug delivery systems, the limitations of conventional formulations as cytosolic delivery vehicles for macromolecules, and potential solutions achieved by the utilization of a bacterial mechanism of cell invasion. The mechanism by which liposomes interact with cells is briefly examined, with a discussion regarding the limitations of the use of conventional liposomes in cytosolic delivery. We then review briefly and describe several approaches to overcoming these limitations, focusing primarily on methods developed in our laboratory that utilize a pore-forming hemolysin within pH-sensitive liposomes.

Liposomes, on interacting with cells, have been demonstrated to be taken up by cells through internalization into the lumen of endocytotic or phagocytotic compartments that are topologically equivalent to the extracellular space. Since the seminal report by Straubinger *et al.* demonstrating that liposomes are internalized by cells through the endocytotic route, it has been well established that liposomes and their contents are delivered to endosomes[1-2a] and that the frequency of liposome fusion with cellular membranes, if it exists at all, is low; thus the extent of liposome-mediated cytosolic delivery is minimal to none. Subsequent reports using the pyranine dye HPTS (8-hydroxypyrene-1,3,6-trisulfonate) provided additional supporting evidence for this notion.[3-5] Later work using the same techniques to monitor poly(ethylene glycol)-phosphatidylethanolamine (PEG-PE)-containing, sterically stabilized immunoliposomes also suggested that neutral or anionic liposomes in general have a tendency to be taken up by cells into endocytotic compartments regardless of their binding mechanism.[6] An exception to these general findings is that direct destabilization or fusion with cellular membranes occurs extensively when liposomes are made of nonnatural, cationic lipids,[7,7a] a description of which is beyond the scope of this chapter. Despite scattered reports suggesting that a fraction of synthetic liposomes can deliver their contents into the cytosol of cells under certain conditions, predominantly in complex *in vivo* cases,[8,9] delivery of liposomal contents that is useful for most therapeutic purposes clearly requires greater efficiency and consistency. Thus, encapsulation of macromolecules inside liposomes, either conventional or sterically stabilized, renders little advantage to favor their delivery into the cytosol. Most macromolecules, typically charged and hydrophilic with extremely low permeability across membranes, remain within the endocytotic compartments surrounded by the membrane barrier, have negligible access to the cytosol, and are ultimately routed to lysosomal compartments. Once routed to

---

[1] R. M. Straubinger, K. Hong, D. S. Friend, and D. Papahadjopoulos, *Cell* **32,** 1069 (1983).
[2] D. Papahadjopoulos, *J. Liposome Res.* **5,** 9 (1995).
[2a] N. Düzgüneş and S. Nir, *Adv. Drug Deliv. Rev.* **40,** 3 (1999).
[3] R. M. Straubinger, D. Papahadjopoulos, and K. Hong, *Biochemistry* **29,** 4929 (1990).
[4] D. L. Daleke, K. Hong, and D. Papahadjopoulos, *Biochim. Biophys. Acta* **1024,** 352 (1990).
[5] K. D. Lee, S. Nir, and D. Papahadjopoulos, *Biochemistry* **32,** 889 (1993).
[6] J. W. Park, K. Hong, P. Carter, H. Asgari, L. Guo, G. Keller, C. Wirth, R. Shalaby, C. Kotts, W. Wood, D. Papahadjopoulos, and C. Benz, *Proc. Natl. Acad. Sci. USA* **92,** 1327 (1995).
[7] D. S. Friend, D. Papahadjopoulos, and R. J. Debs, *Biochin Biophys. Acta* **1278,** 41 (1996).
[7a] P. Pires, S. Simões, S. Nir, R. Gaspar, N. Düzgüneş, and M. C. Pedraso de Lima, *Biochim. Biophys. Acta* **1418,** 71 (1999).
[8] F. Zhou and L. Huang, *Immunomethods* **4,** 229 (1994).
[9] C. R. Alving and N. M. Wassef, *AIDS Res. Hum. Retroviruses* **10,** S91 (1994).

the lysosomes, the macromolecules are degraded quickly by hydrolytic enzymes in these low-pH compartments. Thus, many approaches have been sought to reroute the cargo of endocytosed liposomes to the cytosolic space and away from the lysosomal compartment. One prominent strategy is to incorporate into the liposomes a mechanism to overcome the membrane barrier of endocytotic compartments. Temporarily breaching the endosomal membranes or inducing direct fusion between the internalized liposomal membrane and the endosomal membrane would allow the escape of endocytosed liposomal contents into the cytosolic space before they reach the highly degradative lysosomal environment. Given these considerations, targeted and sterically stabilized liposomes designed to be delivered into the endocytotic compartments of intended cells, while at the same time allowing escape into the cytosol, would be particularly useful for macromolecule delivery.

## Modified Nonconventional Liposome Formulations for Cytosolic Delivery

A common paradigm for intracellular delivery has been to mimic viruses, which efficiently introduce their contents into the cytosol of target cells. Once bound to the cell surface, some viruses promote direct fusion between the viral membrane and the cell plasma membrane. Others are internalized into the endocytotic compartment and trigger membrane fusion between the virus and an endosome on acidification of the endocytotic compartment.[10,11] Two main approaches have been tested in attempting to design liposomes that mimic viruses in their behavior. One approach involves the use of "pH-sensitive" liposomes[12-14]; the other involves incorporation of the functionality of the fusion protein into liposomal membranes. The former approach is achieved by incorporating into the lipid bilayer, made primarily of phosphatidylethanolamine (PE), lipids that have a protonatable moiety with a $pK_a$ near the acidic environment of endosomes. For detailed descriptions of pH-sensitive liposomes, readers are referred to reviews on this subject[14-15a] (and see [20] in this volume[15b]).

[10] M. Marsh and A. Helenius, *Adv. Virus Res.* **36,** 107 (1989).

[11] J. White, M. Kielian, and A. Helenius, *Q. Rev. Biophys.* **16,** 15 (1983).

[12] N. Düzgüneş, R. M. Straubinger, P. A. Baldwin, D. S. Friend, and D. Papahadjopoulos, *Biochemistry* **24,** 3091 (1985).

[12a] R. M. Straubinger, N. Düzgüneş, and D. Papahadjopoulos, *FEBS Lett.* **179,** 148 (1985).

[13] J. Connon and L. Huang, *J. Cell Biol.* **101,** 582 (1985).

[14] R. M. Straubinger, *Methods Enzymol.* **221,** 361 (1993).

[15] D. C. Litzinger and L. Huang, *Biochim. Biophys. Acta* **1113,** 201 (1992).

Indirect evidence suggests that these pH-sensitive liposomes induce a certain extent of leakage of endosomes, partly through the fusion event. Several experiments using antigen, fluorescent dye, toxin, or reporter genes show that cytosolic delivery with pH-sensitive liposomes is greater than with pH-insensitive conventional liposomal formulations.[16–18] However, the reported efficiency of cytosolic delivery indicates less frequent direct fusion between liposomal and endosomal membranes than would be predicted from the design rationale. Another approach is to reconstitute the whole viral fusion protein into liposomes to generate proteoliposomes or to isolate the causative segment of fusion protein for incorporation into liposomes. Several delivery systems, particularly for oligonucleotides and plasmids, have been designed utilizing virus-derived fusion peptides or whole virus.[19–21] In some instances, a viral envelope glycoprotein has been reconstituted into liposomes to make artificial virus-like particles, called "virosomes."[22] Similar formulations have also been generated by fusing liposomes with viruses. Viruses that have been exploited so far include influenza virus, adenovirus, and Sendai virus.[23,24]

## New Liposome Formulation Mimicking the Intracellular Bacterium *Listeria monocytogenes*

A new paradigm that has been exploited is the cytosolic delivery of macromolecules by mimicking the escape of facultative intracellular bacteria from endocytotic compartments into the cytosol. A few parasitic

---

[15a] N. Düzgüneş, R. M. Straubinger, P. A. Baldwin, and D. Papahadjopoulos, in "Membrane Fusion: Mechanisms, Cell Biology and Applications in Biotechnology" (J. Wilschut and D. Hoekstra, eds.), p. 195. Marcel Dekker, New York, 1990.

[15b] R. Tachibana, S. Futaki, H. Harashima, and H. Kiwada, *Methods Enzymol.* **372**, [20], 2003 (this volume).

[16] C. V. Harding, D. S. Collins, O. Kanagawa, and E. R. Unanue, *J. Immunol.* **147**, 2860 (1991).

[17] C.-J. Chu, J. Dijkstra, M.-Z. Lai, K. Hong, and F. C. Szoka, *Pharm. Res.* **7**, 824 (1990).

[18] J. Y. Legendre and F. J. Szoka, *Pharm. Res.* **9**, 1235 (1992).

[19] E. Wagner, C. Plank, K. Zatloukal, M. Cotten, and M. Birnstiel, *Proc. Natl. Acad. Sci. USA* **89**, 7934 (1992).

[20] E. Wagner, K. Zatloukal, M. Cotten, H. Krilappos, K. Mechtler, D. Curiel, and M. Birnstiel, *Proc. Natl. Acad. Sci. USA* **89**, 6099 (1992).

[21] D. T. Curiel, S. Agarwal, E. Wagner, and M. Cotten, *Proc. Natl. Acad. Sci. USA* **88**, 8850 (1991).

[22] R. Bron, A. Ortiz, and J. Wilschut, *Biochemistry* **33**, 9110 (1994).

[23] T. Uchida, J. Kim, M. Yamaizumi, Y. Miyake, and Y. Okada, *J. Cell Biol.* **80**, 10 (1979).

[24] V. J. Dzau, M. J. Mann, R. Morishita, and Y. Kaneda, *Proc. Natl. Acad. Sci. USA* **93**, 11421 (1996).

bacteria or intracellular pathogens that survive by invading and multiplying within the cytoplasm of host cells must have the ability to escape from endocytotic vacuoles. By secreting specialized proteins, *Listeria, Shigella,* and *Trypanosoma cruzi* effectively disrupt endosomal membranes.[25–28] Thus, several bacterial vectors have been developed to transfer exogenous genes, harnessing the bacterial mechanism of escape from endosomes.[29] Facultative intracellular bacteria including *Shigella* and *Listeria* have been utilized in this manner.[30–32] Common to all these bacterial vectors is adequate efficiency in mediating therapeutically significant delivery of genes, as well as weaknesses shared by viral vectors, including pathogenicity, immunogenicity, and difficulty in targeting.

As in the case of certain viral vectors, the minimal component necessary for endosome lysis, that is, hemolysin, has been isolated from bacteria and incorporated into liposomes, formulating a nonviral and nonbacterial delivery vector. The hemolysin of *Listeria monocytogenes,* listeriolysin O (LLO), is a well-characterized protein that is secreted by the intracellular pathogen to escape from the endocytotic compartment.[33–35] LLO is a soluble protein of 58 kDa that is thought to form pores in membranes. It belongs to the thiol-activated, pore-forming hemolysin family, which includes perfringolysin O (PFO), pneumolysin O (PLO), and streptolysin O (SLO). These toxins are homologous ($\sim$70% amino acid similarity, $\sim$40% identity), all bind strongly to cholesterol-containing membranes, and they are thought to oligomerize within lipid bilayers to form pores and lyse membranes.[36,37] LLO is unique among the thiol-activated hemolysins in that it possesses pH-dependent hemolytic activity; LLO is activated by low pH as well as reduction of the unique cysteine (Cys-484) in its amino acid sequence.[33,38] The *Listeria* gene for LLO, *hly,* has been cloned and

[25] N. W. Andrews, *Exp. Parasitol.* **71,** 241 (1990).

[26] N. W. Andrews, C. K. Abrams, S. L. Slatin, and G. Griffiths, *Cell* **61,** 1277 (1990).

[27] N. High, J. Mounier, M. C. Prevost, and P. J. Sansonetti, *EMBO J.* **11,** 1991 (1992).

[28] D. Portnoy, P. Jacks, and D. Hinrichs, *J. Exp. Med.* **167,** 1459 (1988).

[29] D. Higgins and D. Portnoy, *Nat. Biotechnol.* **16,** 138 (1998).

[30] D. Sizemore, A. Branstrom, and J. Sadoff, *Science* **270,** 299 (1995).

[31] G. Ikonomidis, Y. Paterson, F. J. Kos, and D. A. Portnoy, *J. Exp. Med.* **180,** 2209 (1994).

[32] G. Dietrich, A. Bubert, I. Gentschev, Z. Sokolovic, A. Simm, A. Catic, S. Kaufmann, J. Hess, A. Szalay, and W. Goebel, *Nat. Biotechnol.* **16,** 181 (1998).

[33] D. A. Portnoy, T. Chakraborty, W. Goebel, and P. Cossart, *Infect. Immun.* **60,** 1263 (1992).

[34] H. Goldfine, C. Knob, D. Alford, and J. Bentz, *Proc. Natl. Acad. Sci. USA* **92,** 2979 (1995).

[35] C. Geoffroy, J. L. Gaillard, J. E. Alouf, and P. Berche, *Infect. Immun.* **55,** 1641 (1987).

[36] J. Rossjohn, S. C. Feil, W. J. McKinstry, R. K. Tweten, and M. W. Parker, *Cell,* **89,** 685 (1997).

[37] G. Dietrich, J. Hess, I. Gentschev, B. Knapp, S. H. Kaufmann, and W. Goebel, *Trends Microbiol.* **9,** 23 (2001).

[38] S. Jones and D. A. Portnoy, *Infect. Immun.* **62,** 5608 (1994).

was the first virulence gene identified in *Listeria*.[39,40] One *Listeria* mutant, *hly⁻*, cannot invade the cytosol of host cells; instead, it dies in the lysosomes, demonstrating that LLO is necessary for escape of *Listeria* into the cytosol.[39,40] Bielecki and colleagues have demonstrated that when the *hly* gene is introduced into *Bacillus subtilis*, it confers the ability to escape from endocytotic compartments,[41] a property that is not found in the wild-type *B. subtilis*.

## Liposomes Containing LLO: Listeriosomes

The term *listeriosomes* describes a liposomal cytosolic delivery system incorporating the endosomolytic activity of LLO into pH-sensitive liposomes, which destabilize on protonation and leak their contents along with LLO into the endosome lumen. The lipid formulation used thus far retains the LLO molecules inside liposomes at pH 7.4–8.5, such that release is dictated by acidification in the endosomes, where LLO is activated to form pores within the endosomal membranes. LLO is coencapsulated inside liposomes with macromolecular cargo to create an LLO-containing liposome, that is, a listeriosome, and its ability to deliver macromolecules into the cytosol of macrophages has been tested and demonstrated. Testing of listeriosomes to deliver molecules of molecular mass up to 44 kDa has shown dramatic enhancement of the release of fluorescent dyes, proteins,[42a,42b,42d] and 20-mer oligonucleotides[42c] from endosomes into the cytosol. This delivery system can be utilized as a general delivery vehicle for many membrane-impermeant macromolecular drugs, provided they can be encapsulated inside liposomes with reasonable efficiency. The unique ability of listeriosomes to put antigenic proteins into the cytosol makes them optimal for targeting antigens into the MHC class I-mediated cytosolic pathway of antigen presentation and T cell activation for vaccination applications. In addition, targeted listeriosomes have the potential to serve as carriers of various toxins for use in chemotherapy. The versatility of listeriosomes would allow for the incorporation of a variety of cargoes, from proteins to genes, without compromising delivery characteristics.

[39] J. Mengaud, J. Chenevert, C. Geoffroy, J. Gaillard, and P. Cossart, *Infect. Immun.* **55,** 3225 (1987).

[40] P. Cossart, M. F. Vicente, J. Mengaud, F. Baquero, J. C. Perez-Diaz, and P. Berche, *Infect. Immun.* **57,** 3629 (1989).

[41] J. Bielecki, P. Youngman, P. Connelly, and D. Portnoy, *Nature* **345,** 175 (1990).

[42a] M. Mandal and K.-D. Lee, *Biochim. Biophys. Acta.* **1543,** 7 (2002).

[42b] C. J. Provada, F. M. Sher, and K.-D. Lee, *J. Biol. Chem.,* in press (2003).

[42c] E. Mathew, G. E. Hardee, C. F. Bennett, and K.-D. Lee, *Gene Ther.* **10,** 1105 (2003).

## Purification of Recombinant Listeriolysin O

Two different methodologies have been employed thus far to purify hemolytically active LLO for encapsulation inside liposomes: one was used by Lee et al.[42d] for initial experiments that proved the concept of LLO-liposomes, and the other, currently employed and the preferred method, uses bacterially overexpressed recombinant LLO (rLLO). The description of the biochemical purification of LLO from Listeria culture supernatant may be found in Lee et al.[42] Here, the recombinant expression system is described in detail. Recombinant LLO is purified from Escherichia coli strain BL21 ($\lambda$DE3) transformed with the pET29b plasmid carrying the six-histidine (His$_6$)-tagged LLO cDNA[43] (obtained from D. Portnoy, University of California, Berkeley), using one-step affinity chromatography. Recombinant bacteria from frozen glycerol stock are cultured for 24 h at 30° on a Luria broth (LB)–agar plate containing kanamycin (30 $\mu$g/ml). A single colony from the agar plate is typically grown overnight at 30° in 20 ml of LB medium containing kanamycin (30 $\mu$g/ml), with shaking, and a 10-ml volume is inoculated into 250 ml of LB medium (1:25 dilution) containing kanamycin (30 $\mu$g/ml) in a 1-liter flask. The culture is further incubated for 1.5 h, and expression of rLLO is induced by adding 1 m$M$ isopropyl $\beta$-D-thiogalactopyranoside, after which the culture is incubated for an additional 4 h at 30°. The bacterial culture is centrifuged at 5600 g, and the cell pellet is resuspended in 50 m$M$ phosphate, 300 m$M$ NaCl, 20 m$M$ imidazole, pH 8.0 (lysis buffer) and disrupted twice with a French press (Thermo Spectronic, Rochester, NY) at 4° in the presence of 1 m$M$ phenylmethylsulfonyl fluoride (PMSF). The bacterial lysate is centrifuged (16,000 g, 20 min, 4°), and the clear supernatant is adsorbed to Ni-NTA Superflow (Qiagen, Valencia, CA) for 1 h at 4° with gentle agitation. The affinity matrix is washed extensively with lysis buffer containing 1 m$M$ PMSF, and His$_6$-tagged rLLO is eluted with same buffer containing 500 m$M$ imidazole. Protein purity is analyzed by sodium dodecyl sulfate–polyacrylamide gel electrophoresis (SDS–PAGE) followed by Coomassie blue staining, and the concentration is determined by the bicinchoninic acid (BCA) protein assay (Pierce, Rockford, IL).

Affinity-purified, histidine-tagged rLLO migrates at ~58 kDa as a major band in SDS–PAGE and is used for encapsulation in liposomes. The typical yield of rLLO is in the range of 4–8 mg from 1 liter of culture. rLLO is dialyzed against 10 m$M$ HEPES, 140 m$M$ NaCl, 1 m$M$ EDTA, pH 8.4 at 4°, aliquotted, and stored at −80°. The osmolality of the rLLO

[42d] K.-D. Lee, Y. Oh, D. Portnoy, and J. Swanson, J. Biol. Chem. **271**, 7249 (1996).

[43] D. E. Higgins, N. Shastri, and D. A. Portnoy, Mol. Microbiol. **31**, 1631 (1999).

solution is checked with a vapor pressure osmometer (Wescor, Logan, UT) and should be approximately 290 mOsm/kg. The pH of the protein solution is kept above pH 7.4 to facilitate the formation of pH-sensitive liposomes as described below.

## Hemolysis Assay

The membrane pore-forming activity of the purified rLLO is assayed by monitoring hemolysis of sheep red blood cells (SRBCs), which is reflected by the change in 90° light scattering due to the membrane damage caused by rLLO, and is measured in a FluoroMax-2 fluorometer (Jobin Yvon-Spex Instruments, Edison, NJ) with instrument settings at 590 nm for both excitation and emission. SRBCs (ICN, Costa Mesa, CA) are washed three times in 10 m$M$ HEPES, 140 m$M$ NaCl, pH 7.4, resuspended, and diluted to $1 \times 10^8$ cells/ml in a 3-ml final volume containing 1 m$M$ dithiothreitol (DTT). Hemolysis is observed by adding small quantities of rLLO (10, 25, or 50 ng) to the SRBCs with constant mixing, after recording a stable signal for about 200 s. Membrane pore-forming activity of rLLO is monitored up to 800 s by observing a decrease in counts per second (CPS). The rate of hemolysis in this assay depends on the rLLO dose.

## Assays for LLO-Liposome-Mediated Cytosolic Delivery

### Monitoring Cytosolic Delivery by Morphological Assay Using Fluorescent Dye

The ability of LLO to deliver molecules to the cytosol can be monitored in cultured cells by direct visualization of either the movement or steady state distribution of fluorescent cargo. This has been achieved with the membrane-impermeant fluorescent dye 8-hydroxypyrene-1,3,6-trisulfonate (HPTS, pyranine; Molecular Probes, Eugene, OR). HPTS is a relatively small (MW 524) hydrophilic pyrene derivative that has the useful property of exhibiting a pH-dependent excitation spectral shift in the region between 405 and 450 nm (the two predominant absorption maxima).[3,4,44] At low pH, maximum HPTS excitation occurs at 405 nm, whereas maximum excitation at neutral to slightly basic pH occurs at 450 nm; excitation at either wavelength produces emission with a peak at 511 nm. Because of these spectral properties, HPTS has the added advantage of providing quantitative as well as qualitative morphological data regarding its

---

[44] C. C. Overly, K. D. Lee, E. Berthiaume, and P. J. Hollenbeck, *Proc. Natl. Acad. Sci. USA* **92**, 3156 (1995).

subcellular distribution. Thus, the ratio of fluorescence that is produced by excitation at 405 nm to that produced with excitation at 450 nm is indicative of the relative amounts of HPTS in the endocytotic compartments (acidic) versus the cytosol (neutral), respectively.[42] Furthermore, because of its sulfonate groups HPTS is highly negatively charged in all pH environments encountered within the cell, including the endosome and cytosol, and hence exhibits low membrane permeability. Therefore, by being able to exclude diffusion through membranes as a mechanism for cytosolic delivery of HPTS, one can conclude more confidently that delivery has occurred via an aqueous pore such as that formed by LLO. This is reflected in the observation that delivery of HPTS to the cytosol of macrophages is undetectable in cells that have been treated with HPTS-containing liposomes without coencapsulated LLO (Fig. 1). Moreover, if one wishes to use HPTS in a pH-independent way, this can be done conveniently with an excitation wavelength of 413 nm, which excites HPTS approximately at its isobestic point.

*Encapsulation of HPTS in* Listeriosomes

The procedure for making listeriosomes is as follows. Briefly, pH-sensitive liposomes, composed of phosphatidylethanolamine (PE; Avanti Polar Lipids, Alabaster, AL) and cholesteryl hemisuccinate (CHEMS; Sigma, St. Louis, MO) (2:1 molar ratio, typically 15 $\mu$mol of total lipid), are made by drying the lipids under vacuum ($\geq$760 mmHg) in Pyrex glass tubes to a thin film in a Büchi Rotavapor (Brinkmann Instruments, Westbury, NY). The lipids are then rehydrated in 30 m$M$ Tris-Cl, 100 m$M$ NaCl, pH 8.5 in the presence of rLLO at 200 $\mu$g/ml plus 35 m$M$ HPTS and 50 m$M$ $p$-xylene-bispyridinium bromide (DPX) by vigorous vortexing and a combination of four rounds of freeze–thawing, followed either by extrusion four times through 0.2-$\mu$m pore size polycarbonate filters (Whatman Nucleopore, Scarborough, ME) or sonication. The size of the resulting liposomes can be monitored by quasi-elastic light scattering (Nicomp, Santa Barbara, CA). DPX, when coencapsulated at relatively high concentrations, quenches HPTS fluorescence[4,42]; release of both DPX and HPTS from liposomes into either endosomes or the cytosol effectively dilutes DPX, resulting in a dramatic dequenching and thus an increase in HPTS fluorescence. Unencapsulated material is removed by gel-filtration chromatography in a Sepharose CL (Amersham Biosciences, Piscataway, NJ) column.

Murine bone marrow-derived macrophages used in these experiments are harvested and cultured essentially as described.[45] Briefly, bone marrow

---

[45] E. L. Racoosin and J. A. Swanson, *J. Cell Biol.* **121**, 1011 (1993).

405 nm excitation          440 nm excitation

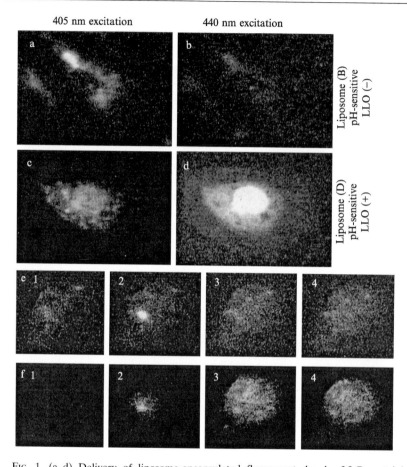

FIG. 1. (a–d) Delivery of liposome-encapsulated fluorescent dye by LLO-containing liposomes. Fluorescence of cells incubated with HPTS/DPX-containing liposomes: (a and b) pH-sensitive liposomes without LLO; (c and d) pH-sensitive liposomes with LLO. Images (a) and (c) were obtained with excitation at 405 nm (20-nm bandpass), and images (b) and (d) were obtained with excitation at 440 nm (20-nm bandpass); all with emission at 520 nm. Liposomes containing HPTS–DPX were incubated with bone marrow-derived macrophages for 15 min at 37°. Cells were washed and chased for an additional 15 min at 37° before viewing. (e and f) A time-lapse sequence of HPTS delivery by pH-sensitive, LLO-containing liposomes. The uptake and fate of HPTS/DPX-containing liposomes was monitored by MetaMorph image analysis (Universal Imaging, Westchester, PA) equipped with alternating excitation filters [(e) excited at 405 nm; (f) excited at 440 nm]. After 8 min of incubation with liposomes, cells were washed with buffer, and the fluorescence at the two excitation wavelengths was collected every minute. Frames represent 1-min intervals, beginning 5 min after the wash. (Adapted, with permission, from Ref. 42.)

cells are extracted from femurs of 4- to 12-week-old mice. Mouse bone marrow precursor-derived macrophages are cultured in complete Dulbecco's modified Eagle's medium (DMEM) containing 10% heat-inactivated fetal bovine serum (FBS), 2 mM L-glutamine, 1 mM sodium pyruvate, penicillin (100 U/ml), streptomycin (100 μg/ml), 50 μM 2-mercaptoethanol (Sigma), plus macrophage colony-stimulating factor in 100-mm petri dishes (Falcon; BD Biosciences Discovery Labware, Franklin Lakes, NJ) for 6 days with two changes of medium. During 6 days of culture, progenitor cells expand and differentiate into adherent bone marrow-derived macrophages.

Listeriosomes containing HPTS and DPX are incubated with the macrophages for 15 min at 37° followed by washing and a 15-min chase before viewing, at which point cells are analyzed for HPTS distribution by recording fluorescence emission at ≥520 nm, using a MetaMorph imaging system (Universal Imaging, Westchester, PA) equipped with alternating excitation filters for 405 and 440 nm (20-nm bandpass) (Fig. 1). For experiments in which release of HPTS is recorded in time-lapse sequences, macrophages are incubated at 37° for 8 min with listeriosomes and washed, and images are recorded at 1-min intervals beginning 5 min after the wash (Fig. 1). A similar method is used to determine the relationship between pH and LLO activity within phagosomes of macrophages that are simultaneously fed HPTS along with *L. monocytogenes* bacteria.[46]

The use of HPTS in these experiments allows conclusions to be drawn on the basis of the subcellular distribution of its fluorescence. Soon after uptake (within approximately the first few minutes of recording images), listeriosomes containing HPTS and DPX display weak fluorescence when excited at 405 or 440 nm, indicating quenching of HPTS by DPX still trapped within the pH-sensitive listeriosomes. Soon thereafter, however, release of HPTS and DPX into the phagosome lumen can be detected, on the basis of the significant increase in fluorescence by excitation at 405 nm, but relatively weak fluorescence by excitation at 440 nm. Within ~1 min of this event, a rapid release of HPTS into the cytosol is observed, on the basis of the emission of 520-nm light produced by excitation at 440 nm (increased) versus that at 405 nm (decreased). This interpretation, based on quantification of fluorescence, is corroborated morphologically in that the pattern of fluorescence recorded at these time points concomitantly changes from punctate (endosomal) to diffuse (cytosolic). It should be noted that HPTS, when delivered at relatively high concentration into the cytosol, displays a tendency to accumulate in the

[46] K. E. Beauregard, K. D. Lee, R. J. Collier, and J. A. Swanson, *J. Exp. Med.* **186,** 1159 (1997).

nucleus (Fig. 1c and d), suggesting the utility of listeriosomes in delivering hydrophilic molecules such as oligonucleotides to the nucleus (see below).

## Coencapsulation with Macromolecular Antigen inside Liposomes

The procedure for coencapsulation of ovalbumin (OVA) in pH-sensitive listeriosomes as a model macromolecule antigen is essentially identical to that for HPTS–DPX described above, with the following modifications. OVA (grade VI, 20 mg/ml; Sigma) with or without rLLO (250 $\mu$g/ml), in 0.75 ml of HEPES-NaCl, pH 8.4, is added to dried lipid films. Tubes are vortexed vigorously to hydrate the lipid films, and the resulting multilamellar vesicles are freeze–thawed for four cycles and sonicated afterward in a bath sonicator (Laboratory Supplies, Hicksville, NY) for 30 s each three times. Depending on the need for a particular size, the liposomes can be further extruded through defined-pore-size polycarbonate membranes (Whatman Nuclepore). However, OVA and rLLO have a tendency to clog up the membranes.

Liposome-encapsulated OVA or OVA–LLO is separated from unencapsulated species by gel-filtration chromatography, using a Sepharose CL-4B column (1 cm in diameter and ∼30 cm in length). The concentrations of encapsulated OVA and LLO in the liposomal fractions are determined by densitometric analysis of samples and known quantities of OVA standards resolved by SDS–PAGE and stained with Coomassie blue, using either MetaMorph imaging system analysis software or other commercially available gel documentation systems. On average, the encapsulation efficiency is between 7% and 15%, and the mean diameter of the liposomes is in the range of ∼240–370 nm as determined by particle size analysis.

## In Vitro Cytosolic Delivery of Macromolecule Antigen in Antigen-Presenting Cells

The ability of rLLO-containing, pH-sensitive liposomes to deliver exogenous macromolecule antigen to the cytosol of antigen-presenting cells (APCs), for cytosolic processing and major histocompatibility complex I (MHC I)-mediated presentation, is examined in vitro with mouse bone marrow-derived macrophages or dendritic cells (DCs). C57BL/6J (B6) mouse bone marrow-derived macrophages used in in vitro antigen presentation assays are cultured as described above.

In vitro antigen delivery experiments using macrophages are performed in triplicate as described by Lee et al.[42] Macrophages are plated ($2 \times 10^5$/ well in 96-well plates) in complete DMEM and are pulsed with liposomal OVA–LLO or liposomal OVA alone in serum-free medium for 1 h at 37°. Afterward, cells are washed three times, incubated in complete medium for

4 h at 37°, and then fixed with 1% paraformaldehyde in phosphate-buffered saline (PBS) (200 $\mu$l/well) for 15 min at room temperature. Cells are washed three times, and any residual paraformaldehyde is quenched with 0.2 $M$ L-lysine in complete DMEM (300 $\mu$l/well) for 20 min at room temperature and washed afterward. Subsequently, $1 \times 10^5$ OVA-specific CD8$^+$ T cells, CD8 OVA T1.3, are added to each well in 0.2 ml of complete DMEM and further incubated for 24 h at 37° in 5% $CO_2$. CD8 OVA T1.3 cells (H-2K$^b$ restricted, recognizing OVA MHC class I peptide, SIINFEKL epitope) are maintained in complete DMEM (kindly provided by C. Harding, Case Western Reserve University, Cleveland, OH). The interleukin 2 (IL-2) concentration in the culture supernatant after 24 h is measured by enzyme-linked immunosorbent assay (ELISA) to monitor the extent of cytosolic delivery and presentation of liposome-encapsulated exogenous antigen in conjunction with MHC class I molecules of APCs to the antigen-specific CD8$^+$ T cells. The Duoset IL-2 assay kit (Genzyme, Cambridge, MA) is used for ELISA according to the manufacturer's instructions. Nunc-Immuno plates (Nunc, Rostilde, Denmark) are coated with the capture anti-IL-2 antibody at 4 $\mu$g/ml in carbonate buffer (0.1 $M$, pH 8.2) overnight at 4°. Plates are washed with PBS containing 0.05% Tween 20 (PBST) and blocked with 2% bovine serum albumin (BSA) in PBST for 2 h at room temperature. Culture supernatants from antigen presentation experiments, and standard murine recombinant IL-2 at a wide range of concentrations, are added to the plate, at 100 $\mu$l/well, and incubated for 1 h at 37°. After extensive washing with PBST, biotin-conjugated anti-IL-2 is added at 0.5 $\mu$g/ml and the plate is incubated for 1 h at 37°. After washing, horseradish peroxidase (HRP)–streptavidin is added, incubated for 40 min at 37°, and washed. HRP substrate Turbo-TMB (Pierce) is added at 100 $\mu$l/well, the reaction is stopped with 2 $N$ $H_2SO_4$ (100 $\mu$l/well), and the plate is read at 450 nm in a microplate reader (Molecular Devices, Sunnyvale, CA). Unknown values are calculated on the basis of recombinant mouse IL-2 standards and expressed as picograms per milliliter IL-2 concentration.

Efficient cytosolic delivery of OVA can be observed in macrophages in an LLO-dependent manner, demonstrating that coencapsulation of LLO inside pH-sensitive liposomes is required for the introduction of exogenous OVA into the cytosol of APCs (Fig. 2). The extent of cytosolic delivery of OVA by pH-sensitive LLO-liposomes is found to be dependent on the dose of OVA, as higher concentrations of OVA generate higher concentrations of IL-2. No IL-2 is detected in the culture supernatant when cells are incubated either with OVA encapsulated in non-LLO-containing, pH-sensitive liposomes or with non-LLO-containing, pH-insensitive liposomes.

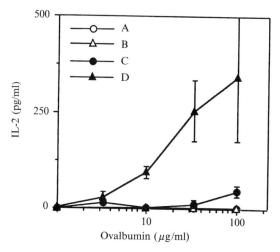

FIG. 2. OVA-specific MHC class I-restricted antigen presentation by macrophages. Various amounts of OVA encapsulated inside liposome formulations: (A) PC/CHEMS (−LLO); (B) PE/CHEMS (−LLO); (C) PC/CHEMS (+LLO); (D) PE/CHEMS (+LLO), are incubated with macrophages. The extent of OVA-specific antigen presentation is documented by measuring interleukin 2 (IL-2) production from CD8 OVA T1.3 cells incubated with macrophages according to the method described in text. The data represent averages of two independent experiments performed in triplicate. (Adapted with permission from Ref. 42.)

LLO-containing, pH-sensitive liposome-mediated efficient cytosolic delivery of OVA can also be observed in murine dendritic cells, using the antigen presentation assay as described for macrophages. DCs are grown in complete RPMI containing granulocyte-macrophage colony-stimulating factor and interleukin 4 in a 24-well plate. The medium is changed every 2 days, and cells are harvested and enriched by passage through a serum column on day 6, according to Inaba et al.[47] These purified DCs are grown in 100-mm petri dishes for an additional 2 days before they are used in antigen presentation assays.

*Cytosolic Delivery of Whole Protein Antigen* in Vivo: *Monitoring Functional Activity or Frequency of CD8⁺ Cytotoxic T Lymphocytes*

[51]*Cr Release Assay for Antigen-Specific Cytotoxic T Lymphocyte Activity.* To examine the efficacy of rLLO-containing, pH-sensitive liposomes in inducing an antigen-specific cytotoxic T lymphocyte (CTL)

[47] K. Inaba, M. Inaba, N. Romani, H. Aya, M. Deguchi, S. Ikehara, S. Muramatsu, and R. M. Steinman, *J. Exp. Med.* **176,** 1693 (1992).

response, B6 mice are immunized via the subcutaneous or intravenous route (four to six mice per group) on day 0 and day 12 with OVA encapsulated in pH-sensitive liposomes, with or without rLLO. Some of the mice are also immunized with soluble OVA as a baseline control group. Splenocytes (responders) are isolated from the immunized mice 9–12 days after the boost and then restimulated *in vitro* by coculturing with mitomycin C (50 $\mu$g/ml, $4 \times 10^7$ cells for 25 min at $37°$)-treated E.G7-OVA (E.OVA) cells (stimulators) at $3 \times 10^7$ responder:$2 \times 10^6$ stimulator cells, in 10 ml of complete RPMI, in upright 25-cm$^2$ flasks (Falcon; BD Biosciences Discovery Labware) as described,[48] without exogenous IL-2 for 5 days at $37°$, 5% $CO_2$. Restimulated viable T cells are separated on a Ficoll (Pharmacia, Uppsala, Sweden) density gradient, washed twice in complete RPMI, and used as effector cells in a standard 4-h $^{51}Cr$ release CTL assay. $^{51}Cr$ (Amersham Biosciences)-labeled E.OVA or EL-4 cells, or EL-4 cells pulsed with SIINFEKL (10 $\mu$g/ml) during $^{51}Cr$ labeling (200 $\mu$Ci/$10^6$ cells) for 1.5 h at $37°$, are used as target cells and plated in U-bottom 96-well plates (Costar, Cambridge, MA) at various effector:target ratios in 200 $\mu$l of complete RPMI. Released $^{51}Cr$ from target cells is measured directly in a $\gamma$ counter (Auto Gamma 5000 series; Packard, Downers Grove, IL) by transferring 100 $\mu$l of supernatant into scintillation vials, and the specific cytolytic activity is determined by the following formula:

$$\% \text{ specific lysis} = \left( \frac{\text{Experimental release} - \text{Spontaneous release}}{\text{Total release} - \text{Spontaneous release}} \right) \cdot 100$$

The spontaneous release from the target cells is routinely between 5% and 15% of the total release of $^{51}Cr$ by detergent (2% Triton X-100) lysis.

The tumor cell lines EL-4 (H-2K$^b$, MHC class II negative), a B6-derived thymoma, and E.OVA, a subclone of EL-4 stably transfected with the pAc-neo-OVA gene,[48] are purchased from the American Type Culture Collection (Manassas, VA). Cells are grown in complete RPMI 1640 medium containing 10% heat-inactivated FBS, 2 m$M$ L-glutamine, 1 m$M$ sodium pyruvate, penicillin (100 U/ml), streptomycin (100 $\mu$g/ml), and 50 $\mu M$ 2-mercaptoethanol (2-ME) (Life Technologies, Grand Island, NY). E.OVA cells are grown in complete RPMI containing G418 sulfate, 400 $\mu$g/ml. Geneticin MHC class I OVA peptide SIINFEKL (amino acid residues 257–264) and vesicular stomatitis virus nucleoprotein peptide RGYVYQGL (amino acid residues 325–332, control peptide) are synthesized (Research Genetics, Huntington, AL), dissolved in PBS, and kept at $-80°$ in aliquots.

[48] M. W. Moore, F. R. Carbone, and M. J. Bevan, *Cell* **54**, 777 (1988).

A significant increase in OVA-specific CTL activity is used as an indicator of enhanced cytosolic delivery of OVA *in vivo*.

*Determination of Antigen-Specific T Cell Frequency as Readout of Cytosolic Delivery of Antigen: Enzyme-Linked Immunospot Assay.* Analysis of cytotoxic T lymphocyte precursor (CTLp) frequency allows a direct comparison of the CTL priming ability between the immunogenic formulations on the basis of quantification of MHC class I peptide-recognizing T cells generated due to vaccination, utilizing the cytosolic delivery strategy. The OVA MHC class I peptide SIINFEKL-specific CTLp frequency is determined by antigen-specific interferon $\gamma$ (IFN-$\gamma$) enzyme-linked immunospot (ELISpot) assay[49] utilizing IFN-$\gamma$-secreting effector T cells generated in immunized mice by employing the paired anti-IFN-$\gamma$ antibodies R4-6A2 for capture and biotinylated XMG1.2 for detection (BD Biosciences PharMingen, San Diego, CA). MAHA-S45 plates (Millipore, Bedford, MA) are coated with the capture antibody in sterile PBS (2 $\mu$g/ml) overnight at 4°, washed with sterile PBS, and blocked with 2% BSA in PBS (sterile) for 2 h at room temperature. Splenocytes from immunized mice are depleted of the red blood cells with Gey's hemolytic solution, washed, and resuspended in HL1 medium (BioWhittaker, Walkersville, MD). Cells are cultured at various cell numbers in triplicate in antibody-coated, blocked plates in the presence of SIINFEKL or RGYVYQGL (10 $\mu$g/ml) for 24 h at 37° without any disturbance. Plates are washed extensively with PBS, and then with PBST. Secreted IFN-$\gamma$ spots on the membrane are detected by adding alkaline phosphatase-conjugated streptavidin (BD Biosciences PharMingen) followed by 5-bromo-4-chloro-3-indolyl phosphate/nitroblue tetrazolium (BCIP/NBT) substrate. The frequencies of IFN-$\gamma$-producing cells in each well are determined with a computerized ImmunoSpot image analyzer (Cellular Technology, Cleveland, OH).

Enhancement in the frequency of SIINFEKL-specific IFN-$\gamma$ spots is a measure of *in vivo* cytosolic delivery of OVA.

*Antigen-Specific Interferon $\gamma$ Production as Measure of Cell-Mediated Immune Response Primed by in Vivo Cytosolic Delivery.* The antigen-specific cytokine (IFN-$\gamma$) response can be monitored as an additional measure of cellular immunity primed by LLO-liposome-mediated cytosolic delivery of antigen. The IFN-$\gamma$ response is generated by culturing splenocytes ($1 \times 10^7$/ml) from immunized mice in the presence of SIINFEKL (10 $\mu$g/ml), control peptide RGYVYQGL (10 $\mu$g/ml), or concanavalin A (Con A, 5 $\mu$g/ml) in 96-well plates in triplicate in 200 $\mu$l of complete RPMI

---

[49] Y. Miyahira, K. Murata, D. Rodriguez, J. R. Rodriguez, M. Esteban, M. M. Rodrigues, and F. Zavala, *J. Immunol. Methods* **181,** 45 (1995).

for 24 or 48 h.[50] The IFN-γ concentration in the culture supernatant is measured by ELISA, using the IFN-γ-specific paired monoclonal antibodies (BD Biosciences PharMingen) as described earlier for the ELISpot assay. Unknown values are calculated on the basis of recombinant mouse IFN-γ (Genzyme) standards and expressed as IFN-γ units per milliliter.

## Cytosolic Delivery of Antisense Oligonucleotides

Various strategies have been employed to improve the delivery of oligonucleotides to the cytosol, with only limited success. Disadvantages associated with these formulations include cytotoxicity, reduced activity in serum-containing media, or the requirement for large doses of drug to obtain a suboptimal effect.[51,52] As with fluorescent dyes and proteins, listeriosomes can be utilized to improve the delivery of antisense oligonucleotides to the cytosol. In this section, we describe several methods to evaluate the efficiency of listeriosome-mediated cytosolic delivery of antisense oligonucleotides: subcellular localization of the delivered oligonucleotide visualized by fluorescence microscopy and antisense efficacy as measured by the expression levels of the target protein and mRNA.

*Encapsulation of Oligonucleotide and LLO into Liposomes.* Listeriosomes are prepared in the same manner as mentioned in the previous section for HPTS encapsulation, with the following modifications. The dried lipid film is hydrated with oligonucleotides (4.3 mg/ml), or LLO (200 μg/ml), or a mixture of both in 140 mM NaCl, 10 mM HEPES, pH 8.5. After freeze–thawing, liposomes are first extruded through 0.4-μm pore size polycarbonate filters (Whatman Nuclepore) and then through 0.2-μm pore size filters four times each to ensure size uniformity. The mean diameters of the liposomal formulations are assessed by quasi-elastic light scattering, using the Zeta Potential/Particle Sizer Nicomp 380 ZLS (Nicomp). After extrusion, liposomes are separated from the unencapsulated contents by size-exclusion chromatography, using a Sepharose CL-4B column. Liposomes are stored under argon at 4° and used in experiments within 7 days of preparation.

Encapsulation efficiencies of oligonucleotide or LLO are determined by quantitative polyacrylamide gel electrophoresis and by measuring the intensities of the bands, using MetaMorph image analysis software. To measure oligonucleotide encapsulation efficiency, liposomes are destabilized with

[50] K. Suzue, X. Zhou, H. N. Eisen, and R. A. Young, *Proc. Natl. Acad. Sci. USA* **94,** 13146 (1997).
[51] I. Lebedeva, L. Benimetskaya, C. A. Stein, and M. Vilenchik, *Eur. J. Pharm. Biopharm.* **50,** 101 (2000).
[52] R. L. Juliano and H. Yoo, *Curr. Opin. Mol. Therapeut.* **2,** 297 (2000).

2% $C_{12}E_8$ (EMD Biosciences, La Jolla, CA) and the samples are resolved in a 15% Tris–borate–EDTA (TBE) polyacrylamide gel (Bio-Rad, Hercules, CA). Oligonucleotide concentration is determined by SYBR Green 1 (Molecular Probes) staining and calculated on the basis of a standard curve generated with known concentrations of oligonucleotide ranging from 6.25 to 50 μg/ml. LLO encapsulation efficiency is measured similarly by staining a 10% polyacrylamide gel with Coomassie Brilliant Blue R-250 (Fisher Scientific, Pittsburgh, PA) and calculated on the basis of known concentrations of LLO ranging from 2.5 to 80 μg/ml.

*Treatment of Cells with Oligonucleotides.* For oligonucleotide delivery experiments, murine bone marrow macrophages are cultured as described above. After 6 days of culture, macrophages are suspended in endotoxin-free PBS at 4° without $CaCl_2$ and $MgCl_2$, centrifuged, and washed in DMEM supplemented with 10% FBS (DME-10F). Cells are seeded at a concentration of $1 \times 10^6$ cells/ml in a 35-mm noncoated Nunc dish in DME-10F for flow cytometric and Northern blot analyses and in a 25-mm no. 1 coverslip (Corning Life Sciences, Acton, MA) for fluorescence microscopic analysis. The cells are incubated at 37° for 2–12 h to allow adherence before treatments. Adherent cells are washed twice with warm DME-10F and treated with oligonucleotides in various delivery formulations diluted in DME-10F at 37° for 2 h. Oligonucleotides encapsulated in LLO-containing liposomes are tested and compared with control formulations including naked oligonucleotides, blank liposomes, blank listeriosomes, oligonucleotides in liposomes containing no LLO, and random control oligonucleotide sequences in listeriosomes. After a 2-h incubation with the formulations, cells are washed three times, chased in DME-10F, typically for 0.5 h at 37°, and activated by adding IFN-γ (100 units/ml; Genzyme) and lipopolysaccharide (LPS, 100 ng/ml; Sigma). Numerous cell surface receptors and cytokines expressed in bone marrow-derived macrophages, including intercellular adhesion molecule-1 (ICAM-1), IFN-β, and B7 costimulatory molecules, are upregulated by the activation procedure, many of which constitute the targets for inhibition using the cytosolic delivery of specific antisense oligonucleotides. In the specific example of ICAM-1, activating cells for 8 and 2 h induces the maximal upregulation of ICAM-1 protein and mRNA, respectively.

*Determination of Protein Expression by Flow Cytometric Analysis.* The cell surface expression of the target protein is measured by flow cytometric analysis after treatment with various oligonucleotide formulations. Macrophages are washed with DMEM supplemented with 2% FBS and 0.01% sodium azide (DME-2F), and treated with 0.05% trypsin and 0.53 m$M$ EDTA for 1 min at 37° to detach cells. Cells are detached with minimal damage and with no apparent reduction in the cell surface target protein, as determined by flow cytometry. The cell suspension is washed twice with DME-2F and then incubated with a fluorescein-labeled monoclonal antibody (mAb) specific for the target molecule at 4° for 45 min in the dark. Fluorescein-labeled polyclonal nonspecific IgG (BD Biosciences PharMingen) serves as the control antibody for monitoring nonspecific mAb labeling. After staining, cells are kept at 4°, washed twice with DME-2F and PBS, and resuspended in 0.5 ml of PBS. The mean fluorescence intensity for the cell suspension is measured with a Coulter Epics Elite ESP flow cytometer equipped with Elite acquisition software (Beckman Coulter, Fullerton, CA).

*Determination of mRNA Levels by Northern Blot Analysis.* Northern blot analysis is performed to determine whether oligonucleotides delivered in LLO-liposomes reach the cytosolic targets and specifically inhibit the upregulated expression of target mRNA in activated macrophages. After treatment, cells are washed twice with Hanks' balanced salt solution (HBSS; GIBCO-BRL, Grand Island, NY) and lysed directly in a 35-mm Nunc plate, using an RNAqueous phenol-free total RNA isolation kit (Ambion, Austin, TX). The total extracted cellular RNA concentration and relative purity are determined spectrophotometrically by measuring the absorbance at 260 nm and by the 260 nm/280 nm quotient, respectively. RNA samples are mixed with ethidium bromide and resolved electrophoretically on a 1% agarose–formaldehyde gel in morpholinepropanesulfonic acid (MOPS) buffer. The RNA bands are transferred to a nylon membrane (Hybond-N; Amersham Biosciences) for 2 h by capillary diffusion, using Ambion transfer buffer, and cross-linked by ultraviolet light. The membrane is prehybridized with 10 ml of Zip/Hyb solution (Ambion) at 42° for 2 h to block nonspecific binding and then hybridized with the radioactive probe for 4 h. The membrane is then probed with a $^{32}$P-labeled cDNA typically corresponding to one of two genes: (1) the target gene of interest or (2) a housekeeping gene such as murine glyceraldehyde-3-phosphate dehydrogenase (GAPDH), which serves as a control for variability in the assay. To make the probe, the cDNA fragment of the target gene is labeled with $[^{32}P]dCTP$, utilizing a Rediprime II random prime labeling kit (Amersham Biosciences). The labeled probe is removed from unincorporated nucleotides, using a G-25 Sephadex column (Roche Molecular Biochemicals, Indianapolis, IN). The concentrated probe is heat denatured at 100° for 5 min and added immediately to the hybridization solution at a specific activity of $1–2 \times 10^6$ cpm/ml. After hybridization, the membrane is washed twice with 20 ml of 2× saline–sodium phosphate–EDTA (SSPE) solution for 10 min at 42° and then washed two more times with 20 ml of 0.1× SSPE for 20 min at 42°. The probed membranes are exposed to X-Omat film (Eastman Kodak, Rochester, NY) with a Kodak

BioMax TranScreen-HE intensifying screen at $-80°$ for several days and examined by autoradiography. The bands on the film are quantified by MetaMorph image analysis software. After analysis, the probe is stripped overnight at $75°$ in a solution of 1 m$M$ Tris, 1 m$M$ EDTA, and $0.1\times$ Denhardt's solution (Sigma) and then washed in $0.1\times$ SSPE at room temperature for 1 h. The membrane is reprobed with GAPDH-mouse DECAtemplate (Ambion) labeled with $[^{32}P]dCTP$, and the GAPDH bands are quantified as mentioned previously.

*Intracellular Localization of Oligonucleotides by Fluorescence Microscopy.* To detect the subcellular distribution of delivered oligonucleotides, cells are incubated with fluorescein-labeled oligonucleotides coencapsulated with rhodamine-labeled dextrans in pH-sensitive liposomes with or without LLO for 2 h at $37°$ in DME-10F. After incubation, cells are washed three times with Ringer's solution, pH 7.4, and placed on a temperature-controlled stage set to $37°$. Cells are analyzed with an Axiovert 135 TV inverted microscope (Carl Zeiss, Thornwood, NY), equipped with a 100-W mercury arc and halogen lamp, $\times60$ and $\times100$ oil immersion objective lenses, an excitation filter wheel alternating between 490 and 570 nm, and a long-bandpass $\geq$520-nm emission filter. Cells are photographed with a cooled CCD camera (Roper Scientific, Trenton, NJ). Three images are taken of each cell: a phase-contrast image, a 490-nm image, and a 570-nm image. Images are analyzed by MetaMorph image analysis software.

Concluding Remarks

LLO-containing, pH-sensitive liposomes have demonstrated their ability to deliver exogenous macromolecules into the cytosol of macrophages *in vitro*. As this method can deliver proteins into the cytosol of DCs and other cell lines in culture, the cytosolic delivery results observed so far can be extended to other proteins such as toxins or various nucleic acid-based drugs, as well as to other cell types. This nonviral, nonbacterial liposomal delivery system is relatively easy to produce and simple to manipulate, and it has the ability to encapsulate various macromolecular drug candidates for cytosolic delivery. Because recombinant LLO can be obtained in reasonable quantities and is fairly stable during storage, the difficulties associated with manufacturing LLO-liposome formulations, relative to conventional liposomes, should be outweighed by their superior efficiency in cytosolic delivery. One caveat regarding the use of rLLO from an *E. coli* expression system is the potential for contamination with endotoxin, which is difficult to remove once introduced.

In this chapter, we have presented several methods for LLO-liposomal formulation designed to deliver macromolecules to the cytosol of cells. We have also described several methods tailored specifically to monitor either protein antigen delivery into the cytosolic compartment of antigen processing and presentation, or antisense oligonucleotide delivery. Most of these methods, except for the morphological method, rely on indirect, amplified readouts of the downstream effects on cytosolic delivery of the relevant molecules. In the case of LLO-liposomes, proper controls including non-LLO-liposomes are important in validating the extent of cytosolic delivery augmented by the LLO-mediated mechanism.

The ability to deliver a variety of drugs and macromolecules into the cytosol is essential to many areas of basic biology and pharmaceutics. Liposomes have gone through several transformations and will need significant further investigation before use as a general delivery system for a variety of macromolecular therapeutic agents. In addition, modifications and customization for each application, depending on particular drugs and target cells, will still be necessary. Among the promising new strategies, the more recent approach using the LLO-based mechanism presents an exciting solution to the limitations of liposomal cytosolic delivery.

## Acknowledgments

The authors are thankful for financial support from the NIH (F32 AI10571 to C.J.P.; R29AI42084 and R01AI47173 to K.-D.L.). This work was also supported in part by the Vahlteich Research Fund from the College of Pharmacy at the University of Michigan.

# [19]  Tat Peptide-Mediated Intracellular Delivery of Liposomes

By Tatiana S. Levchenko, Ram Rammohan, Natalia Volodina, and Vladimir P. Torchilin

## Introduction

Intracellular drug delivery for most purposes is hampered by problems related to the efficacy of delivery as well as to endocytotic processing and degradation. Numerous attempts have been made to develop an efficient intracellular drug and DNA delivery system bypassing the endocytotic pathway. In this context, Tat peptides, protein transduction domains (PTDs) derived from the human immunodeficiency virus type 1 (HIV-1) Tat protein,[1] have been shown to facilitate the intracellular delivery of proteins,[2] DNA,[3] and small colloidal particles.[4] Although the exact mechanism(s) of this uptake/translocation has not yet been clearly established, it is suspected that some form of ligand–lipid interaction plays a role, and direct contact between the translocating moiety and cell membrane is required.[5] A challenging task is to use PTDs for intracellular delivery of various drug carriers, such as micelles, liposomes, and nanoparticles. So far, dextran-coated iron oxide colloidal particles, about 40 nm in size and containing several attached molecules of Tat peptide per particle, have been delivered into lymphocytes much more efficiently than free particles.[6] One may assume that even bigger particulate drug carriers could be translocated into cells by Tat peptides if a sufficient number of these molecules are attached to the particle surface. Relatively large (200 nm) plain and PEGylated liposomes can be delivered into various cells by multiple Tat peptide molecules attached to the liposome surface via the spacer p-nitrophenylcarbonyl-PEG-phosphatidylethanolamine (pNP-PEG-PE).[7] This covalent coupling of Tat peptides to microparticulate drug carriers, via a single terminus-reactive spacer group, may provide an efficient tool for cytosolic delivery of drug- and DNA-loaded plain and sterically protected nanoparticles *in vitro* and *in vivo*.[8]

Methods

*Liposomes*

*Reagents.* Egg phosphatidylcholine (PC), cholesterol (Ch), polyethylene glycol (PEG)-phosphatidylethanolamine (PE) with differently sized PEG units (MW 2000 and 5000 Da; PEG2000-PE and PEG5000-PE, respectively), positively charged dioleoyl phosphatidylethanolamine (DOPE), N-glutaryl-PE (NGPE), and the amphiphilic fluorescent dye tetramethyl-rhodamine-PE (Rh-PE) are obtained from Avanti Polar Lipids (Alabaster,

[1] M. Green and P. M. Loewenstein, *Cell* **55,** 1179 (1988).
[2] S. R. Schwarze, A. Ho, A. Vocero-Akbani, and S. F. Dowdy, *Science* **285,** 1569 (1999).
[3] A. Eguchi, T. Akuta, H. Okuyama, T. Senda, H. Yoloi, H. Inokuchi, S. Fujita, T. Hayakawa, K. Takeda, M. Hasegawa, and M. Nakanishi, *J. Biol. Chem.* **276,** 26204 (2001).
[4] R. Bhorade, R. Weissleder, T. Nakakoshi, A. Moore, and C.-H. Tung, *Bioconjug. Chem.* **11,** 301 (2000).
[5] D. A. Mann and A. D. Frankel, *EMBO J.* **10,** 1733 (1991).
[6] M. Lewin, N. Carlesso, C.-H. Tung, X.-W. Tang, D. Cory, D. T. Scadden, and R. Weissleder, *Nat. Biotechnol.* **18,** 410 (2000).
[7] V. P. Torchilin, T. S. Levchenko, A. N. Lukyanov, B. A. Khaw, A. L. Klibanov, R. Rammohan, G. P. Samokhin, and K. R. Whiteman, *Biochim. Biophys. Acta* **1511,** 397 (2001).
[8] V. P. Torchilin, R. Rammohan, V. Weissig, and T. S. Levchenko, *Proc. Natl. Acad. Sci. USA* **98,** 8786 (2001).

AL). pNP-PEG-DOPE is synthesized as described below. Diethylene triamine pentaacetic acid anhydride (DTPA), DTPA-PE, PEG-di(nitrophenylcarbonyl) [PEG-(pNP)$_2$], triethylamine (TEA), octyl glucoside (OG), 1-ethyl-3-(3-dimethylaminopropyl)carbodiimide (EDC), N-hydroxysulfosuccinimide (sulfo-NHS), fluorescent FITC-dextran (MW 4400 Da), sodium azide, and iodoacetamide (IAA) are from Sigma-Aldrich (Milwaukee, WI). Tat peptide (11-mer, TyrGlyArgLysLysArgArgGlnArgArgArg; MW 1560 Da; three reactive amino groups) is prepared by ResGen Invitrogen (Huntsville, AL). Indium-111 chloride (40–4000 Ci/mg) is from NEN. Cell culture media and fetal bovine serum (FBS) (Cellgro) are from Mediatech (Herndon, VA). All solvents and buffer components are analytical-grade preparations.

*Synthesis of pNP-PEG-DOPE.* We have introduced a new amphiphilic single terminus-reactive PEG derivative, *p*-nitrophenylcarbonyl-PEG-PE (pNP-PEG-DOPE),[7] which is stable at slightly acidic pH, firmly incorporates into the liposomal membrane via the phosphatidylethanolamine (PE) residue, and binds easily to primary amino group-containing ligands via water-exposed pNP groups, forming stable and nontoxic urethane (carbamate) bonds. The reaction proceeds at pH 7.5–8.0 (Fig. 1), and nonreacted pNP groups are eliminated by spontaneous hydrolysis. In our experiments, we have used 0.5 mol% of pNP-PEG-DOPE in the lipid mixture (preliminary experiments have shown that as little as 0.5 to 1 mol% of

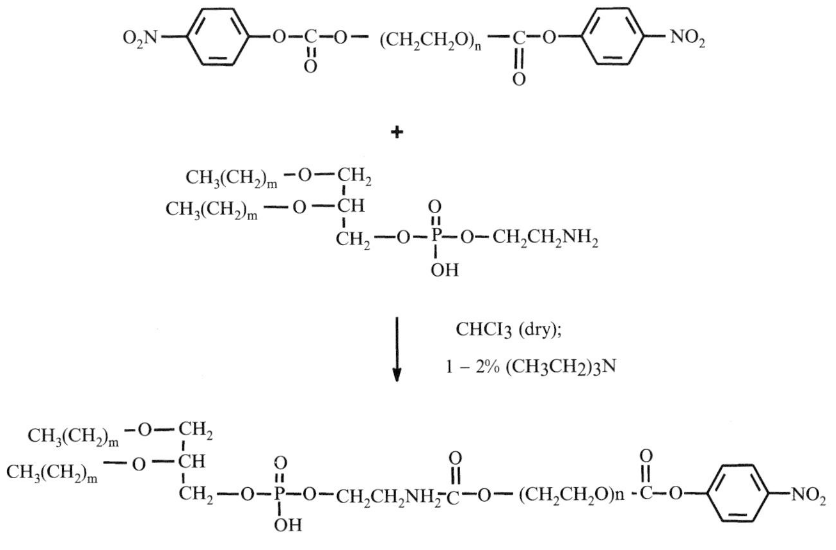

FIG. 1. Reaction scheme for the synthesis of pNP-PEG-DOPE.

total lipid of pNP-PEG-DOPE provides a sufficient number of reactive groups on the liposome surface to bind several hundred protein molecules to a single 200-nm liposome).

Twenty-four milligrams (32.2 $\mu$mol) of DOPE is dissolved in chloroform to obtain a 50-mg/ml solution. Eighty microliters of tetraethylammonium (TEA) [~2-fold molar excess over PEG-(pNP)$_2$] is added. One gram (~10-fold molar excess over DOPE) of PEG-(pNP)$_2$ is dissolved in 5 ml of chloroform and added to the mixture, which is then incubated overnight at room temperature under argon with stirring. The organic solvents are then removed, using a rotary evaporator. To the dry film is added 0.01 $M$ HCl–0.15 $M$ NaCl, and the mixture is sonicated with a bath sonicator. Micelles are separated from unbound PEG and released pNP on a Sepharose CL-4B (Bio-Rad, Hercules, CA) column, using 0.01 $M$ HCl–0.15 $M$ NaCl as the eluent. The pooled fractions containing pNP-PEG-DOPE are freeze-dried and then chloroform is added to extract pNP-PEG-DOPE. The latter procedure is repeated twice to ensure the complete removal of NaCl from the preparation (Fig. 1). pNP-PEG-DOPE is stored as a chloroform solution at $-80°$.

*Preparation of Liposomes.* Liposomes are prepared by making a lipid film from a mixture of PC, Ch, and pNP-PEG-DOPE (7:3:0.05, molar ratio) in chloroform by rotary evaporation followed by freeze–drying. The film is rehydrated in citrate-buffered saline (pH 5.4), and the suspension is extruded 30–40 times through a polycarbonate filter (pore size 200 nm), using a syringe extruder (Avestin, Ottawa, ON, Canada). For visualization, all liposomes are labeled with trace amounts of Rh-PE ($\lambda_{ex}$ 550 nm, $\lambda_{em}$ 590 nm). To prepare FITC-dextran-loaded liposomes, FITC-dextran (45 mg/ml) is added to the citrate buffer during hydration. Nonincorporated FITC-dextran is removed by size-exclusion column chromatography on a BioGel A1.5m (0.7 × 25 cm) column (Bio-Rad). All the liposome preparations used in our experiments have approximately the same average diameter of 150 to 200 nm, as measured by dynamic light scattering with a Coulter N4 Plus submicron particle analyzer (Beckman Coulter, Fullerton, CA).

*Attachment of Tat Peptide to Liposomes.* One milligram of Tat peptide in borate buffer (0.1 $M$ sodium tetraborate, 150 m$M$ NaCl, pH 9.2) is added to 10 mg of pNP-PEG-DOPE-containing liposomes in citrate buffer (10 m$M$ sodium citrate, 0.150 $M$ NaCl, pH 5.0), and incubated overnight at room temperature. Tat-liposomes are purified from unbound Tat peptide and released pNP by gel filtration on a BioGel A1.5M column (0.7 × 25 cm).

Initial attachment of Tat peptide via pNP groups at the distal ends of liposome-grafted pNP-PEG-DOPE chains must be done at or above pH

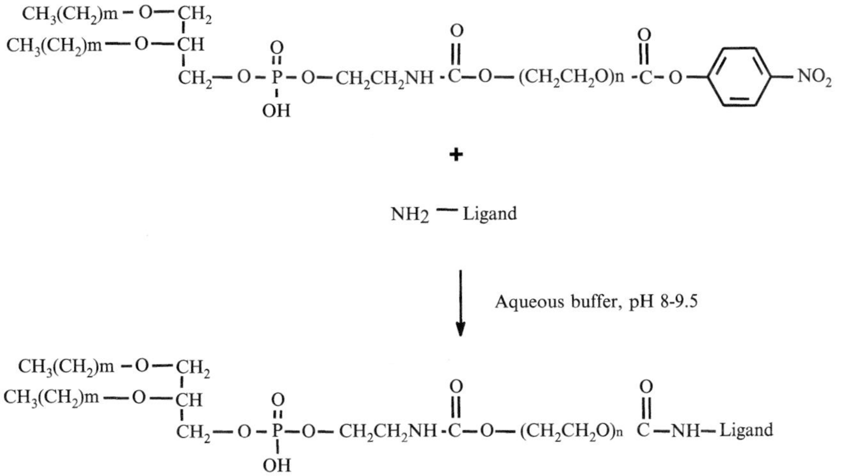

CH$_3$(CH$_2$)m – O—CH$_2$
CH$_3$(CH$_2$)m —— O—CH
CH$_2$— O - P—O—CH$_2$CH$_2$NH ·C—O— (CH$_2$CH$_2$O)n ·C - O—⟨benzene ring⟩—NO$_2$
OH

+

NH$_2$ —— Ligand

Aqueous buffer, pH 8-9.5

CH$_3$(CH$_2$)m – O—CH$_2$
CH$_3$(CH$_2$)m —— O—CH
CH$_2$— O - P—O— CH$_2$CH$_2$NH ·C—O— (CH$_2$CH$_2$O)n  C —NH— Ligand
OH

FIG. 2. Reaction scheme for the attachment of amino group-containing ligand to the pNP group of pNP-PEG-PE.

8.5 (Fig. 2). Incubation under such slightly alkaline conditions allows for both the relatively fast coupling reaction and slower hydrolysis of the non-reacted pNP residues. The average number of Tat peptide molecules attached per liposome had earlier been estimated from radioactivity data[8] (assuming that practically all Tat peptides are labeled with indium-111 via the attached DTPA residues, and $\sim 3 \times 10^5$ molecules of a lipid form a liposome with an average diameter of 200 nm,[9] approximately 500 Tat peptide molecules were found to be coupled to a single liposome). This corresponds well to the number of reactive pNP groups available on the liposome surface, assuming an even distribution of pNP-PEG-DOPE between the inner and outer lipid monolayers. This also corresponds well to the range of protein molecules bound to liposomes via pNP-PEG-DOPE, and provides a possibility that even such large particles as liposomes can be coaxed to translocate through the cytoplasmic membrane due to the multipoint Tat peptide-to-membrane interaction.

This estimate of the number of attached Tat peptide molecules is also true for PEGylated long-circulating liposomes (liposomes with an additional 2 to 6 mol% of PEG-PE[10] were used in these experiments). The only caveat to be kept in mind in this case is that the length of the

[9] H. G. Enoch and P. Strittmater, *Proc. Natl. Acad. Sci. USA* **76,** 145 (1979).
[10] A. L. Klibanov, K. Maruyama, V. P. Torchilin, and L. Huang, *FEBS Lett.* **268,** 235 (1990).

liposome-attached PEG-PE chains used for the steric protection of liposomes in the circulation should be less than the length of Tat-pNP-PEG-PE. Otherwise steric hindrance of the Tat peptide-to-membrane interaction created by protruding long free PEG chains may abolish liposome internalization (illustrating the key importance of direct contact between the Tat peptide and the cell surface for translocation; see Fig. 3).

## Cell Uptake Studies

*Cell Lines and Reagents.* Human breast tumor (BT20) cells are maintained in Eagle's minimal essential medium (EMEM, with 10 m$M$ pyruvate, nonessential amino acids, L-glutamine, and 10% FBS). Lewis lung carcinoma (LLC) cells are maintained in RPMI 1640 medium (with 10% FBS). Rat embryonic cardiac myocytes (H9C2 cells) are maintained in Dulbecco's modified Eagle's medium (DMEM, with 10% FBS). RPMI 1640, DMEM, EMEM, and serum-free medium (complete, serum-free) are from CellGro. Fluorescence-free glycerol-based mounting medium (Fluoromount-G) is from Southern Biotech (Birmingham, AL). Heat-inactivated FBS is from CellGro. Iodoacetamide, sodium azide, and all other cell culture supplements and reagents are from Sigma-Aldrich.

*Intracellular Localization and Trafficking of Tat-Liposomes.* We have tested Tat peptide-mediated uptake of liposomes *in vitro,* using three different cell lines. After initial passage in tissue culture flasks, cells are grown on coverslips in six-well tissue culture plates. After the cells reach a confluence of 60% to 70%, the plates are washed twice with sterile PBS, pH 7.4, and then treated with liposome samples in serum-free medium (2 ml/well, 16 $\mu$g of total lipid per milliliter). After various incubation times (we have used time points of 1, 2, 4, 6, 9, and 24 h), the medium is removed and the plates are washed with sterile PBS three times. Individual coverslips are mounted cell side down onto fresh glass slides with fluorescence-free glycerol-based mounting medium. The cells are viewed with a Nikon Eclipse E400 microscope under bright light, or under epifluorescence with rhodamine/TRITC and fluorescein/FITC filters. For the additional control of the intracellular localization of internalized liposomes, a differential interference contrast microscope (DIC) is used, such as an Axioplan 2 (Zeiss, Thornwood, NY) with deconvolution and pseudo-coloring for fluorescence imaging.

The interaction of various liposomal preparations with cells at 37° with no metabolic inhibitors added yields practically identical results for all cell lines used (LLC, BT20, and H9C2). All cells treated under similar conditions with Rh-PE-labeled Tat-liposomes or PEGylated Tat-liposomes (Tat-liposomes are prepared with pNP-PEG3000-PE, whereas in the case

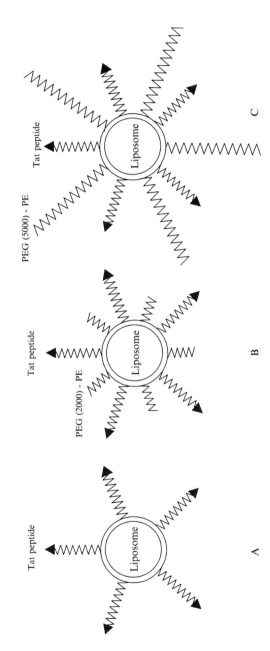

FIG. 3. Structures of various Tat-liposomes. In the case of PEGylated liposomes, free PEG chains, if longer than the PEG chains to which Tat peptide is attached, may create steric hindrance for Tat exposure and direct interaction with the cell membrane [compare (B) and (C)]. (A) Liposomes with Tat-pNP-PEG3000-DOPE; (B) PEGylated liposomes (Tat-pNP-PEG3000-DOPE and PEG2000-PE are used); (C) PEGylated liposomes (Tat-pNP-PEG3000-DOPE and PEG5000-PE are used).

of PEGylated Tat-liposomes PEG2000-PE is used) demonstrate high intracellular fluorescence, that is, liposome internalization (Fig. 4). The results
of fluorescence microscopy with differential interference contrast confirm
the intracellular localization of Tat-liposomes, because the majority of
the liposome-associated fluorescence can be seen inside the cell (when a
single cell is viewed at the midpoint of cell thickness), and not on the cell
surface.[8] After 1 h of incubation, Tat-liposomes are seen mainly in the
cytoplasm, with minimal nuclear/perinuclear association (whereas free
Tat peptide was shown earlier to have a predominantly nuclear localization[1]). Our preliminary data have shown that with time, Tat-liposomes
eventually accumulate in the perinuclear zone and gradually disintegrate
(see further in Fig. 5).

To investigate whether Tat peptide-mediated internalization results in
the intracellular delivery of intact liposomes, Rh-PE-labeled liposomes
with entrapped FITC-dextran are used. After incubation with such liposomes, cells are studied for both Rh and FITC fluorescence. As can be seen
from the typical data presented in Fig. 5, incubation of cells with Rh-PE-
labeled Tat-liposomes loaded with FITC-dextran results in the intracellular
appearance of colocalized Rh and FITC fluorescence; that is, both the
membrane and internal space markers are delivered into the cytoplasm
and the majority of liposomes remain intact during the first 1 h of observation time (free FITC-dextran shows only minor intracellular accumulation[8]). By 2 to 4 h, the Tat-liposomes are seen clustered in the perinuclear

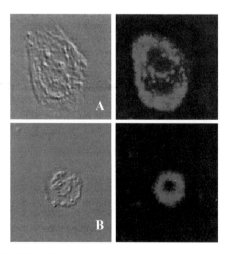

FIG. 4. Bright-field (*left*) and fluorescence *(right)* microscopy of cell cultures 1 h after
treatment with Tat-liposomes. (A) BT20; (B) Lewis lung carcinoma. (See color insert.)

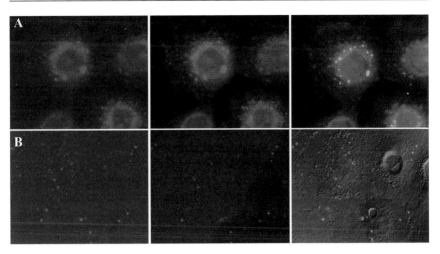

FIG. 5. BT20 cells viewed after incubation with Tat-liposomes loaded with FITC-dextran: DIC with rhodamine filter *(left)*; DIC with FITC *(middle)*; composite of all layers *(right)*. (A) After 2 h of incubation; (B) after 9 h of incubation. (See color insert.)

region with the reduced cytoplasmic distribution. Still, the red fluorescence of Rh-PE-liposomes and green fluorescence of the intraliposomal FITC-dextran can be seen as colocalized in this region. However, from 9 h onward, the liposome quantity in the perinuclear zone is seen to progressively reduce. There is a minimal cytoplasmic presence of liposomes, and the FITC-dextran has by this time been mostly released from the liposomes. The degradation of the liposomes is evidenced by the diffuse green and red/orange fluorescence in the cytoplasm as well as the nucleus. The data produced by DIC microscopy support this conclusion, because the superposition of Rh- and FITC-based images also yielded identical patterns (Fig. 5).

*Temperature Dependence of Liposome Uptake.* To study the temperature dependence of Tat-mediated translocation of liposomes into cells, Tat-liposomes are labeled with indium-111 (In-111) and their uptake is tested with BT20 cells. Tat-liposomes are prepared as described above. For radiolabeling with In-111, DTPA-PE-containing Tat-liposomes are incubated with indium-111 citrate complex (25–60 $\mu$Ci/ml) for 1 h at room temperature. The product is then dialyzed against HEPES-buffered saline (HBS), pH 7.4, overnight at 4°. The samples are then added to BT20 cells in serum-free medium, and the plates are incubated at 4, 15, 25, or 37° for 1 h. After the incubation, cells are washed twice with sterile PBS, pH 7.4, and then trypsinized to harvest and transfer to test tubes for $\gamma$ counting.

TABLE I
INTERNALIZATION OF LIPOSOMES AT VARIOUS TEMPERATURES[a,b]

| | Temperature | | | |
| --- | --- | --- | --- | --- |
| | 4° | 15° | 25° | 37° |
| Relative intracellular uptake (%) | 100 ± 0.11 | 103 ± 0.21 | 98 ± 0.12 | 104 ± 0.21 |

[a] Tested on BT20 human breast carcinoma cells.
[b] Assuming binding at 4° as 100%.

Tat-liposomes may be radiolabeled with indium-111 by two commonly used methods. In the first method, an amphiphilic chelating agent, that is, a trace amount of DTPA-PE, is included in the lipid mixture before the preparation of liposomes. Subsequently, the liposomal suspension is radiolabeled as described above. Alternatively, the Tat peptide itself may be modified directly with DTPA, attached to pNP-PEG-PE-containing liposomes, and then labeled with indium-111 by transchelation, using the same condition as in liposome labeling via DTPA-PE.

Although the exact mechanism by which Tat peptide-mediated uptake occurs is still to be determined, it has been suggested that it is neither receptor mediated nor energy dependent.[5] Preliminary results of the incubations carried out at 4 and 37°, as well as in the presence of metabolic inhibitors such as iodoacetamide or sodium azide, indicated that these factors only minimally affect the binding and uptake of Tat-liposomes.[8] Further experiments carried out at 4, 15, 25, and 37° show that there is little, if any, difference in the internalization of Tat-liposomes (assuming binding at 4° as 100%) at these various temperatures. This confirms the energy-independent mechanism of Tat peptide-mediated binding and uptake of liposomes (Table I).

Concluding Remarks

Our experiments provide a proof-of-principle for the efficient Tat peptide-mediated "transduction" of conventional as well as sterically protected liposomes and their contents into cells. Tat peptides attached to the liposome surface need to be nonshielded and accessible for cell membrane interaction. Our results seem to favor the hypothesis that spontaneous formation of certain membrane structures (reverse micelles?) sharply, albeit temporarily, increases cell membrane permeability under the action of the Tat peptide. No positive dependence on temperature has been found. Initial cytoplasmic and later perinuclear localization of

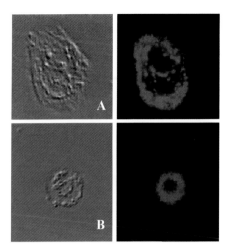

LEVCHENKO *ET AL.,* CHAPTER 19, FIG. 4. Bright-field (*left*) and fluorescence (*right*) microscopy of cell cultures 1 h after treatment with Tat-liposomes. (A) BT20; (B) Lewis lung carcinoma.

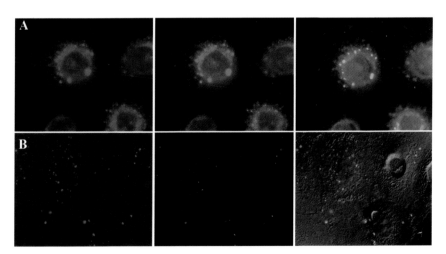

LEVCHENKO *ET AL.,* CHAPTER 19, FIG. 5. BT20 cells viewed after incubation with Tat-liposomes loaded with FITC-dextran: DIC with rhodamine filter (*left*); DIC with FITC (*middle*); composite of all layers (*right*). (A) After 2 h of incubation; (B) after 9 h of incubation.

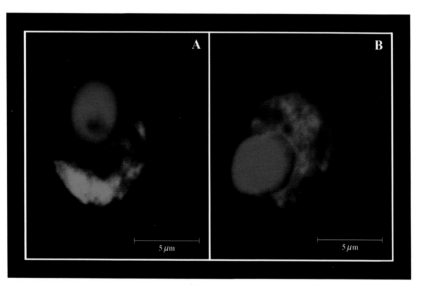

TACHIBANA ET AL., CHAPTER 20, FIG. 3. Release of FITC-alb into the cytosol from endosomes by pH-sensitive liposomes. Macrophages were incubated in 5% $CO_2$ at $37^\circ$ with liposomes for 30 min. Cells were then washed and fixed. Liposomes were composed of HEPC–CHEMS (A) or DOPE–CHEMS (B). Cells were observed by confocal laser microscopy.

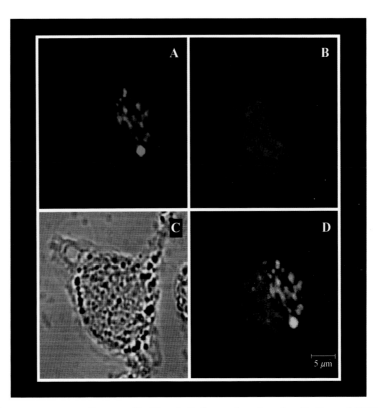

TACHIBANA *ET AL.*, CHAPTER 20, FIG. 5. Failure of nuclear localization of FITC-alb without NLS by pH-sensitive liposomes. Macrophages were incubated in 5% $CO_2$ at $37^\circ$ with liposomes for 1 h. After washing the liposomes in medium, cells were incubated for periods of up to 3 h. The cells were then washed and fixed. Liposomes were composed of DOPE–CHEMS. Cells were observed by confocal laser microscopy. (A) FITC-alb (green); (B) nucleus labeled with propidium iodide (red); (C) the entire shape of a macrophage by phase-contrast microscopy; (D) a superimposed image of (A) and (B).

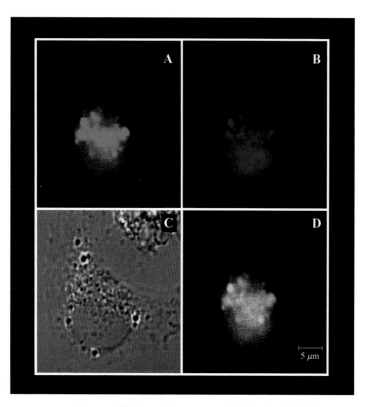

TACHIBANA *ET AL.*, CHAPTER 20, FIG. 6. Transport of FITC-alb with NLS to the nuclear compartment by pH-sensitive liposomes. Macrophages were incubated in 5% $CO_2$ at 37° with liposomes for 1 h. After washing the liposomes in medium, cells were incubated for periods of up to 3 h and the cells were then washed and fixed. Liposomes were composed of DOPE–CHEMS. Cells were observed by confocal laser microscopy. (A) FITC-alb (green); (B) nucleus labeled with propidium iodide (red); (C) the entire shape of a macrophage by phase-contrast microscopy; (D) a superimposed image of (A) and (B).

Tat-liposomes was observed. The approach described may have important implications for drug and DNA delivery directly into the cytoplasm and the nuclear space, bypassing the endocytotic pathway. This approach may be utilized for *ex vivo* cell treatment, as well as in certain protocols of local drug application.

## [20]  pH-Sensitive Liposomes in Nuclear Targeting of Macromolecules

*By* Rieko Tachibana, Shiroh Futaki, Hideyoshi Harashima, and Hiroshi Kiwada

### Introduction

Macromolecular targeting based on molecular mechanisms of intracellular trafficking represents a potentially powerful tool for a rational drug delivery system that would be of great therapeutic value.[1,2] For the active targeting of macromolecules to the cell, drug carriers such as liposomes, microspheres, or nanospheres are required, most of which are taken up via endocytosis. Receptor-mediated endocytosis, in particular, is a selective internalizing pathway for macromolecules.[3] However, to accomplish selective targeting to organelles, macromolecules must escape from the endosome to the cytosol before their degradation by lysosomal enzymes. Furthermore, regulation of the selective targeting of macromolecules, which have been released into the cytosol to the organelle, is required.

In this chapter we review methods of delivery of macromolecules from the endosome to the cytosol and the selective targeting of macromolecules to the nucleus, using nuclear localization signals.

### Endosomal Escape

pH-sensitive liposomes were first designed by Yatvin *et al.*[4] for the site-specific delivery of lower pH, encapsulated compounds to primary or metastasized tumors, or to sites of inflammation and infection. At the intracellular level, the pH of endosomes and lysosomes is lower than that of

---

[1] H. Harashima, Y. Shinohara, and H. Kiwada, *Eur. J. Pharm. Sci.* **13**, 85 (2001).

[2] R. Tachibana, H. Harashima, Y. Shinohara, and H. Kiwada, *J. Biol. Macromol.* **1**, 60 (2001).

[3] C. M. Varga, T. J. Wickham, and D. A. Lauffenburger, *Biotechnol. Bioeng.* **70**, 593 (2000).

[4] M. B. Yatvin, W. Kreutz, B. A. Horwitz, and M. Shinitzky, *Science* **210**, 1253 (1980).

other organelles. The interior of the endosomal compartment is kept acidic by ATP-driven $H^+$ pumps in the endosomal membrane that pump $H^+$ into the lumen from the cytosol.[5] Conventional liposomes are taken up by endocytosis. Therefore, pH-sensitive liposomes can be used for the cytosolic delivery of liposomally encapsulated substances after internalization and acidification of the endosomal pH.[6-6b] We utilized a pH-sensitive liposome composed of dioleoylphosphatidylethanolamine (DOPE) and cholesterylhemisuccinate (CHEMS).[7] DOPE, which is generally considered to be a fusogenic lipid,[8] has a small polar group within the molecule and can adopt an inverted cone shape, which confers fusogenic properties with respect to the endosomal membrane. Most pH-sensitive liposomes take advantage of this property. CHEMS is a weak acid and plays the role of a pH sensor in this system. CHEMS has a negative charge at neutral pH but is uncharged in the acidic range, because it is protonated. Uncharged CHEMS cannot maintain the stability of a bilayer structure, but DOPE tends to undergo a structural change from a bilayer structure to that of an inverted hexagonal ($H_{II}$) structure.[9] In addition to CHEMS, other acidic amphiphiles such as dipalmitoylsuccinylglycerol, dioleoylsuccinylglycerol, oleic acid, and palmitoyl homocysteine can be used in such systems.[10]

Several peptides are used to enhance endosomal escape in the gene delivery system, and include GALA,[11] HA2,[12] as well as others.[13] GALA is a synthetic pH-dependent fusogenic peptide, which undergoes a conformational change to an amphipathic $\alpha$ helix when the pH is reduced. GALA has been shown to increase the transfection efficiency of luciferase, when complexed with cationic liposomes and transferrin in macrophages.[14]

[5] B. Alberts, D. Bray, J. Lewis, M. Raff, K. Roberts, and J. D. Watson, "Molecular Biology of the Cell," 3rd Ed., p. 622. Garland Publishing, New York, 1994.

[6] D. C. Drummond, M. Zignani, and J. C. Leroux, *Prog. Lipid Res.* **39**, 409 (2000).

[6a] R. M. Straubinger, N. Düzgüneş, and D. Papahadjopoulos, *FEBS Lett.* **179**, 148 (1985).

[6b] N. Düzgüneş, R. M. Straubinger, P. A. Baldwin and D. Papahadjopoulos, *in* "Membrane Fusion" (J. Wilschut and D. Hoekstra, eds.), p. 713. Marcel Dekker, New York, 1991.

[7] R. Tachibana, H. Harashima, M. Shono, M. Azumano, M. Niwa, S. Futaki, and H. Kiwada, *Biochem. Biophys. Res. Commun.* **251**, 538 (1998).

[8] D. P. Siegel, *Biophys. J.* **49**, 1155 (1986).

[9] D. C. Litzinger and L. Huang, *Biochim. Biophys. Acta* **1113**, 201 (1992).

[10] V. P. Torchilin, F. Zhou, and L. Huang, *J. Liposome Res.* **3**, 201 (1993).

[11] R. A. Parente, S. Nir, and F. C. Szoka, Jr., *Biochemistry* **29**, 8720 (1990).

[12] E. Wagner, C. Plank, K. Zathoukal, M. Cotton, and M. L. Bimstiel, *Proc. Natl. Acad. Sci. USA* **89**, 7934 (1992).

[13] C. Plank, B. Oberhauser, K. Mechtler, C. Koch, and E. Wagner, *J. Biol. Chem.* **269**, 12919 (1994).

[14] S. Simoes, V. Slepushkin, E. Pretzer, P. Dazin, R. Gaspar, M. C. Pedroso de Lima, and N. Düzgüneş, *J. Leukoc. Biol.* **65**, 270 (1999).

The utilization of these specific peptides will also be a key factor in the optimization of a carrier system.

## Preparation of Liposomes

Multilamellar vesicles (MLVs) are prepared by the hydration method. pH-sensitive liposomes are composed of DOPE–CHEMS at a molar ratio of 3:2, whereas hydrogenated egg phosphatidylcholine (HEPC) or hydrogenated egg phosphatidylserine (HEPS) is used in place of DOPE in control liposomes. Chloroform solutions of lipids are placed into a flask, and the solvent is evaporated. Residual solvent is removed with a vacuum pump. Lipid films are hydrated with a calcein solution (40 mM calcein–1 mM EDTA, pH 8.0) as an aqueous phase marker or with fluorescein isothiocyanate-labeled bovine serum albumin (FITC-alb, 100 mg/ml PBS, pH 8.0; Sigma, St. Louis, MO) for use as a model macromolecule. The final concentration is 40 μmol of lipid per milliliter. Liposomes are sized by extrusion through a polycarbonate membrane (Whatman Nuclepore, Pleasonton, CA) with 800-nm-diameter pores. The average diameter of the liposomes is measured by dynamic light scattering (NICOMP 370; Particle Sizing Systems, Santa Barbara, CA). Unincorporated calcein is removed by dialysis in a cellulose tube in PBS (pH 8.0). Unincorporated FITC-alb is removed by three ultracentrifugations (9000 g for 1 h).

## Effect of DOPE on pH Profiles of Calcein Release from Liposomes

Ten microliters of liposomes, diluted in PBS, is incubated with 2.99 ml of citrate buffer at various pH values at room temperature. After 15 min, the pH of the mixture is adjusted to pH 8.0 by adding an appropriate amount of NaOH solution. Calcein fluorescence is measured at $\lambda_{ex} = 490$ nm and $\lambda_{em} = 520$ nm, before and after the addition of 300 μl of 5% Triton X-100. The percent release of calcein is calculated by the equation:

$$\% \text{ release} = [F \times (3 + V)/F_{total} \times (3.3 + V)] \times 100$$

where $F$ represents the fluorescence intensity before the addition of Triton X-100, and $F_{total}$ is the fluorescence intensity afterward. $V$ represents the volume of the 2 N NaOH solution added.

## Isolation of Rat Peritoneal Macrophages

Male Wistar rats weighing 180–250 g (Inoue Experimental Animal, Kumamoto, Japan) are stimulated by an intraperitoneal injection of about 20 ml of thioglycolic acid. After a 3-day period, the peritoneal macrophages are isolated by peritoneal lavage. Hanks' solution (total 50 ml) is injected into the peritoneal cavity, and the lavaged fluid is collected. The resulting cells are washed three times by centrifugation at $4°$ (100 $g$, 5 min) and are then resuspended in RPMI 1640 medium and plated at $2 \times 10^6$ cells/35-mm dish. After incubation for 1 h in 5% $CO_2$ at $37°$, the medium is replaced by RPMI 1640 medium supplemented with 5% heat-inactivated fetal bovine serum, followed by culturing for 20 h.

### Intracellular Degradation of Liposomal FITC-alb in Peritoneal Macrophages

Macrophages are incubated in 5% $CO_2$ at $37°$ with liposomes (final concentration, 1 $\mu$mol of lipid/ml) for 1 h in Hanks' balanced salt solution and then washed five times with cold Hanks' solution. The cells are further incubated for 1 or 3 h. At the end of the incubation period, the cells are washed and lysed by treatment with 0.5% Triton X-100. The intact and degraded FITC-alb is isolated and separated by equilibrium dialysis against 0.5% Triton X-100. The molecular weight cutoff (MWCO) is 50,000 (Spectra/Por membrane, MWCO 50,000; Spectrum Laboratories, Rancho Dominguez, CA). The dialysis is continued for 24 h at $4°$. The fluorescence intensity of FITC in each compartment is then measured at $\lambda_{ex} = 485$ and $\lambda_{em} = 538$ nm. The percentage of intact FITC-alb is calculated after correction for background intensity in the absence of liposomes.

### Intracellular Trafficking of Macromolecules Encapsulated in Liposomes

Confocal microscopy is used to examine the effect of pH-sensitive liposomes on the cytosolic delivery of encapsulated FITC-alb, and compared with control liposomes.

Macrophages are cultured on stenle coverslips ($4 \times 10^5$ cells per coverslip) that have been placed in 35-mm dishes and incubated with liposomes at a final lipid concentration of 1 $\mu$mol/ml for various time(s) in 5% $CO_2$ at $37°$ and then washed with PBS. After incubation of the cells and liposomes for 30 min, they are fixed with 2% paraformaldehyde in PBS for 30 min. The cells are washed with PBS and immersed in PBS for 5 min, and this procedure is repeated twice. Nuclei are stained with a 20-$\mu$g/ml solution of propidium iodide (PI). After incubation for 10 min with the PI solution, the cells are washed twice with PBS. After drying, the coverslip is mounted on a glass slide with one drop of 2% $n$-propyl gallate. Drops of nail polish are applied to the edge of the coverslip for shielding purposes. The intracellular localization of FITC-alb is imaged by confocal laser microscopy (FITC, $\lambda_{ex} = 488$ nm and $\lambda_{em} = 530$ nm; PI, $\lambda_{ex} = 540$ nm and $\lambda_{em} = 580$ nm).

*Effect of DOPE on pH Profiles of Calcein Release from Liposomes*

The pH profiles for the release of calcein from liposomes are shown in Fig. 1. Liposomes composed of DOPE–CHEMS (3:2, molar ratio) release calcein in a pH-dependent manner, with a dramatic increase being observed between pH 6.0 and 5.0. On the other hand, liposomes including HEPS show a gradual release, which is pH dependent. However, liposomes without DOPE (HEPC–CHEMS) release only slight amounts. The pH profile of liposomes composed of DOPE–CHEMS (3:2, molar ratio) indicates that they would be ideal for the release of incorporated substances from the endosome to the cytosol, because the endosomal pH is between pH 6.0 and 5.0. On the basis of these results, liposomes composed of DOPE–CHEMS are employed as pH-sensitive liposomes and HEPC–CHEMS liposomes are used as controls.

*Effect of pH-Sensitive Liposomes on Intracellular Degradation of Encapsulated Albumin in Rat Peritoneal Macrophages*

Liposomes are taken up by cells via endocytosis.[15] The endocytosed liposomes are then localized in endosomes, where the pH decreases from pH 7.4 to ~5.0.[16] On the basis of these facts, pH-sensitive liposomes would

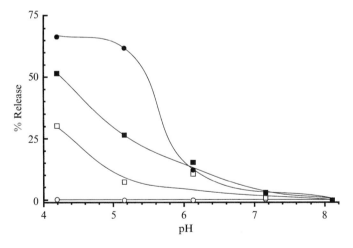

FIG. 1. Acid sensitivity of pH-sensitive liposomes. Percent calcein release from liposomes is plotted against the pH of the incubation buffer. Liposomes were composed of (●) DOPE–CHEMS (3:2, molar ratio); (■) HEPS–DOPE–CHEMS (3:4.2:2.8); (○) HEPC–CHEMS (3:2); (□) HEPS–CHEMS (3:2).

[15] R. M. Straubinger, K. Hong, D. S. Friend, and D. Papahadjopoulos, *Cell* **32**, 1069 (1983).
[16] T. Yoshimura, M. Shono, K. Imai, and K. Hong, *J. Biochem.* **117**, 34 (1995).

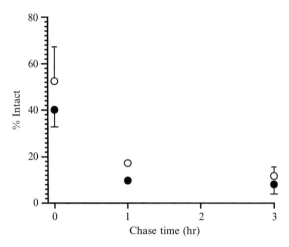

FIG. 2. Intracellular degradation of liposomal FITC-alb. The levels of intact and degraded FITC-alb were measured separately by equilibrium dialysis. Macrophages were incubated in 5% $CO_2$ at 37° with liposomes for 1 h and then chased. The cells were lysed by treatment with 0.5% Triton X-100. Intact and degraded FITC-alb were separated by membrane equilibrium dialysis (MWCO 50,000), and the fluorescence intensity of each compartment was measured. The fraction of degraded FITC-alb was calculated by dividing the degraded FITC-alb by the total FITC-alb. (○) pH-sensitive liposomes; (●) conventional liposomes.

be expected to avoid lysosomal degradation by fusing with the endosomal membrane at the lower pH. Therefore, we examined the time course for the degradation of liposomally encapsulated FITC-alb in macrophages (Fig. 2). The degradation of FITC-alb in pH-sensitive liposomes is less extensive than that in conventional liposomes at each time point, whereas for the case of degradation in liposomes composed of HEPS–DOPE–CHEMS, which have pH sensitivity lower than that of pH-sensitive liposomes (DOPE–CHEMS), little difference is observed from the control liposomes. The difference between pH-sensitive liposomes and control liposomes tends to gradually decrease [12.1% (0 h), 7.5% (1 h), 3.9% (3 h)]. From these results, it would be predicted that FITC-alb is delivered to the cytosol from endosomes in macrophages within several hours, and the degradation of albumin in the cytosol is considerably slower than in lysosomes.[17] The extent of cytosolic delivery of conventional pH-sensitive liposomes such as DOPE–CHEMS reaches a maximum of 10% after endocytosis, which is consistent with observations reported by Chu et al.[18]

[17] S. Bigelow, R. Hough, and M. Rechsteiner, *Cell* **25,** 83 (1981).
[18] C. J. Chu, J. Dijkstra, M. Z. Lai, K. Hong, and F. C. Szoka, *Pharm. Res.* **7,** 824 (1990).

It would be expected that the various synthetic peptides previously mentioned would increase the efficiency of cytosolic delivery.

## Confocal Microscopic Analysis of Manipulation of Intracellular Trafficking with pH-Sensitive Liposomes

FITC-alb is used as a model macromolecule. After internalization of the liposomes, the pH-sensitive liposomes are compared with conventional liposomes by determining the distribution of FITC-alb in the cytosol, as shown in Fig. 3. Green fluorescence and red fluorescence represent FITC bound to alb (FITC-alb) and the nucleus stained with propidium iodide, respectively. FITC-alb encapsulated in conventional liposomes is localized in the vesicular compartment, possibly endosomes/lysosomes. In the case of pH-sensitive liposomes, FITC-alb is distributed homogeneously in the cytosol and surrounds the nuclear surface. These results indicate that the use of pH-sensitive liposomes successfully permits the cytosolic delivery of liposomal FITC-alb from endosomes.

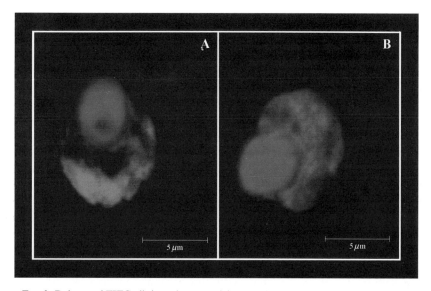

FIG. 3. Release of FITC-alb into the cytosol from endosomes by pH-sensitive liposomes. Macrophages were incubated in 5% $CO_2$ at 37° with liposomes for 30 min. Cells were then washed and fixed. Liposomes were composed of HEPC–CHEMS (A) or DOPE–CHEMS (B). Cells were observed by confocal laser microscopy. (See color insert.)

Nuclear Transport of Macromolecules

To achieve selective and efficient nuclear delivery of FITC-alb normally released to the cytosol by pH-sensitive liposomes, we have used a nuclear localization signal (NLS) peptide derived from the simian virus 40 (SV40) large tumor antigen, for conjugation to FITC-alb.

The nucleus contains a nuclear envelope, in the form of a double membrane. Nuclear pore complexes (NPCs) are present in the nuclear envelope, and each contains an aqueous diffusion channel, approximately 9 nm in diameter, which is involved in the transport of a variety of molecules. Molecules smaller than 40–45 kDa are apparently able to diffuse freely into and out of the nucleus. Karyophilic proteins larger than 45 kDa are actively transported through the nuclear pores. Their selective and active transport is mediated by an NLS as a part of the karyophilic protein sequence.[19,20] NLSs are short peptides, but no general consensus sequence exists for NLSs. However, nearly all NLSs identified to date have been found to contain a sequence of basic amino acids. Proteins that contain the NLS peptide form complexes with cytosolic transport factors such as importin $\alpha/\beta$ or transportin.[21] The complex is docked at the NPC and then translocated into the nucleus. Moreover, regarding NLSs, it is noteworthy that a synthetic peptide containing the NLS sequence provides a pathway for exogenous protein to migrate into the nucleus.

As stated above, typical NLS segments are basic peptide segments, which consist of four to six lysine and arginine residues in a hexapeptide.[22] These segments are often found in transcription factors and viral proteins, which are recognized by the nuclear transport system comprising importin $\alpha/\beta$.

In earlier studies directed at identifying the roles of the NLS peptide, SV40 NLS peptides were attached to bovine serum albumin (BSA).[23] BSA has a molecular mass of 69 kDa and usually is not able to pass through nuclear pores. However, the microinjection of an NLS–BSA conjugate resulted in translocation of the conjugate into the nucleus. This finding serves as an indication of the great potential of NLSs for use in the nuclear targeting of macromolecules. The NLS from SV40 has been reported to promote highly efficient nuclear localization.[24] Thus, the

[19] D. A. Jans and S. Hubner, *Physiol. Rev.* **76,** 651 (1996).

[20] Y. Yoneda, *J. Biochem. Tokyo* **121,** 811 (1997).

[21] N. Imamoto, *Cell Struct. Funct.* **25,** 207 (2000).

[22] T. Boulikas, *J. Cell Biochem.* **55,** 32 (1994).

[23] Y. Yoneda, N. Imamoto-Sonobe, M. Yamaizumi, and T. Uchida, *Exp. Cell Res.* **173,** 586 (1987).

[24] D. Gorlich and I. W. Mattaj, *Science* **271,** 1513 (1996).

utilization of this segment for the delivery of DNAs into the nucleus has also been a subject of great interest.[25–33]

Conjugation of the NLS peptide with target proteins can be accomplished both chemically and genetically, although this chapter deals with only the former approach. For chemical conjugation, the use of heterobifunctional cross-linking agents is preferable, because this avoids linkage formation between the cargo molecules. Among typical cross-linking agents, both succinimide and maleimide moieties have been employed frequently. The former reacts with $\varepsilon$-amino moieties of lysine and the N-terminal $\alpha$-amino moiety, whereas the latter reacts with sulfhydryl moieties of cysteine.

## Conjugation of NLS Segment to Cargo Protein

To an ice-chilled solution of FITC-alb (158 mg) in 5 ml of 0.145 $M$ NaCl–10 m$M$ PBS (pH 6.3), $N$-(6-maleimidocaproyloxy) succinimide ester (Dojin) (158 mg, 30 Eq) in dimethylformamide (0.5 ml) is added dropwise (Fig. 4). The mixture is stirred gently at room temperature for 1 h. To remove the excess cross-linking reagents, the reaction mixture is subjected to gel filtration on a Sephadex G-10 column (1.6 × 17 cm; Pharmacia, Uppsala, Sweden), using PBS (pH 6.3) as the eluate. The first main peak detected by UV (280 nm) is collected. The SV40 NLS peptide (Cys-Pro-Lys-Lys-Lys-Arg-Lys-Val-Glu-Asp-Pro-amide) (93 mg, 33 Eq) is added directly to this solution and allowed to react for 1 h with gentle stirring. The mixture is dialyzed against PBS (pH 8.0) three times (4–6 h each time) to remove the unreacted SV40 peptide. On the basis of an estimate of its molecular weight by sodium dodecyl sulfate–polyacrylamide gel electrophoresis (SDS–PAGE), four or five molecules of basic peptides are estimated to be incorporated into each protein molecule.

The NLS peptides should be designed so as to have a cysteine residue at the N terminus or the C terminus. Such peptides can be obtained by means of a solid-phase peptide synthesizer or by custom synthesis.

[25] M. G. Sebestyén, J. J. Ludtke, M. C. Bassik, G. Zhang, V. Budker, E. A. Lukhtanov, J. E. Hagstrom, and J. A. Wolff, *Nat. Biotechnol.* **16,** 80 (1998).

[26] C. Neves, V. Escriou, G. Byk, D. Scherman, and P. Wils, *Cell Biol. Toxicol.* **15,** 193 (1999).

[27] M. A. Zanta, P. Belguise-Valladier, and J. P. Behr, *Proc. Natl. Acad. Sci. USA* **96,** 91 (1999).

[28] X. Gao and L. Huang, *Biochemistry* **35,** 1027 (1996).

[29] J. D. Fritz, H. Herweijer, G. Zhang, and J. A. Wolff, *Hum. Gene Ther.* **7,** 1395 (1996).

[30] A. I. Aronsohn and J. A. Hughes, *J. Drug Target.* **5,** 163 (1998).

[31] P. Collas and P. Aleström, *Biochem. Cell Biol.* **75,** 633 (1997).

[32] L. J. Brandén, A. J. Mohamed, and C. I. E. Smith, *Nat. Biotechnol.* **17,** 784 (1999).

[33] A. Subramanian, P. Ranganathan, and S. L. Diamond, *Nat. Biotechnol.* **17,** 873 (1999).

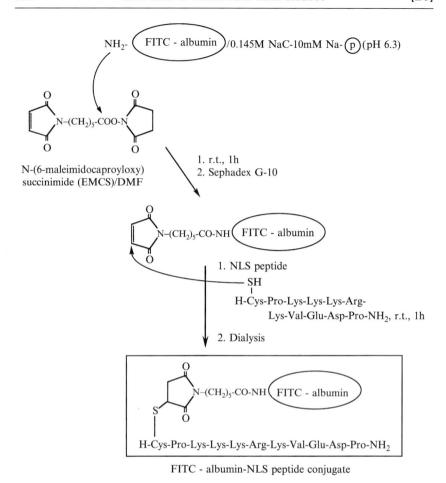

FITC - albumin-NLS peptide conjugate

Fig. 4. Preparation of a conjugate of the NLS peptide and FITC-alb.

*Intracellular Trafficking of Nuclear-Targeted Macromolecules*
*Encapsulated into Liposomes*

FITC-alb-NLSs (or FITC-alb) are encapsulated into pH-sensitive liposomes. Macrophages are incubated with liposomes at a final lipid concentration of 1 $\mu$mol/ml for 1 h in 5% $CO_2$ at 37° and then washed with PBS. The medium is replaced with RPMI supplemented with 5% heat-inactivated fetal calf serum, and the cells are then incubated further for 3 h. Samples for observation by confocal microscopy are prepared as described above.

Figure 5 shows a micrograph of FITC-alb delivered to the cells by pH-sensitive liposomes. Figure 5C shows the phase-contrast micrograph of a complete cell. The green regions in Fig. 5A represent the FITC, and the red region in Fig. 5B represents the nucleus, which is stained with propidium iodide. Figure 5D shows a micrograph in which Fig. 5A is superimposed on Fig. 5B. The yellow/orange region in Fig. 5D represents the region of colocalization of the FITC and propidium iodide. In the case of FITC-alb without the NLS, the green and red regions are separated and little colocalization between FITC-alb and nucleus is detected. On the other hand, Fig. 6 shows a micrograph of FITC-alb-NLS delivered to the

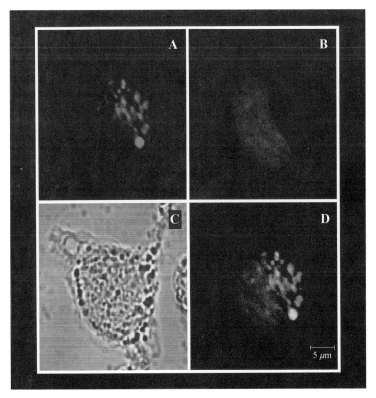

FIG. 5. Failure of nuclear localization of FITC-alb without NLS by pH-sensitive liposomes. Macrophages were incubated in 5% $CO_2$ at 37° with liposomes for 1 h. After washing the liposomes in medium, cells were incubated for periods of up to 3 h. The cells were then washed and fixed. Liposomes were composed of DOPE–CHEMS. Cells were observed by confocal laser microscopy. (A) FITC-alb (green); (B) nucleus labeled with propidium iodide (red); (C) the entire shape of a macrophage by phase-contrast microscopy; (D) a superimposed image of (A) and (B). (See color insert.)

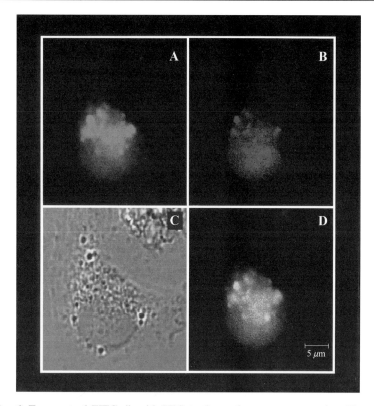

Fig. 6. Transport of FITC-alb with NLS to the nuclear compartment by pH-sensitive liposomes. Macrophages were incubated in 5% $CO_2$ at 37° with liposomes for 1 h. After washing the liposomes in medium, cells were incubated for periods of up to 3 h and the cells were then washed and fixed. Liposomes were composed of DOPE–CHEMS. Cells were observed by confocal laser microscopy. (A) FITC-alb (green); (B) nucleus labeled with propidium iodide (red); (C) the entire shape of a macrophage by phase-contrast microscopy; (D) a superimposed image of (A) and (B). (See color insert.)

cells by pH-sensitive liposomes. The fluorescence of FITC-alb and PI is colocalized in Fig. 6D. This result suggests that FITC-alb is selectively delivered from the cytosol to the nucleus by means of the NLS conjugated to albumin.

The nuclear transport of plasmid DNA, a typical macromolecule, is thought to be a major barrier, especially for nonviral gene delivery systems. NLSs, either covalently[25–27] or noncovalently[28–33] associated with the plasmid DNA, have been extensively utilized to enhance nuclear transfer and/ or expression. It would be expected that the efficient nuclear transport of genes would be difficult compared with proteins because the molecular weight of plasmid DNA is much greater than that of typical proteins. However, the effective utilization of an NLS may be possible by developing innovative methods for attaching NLS peptides to plasmid DNA.[27]

## Acknowledgments

This study was supported by fellowships and grants from Research Fellowships of the Japan Society for the Promotion of Science for Young Scientists, and by Grants-in-Aid for Scientific Research on Priority Areas from the Ministry of Education, Science, Sports, and Culture of Japan, respectively.

# [21]   Liposomes in Identification and Characterization of Viral Fusogenic Peptides

By SERGIO GERARDO PEISAJOVICH and YECHIEL SHAI

## Introduction

Infection of cells by enveloped viruses requires fusion of the viral and cellular plasma or endosomal membranes. As a result of membrane fusion, the viral contents (i.e., the genome-containing capsid) are transferred to the cytoplasm of the newly infected cell. Energy barriers hinder spontaneous fusion between two apposing membranes. During viral infection, however, these barriers are overcome by specific viral "fusion proteins" present in the viral envelope.[1] Despite the lack of sequence homology between fusion proteins from distantly related viruses, most of them share a common motif, termed the "fusion peptide," formed mostly of hydrophobic residues, sometimes with an unusually high proportion of alamine and glycine.[2] Fusion peptides are located either at the N terminus or in the interior of viral fusion proteins and are thought to penetrate and destabilize the target cell membrane, thus facilitating fusion. Numerous studies have demonstrated that fusion peptides play a crucial role in membrane fusion. Mutational analysis revealed that substituting critical residues within the fusion peptides could abolish the fusogenic activity of the viral protein.[3] However, mutations are difficult to analyze, because, in addition to their local effects, the overall folding of the fusion protein can be affected.

[1] T. Stegmann, R. W. Doms, and A. Helenius, *Annu. Rev. Biophys. Biophys. Chem.* **18,** 187 (1989).
[2] J. White, M. Kielian, and A. Helenius, *Q. Rev. Biophys.* **16,** 151 (1983).
[3] M. J. Gething, R. W. Doms, D. York, and J. White, *J. Cell Biol.* **102,** 11 (1986).

Accordingly, an alternative strategy was developed that complements mutational studies with the full-length proteins. This consists of studying the interaction between liposomes and synthetic peptides that mimic the corresponding region in their parent proteins. The significance of this approach has been demonstrated by three observations: (1) there is a direct correlation between the effect of mutations in the intact protein and the peptide analogs[4–6]; (2) the fusion activity of the virus-derived synthetic peptides, measured *in vitro,* is sensitive to factors (such as pH or the addition of inhibitory agents) that affect the infectivity of the virus *in vivo*[7,8]; and (3) some virus-derived fusion peptides show antiviral activity, suggesting that they can model accurately and interact with functional domains of the viral proteins.[9,10] Indeed, the fact that a small fragment of a much longer protein can mimic partially the activity of the parent protein provides us with a unique opportunity to identify unequivocally specific functional regions.

This chapter focuses primarily on the methodology used to study the activity of virus-derived fusion peptides, using lipid vesicles. We analyze only a few examples to help clarify the techniques described. Therefore, readers interested in more extensive reviews on virally induced membrane fusion are referred to other works.[11–14]

## Preparation of Liposomes

The experiments detailed in this chapter can be done either with large unilamellar vesicles (LUVs) or with small unilamellar vesicles (SUVs). However, SUVs possess higher curvature and are therefore easier to fuse than LUVs. Moreover, it has been observed that some peptides are able

[4] M. Rafalski, J. D. Lear, and W. F. DeGrado, *Biochemistry* **29,** 7917 (1990).

[5] D. Rapaport and Y. Shai, *J. Biol. Chem.* **269,** 15124 (1994).

[6] N. Düzgüneş, and F. Gambale, *FEBS Lett.* **227,** 110 (1988).

[7] S. A. Wharton, S. R. Martin, R. W. Ruigrok, J. J. Skehel, and D. C. Wiley, *J. Gen. Virol.* **69,** 1847 (1988).

[8] F. B. Pereira, F. M. Goni, and J. L. Nieva, *AIDS Res. Hum. Retroviruses* **13,** 1203 (1997).

[9] Y. Kliger, A. Aharoni, D. Rapaport, P. Jones, R. Blumenthal, and Y. Shai, *J. Biol. Chem.* **272,** 13496 (1997).

[10] M. Pritsker, P. Jones, R. Blumenthal, and Y. Shai, *Proc. Natl. Acad. Sci. USA* **95,** 7287 (1998).

[11] S. R. Durell, I. Martin, J. M. Ruysschaert, Y. Shai, and R. Blumenthal, *Mol. Membr. Biol.* **14,** 97 (1997).

[12] R. M. Epand, *Biochim. Biophys. Acta* **1376,** 353 (1998).

[13] E. I. Pecheur, J. Sainte-Marie, A. Bienvenüe, and D. Hoekstra, *J. Membr. Biol.* **167,** 1 (1999).

[14] Y. Shai, *Biosci. Rep.* **20,** 535 (2000).

to cause fusion of SUVs but not of LUVs. On the contrary, to our knowledge, any peptide able to fuse LUVs is also able to cause the fusion of SUVs to a higher extent. Thus, for the experiment to be relevant it is recommended that LUVs be used exclusively.

To prepare LUVs it is necessary to make lipid films of the desired composition (e.g., phosphatidylcholine–cholesterol, 10:1 molar ratio). For this purpose, the lipids are dissolved in chloroform–methanol (2:1, v/v) and 4-mg (total lipid) aliquots are transferred to small glass vials. The films are prepared by removing the organic solvents under a stream of nitrogen in a rotary evaporator. After the last traces of organic solvents are removed by vacuum overnight, the dried films are hydrated with the desired buffer at a starting concentration of 10 mg/ml and are then vortexed to obtain a milky suspension of multilamellar vesicles (MLVs). MLVs are then freeze–thawed five times in liquid nitrogen and extruded (1) 5 times through two polycarbonate membranes with 0.4-$\mu$m-diameter pores and (2) 10 times through two polycarbonate membranes with 0.1-$\mu$m-diameter pores. Although filtration through 0.1-$\mu$m-diameter pore membranes is expected to yield LUVs with an average diameter of 0.1 $\mu$m, the aggregation of lipids may reduce the actual size of the pores; therefore it is recommended that the final size of the vesicles be checked either by light scattering or electron microscopy. Usually, the final lipid concentration obtained after the steps mentioned is about 7 mg/ml (presumably because of some loss of lipids during the extrusion steps). LUVs at this concentration look almost transparent (depending on the amount of cholesterol). They can be stored at 4° and used within 24 h. Special attention should be given to the composition of the buffer used, because low amounts of divalent cations [such as $Ca^{2+}$ or $Mg^{2+}$, normally present in phosphate-buffered saline (PBS)] can cause LUV aggregation and facilitate fusion. Thus, PBS$^=$ (without $Ca^{2+}$ and $Mg^{2+}$) should be used.

### Determining Ability of Virus-Derived Peptide to Cause Fusion of Liposomes

Virus-induced membrane fusion is thought to proceed through several consecutive steps: (1) close apposition of the involved bilayers, (2) mixing between the outer leaflets of the opposing membranes (also called hemifusion), (3) mixing of both the outer and inner monolayers, and (4) fusion pore formation (consequently mixing the aqueous contents). In principle, by using a simple model such as virus-derived fusion peptides and LUVs, it is possible to analyze the four different steps with a combination of complementary experiments. Note that although in principle it is possible to determine the ability of fusion peptides to induce the mixing of liposome contents,[15,16] most fusion peptides also cause a high degree of membrane destabilization. This results in leakage of the liposome contents; thus it is difficult, if not impossible, to detect real content mixing. This section describes these experiments, as well as their limitations, particularly when working with peptides that cause leakage of liposome contents.

### Determination of Membrane Destabilization

Like other "membrane-active" peptides, such as antimicrobial peptides or some signal sequences, most fusion peptides destabilize the membrane structure. Whether this destabilization is a prerequisite for the fusogenic activity or only a by-product remains to be determined. Indeed, there are some reports that infection with some viruses, such as influenza and human immunodeficiency virus type 1 (HIV-1), may result in initial membrane destabilization that causes some leakage of the cell contents.[17,18] In more practical terms, it is important to determine the ability of fusion peptides to increase membrane permeability, because this may affect some of the experiments described next and thus needs to be taken into account. Two different methods to determine the extent of membrane destabilization are discussed: dissipation of a diffusion potential and calcein release.

*Dissipation of Diffusion Potential.* Membrane destabilization can be assessed easily by monitoring fluorimetrically the collapse of a diffusion potential caused by the addition of the peptides.[19,20] First, LUVs are prepared in $K^+$ buffer (50 m$M$ $K_2SO_4$, 25 m$M$ HEPES-sulfate, pH 6.8), as described previously. Next, the LUVs, which have an initial concentration of about 8–9 m$M$ (depending on the particular lipid composition), are diluted to a final concentration of 35 $\mu M$ in an isotonic $K^+$-free buffer (50 m$M$ $Na_2SO_4$, 25 m$M$ HEPES-sulfate, pH 6.8), and a 1-$\mu$l/ml concentration of the fluorescent dye diS-$C_2$-5 (Sigma, St. Louis, MO) is added (the stock of diS-$C_2$-5 is 0.5 $M$ in dimethylsulfoxide). Subsequent addition of a 1.7-$\mu$l/ml concentration of valinomycin (from a 100 n$M$ stock in ethanol) creates a negative diffusion potential inside the vesicles by the selective efflux of $K^+$ ions, which results in a quenching of the dye fluorescence. This quenching should be monitored in a fluorometer (in a glass cuvette) with

[15] J. Wilschut and D. Papahadjopoulos, *Nature* **281,** 690 (1979).

[16] R. F. Epand, J. C. Macosko, C. J. Russell, Y. K. Shin, and R. M. Epand, *J. Mol. Biol.* **286,** 489 (1999).

[17] P. Bonnafous and T. Stegmann, *J. Biol. Chem.* **275,** 6160 (2000).

[18] A. S. Dimitrov, X. Xiao, D. S. Dimitrov, and R. Blumenthal, *J. Biol. Chem.* **276,** 30335 (2001).

[19] L. M. Loew, I. Rosenberg, M. Bridge, and C. Gitler, *Biochemistry* **22,** 837 (1983).

[20] Y. Shai, D. Bach, and A. Yanovsky, *J. Biol. Chem.* **265,** 20202 (1990).

the excitation set at 620 nm and the emission measured at 670 nm, until the fluorescence of diS-C$_2$-5 is stable. The fluorescence should be monitored as the peptide to be analyzed is added. If the peptide can destabilize the membrane, an increase in membrane permeation for all the ions in the solution will cause a dissipation of the diffusion potential. This will be accompanied by an increase in the fluorescence of the dye. The percentage of fluorescence recovery, $F_t$, is defined by

$$F_t(\%) = [(I_t - I_o)/(I_f - I_o)] \times 100$$

where $I_t$ is the fluorescence observed after adding a peptide at time $t$, $I_o$ is the fluorescence after adding valinomycin, and $I_f$ is the total fluorescence before adding valinomycin. It is recommended that the dose response be analyzed in a range of peptide:lipid molar ratios of up to 0.1. Note that fusion peptides may not be soluble in aqueous solvents and thus might need to be dissolved in organic solvents, such as dimethylsulfoxide. In these cases the effect of the solvent on membrane permeability should be determined. Figure 1 shows an example of the fluorescence recovery as a function of the peptide:lipid molar ratio for the 33-residue Sendai F protein fusion peptide and its G12A mutant. As shown, changing Gly-12 to alanine increases the membrane-permeating activity of the peptide. Interestingly, the same mutation was reported to increase the cell–cell fusion activity of the full-length F protein from SV5, a virus from the Sendai family.

*Calcein Release.* Calcein, a membrane-impermeable fluorescent probe, is self-quenched at high concentrations.[21] Calcein is entrapped at a self-quenching concentration (60 m$M$) into LUVs prepared in 10 m$M$ HEPES, 150 m$M$ NaCl, pH 7.4. The nonencapsulated calcein is removed from the LUV suspension by gel filtration, using a Sephadex G-50 column. In a typical run, 50 $\mu$l of the LUV suspension is loaded onto a column preequilibrated in 10 m$M$ HEPES, 0.15 $M$ NaCl, pH 7.4. Elution is monitored by absorbance at 280 nm, and the collected LUV peak is diluted to a final volume of 2 ml in the same buffer (to a concentration of 225 $\mu M$, which can be stored at 4° for about 24 h). The LUVs are further diluted to a final lipid concentration within the range of 2–10 $\mu M$, depending on the sensitivity of the fluorometer used, and the changes in calcein fluorescence (with excitation at 485 nm and emission at 515 nm) are monitored as different amounts of the peptides to be tested are added. Peptide-induced calcein leakage results in dequenching and a concomitant increase in the calcein fluorescence. Complete dye release is obtained after disrupting the vesicles

[21] Y. Pouny, D. Rapaport, A. Mor, P. Nicolas, and Y. Shai, *Biochemistry* **31**, 12416 (1992).

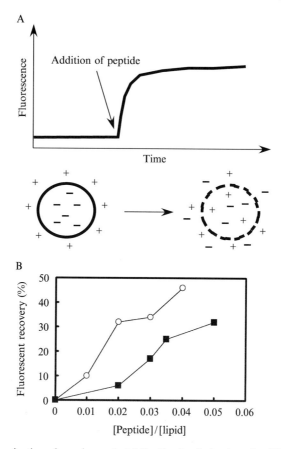

FIG. 1. Determination of membrane destabilization by dissipation of a diffusion potential. (A) Valinomycin creates a negative diffusion potential inside the vesicles by a selective efflux of $K^+$ ions, thus quenching the fluorescence of diS-$C_2$-5; addition of a peptide able to increase membrane permeability toward all the ions in the solution causes dissipation of the diffusion potential with a concomitant increase in the fluorescence of the dye. (B) As an example, fluorescence recovery (measured 1 min after adding the peptides) as a function of the peptide:lipid molar ratio is shown for a 33-residue segment corresponding to Sendai F protein N-terminal fusion peptide (■) and its G12A mutant (○). Adapted with permission from *J. Biol. Chem.* **269,** 15124–15131. Copyright 1994 American Society for Biochemistry and Molecular Biology, Inc.

with Triton X-100 [0.1% (w/v) final concentration]. The percentage of fluorescence recovery is determined by comparing the peptide-induced calcein release with the value obtained after adding Triton X-100, which corresponds to 100%.

## Determination of Aggregation and/or Fusion of Vesicles by Visible Light Absorbance

Changes in particle size can be monitored by measuring light scattering at visible wavelengths. Addition of a fusion peptide to an LUV suspension may result in an increase in light scattering at 405 nm because of aggregation and/or fusion of the LUVs.[22] To optimize the signal-to-noise ratio, it is important to keep the LUV and peptide concentrations within appropriate values. Regularly, a 100–200 $\mu M$ lipid concentration and peptide:lipid molar ratios up to 0.05 are used. Increasing amounts of peptide (from stock solutions) are added to 200-$\mu l$ suspensions of LUVs in PBS$^=$ preloaded in 96-well plates. Each desired peptide:lipid molar ratio should be analyzed at least in triplicate. The solvent used to dissolve the peptides may have an effect on the LUV suspension, and thus controls with equal amounts of solvents but without peptide should be considered. The 96-well plates are then incubated on a rotary shaker at the desired temperature (usually 37°). LUV aggregation is a fast process, so if the kinetics need to be analyzed, it is recommended that measuring (in a 96-well plate reader) be started as soon as the peptides are added. Usually, 10–15 min is enough to reach maximum activity. In a study done to determine the role of the Sendai virus F protein N-terminal repeat in the process of membrane fusion, different constructs corresponding to the first 70 N-terminal residues of Sendai F protein were compared.[22] It was observed that, whereas the wild-type peptide caused fusion of LUVs even at low peptide:lipid ratios (0.01–0.03), two mutants (one in which Gly-119 was changed to lysine and the other in which Ile-154 was changed to lysine) were virtually inactive. However, the three peptides induced aggregation of LUVs to similar extents (Fig. 2). Thus, the failure of the mutants to induce membrane fusion occurred at some stage after membrane aggregation.

## Determination of Bilayer Mixing by Using Fluorescence Probe Dilution Assay

The most widely used method to detect lipid mixing of LUVs is the fluorescent probe dilution assay, based on resonance energy transfer measurements.[23] In this method, two different LUV populations are prepared. One consists of regular lipids (e.g., phosphatidylcholine–cholesterol, 10:1 molar ratio) and the other has, in addition, 0.6 mol% each of 7-nitrobenz-2-oxa-1,3-diazole (NBD)-phosphatidylethanolamine (energy donor) and Rhodamine-phosphatidylethanolamine (energy acceptor). The fluorescent

[22] J. K. Ghosh and Y. Shai, *J. Mol. Biol.* **292**, 531 (1999).
[23] D. K. Struck, D. Hoekstra, and R. E. Pagano, *Biochemistry* **20**, 4093 (1981).

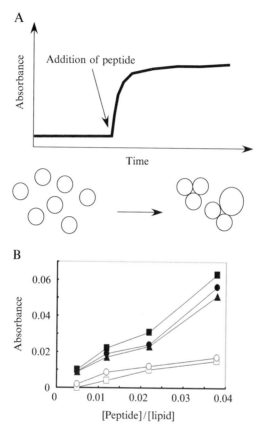

Fig. 2. Detection of aggregation and/or fusion of vesicles by measuring changes in light scattering at 405 nm. (A) The initial suspension of LUVs has a basal level of light scattering; the addition of a fusogenic peptide causes aggregation and/or fusion of liposomes, increasing the scattering of light at 405 nm. (B) As an example, changes in light scattering as a function of the peptide:lipid molar ratio are shown, for a family of five peptides corresponding to different regions of the Sendai virus F protein N-terminal domain. (■) Sendai virus F1 protein residues 1–70; (●) 1154K 1–70; (▲) G119K; (□) 1–33; (○) an equimolar mixture of 1–33 and 34–70. The LUVs were composed of phosphatidylcholine–phosphatidylglycerol (1:1, molar ratio) and used at a concentration of 116 $\mu M$ in PBS$^=$. Reproduced with permission from *J. Mol. Biol.* **292,** 531–546. Copyright 1999 Academic Press.

phospholipids are added when preparing the lipid films, and thus the final LUVs contain fluorescent probes on both the inner and outer leaflets. Then, both LUV populations are mixed in PBS$^=$, at a total lipid concentration of 110 $\mu M$, with a 1:4 ratio of labeled and unlabeled vesicles. For a typical experiment, 400-$\mu l$ aliquots are placed in a fluorescence cuvette and the excitation is set at 467 nm (maximum of NBD absorbance), while the emission is recorded at 530 nm (maximum of NBD emission). Initially, the average distance between an NBD-Rho pair is such that NBD fluorescence is partially quenched and, therefore, the intensity of NBD fluorescence is low. Addition of a fusogenic peptide may result in fusion between labeled and unlabeled vesicles, with a concomitant dilution of the fluorescent probes on both the internal and external surfaces of the fused LUVs. This causes an increase in the average distance between NBD-Rho pairs, which in turn results in a dequenching of NBD fluorescence, observed as an increase in the fluorescence intensity. The fluorescence before the addition of the peptide is referred to as 0% lipid mixing, and the fluorescence intensity on the addition of reduced Triton X-100 (Sigma) [0.05% (v/v) final concentration] is referred to as 100% lipid mixing. Note that regular Triton X-100 can quench NBD fluorescence; thus, it should be avoided. Regularly, increasing peptide:lipid molar ratios (of up to 0.1) are analyzed. As mentioned before, the solvents in which the peptides are dissolved might affect the LUV (or NBD-Rho fluorescence); thus it is necessary to perform controls with similar amounts of solvent but without peptides.

We detected an internal fusion peptide in Sendai virus F protein. Unlike N-terminal fusion peptides, the N-terminal boundary of which is explicitly defined, the exact location within the protein sequence of internal fusion peptides should be determined by comparing the activity of peptides of different lengths, shifted either to the N or the C terminus. Accordingly, the fluorescent probe dilution assay was used to study a family of peptides derived from the internal fusogenic region of Sendai F protein. It was observed that, when using LUVs composed of phosphatidylcholine, the segment corresponding to residues 201–229 (SV-201) was itself sufficient to cause fusion (Fig. 3), whereas a shorter peptide (residues 208–229, SV-208) showed no activity. Furthermore, shifting the peptide to the N terminus (residues 178–210, SV-178), as well as introducing mutations within residues 201–208 (Mu-SV-201 and Mu-SV-208), completely abolished the activity. Thus, it was concluded that the region responsible for the fusogenic activity is contained within residues 201–229, and that the N-terminal residues are important for this activity. Interestingly, N-terminal elongation (residues 198–201, SV-197) resulted in a significant increase in fusion activity, suggesting that the internal fusogenic region can have a similar role in the full-length F protein.

It is important to note that a peptide that causes micellization but not membrane fusion will have an effect on NBD fluorescence indistinguishable from that of a true fusogen. Thus, lipid-mixing experiments should be confirmed by an independent method, such as examination of the

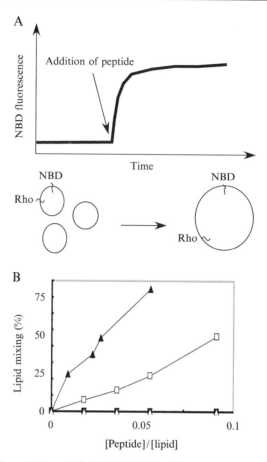

Fig. 3. Detection of intervesicular lipid mixing by a fluorescent probe dilution assay. (A) The LUV suspension has a basal fluorescence resulting from the average distance between NBD-Rho pairs in the labeled population of vesicles. Addition of an active peptide causes fusion between labeled and unlabeled liposomes, with a concomitant dilution of the fluorescent probes. Consequently, the average distance between NBD-Rho pairs is greater, thus increasing the NBD fluorescence. (B) Dose dependence of lipid mixing of phosphatidylcholine LUVs induced by a family of peptides corresponding to the Sendai F protein internal fusogenic region. (□) SV-201; (■) SV-208; (▲) SV-197; (△) SV-178. Adapted with permission from *Biochemistry* **39,** 11581–11592. Copyright 2000 American Chemical Society.

LUVs, before and after adding the tested peptides, by electron microscopy (as described next). Some virus-derived peptides fail to cause aggregation of vesicles (no increase in light scattering of LUVs at 405 nm is detected), and consequently they do not induce membrane fusion. However, they can be readily active if the vesicles are preaggregated by addition of 1–5 m$M$ $Mg^{2+}$ (a concentration that does not cause fusion per se).[24] For example, in the absence of $Mg^{2+}$, a 33-residue peptide corresponding to HIV-1 gp41 N-terminal fusion peptide causes fusion of zwitterionic phosphatidylcholine LUVs, whereas a shorter peptide comprising only the first 16 residues is inactive. However, when the LUVs are preaggregated with $Mg^{2+}$, the short peptide retains 60% of the activity of the longer one.[25]

## Determination of Hemifusion by Fluorescence Probe Dilution of Liposomes Pretreated with Dithionite

As mentioned previously, virus-induced membrane fusion is thought to proceed through an intermediate in which only fusion between the outer monolayers occurs. Interestingly, an assay originally developed to study membrane asymmetry can be useful to detect such intermediates.[26] The method is, in essence, similar to the probe dilution assay described previously. The only difference is that the labeled LUVs are pretreated with sodium dithionite ($S_2O_4^{2-}$), an agent that reduces NBD, irreversibly quenching its fluorescence. Dithionite cannot permeate lipid membranes, and thus its addition to already prepared LUVs causes reduction of the NBD present on the outer leaflet only. Next, the dithionite is separated from the liposomes by size-exclusion chromatography, and the labeled LUVs can be used as mentioned in the previous section. An increase in NBD fluorescence, detected on the addition of the tested peptides, can be interpreted as lipid mixing of the inner (and therefore also the outer) monolayers. If no increase in NBD fluorescence is observed, although such an increase was observed when the vesicles were not pretreated with dithionite, it is possible to conclude that only mixing of the outer monolayers (hemifusion) occurred. Labeled LUV stocks prepared as indicated previously have a total lipid concentration of about 8–9 m$M$ (0.6% molar NBD-labeled phospholipid). To completely reduce outer leaflet NBD, dithionite should be added at a final concentration of 80 m$M$ (from a freshly prepared stock of 1 $M$ dithionite in 1 $M$ Tris, pH 10.0). The reaction is usually complete after 30–45 min of incubation in ice. It is important to mention that Meers et al.[26] have reported that NBD-phosphatidylserine has a low translocation rate across the lipid bilayers and therefore is preferable to NBD-phosphatidylethanolamine.

[24] D. Hoekstra and N. Düzgüneş, Biochemistry 25, 1321 (1986).

[25] M. Pritsker, J. Rucker, T. L. Hoffman, R. W. Doms, and Y. Shai, Biochemistry 38, 11359 (1999).

[26] P. Meers, S. Ali, R. Erukulla, and A. S. Janoff, Biochim. Biophys. Acta 1467, 227 (2000).

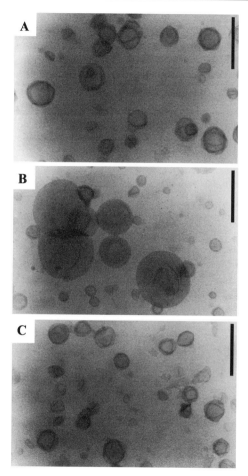

Fig. 4. Detection of fusion of liposomes by electron microscopy. Phosphatidylcholine–phosphatidylglycerol (1:1) LUVs were incubated for 15 min before staining and visualization with the corresponding amount of dimethylsulfoxide (A), SV-208 (B), and Mu-SV-208 (C); the bars represent 200 nm. As observed, SV-208, the wild-type peptide, causes an increase in the average diameter of the vesicles, compared with the inactive peptide Mu-SV-208 and the vesicles incubated with solvent only. Reproduced with permission from *J. Mol. Biol.* **296,** 1353–1365. Copyright 2000 Academic Press. (See color insert.)

## Determination of Fusion of Liposomes by Electron Microscopy

Direct visualization of LUVs before and after incubation with the peptides could be, in principle, the best method to detect fusion (or aggregation) of liposomes. Indeed, by comparing the average diameter of liposomes not treated with the peptides with that of the treated vesicles,

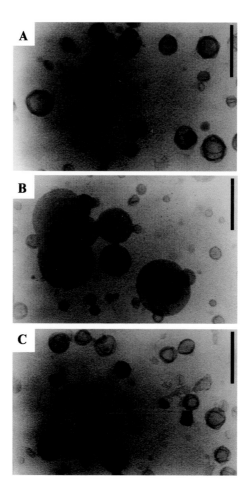

PEISAJOVICH AND SHAI, CHAPTER 21, FIG. 4. Detection of fusion of liposomes by electron microscopy. Phosphatidylcholine–phosphatidylglycerol (1:1) LUVs were incubated for 15 min before staining and visualization with the corresponding amount of dimethylsulfoxide (A), SV-208 (B), and Mu-SV-208 (C); the bars represent 200 nm. As observed, SV-208, the wild-type peptide, causes an increase in the average diameter of the vesicles, compared with the inactive peptide Mu-SV-208 and the vesicles incubated with solvent only. Reproduced with permission from *J. Mol. Biol.* (2000) **296,** 1353–1365. Copyright 2000 Academic Press.

it is possible to obtain a qualitative estimation of the extent of fusion, as well as to differentiate fusion from vesicle aggregation. Moreover, electron microscopy is a useful tool for confirming that lipid mixing is the result of membrane fusion and not of micellization, because fused liposomes have an average diameter larger than that of the untreated vesicles, whereas micelles appear concomitantly with an abrupt reduction in LUV diameter. However, it is difficult to analyze electron microscopy results quantitatively; thus we prefer to use it not exclusively, but rather combined with more quantitative techniques. Furthermore, electron micrographs are snapshots of the fusion process, whereas the fluorescence techniques described previously can be used for kinetic analysis as well. In a typical electron microscopy experiment, a 10- to 15-$\mu$l aliquot from a suspension of LUVs in PBS$^=$, at a concentration of 4–5 m$M$, is incubated with the desired amount of peptide (peptide:lipid molar ratios similar to those used in the fluorescent probe dilution assay described previously) or with an equivalent amount of solvent for 30 min (the actual time of incubation to obtain maximal activity can be deduced from the probe dilution assay). Then, a 10-$\mu$l aliquot is deposited onto a carbon-coated grid and negatively stained with 2% (w/v) uranyl acetate. After 1 min, the excess dye is removed with filter paper and the grid can be examined in an electron microscope. The dried grids can be stored at room temperature for months without any visible alteration. Figure 4 shows electron micrographs of negatively stained phosphatidylcholine–phosphatidylglycerol (1:1 molar ratio) LUVs incubated with the corresponding amount of solvent alone (dimethylsulfoxide in this case; Fig. 4A), with a peptide corresponding to Sendai virus F protein internal fusion peptide (Fig. 4B), or with an inactive mutant (Fig. 4C).[27] Clearly, the wild-type peptide causes an increase in the average size of the vesicles as a result of membrane fusion, compared with the vesicles incubated with dimethylsulfoxide or with the inactive mutant.

## Concluding Remarks

The use of liposomes to identify and characterize viral fusogenic peptides has enabled rapid advances in the field of virus-mediated membrane fusion. The development of sensitive, and concomitantly versatile, fluorescence-based assays has facilitated the characterization of the different stages involved in the process of membrane fusion. To date, most liposome-based *in vitro* studies of virus-mediated membrane fusion have focused on the role of short fragments derived from large viral proteins. Extending these studies to analyze the full-length fusogenic proteins seems the most logical step for the coming years.

[27] S. G. Peisajovich, O. Samuel, and Y. Shai, *J. Mol. Biol.* **296,** 1353 (2000).

## [22] Liposomes as Target Membranes in the Study of Virus Receptor Interaction and Membrane Fusion

By Jolanda M. Smit, Barry-Lee Waarts,
Robert Bittman, and Jan Wilschut

### Introduction

The genome of an enveloped virus gains access to the cell cytoplasm by a membrane fusion process. Enveloped viruses can infect their host cell via two different pathways: by direct fusion with the plasma membrane or by cell entry through receptor-mediated endocytosis and subsequent fusion from within acidic endosomes.[1] Fusion with the plasma membrane is triggered by the interaction of a virus particle with a receptor on the cell surface. In the process of receptor-mediated endocytosis, fusion is initiated through exposure of the virus to low pH within the lumen of the endosomal compartment of the cell. The route of cell entry depends on the family to which the virus belongs. For example, alphaviruses[2] and orthomyxoviruses[3] infect their host cells via receptor-mediated endocytosis, whereas lentiviruses[4] and paramyxoviruses[5] infect cells via fusion with the plasma membrane.

The basic features underlying viral membrane fusion processes have been studied extensively in virus–cell and virus–liposome model systems. In most virus–cell systems, membrane fusion has been evaluated in an indirect manner, for example, on the basis of cellular infection, hemolysis, or polykaryon formation.[2] To obtain a more detailed insight into the molecular mechanisms underlying viral membrane fusion processes, virus–liposome systems have been developed.[6] Studies of virus fusion with liposomes offer a number of advantages over virus–cell fusion assays, including higher sensitivity and the possibility of obtaining quantitative data. Furthermore, virus–liposome model systems give the opportunity to vary the composition of the target membrane so that specific components required for virus–receptor interaction and membrane fusion may be identified.

[1] M. March and A. Pelchen-Matthews, *Rev. Med. Virol.* **3,** 173 (1993).

[2] M. Kielian, *Adv. Virus Res.* **45,** 113 (1995).

[3] L. D. Hernandez, L. R. Hoffman, T. G. Wolfsberg, and J. M. White, *Annu. Rev. Cell. Dev. Biol.* **12,** 627 (1996).

[4] R. Blumenthal and D. S. Dimitrov, *in* "Handbook of Physiology" (J. F. Hoffman and J. C. Jamieson, eds.), "Section on Cell Physiology," p. 564. Oxford University Press, New York, 1996.

[5] C. J. Russell, T. S. Jardetzky, and R. A. Lamb, *EMBO J.* **20,** 4924 (2001).

[6] D. Hoekstra and K. Klappe, *Methods Enzymol.* **220,** 261 (1993).

The first step in the membrane fusion process of a virus with a liposome involves binding of the virus to the target membrane. Then, fusion of the viral membrane with the liposomal membrane occurs, which may be subdivided into two steps: merging of the two membranes and mixing of the interior of the virus with the liposomal lumen. Accordingly, assays have been developed to determine the extent of virus–liposome binding and to evaluate fusion on the basis of either *lipid mixing* or *content mixing*. In lipid-mixing assays, fluorescent probes are generally used to monitor membrane merging in a sensitive and continuous fashion, thus providing information about the kinetics and the extent of the membrane fusion process. Content-mixing assays do not normally permit kinetic analysis of the membrane fusion process, but do meet a stringent criterion for fusion, that is, the delivery of the internal content of the virus to the aqueous lumen of the target liposome. In this chapter, the procedures and applications of binding, lipid-mixing, and content-mixings assays are described, with an emphasis on the analysis of the role of target membrane lipid composition on virus binding and fusion. All the examples presented in this chapter relate to receptor interaction and membrane fusion of two alphaviruses, Semliki Forest virus (SFV) and Sindbis virus (SINV). However, the approaches can be readily adapted to the study of other enveloped viruses.

Alphaviruses represent a separate genus within the family of the Togaviridae. Alphaviruses are simple, highly symmetrical enveloped viruses with a diameter of approximately 70 nm.[7] An alphavirus virion contains one single-stranded positive-sense RNA molecule of approximately 11 kb, complexed to 240 copies of the core protein C. The viral envelope contains the glycoproteins E1 and E2, organized in 80 spike-like protrusions. Each spike consists of three E1–E2 heterodimers. The E2 component of the alphavirus spike is responsible for virus-receptor binding, whereas the E1 protein mediates the low-pH-dependent fusion process from within the target cell endosomes.

## Experimental Procedures

### Background

*Binding Assays.* To study the binding characteristics of virus particles to liposomal membranes, assays have been developed that rely on flotation of virus–liposome complexes on density gradients.[8–10] Briefly,

[7] J. H. Strauss and E. G. Strauss, *Microbiol. Rev.* **58,** 491 (1994).
[8] M. Kielian and A. Helenius, *J. Virol.* **52,** 281 (1984).
[9] R. Bron, J. M. Wahlberg, H. Garoff, and J. Wilschut, *EMBO J.* **12,** 693 (1993).
[10] J. M. Smit, R. Bittman, and J. Wilschut, *J. Virol.* **73,** 8476 (1999).

after incubation of the virus (preferably radioactively labeled) with liposomes, the virus–liposome suspension is mixed with high-density sucrose and overlaid with a sucrose density gradient. Bound virus is then separated from nonbound virus by flotation of virus–liposome complexes to the top of the gradient.

*Lipid-Mixing Assays.* Many fluorescence assays have been developed for monitoring the course of viral membrane fusion processes (for a review, see Hoekstra and Klappe[6]). To study virus fusion with liposomes, it is preferable to use an assay that involves labeling of the virus with a fluorescent probe. Fusion of the virus with the larger liposome results in a substantial dilution of the probe into the target membrane, which is monitored as a change in the fluorescence properties or intensity of the probe. For a comparatively small virus, this is the preferred method by which to carry out the fusion assay. It is also possible, particularly with larger viruses, to label the target liposomes with a fluorescent probe or a combination of probes and subsequently monitor fusion as a dilution of the probe(s) from the target liposomes into the viral membrane. In this case, the liposomes should be comparatively small. This chapter presents both assay variants. In the first variant of the assay, virus particles are labeled with pyrene fatty acids, which are biosynthetically incorporated into the envelope phospholipids of the virus. In the second variant, pyrene-labeled phosphatidylcholine (pyrPC) is incorporated into small liposomes. In either case, fusion is followed as a change in pyrene fluorescence intensity. The assays rely on the capacity of the pyrene probe to form excited state dimers (excimers), which consist of a pyrene probe molecule in the excited state and a probe molecule in the ground state. The fluorescence emission of the excimer is shifted to longer wavelengths relative to that of the pyrene monomer. Excimer formation is dependent on the average distance between the probe molecules. Thus, within the membrane, the intensity of excimer fluorescence is proportional to the surface density of pyrene-labeled molecules. After fusion of a pyrene-labeled membrane with an unlabeled membrane (containing no pyrene-conjugated lipids), the pyrene probe diffuses into the target membrane. As a result, a decrease in pyrene excimer fluorescence intensity is observed, which can be monitored in a continuous fashion.[9–11]

Other lipid-mixing assays have been developed that involve labeling of virus particles with a fluorescent probe, notably octadecylrhodamine (R18).[6] The R18 assay involves insertion of the probe into the viral membrane to such an extent that its fluorescence is self-quenched. After fusion of R18-labeled virus with unlabeled target membranes such as liposomes,

[11] J. Corver, A. Ortiz, S. L. Allison, J. Schalich, F. X. Heinz, and J. Wilschut, *Virology* **269,** 37 (2000).

the probe is diluted, which is monitored directly as a relief of fluorescence self-quenching. We prefer the pyrene assay over the R18 assay, mostly because it has been demonstrated that under certain conditions spontaneous transfer of the R18 probe from the viral membrane into the target membrane may occur (without membrane fusion).[12]

*Content-Mixing Assays.* A stringent criterion for virus–liposome fusion involves mixing of the interior of the virus with the liposomal lumen. To monitor content mixing, liposomes are prepared with an encapsulated enzyme, such as trypsin or RNase; on fusion, the viral nucleocapsid is degraded.[13,14] The trypsin variant of these assays is described in detail below.

## Detailed Procedures

*Production and Characterization of SINV or SFV.* Alphaviruses can be grown to high titers in baby hamster kidney (BHK-21) cells. BHK-21 cells are cultured in Glasgow's modification of Eagle's minimal essential medium (e.g., from GIBCO-BRL, Gaithersburg, MD), supplemented with 5% fetal bovine serum, 10% tryptose phosphate broth, 200 m$M$ glutamine, 25 m$M$ HEPES, and 7.5% sodium bicarbonate. On day 1, BHK-21 cells (grown to confluence in a 162.5-cm$^2$ flask) are trypsinized and resuspended in 10 ml of cold culture medium. Subsequently, the BHK-21 cells are added to a roller bottle containing 90 ml of cold culture medium. The roller bottle is placed in a 37° incubator with a rotation speed of 0.4 rpm. On day 4, the subconfluent cell monolayer is infected with SINV or SFV at a multiplicity of infection (MOI) of 4. At 24 h postinfection, the progeny virus is harvested from the medium. First, a low-speed centrifugation (350 $g$, 15 min) min) in the cold is carried out to remove cell debris. Then, the virus is sedimented from the supernatant by ultracentrifugation in a Beckman Coulter (Fullerton, CA) type 19 rotor at 54,000 $g$ for 2.5 h at 4°. The virus pellet is resuspended overnight in 1 ml of HNE (5 m$M$ HEPES, 150 m$M$ NaCl, 0.1 m$M$ EDTA, pH 7.4). Finally, the virus suspension is subjected to sucrose gradient centrifugation [20–60% (w/v), linear sucrose gradient in HNE] in a Beckman Coulter SW 41 rotor at 100,000 $g$ for 16 h at 4°. SINV or SFV equilibrates as a single band at approximately 45% (w/v) sucrose. After isolation of the virus band, the virus is frozen in small aliquots in liquid N$_2$ and stored at −80°. The purity of the viruses is checked by sodium dodecyl sulfate–polyacrylamide gel electrophoresis

---

[12] T. Stegmann, P. Schoen, R. Bron, J. Wey, I. Bartoldus, A. Ortiz, J. L. Nieva, and J. Wilschut, *Biochemistry* **32,** 11330 (1993).

[13] J. White and A. Helenius, *Proc. Natl. Acad. Sci. USA* **77,** 3273 (1980).

[14] A. Helenius, J. Kartenbeck, K. Simons, and E. Fries, *J. Cell Biol.* **84,** 404 (1980).

(SDS–PAGE).[15] SINV or SFV particles are further characterized by plaque assay on BHK-21 cells,[16] phosphate analysis,[17] and protein determination.[18] Then, the ratio of physical virus particles to plaque-forming units (particle-to-PFU ratio) may be calculated to assess the specific infectivity of the virus preparation. For the calculation, a theoretical amount of $5.45 \times 10^{-17}$ g (SINV) of protein or $4.6 \times 10^{-20}$ g (SFV) protein mol of phosphate per SINV particle is used. Typically, the particle-to-PFU ratio is about 5–10 for a SINV or SFV preparation.[9,10]

*Production of Pyrene-Labeled Virus.* Pyrene-labeled SINV or SFV is produced in essentially the same fashion as unlabeled virus, with the exception that, after the BHK-21 cells have been cultured for 12 to 15 h in a roller bottle as described above, an aliquot of a solution of 2.5 mg of 16-(1′-pyrenyl)hexadecanoic acid (Molecular Probes, Eugene, OR) per milliliter of dimethyl sulfoxide (DMSO) is added to the medium to give a final concentration of 15 $\mu$g/ml. We have observed that concentrations higher than 15 $\mu$g/ml become toxic for the BHK-21 cells; the precise toxic level probably depends on the cell type. As described above, on day 4 the cells are infected with virus, and at 24 h postinfection the pyrene-labeled virus is purified from the medium and characterized as described above. To assess the extent of labeling of the viral envelope phospholipids, the pyrene excimer fluorescence intensity of the labeled virus is determined. Pyrene-labeled SINV or SFV (0.5 $\mu M$ viral phospholipid) in 700 $\mu$l of HNE buffer is added to a quartz cuvette, thermostatted at 37°. The pyrene emission spectrum is measured from 355 to 550 nm with excitation at 345 nm, in a fluorometer (e.g., an SLM-Aminco AB2 instrument from Thermo Spectronic, Rochester, NY). The emission spectrum is characterized by two sharp monomer peaks at 377 and 397 nm. A concentration-dependent broad excimer peak is observed at 480 nm. The ratio of the fluorescence intensity of excimer$_{480 \text{ nm}}$ to that of monomer$_{377 \text{ nm}}$ in the viral membrane should be approximately 0.20 to 0.40 at 37°. Figure 1 shows a typical example of an emission spectrum of pyrene-labeled SINV.

*Production of* $^{35}S$*methionine-Labeled Virus.* On day 1, BHK-21 cells (grown to confluence in a 75-cm$^2$ flask) are infected with SINV or SFV in 2.5 ml of culture medium as described above. At 2.5 h postinfection, the virus-containing medium is replaced by 5 ml of label medium, consisting of methionine-free Dulbecco's modified Eagle's medium (e.g., from GIBCO-BRL), supplemented with 5% fetal bevine serum and 200 m$M$

[15] U. K. Laemmli, *Nature* **227,** 680 (1970).
[16] W. B. Klimstra, K. D. Ryman, and R. E. Johnston, *J. Virol.* **72,** 7357 (1998).
[17] C. J. F. Böttcher, C. M. van Gent, and C. Fries, *Anal. Chim. Acta* **24,** 203 (1961).
[18] G. L. Peterson, *Anal. Biochem.* **83,** 346 (1977).

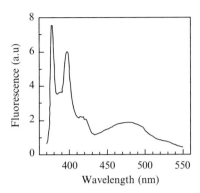

Fig. 1. Fluorescence emission spectrum of pyrene-labeled SINV. The spectrum was recorded at 37°, with SINV at a concentration of 0.5 $\mu M$ phospholipid in HNE-buffer, between 355 and 550 nm using an excitation wavelength of 345.

glutamine. After 2 h of starvation, 200 $\mu$Ci of [35]S-labeled methionine (e.g., from Amersham Biosciences, Piscataway, NJ) is added to the medium and incubation is continued overnight. At 24 h postinfection, [35]S-labeled viral particles are harvested from the medium. After a low-speed centrifugation (350 $g$, 15 min), the virus-containing supernatant is subjected to centrifugation on a discontinuous sucrose gradient [5–20–50–60% (w/v) sucrose gradient in HNE] at 150,000 $g$ in a Beckman Coulter SW 41 rotor for 2 h at 4°. The virus bands at the 20–50% sucrose interface. After isolation of the virus band, the purified viral particles are frozen in small aliquots at −20°. The purity of the viruses is checked by SDS–PAGE.[15] The virus preparations may be further characterized by plaque assay on BHK-21 cells,[16] to determine the specific infectivity (PFU/cpm).

*Preparation of Liposomes.* Large unilamellar vesicles (referred to as liposomes in this chapter) are typically prepared from mixtures of phospholipids and cholesterol. The phospholipids and cholesterol are available from several vendors (e.g., Avanti Polar Lipids, Alabaster, AL). We routinely use a freeze/thaw–extrusion procedure, which allows the preparation of liposomes having different size ranges. Typically, in the fusion assays involving a pyrene-labeled or radioactive virus, we use liposomes with an average diameter of 200 nm. Lipid mixtures (2–10 mg) are dried from stock solutions in chloroform or chloroform–methanol under a stream of $N_2$ and further dried under vacuum for at least 1 h. Then, the lipids are hydrated in 1.0 ml of HNE and subjected to five cycles of freezing in liquid $N_2$ and thawing in lukewarm water. Subsequently, the liposomes are sized by extrusion (21 times) through a Unipore polycarbonate filter with a pore size of 0.2 $\mu$m (e.g., from Whatman Nuclepore, Scarborough, ME) in a

LiposoFast miniextruder (Avestin, Ottawa, ON, Canada). The sizes of the liposomes are determined by quasi-elastic light-scattering analysis (e.g., with a submicron particle sizer, NICOMP model 370; Particle Sizing Systems, Santa Barbara, CA). The phospholipid concentration of the liposomes is determined by phosphate analysis.[17]

Trypsin-containing liposomes are prepared in the same fashion, except that the lipids are hydrated in HNE containing trypsin (10 mg/ml; e.g., from Boehringer Mannheim). After the extrusion step, the trypsin-containing liposomes are separated from free (untrapped) trypsin by gel filtration on Sephadex G-100 (20-cm-long column) in HNE.

*Preparation of pyrPC-Labeled Liposomes.* As indicated above, pyrPC-containing liposomes can be used to monitor fusion with unlabeled virus. The liposomes must be small, so that fusion of a liposome with a virion will result in significant dilution of the probe. In our assay, we use liposomes with an average diameter of 70 nm. A solution of pyrPC [1-palmitoyl-2-(1'-pyrenedecanoyl)-*sn*-glycero-3-phosphocholine; Molecular Probes] in chloroform (at a concentration of 5 mol% relative to the total phospholipids) is added to the lipid mixture used to prepare the liposomes. Liposomes are prepared as described above, except that the preparation is extruded an additional 81 times through two stacked Unipore polycarbonate filters with a pore size of 0.05 $\mu$m. The liposomes thus produced have an average diameter of $\sim$70 nm.

*Protocol for Binding of $^{35}$Smethionine-Labeled Virions to Liposomes.* Virus–liposome binding is assessed by flotation analysis. $^{35}$Smethionine-Labeled SINV or SFV particles ($10^5$ to $10^6$ CPM supplemented with unlabeled virus to a total concentration of 0.5 $\mu M$ viral phospholipid) are mixed with liposomes (100-200 $\mu M$ phospholipid) in a final volume of 0.125 ml in HNE buffer. After a 1-min incubation, typically at 37°, the mixture is acidified, with continuous stirring, by adding 6.5 $\mu$l of 0.1 $M$ morpholinoethanesulfonic acid (MES), 0.2 $M$ acetic acid, pretitrated with NaOH to achieve the final desired pH. After 60 s, the samples are neutralized by the addition of a pretitrated volume of 0.1 $M$ NaOH and placed on ice. Subsequently, 0.1 ml of the virus–liposome mixture is added to 1.4 ml of 50% (w/v) sucrose in HNE and overlaid with a discontinuous (50–35–20–5%, w/v) sucrose gradient. After centrifugation in a Beckman Coulter SW 50 rotor at 150,000 $g$ for 2 h at 4°, the gradient is fractionated into 10 samples, starting from the top. The distribution of the viral radioactivity is quantified by liquid scintillation analysis. The radioactivity in the top four fractions, relative to the total amount of radioactivity, is taken as a measure of virus–liposome binding.

Figure 2 shows an example of binding of $^{35}$Smethionine-labeled SINV with liposomes at 37°. At pH 5.0, most of the viral particles float together

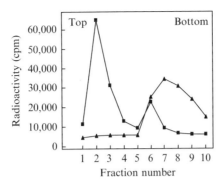

FIG. 2. Binding of [$^{35}$S]methionine-labeled SINV to liposomes. Binding of SINV to liposomes was determined at 60 s postacidification at 4° by coflotation analysis on sucrose density gradients as described in text. The final concentrations of [$^{35}$S]methionine-labeled virus and PC/PE/SPM/Chol (molar ratio, 1:1:1:1.5) liposomes correspond to 0.5 and 100 $\mu M$ membrane phospholipid, respectively. (■) pH 5.0; (▲) pH 7.4.

with the liposomes to the top of the gradient (Fig. 2, squares). Quantification of the flotation results showed that, in this particular experiment, 76% of the virus was associated with the liposomes. At neutral pH, there was a negligible interaction between the virus and the liposomes (Fig. 2, triangles).

*Protocol for Fusion of Pyrene-Labeled SINV or SFV with Liposomes.* On-line fusion of pyrene-labeled SINV or SFV with an excess of liposomes is monitored as a decrease of pyrene excimer fluorescence intensity. In this assay, liposomes with an average diameter of 200 nm are used. The diameter of the viral membrane (excluding the glycoproteins) is 50 nm. Therefore, on fusion of a viral particle with a liposome, the pyrene-labeled phospholipids are diluted by at least an order of magnitude. The fusion reaction is carried out in a fluorometer (e.g., SLM/Aminco), the excitation and emission wavelengths being set at 345 and 480 nm, respectively. Pyrene-labeled SINV or SFV (0.5 $\mu M$ viral phospholipid) is mixed with liposomes (100-200 $\mu M$ phospholipid) in a quartz curvette in a volume of 0.665 ml of HNE. The contents of the cuvette are stirred magnetically and thermostatted at the desired temperature. After a 1-min incubation, fusion is triggered by the addition of 35 $\mu l$ of 0.1 $M$ MES–0.2 $M$ acetic acid, pretitrated with NaOH to achieve the final desired pH. The extent of fusion ($F$) is calculated as follows:

$$F(\%) = [(0.95 \times E_o) - E_m]/[(0.95 \times E_o) - (1.05 \times E_{C_{12}E_8})] \times 100$$

where $E_0$ represents the initial excimer fluorescence intensity, $E_m$ is the measured excimer fluorescence intensity, and $E_{C_{12}E_8}$ is the fluorescence intensity at infinite dilution of the pyrene probe. The latter value is obtained after the addition of 35 $\mu$l of 0.2 $M$ octaethyleneglycol-$n$-dodecyl monoether ($C_{12}E_8$) (e.g., from Fluka, Buchs, Switzerland). $E_0$ and $E_{C_{12}E_8}$ are corrected for the dilution.

An example of an on-line fusion measurement of pyrene-labeled SINV with liposomes at 37° is shown in Fig. 3. At pH 5.0, a decrease in pyrene excimer fluorescence intensity of 60% was observed, which corresponds to 60% fusion (Fig. 3, curve a). The initial rate of fusion, as determined from the tangent to the first part of the curve, was 25%/s (calculated using an expanded scale), and fusion was complete after ~10 s. There was no fusion at pH 7.4 (Fig. 3, curve b). Furthermore, the pyrene excimer fluorescence intensity remained constant when SINV was incubated at pH 5.0 in the absence of liposomes (Fig. 3, curve c), indicating that the decrease of pyrene fluorescence intensity in the presence of target membranes (Fig. 3, curve a) was a result of dilution of the pyrene probe into the liposomal membrane. Clearly, fusion of SINV with liposomes requires a mildly acidic pH, consistent with cellular entry of the virus through endosomes or lyosomes.

*Protocol for Fusion of Unlabeled SINV or SFV with pyrPC-Labeled Liposomes.* In the reverse variant of the pyrene assay, fusion of pyrPC-labeled liposomes with an excess of unlabeled SINV or SFV is measured on the basis of dilution of the pyrene probe from the liposomal to the viral membrane. This assay uses liposomes with an average diameter of 70 nm (see above). Fusion of a 70-nm liposome with a viral membrane 50 nm in diameter results in an increase in liposomal membrane surface area of only 33%, with a concomitant decrease in pyrene excimer fluorescence intensity

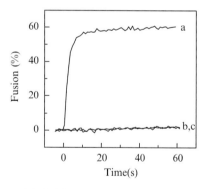

FIG. 3. Fusion of pyrene-labeled SINV with liposomes. Fusion of pyrene-labeled SINV (0.5 $\mu M$ viral phospholipid) with liposomes (200 $\mu M$ membrane phospholipid) was measured on-line at 37° as described in text. Liposomes consisted of PC/PE/SPM/Chol (molar ratio, 1:1:1:1.5). Curve a, pH 5.0; curve b, pH 7.4; curve c, pH 5.0, but in the absence of liposomes. (Adapted from ref. 10, with permission.)

of 33%. Unlabeled SINV or SFV (10 $\mu M$ viral phospholipid) is mixed with pyrPC-liposomes (2 $\mu M$ phospholipid) in 0.665 ml of HNE. The mixture is acidified to the desired pH with 0.1 $M$ MES–0.2 $M$ acetic acid, as indicated above. The fusion scale may be calibrated such that the initial excimer fluorescence intensity represents 0% fusion, and a 33% decrease of the excimer intensity represents 100% fusion (in which case 100% fusion corresponds to fusion of each liposome with a single virus particle). Again, 0% excimer fluorescence is set by the intensity after complete dilution of the probe in the presence of $C_{12}E_8$ as above.

An example of fusion of unlabeled SINV with pyrPC-containing liposomes at 37° shows a decrease in pyrene excimer fluorescence intensity of approximately 38% after 60 s of acidification to pH 5.0 (Fig. 4, curve a). Because 100% fusion (one liposome fusing with one virus particle) corresponds to a decrease in pyrene excimer fluorescence intensity of 33%, under the conditions of the experiment a liposome fused on average with more than one virus particle. There was no fusion at pH 7.4 (Fig. 4, curve b). Furthermore, pyrPC-labeled liposomes in the absence of virus did not exhibit any change in fluorescence intensity on acidification to pH 5.0 (Fig. 4, curve c), indicating that the decrease in pyrene fluorescence in the presence of virus was the result of dilution of the pyrene probe into the viral membrane.

*Protocol for Fusion of [35]Smethionine-Labeled SINV or SFV Particles with Trypsin-Containing Liposomes.* Transfer of viral nucleocapsid to the liposomal lumen during fusion is measured as degradation of the viral capsid protein by trypsin, initially encapsulated in liposomes. In the fusion reaction, [[35]S]methionine-labeled viral particles ($10^5$ to $10^6$ CPM

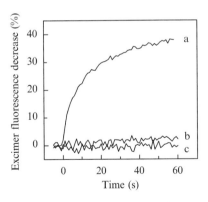

FIG. 4. Fusion of SINV with pyrPC-labeled liposomes. Fusion was monitored on-line at 37° as a decrease in liposomal excimer fluorescence intensity as described in text. The final virus and liposome concentrations were 10 and 2 $\mu M$ (membrane phopholipid), respectively. Liposomes consisted of PC/pyrPC/PE/SPM/Chol (molar ratio, 0.85:0.15:1:1:1.5). Curve a, pH 5.0; curve b, pH 7.4; curve c, pH 5.0 without SINV. (From ref. 10, with permission.)

supplemented with unlabeled virus to a total concentration of 0.5 $\mu M$ viral phospholipid) are mixed with trypsin-containing liposomes (100 $\mu M$ phospholipid) in the presence of trypsin inhibitor (125 $\mu g/ml$; e.g., from Boehringer) in the external medium at 37°. Subsequently, the mixture is acidified to the desired pH, as described for the pyrene assay. After 30 s, the reaction mixture is neutralized by the addition of a pretitrated volume of NaOH, and further incubated for 1 h at 37°. Subsequently, the samples are analyzed by SDS–PAGE. The gels are incubated for 30 min in 1 $M$ sodium salicylate and dried. The protein bands are visualized and quantified by phosphorimaging analysis using Image Quant 3.3 software (Molecular Dynamics, Sunnyvale, CA). Capsid degradation is determined by relating the intensity of the capsid protein band to the intensities of E1 and E2, in the reaction samples and in a virus sample incubated with empty liposomes, the latter providing the initial amount of capsid protein. To exclude the possibility that the amount of trypsin is limiting under the conditions of the experiment, a control experiment is performed in which Triton X-100 (TX-100; Sigma) is added to the reaction mixture in the absence of trypsin inhibitor.

Figure 5 shows an example of capsid degradation on fusion of [35]Smethionine-labeled SINV with an excess of trypsin-containing liposomes in the presence of trypsin inhibitor in the external medium. Incubation of SINV with trypsin-containing liposomes at pH 5.0 resulted in the degradation of a substantial fraction of the capsid protein (Fig. 5, lane a). The extent of capsid degradation was quantified by phosphorimaging analysis. At pH 5.0, approximately 80% of the capsid protein was

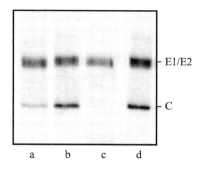

FIG. 5. Fusion of [35S]methionine-labeled SINV with trypsin-containing PC/PE/SPM/Chol liposomes. SINV (containing a trace of [35S]methionine-labeled virus and unlabeled virus supplemented with unlabeled virus to 0.5 $\mu M$ viral phospholipid) was incubated with trypsin-containing liposomes (200 $\mu M$ liposomal phospholipid) in the presence of trypsin inhibitor in the external medium at 37°, and viral capsid protein degradation was determined as described in text. Lane a, trypsin-containing liposomes at pH 5.0; lane b, empty liposomes at pH 5.0; lane c, trypsin-containing liposomes in the presence of TX-100 and in the absence of trypsin inhibitor in the medium at pH 5.0; lane d, trypsin-containing liposomes at pH 7.4. (See also ref. 10.)

degraded. At neutral pH, no degradation of the capsid protein was observed (Fig. 5, lane d). Furthermore, there was no capsid degradation when SINV was incubated at pH 5.0 with empty liposomes (containing no trapped trypsin) (Fig. 5, lane b). Complete degradation of the viral capsid protein was observed when TX-100 was added to the reaction mixture in the absence of trypsin inhibitor (Fig. 5, lane c).

## Selected Studies of Alphavirus Receptor Interaction and Fusion

### Identifying Target Membrane Lipids Involved in Alphavirus Fusion

Virus–liposome systems provide an excellent tool for identification of target membrane components involved in viral fusion, since the lipid composition of liposomes can be varied at will. While early studies employing a content-mixing assay in a liposomal model system had revealed that fusion of SFV was dependent on the presence of cholesterol in the target membrane,[13] we established subsequently that SFV fusion also requires the presence of sphingolipids.[19] More recently, it was found that SINV fusion is similarly dependent on the simultaneous presence of cholesterol and sphingolipids in the target membrane.[10] This observation is illustrated in Fig. 6. Liposomes composed of phosphatidylcholine (PC) from egg yolk, phosphatidylethanolamine (PE) prepared by transphosphatidylation of egg PC, sphingomyelin (SPM) from egg yolk, and cholesterol (Chol) at a molar ratio of 1:1:1:1.5 supported rapid and extensive fusion of SINV at pH 5.0 (Fig. 6, curve a), but liposomes prepared without either cholesterol or SPM were ineffective at pH 5.0 (Fig. 6, curves b and c, respectively).

Figure 7 shows that cholesterol is required for the binding (shaded columns) of SINV to the liposomal membrane, whereas SPM is important for the subsequent fusion process (open columns).[10] When both cholesterol and SPM are present in the bilayer (Fig. 7, column A), extensive binding and extensive fusion were observed. In the absence of SPM (Fig. 7, columns B and D), SINV failed to undergo fusion with the liposomes. In the absence of cholesterol (Fig. 7, columns C and D) there was little binding. Similar results were reported earlier for SFV.[19] Therefore, for both SFV and SIN, it appears that cholesterol is required for virus-liposome binding whereas SPM is involved in the subsequent fusion process.[10,19]

The principal structural features needed in the cholesterol molecule for optimal interaction with phospholipids in bilayers, as established from a wide variety of biophysical studies, are the planar ring structure, the isooctyl side chain, and the $3\beta$-hydroxyl group (for a review, see Bittman[20]). The

[19] J. L. Nieva, R. Bron, J. Corver, and J. Wilschut, *EMBO J.* **13**, 2797 (1994).

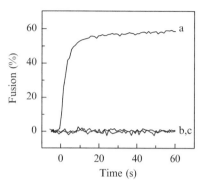

FIG. 6. Fusion of pyrene-labeled SINV with liposomes of different compositions. On-line fusion experiments were performed as described in the legend to Fig. 3. Curve a, PC/PE/SPM/Chol (molar ratio, 1:1:1:1.5) liposomes; curve b, PC/PE/SPM (molar ratio, 1:1:1) liposomes; curve c, PC/PE/Chol (molar ratio, 1:1:1) liposomes. (From ref.10, with permission.)

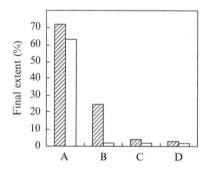

FIG. 7. Effect of target membrane composition on binding and fusion of SINV with liposomes at pH 5.0. The extents of binding (shaded columns) and fusion (open columns) of SINV with liposomes are shown at 60 s postacidification. SINV-liposome binding was determined as described in the legend to Fig. 2. Fusion was determined as described in the legend to Fig. 3. Column A, liposomes prepared with PC/PE/SPM/Chol (molar ratio, 1:1:1:1.5); column B, PC/PE/Chol (molar ratio 1:1:1); column C, PC/PE/SPM (molar ratio, 1:1:1); column D, PC/PE (molar ratio, 1:1). (Adapted from ref.10, with permision.)

$3\beta$-hydroxyl group of cholesterol was essential for fusion of SFV, as revealed by the lipid-mixing and content-mixing assays.[8,9,21,22] However, studies using virus–cell and virus–liposome systems revealed that the planar steroid ring structure, the aliphatic side chain, and the C5–C6 double bond

[20] R. Bittman, in "Subcellular Biochemistry" (R. Bittman, ed.), vol. 28, p. 145. Plenum Press, New York, 1997.

[21] T. Phalen and M. Kielian, J. Cell Biol. 112, 615 (1993).

[22] J. M. Wahlberg, R. Bron, J. Wilschut, and H. Garoff, J. Virol. 66, 7309 (1992).

are not required for SFV fusion.[8] This finding, together with the observations[13] that high amounts of cholesterol (~30 mol%) but much lower amounts of SPM (~2 mol%) are needed[19] for maximal fusion of SFV fusion with liposomes, suggests that complex formation between the sphingolipid and cholesterol in the target membrane is not required for alphavirus fusion.

The structural features required in the sphingolipid molecule to mediate fusion of SFV with liposomes have also been examined in some detail. When the liposomes were prepared with various sphingolipids in addition to cholesterol, PC, and PE, it was found that the phosphocholine head group of sphingomyelin is not essential for fusion of SFV, because glycosphingolipids also supported fusion.[19] An N-acylsphingosine structure, as present in ceramide, is minimally required to support SFV fusion. Ceramide was almost as effective as SPM, whereas sphingosine, which is a free amine, was inactive.[19] The length of the N-acyl chain in ceramide or SPM had no detectable effect on the capacity of the sphingolipid to support fusion of SFV with liposomes.[23] In contrast, the hydroxyl group at C-3 of the sphingoid backbone of ceramide is required, because synthetic ceramide analogs in which this site is modified did not support fusion.[23] Moreover, stereochemical configurations at the C-2 and C-3 positions of the sphingoid backbone play a critical role, because the unnatural D-threo, L-threo, and L-erythro ceramides were found to be fusion-inactive.[24] The position of the double bond in the sphingoid backbone was also critical; relocation of the 4,5-trans double bond of the sphingosine backbone of D-*erythro*-ceramide to a 5,6-trans double bond resulted in the complete loss of SFV fusion with liposomes.[25] In conclusion, studies with liposomes prepared with a wide variety of sphingolipids have demonstrated that the sphingolipid exerts a strong molecular specificity in the membrane fusion process of SFV. On the other hand, SINV fusion with liposomes exhibits a less stringent dependence on the structural features in the sphingolipid molecules. More work is needed to establish the molecular basis for the sphingolipid dependence of fusion of these viruses.

*Lack of Involvement of Sphingolipid/Cholesterol Microdomains in Alphavirus Membrane Fusion*

The dependence of SFV and SINV fusion on the presence of both cholesterol and sphingolipids in the target membrane suggests a possible involvement of sphingolipid–cholesterol microdomains or "lipid rafts,"

[23] J. Corver, L. Moesby, R. K. Erukulla, K. C. Reddy, R. Bittman, and J. Wilschut, *J. Virol.* **69**, 3220 (1995).

[24] L. Moesby, J. Corver, R. K. Erukulla, R. Bittman, and J. Wilschut, *Biochemistry* **34**, 10319 (1995).

[25] L. L. He, H.-S. Byun, J. M. Smit, J. Wilschut, and R. Bittman, *J. Am. Chem. Soc.* **121**, 3807 (1999).

which are present as small entities in many mammalian cellular membranes and participate in a variety of protein-sorting events and cell-signalling cascades.[26,27] Lipid rafts appear to be insoluble in TX-100 at 4° and can be isolated from cells as "detergent-insoluble complexes."

We prepared liposomes containing synthetic (SPMs) and sterols that vary in their ability to form DICs.[28] The analysis of detergent-insoluble formation in Liposomes prepared from PC/PE/SPM/Chol (molar ratio, 1:1:1:1.5) involves treatment of the liposomes with TX-100 at 4°, followed by density gradient flotation analysis. DICs floating to the top of a 45–10–5% density gradient were detected by liquid scintillation counting of the [$^3$H]cholesterol initially incorporated in the liposomes. With PC/PE/SPM/Chol liposomes, as much as 30% of the total amount of cholesterol present in the gradient floated to low density, but when liposomes were prepared without SPM, no radiolabeled cholesterol was detected at the top of the gradient. About 90% of the total amount of phospholipid was SPM in the DIC fractions isolated from liposomes prepared from PC/PE/SPM/Chol, as determined by phospholipid analysis of the lipid spots separated by thin-layer chromatography (TLC). Thus, PC and PE were largely excluded from these complexes.

The flotation assay with [$^3$H]Chol was used to assess the ability of synthetic SPMs with different $N$-acyl chains to form DICs. These synthetic SPMs were incorporated into liposomes with PC, PE, and Chol (PC/PE/SPM/Chol, molar ratio 1:1:1:1.5). A high extent of DIC formation was observed with $N$-stearoyl- and bovine brain SPMs. On the other hand, when $N$-oleoyl- or $N$-linoleoyl-SPM was incorporated into PC/PE/Chol liposomes, no flotation of [$^3$H]Chol occurred.

The ability of various sterols to support formation of DICs was also analyzed.[28] The sterols were incorporated into liposomes (PC/PE/SPM/sterol, molar ratio 1:1:1:1.5). Flotation analysis after extraction with TX-100 indicated that DICs were formed to a greater extent with stigmasterol and ergosterol than with cholesterol, whereas androstenol (which lacks the side chain) failed to support significant DIC formation.

To assess the potential correlation between the fusion of SFV and DIC formation in target liposomes, we used pyrene-labeled virus and liposomes containing different SPMs or sterols. As shown in Fig. 8 (shaded columns A–D), SFV underwent efficient fusion with liposomes containing cholesterol and SPM, irrespective of the degree of unsaturation of the $N$-acyl

[26] K. Simons and E. Ikonen, *Nature* **387,** 569 (1997).

[27] A. Pralle, P. Keller, E.-L. Florin, K. Simons, and J. K. H. Hörber, *J. Cell Biol.* **148,** 997 (2000).

[28] B.-L. Waarts, R. Bittman, and J. Wilschut, *J. Biol. Chem.* **277,** 38141 (2002).

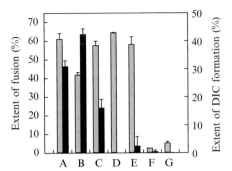

Fig. 8. Lack of correlation between the extent of SFV fusion with liposomes and the extent of detergent-insoluble complex (DIC) formation in liposomal membranes. SFV fusion with liposomes was measured as described in the legend to Fig. 3. Shaded columns: extent of fusion. Solid columns: Extent of detergent-insoluble complex formation in the liposomal membranes, as determined by TX-100 treatment of the liposomes at 4° followed by density gradient flotation analysis. Columns A–D, liposomes were prepared from PC/PE/Chol mixtures plus different SPMs (molar ratio, 1:1:1.5:1): column A, brain SPM; column B, *N*-stearoyl-SPM; column C, *N*-elaidoyl-SPM; column D, *N*-oleoyl-SPM. Column E, PC/PE/brain SPM/androstenol (molar ratio, 1:1:1:1.5) liposomes; column F, PC/PE/brain SPM (molar ratio, 1:1:1) liposomes; column G, PC/PE/Chol (molar ratio, 1:1:1.5) liposomes. (From ref. 28, with permission.)

chain in the SPM. For example, similar kinetics and extents of fusion were observed with liposomes containing either brain SPM (Fig. 8, lane A, shaded column) or *N*-oleoyl SPM (Fig. 8, lane D, shaded column), whereas there was no fusion with liposomes prepared from PC/PE/Chol but lacking SPM (Fig. 8, lane G). Figure 8 also presents the extents of DIC formation in these liposomes (solid columns). Clearly, fusion of SFV was rapid and extensive with liposomes containing synthetic SPMs that lack the ability to assemble into DIC with cholesterol. It is noteworthy that fusion supported by *N*-stearoyl-SPM was slower and 20% less extensive (Fig. 8, lane B) than that supported by brain SPM (Fig. 8, lane A), although *N*-stearoyl-SPM was more effective than brain SPM in promoting formation of detergent-resistant membranes.

SFV fusion with liposomes containing sterols that have different capacities to assemble in DICs, together with PC, PE, and brain SPM, was also studied.[28] The extents as well as the kinetics of fusion of SFV with liposomes containing stigmasterol or androstenol (Fig. 8, lane E) were virtually indistinguishable. As expected, no fusion and no DIC formation were found with liposomes prepared without cholesterol (Fig. 8, lane F). In conclusion, we found that the fusion of SFV is rapid and extensive in the absence or presence of DICs, indicating that SFV fusion does not require

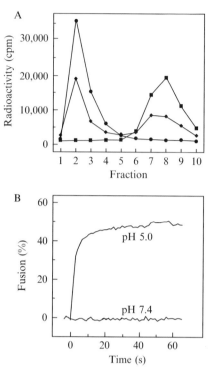

FIG. 9. Binding and fusion of HS-adapted SINV with PC/PE/SPM/Chol liposomes supplemented with various concentrations of HepPE. [$^{35}$S]Methionine-labeled SINV-liposome binding was determined as described in the legend to Fig. 2. Fusion of pyrene-labeled SINV with liposomes was determined as described in the legend to Fig. 3. (A) Binding at 1 h, 4°, pH 7.4: (■) liposomes lacking HepPE; (◆) liposomes supplemented with 0.01 mol% HepPE; (●) liposomes supplemented with 0.02 mol% HepPE. (B) Pyrene-labeled SINV fusion with liposomes supplemented with 0.01 mol% HepPE at pH 5.0; lack of fusion at pH 7.4 is also shown. (Adapted from ref. 30, with permission.)

target membrane sphingolipids to reside in raft-like microdomains with cholesterol.

*Virus–Receptor Interaction and Its Influence on Membrane Fusion Activity*

Besides membrane fusion, virus–receptor interactions can also be studied using the binding and membrane fusion assays described above. It has been reported that SINV readily adapts to interaction with the cellular receptor heparan sulfate (HS), a highly sulfated polysaccharide, and heparin.[16,29] We characterized interaction using lacking HepPE liposomes supplemented with a heparin-conjugated lipid (PE-heparin conjugate, HepPE), which served as a heparin sulfate receptor analog.[30] No binding of [$^{35}$S]methionine-labeled SINV was detected at 4° to liposomes prepared with PC/PE/SPM/Chol lacking HepPE pH 7.4, but efficient binding was observed with liposomes supplemented with remarkably low levels (0.01 and 0.02 mol%) of HepPE (Fig. 9A). Incubation of SINV with soluble heparin inhibited the binding of the virus to HepPE-containing liposomes, indicating that SINV binds specifically to the heparin moiety of HepPE in the target membrane. Despite the efficient interaction of SINV with HepPE-containing liposomes at neutral pH, there was no fusion under these conditions, as revealed by both the pyrene excimer and content-mixing assays. Fusion of pyrene-labeled SINV with liposomes containing 0.01 mol% HepPE was observed, however, at mildly acidic pH (Fig. 9B), consistent with the cell entry of SINV via acidic endosomes. In conclusion, the interaction of SINV with HepPE in the liposomal membrane is efficient and specific, but has no influence on the subsequent fusion process at low pH.

### Concluding Remarks

Liposomes represent an excellent model system for the study of virus-receptor interactions and viral membrane fusion processes. The liposomal model system involving lipid-conjugated heparin, described in this chapter, provides a useful tool for the study of virus-receptor interactions. As many viruses have the capacity to interact with HS, the system may well be adapted to the study of viruses other than alphaviruses. The assay methods for studying fusion of enveloped viruses with liposomes lacking a specific protein or carbohydrate receptor, described in this chapter, have allowed investigators to examine in detail the role of lipids in both the binding and fusion of alphaviruses with target membranes, as well as the pH dependence and the kinetics and extents of those processes. In fact, these assays have already provided a great deal of information about the structural features needed in both the cholesterol and sphingolipid molecules for the low-pH-dependent fusion of SFV and SINV with liposomes. The methods can be readily adapted to the study of fusion of other enveloped viruses, such as flaviviruses[11].

The stringency of the assay methods, including membrane lipid-mixing as well as content-mixing assays, ensures that the virus-liposome fusion

[29] A. P. Byrnes and D. E. Griffin, *J. Virol.* **72**, 7349 (1998).
[30] J. M. Smit, B.-L. Waarts, K. Kimata, W. B. Klimstra, R. Bittman, and J. Wilschut, *J. Virol.* **76**, 10128 (2002).

reactions observed represent genuine processes of membrane merging. Thus, virus-liposome fusion is not a laboratory artifact but rather provides a close reflection of the physiological processes occurring during entry of enveloped viruses to their host cells. Indeed, for several mutants of SINV with varying degrees of infectivity, we have demonstrated that there is a close correlation between infectivity and membrane fusion capacity in the liposomal model system[31].

## Acknowledgments

We acknowledge the support of The Netherlands Organization for Scientific Research (NOW), under the auspices of the Foundation for Chemical Research (CW), and the US National Institutes of Health (Grant HL-16660).

[31] J. M. Smit, W. B. Klimstra, K. D. Ryman, R. Bittman, R. E. Johnston, and J. Wilschut, *J. Virol.*, **75**, 11196 (2001).

## [23]  Pathway of Virus-Induced Membrane Fusion Studied with Liposomes

*By* Stéphane Roche and Yves Gaudin

## Introduction

Enveloped viruses enter cells by protein-mediated membrane fusion. This step is mediated by virally encoded fusogenic glycoproteins. Activation of the fusion capacity involves structural rearrangements of these proteins induced by specific triggers. For viruses fusing at the plasma membrane, the conformational change is triggered by interaction with cellular receptors, whereas for viruses that are internalized and fuse in the endosome, the conformational change is triggered by the low pH of this internal compartment. It is generally assumed that these conformational changes result in the exposure of a fusion peptide or domain, which then interacts with and destabilizes one or both of the participating membranes. In the case of viruses having low pH-induced membrane fusion activity, the use of liposomes as target membranes has given kinetic and quantitative insight into the fusion process. Many assays based on the use of fluorescent probes have proved to be simple and valuable approaches in the elucidation of the pathway and the mechanisms of membrane merging.

The fusion process involves the formation of several intermediate lipidic structures.[1] The lipidic monolayers must thus locally adopt highly positive or negative curvatures, and the stability of these intermediates is dependent on the composition of each monolayer. Lipidic compound such as lysophosphatidylcholine (LPC), inducing positive curvature, or unsaturated fatty acids, inducing negative curvature, will strongly affect the stability of these intermediates and modify the kinetics and the extent of the fusion process.[2,3] The sensitivity of the different intermediates to LPC or unsaturated fatty acids may give information about their topologies. Before discussing in detail the use of these lipids, we briefly describe some assays that have been used to study virus membrane fusion.

Technical Approaches

*Lipid-Mixing Assays*

Many fluorescence assays have been developed to study virus–liposome fusion and have been described previously.[4] One of the most popular assays is based on a resonance energy transfer assay. This assay involves the nonexchangeable probes N-(7-nitro-2,1,3-benzoxadiazol-4-yl)-phosphatidylethanolamine (NBD-PE) and N-(Lissamine rhodamine B sulfonyl)-phosphatidylethanolamine (Rho-PE) as fluorescent donor and acceptor, respectively.[5] When these probes are incorporated in liposomes at a concentration of about 0.6% of the total lipid concentration, there is an efficient fluorescence resonance energy transfer (FRET) from NBD to rhodamine. On fusion of the liposomes with the virus membrane, the fluorophore density decreases, which causes a decrease in transfer efficiency and, hence, an increase in the fluorescence intensity of NBD. This increase can be monitored continuously with a spectrofluorometer and allows an accurate quantitation of the fusion process. This assay has been used successfully in the case of many viruses such as influenza virus,[6] rabies virus,[7] and lymphocytic choriomeningitis virus.[8]

[1] L. V. Chernomordik, M. Kozlov, and J. Zimmerberg, *J. Membr. Biol.* **146,** 1 (1995).
[2] L. V. Chernomordik, E. Leikina, V. Frolov, P. Bronk, and J. Zimmerberg, *J. Cell Biol.* **136,** 81 (1997).
[3] V. I. Razinkov, G. B. Melikyan, R. M. Epand, R. F. Epand, and F. S. Cohen, *J. Gen. Physiol.* **112,** 409 (1998).
[4] D. Hoekstra and K. Klappe, *Methods Enzymol.* **220,** 261 (1993).
[5] D. K. Struck, D. Hoekstra, and R. E. Pagano, *Biochemistry* **20,** 4093 (1981).
[6] T. Stegmann, J. M. White, and A. Helenius, *EMBO J.* **9,** 4231 (1990).
[7] Y. Gaudin, R. W. H. Ruigrok, M. Knossow, and A. Flamand, *J. Virol.* **67,** 1365 (1993).
[8] C. Di Simone, M. A. Zandonatti, and M. Buchmeier, *Virology* **198,** 455 (1994).

A similar assay that uses cholesteryl anthracene-9-carboxylate (CAC) as fluorescent donor and NBD-PE as fluorescent acceptor has also been developed[9] and used to study influenza virus-induced membrane fusion.[10] More recently, Malinin et al.[11] have proposed a new assay for monitoring lipid transfer during membrane fusion. This assay is also based on FRET between the donor 2-(4,4-difluoro-5,7-diphenyl-4-bora-3a,4a-diaza-s-indacene-3-dodecanoyl)-1-hexadecanoyl-sn-glycero-3-phosphocholine (BODIPY 500-PC) and the acceptor 2-(4,4-difluoro-5,7-diphenyl-4-boro-3a,4a-diaza-s-indacene-3-dodecanoyl)-1-hexadecanoyl-sn-glycero-3-phos-phoethanolamine (BODIPY 530-PE). These probes are incorporated in vesicles at concentrations of 0.5 mol% each. The advantage of this assay is that the fluorescent probes are attached to the acyl chain of the lipid instead of the polar head group in the case of Rho-PE and NBD-PE. This renders the fluorescent group less sensitive to environmental properties such as ionic strength or the presence of quenchers.

Other approaches have been developed to incorporate a fluorescent probe in the viral membrane. Among them, two principal assays can be listed. The first method relies on the relief of fluorescence self-quenching of octadecylrhodamine (R18).[12] Octadecyl rhodamine B can be readily inserted into native viral membranes by addition of an ethanolic solution of the probe.[12] For labeling of viral membranes, about 15 nmol of R18 (in 10 $\mu$l of ethanol solution) is injected rapidly into the virus solution (about 1.5 mg of protein in 1 ml) under vigorous vortexing. The mixture is then incubated for 30 min at room temperature in the dark. The noninserted probe is removed on a column of Sephadex G-75. On fusion with nonlabeled target membranes, the decrease in surface density of the fluorophore results in a concomitant increase in fluorescence intensity. Although, in theory, this technique should allow kinetic and quantitative measurements of the fusion process, several aspects of the R18 assay deserve further attention. First, there is evidence that the probe can exchange spontaneously between membranes in the absence of fusion. Second, many results suggest an interaction between the probe and the viral proteins that affects its behavior.[13] For these reasons, another approach has been developed, one that uses viruses that have pyrene-labeled lipids in their membranes because they have been produced in cells cultured in the

[9] P. S. Uster and D. W. Deamer, Biochemistry 24, 1 (1985).
[10] S. A. Wharton, J. J. Skehel, and D. C. Wiley, Virology 149, 27 (1986).
[11] V. S. Malinin, M. E. Haque, and B. Lentz, Biochemistry 40, 8292 (2001).
[12] D. Hoekstra, T. de Boer, K. Klappe, and J. Wilschut, Biochemistry 23, 5675 (1984).
[13] T. Stegmann, P. Schoen, R. Bron, J. Wey, I. Bartoldus, A. Ortiz, J. L. Nieva, and J. Wilschut, Biochemistry 32, 11330 (1993).

presence of pyrene-labeled fatty acids.[14,15] Detailed procedures have already been published[16] but must be adapted depending on the virus studied. The pyrene fluorophore forms fluorescent excimers, representing dimers of a molecule in the ground state and a molecule in the excited state.[17] As the formation of excimers depends on the concentration of pyrene, dilution of viral envelope lipids into the liposomal fusion target leads to a decrease in excimer fluorescence intensity.

The choice of an assay depends on the virus and its properties. Fluorescence variation must reflect a true membrane-merging step and should not arise from an interaction between the fusogenic glycoproteins and the fluorescent probes, nor from unspecific exchange of the probe between membranes. Finally, whatever the assay used, it is important to be able to determine the number of fusion events (i.e., the number of viruses that have fused with liposomes). In general, when liposomes are in excess in the assay, changes in fluorescence intensity are proportional to the number of fusion events.

*Content-Mixing Assays*

For virus–liposome fusion, monitoring content mixing is not possible. However, it has been reported that viral fusion is often leaky, meaning that the aqueous content of the phospholipid vesicles is released into the medium as a by-product of fusion.[18] Furthermore, in experiments in which LPC is used to investigate the fusion pathway, an eventual leakage of liposome membranes induced by this lipid must be checked.[19] Here again, many assays have been developed to monitor leakage and have already been described in detail.[20] Most of them are based on the encapsulation of a fluorophore at self-quenching concentrations. Carboxyfluorescein (sodium salt at a concentration between 50 and 100 m$M$) or tetramethylrhodamines coupled to dextran (TMRD) of different molecular masses (3000 or 10,000 Da at concentrations of 20 mg/ml) have been used successfully.[18,21] Before use, liposomes must be separated from unencapsulated fluorophores by passage through a Sephadex G-75 gel-filtration column. Maximal fluorescence can be established after lysis of the vesicles with a detergent.

[14] R. Pal, Y. Barenholz, and R. R. Wagner, *Biochemistry* **27**, 30 (1988).

[15] R. Bron, J. M. Wahlberg, H. Garoff, and J. Wilschut, *EMBO J.* **12**, 693 (1993).

[16] Y. Barenholz, R. Pal, and R. R. Wagner, *Methods Enzymol.* **220**, 288 (1993).

[17] H. J. Galla and W. Hartmann, *Chem. Phys. Lipids* **27**, 199 (1980).

[18] T. Shangguan, D. Alford, and J. Bentz, *Biochemistry* **35**, 4956 (1996).

[19] Y. Gaudin, *J. Cell Biol.* **150**, 601 (2000).

[20] N. Düzgüneş and J. Wilschut, *Methods Enzymol.* **220**, 3 (1993).

[21] P. Bonnafous and T. Stegmann, *J. Biol. Chem.* **275**, 6160 (2000).

## Hydrophobic Photolabeling

To investigate the interactions between the fusogenic glycoproteins and the membrane, the most direct approach is hydrophobic photolabeling, a technique that makes use of the ability of photoactivatable lipids to covalently modify the polypeptide segments that interact with the hydrophobic core of the membrane.[22–24] The reagents used for such studies are hydrophobic molecules, some of them, such as [$^3$H]PTPC/11 or [$^{125}$I]TID-PC/16, being analogs of phospholipids. When irradiated by ultraviolet (UV) light, these reagents generate in the inner core of the membrane highly reactive and short-lived groups, such as carbene, capable of reacting even with saturated C–H bonds. Such analogs of phospholipids can easily be incorporated in liposomes. These probes have been used largely to study the interaction between fusogenic glycoproteins and the liposomes under various conditions of temperature and pH.[25–28] This approach has allowed the identification of the rhabdovirus fusion peptide.[27] It has also proved to be a powerful tool to study early events preceding membrane merging.[25,26,29]

## Pathway of Virus-Induced Membrane Fusion

### Kinetics of Fusion Studied with Liposomes

Studies on the kinetics of the fusion process revealed striking differences between viruses in the speed of the process and its sensitivity to temperature variation. In the case of tick-borne encephalitis virus (TBEV), the rate of fusion is high: at optimum pH and 37°, the fusion reaction occurs within a few seconds. Lowering of the temperature does not significantly decrease the rate of fusion.[30] On the other hand, for rabies[7] and influenza

[22] J. Brunner, Methods Enzymol. **172**, 628 (1989).

[23] J. Brunner, Annu. Rev. Biochem. **62**, 483 (1993).

[24] Y. Gaudin, R. W. H. Ruigrok, and J. Brunner, J. Gen. Virol. **76**, 1541 (1995).

[25] T. Stegmann, J. M. Delfino, F. Richards, and A. Helenius, J. Biol. Chem. **266**, 18404 (1991).

[26] M. Tsurudome, R. Glück, R. Graf, R. Falchetto, U. Schaller, and J. Brunner, J. Biol. Chem. **267**, 20225 (1992).

[27] P. Durrer, Y. Gaudin, R. W. H. Ruigrok, R. Graf, and J. Brunner, J. Biol. Chem. **270**, 17575 (1995).

[28] P. Durrer, C. Galli, S. Hoenke, C. Corti, R. Gluck, T. Vorherr, and J. Brunner, J. Biol. Chem. **271**, 13417 (1996).

[29] T. Weber, G. Paesold, R. Mischler, G. Semenza, and J. Brunner, J. Biol. Chem. **269**, 18353 (1994).

[30] J. Corver, A. Ortiz, S. L. Allison, J. Schalich, F. X. Heinz, and J. Wilschut, Virology **30**, 269 (2000).

virus,[6] the fusion process is much slower and its rate is highly sensitive to temperature. For influenza virus, complete fusion occurs within 1 min at pH 5.1 and 37° but takes up to 60 min at the same pH and 0°.

These studies have also revealed that fluorescence variation is preceded by a lag phase, indicating that membrane fusion is a multistep process. The duration of the lag time increases with pH up to the pH threshold for fusion and decreases with higher temperature.[6,7,15,30] Many experiments have indicated that this lag time is not due to slow binding of the viruses to the target membrane, the most convincing experiments being the following.

1. For influenza virus (strain X31), at pH 5.1 and 0°, all viruses are bound to the liposomes in less than 15 s, whereas, under the same conditions, the lag time preceding fusion is longer than 4 min.[6]
2. For both influenza and Semliki Forest virus, the duration of the lag time is independent of the liposome concentration.[6,15]
3. For rabies virus, the lag time preceding fusion is no longer detected when small amounts of short-chain (dodecyl and myristoyl) lysophosphatidylcholines (LPCs) are incorporated in the liposomes.[19] This suggests that for this virus, the rate-limiting step responsible for this lag time is the formation of a lipidic structural intermediate (see below).

These observations have suggested that, for these viruses, the lag time is due to postbinding interactions between fusogenic glycoproteins to form the fusion machinery or to the local reorganization of lipids before their diffusion from one membrane to another. It must be kept in mind, however, that the rate-limiting step responsible for the lag time may differ from one virus to another and even for the same virus when different experimental approaches and conditions (pH, temperature, lipid composition) are used.

### Characterization of Fusion Pathway Using Lysophosphatidylcholine and Cis-Unsaturated Fatty Acids

One major question in the fusion field is the structure of the lipidic intermediates formed during the fusion process. It has been proposed that membrane fusion may proceed via the formation of stalk intermediates that are local lipidic connections with negative curvatures between contacting monolayers of fusing membranes. This step would be followed by the formation of a transient hemifusion diaphragm in which a pore would form and enlarge, leading to complete fusion (Fig. 1).[1]

This model can be challenged by altering membrane composition, particularly by addition of lipids that affect the curvature of lipid monolayers

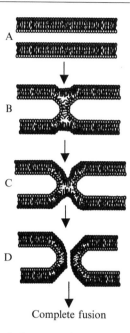

Complete fusion

Fig. 1. Stages of membrane fusion according to the stalk–pore model. Two opposed membranes (A) form a stalk (B). Stalk expansion leads to the formation of a local hemifusion diaphragm (C). Pore formation (D) and its enlargement in the hemifusion diaphragm lead to complete fusion.

(Fig. 2). Of particular interest is the inverted cone-shaped LPC that promotes a spontaneous positive monolayer curvature and the cone-shaped cis-unsaturated fatty acids that favor the hexagonal $H_{II}$ phase. These lipids have been used largely to characterize the fusion pathway in model systems.[2,18,19,31–35]

The effect of these lipids can also be tested in virus–liposome fusion assays. The assay of Struck et al.,[5] using NBD-PE and Rho-PE, is particularly suitable for this kind of study.[18,19] If possible, it seems reasonable to study the effect of various concentrations of LPC with saturated aliphatic chains of various lengths (e.g., from 12 to 18 hydrocarbon groups). Stock

[31] L. V. Chernomordik, S. S. Vogel, A. Sokoloff, H. O. Onaran, E. A. Leikina, and J. Zimmerberg, FEBS Lett. 318, 71 (1993).

[32] S. S. Vogel, E. A. Leikina, and L. V. Chernomordik, J. Biol. Chem. 268, 25764 (1993).

[33] L. Chernomordik, E. Leikina, M. S. Cho, and J. Zimmerberg, J. Virol. 69, 3049 (1995).

[34] L. V. Chernomordik, V. A. Frolov, E. Leikina, P. Bronk, and J. Zimmerberg, J. Cell Biol. 140, 1369 (1998).

[35] P. L. Yeagle, F. T. Smith, J. E. Young, and T. D. Flanagan, Biochemistry 33, 1820 (1994).

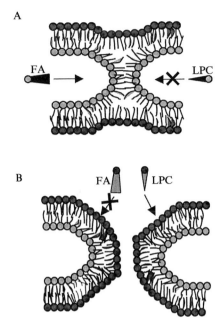

FIG. 2. (A) Formation of the stalk is inhibited (favored) in the presence of LPC [cis-unsaturated fatty acid (FA)] in the outer leaflets of fusing membranes. (B) Pore formation is favored (inhibited) in the presence of LPC (cis-unsaturated fatty acid) in the inner leaflets of fusing membranes.

solutions of lauroyl ($C_{12}$), myristoyl ($C_{14}$), and palmitoyl ($C_{16}$) LPCs can be prepared as aqueous dispersions at concentrations of 5 m$M$ (for $C_{12}$ LPC) or 1 m$M$ (for $C_{14}$ and $C_{16}$ LPC). Stearoyl LPC ($C_{18}$) must be prepared as an ethanolic solution at a concentration of 1 m$M$. In fusion experiments, the lipids can be added directly to the cuvette containing the liposomes in the buffer at the required pH. It is necessary to wait for their partition equilibrium between the aqueous medium and the liposome membrane, before addition of the virus. In fact, LPC incorporation into liposomes induces a slight increase in NBD fluorescence (due to dilution of Rho-PE and NBD-PE and, consequently, a decrease in FRET). This increase reaches a plateau after a few seconds, indicating that the partition equilibrium is also reached (Fig. 3A). It must be kept in mind that LPCs are efficient detergents. Thus, the fluorescence increase should be much smaller than the increase in NBD fluorescence observed on addition of solubilizing concentrations of LPC or Triton X-100. Experimentally at 23°, in buffers prepared from 100 m$M$ citric acid and 200 m$M$ dibasic sodium phosphate solutions, target liposomes containing about 10 $\mu M$ lipids [PC/PE (7:3, w/w) PC/PE/

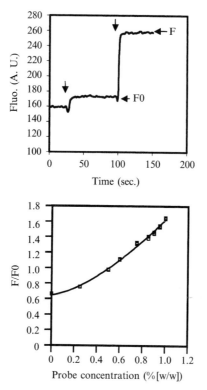

Fig. 3. (A) NBD fluorescence increases on LPC incorporation in liposomes. Palmitoyl LPC was added to a spectrofluorometer cuvette (final concentration, 6 $\mu M$) containing liposomes (10 $\mu M$ phospholipids) (first vertical arrow). Triton X-100 was added (final concentration, 0.8%) (second vertical arrow). The values of $F_0$ (fluorescence at the plateau after addition of $C_{16}$ LPC) and $F$ (fluorescence at the plateau after addition of Triton X-100) can be used to estimate the incorporation of LPC into the liposomes by using the calibration curve obtained in (B). (B) Plot of $F_0/F$ against fluorescent probe concentration. Both NBD-PE and Rho-PE were incorporated in liposomes at the indicated concentration. Fluorescence of the liposomes (10 $\mu M$ total lipids) was measured before ($F_0$) and after ($F$) 0.8% Triton X-100 addition. The experimental point with the abscissa equal to 0 corresponds to liposomes containing only NBD-PE. The experimental data (results of two experiments) were fitted using the equation $y = a + bx^n$. Best fit was obtained with $a = 0.649, b = 0.974, n = 1.505$ ($R^2 = 0.995$). Note that $a$ is the ratio of the quantum yields of NBD-PE in the Triton X-100 micelles and in the liposomes; this value may depend on Triton X-100 purity. This equation can be used to estimate the dilution of the fluorescent probe after addition of LPC.

gangliosides (7:3:1, w/w)] are not solubilized, when incubated with 400 $\mu M$ $C_{12}$ LPC, 40 $\mu M$ $C_{14}$ LPC, 10 $\mu M$ $C_{16}$ LPC, or 8 $\mu M$ $C_{18}$ LPC.[19] However, as this process depends on temperature, buffer, and membrane composition, potential solubilization must be checked in the chosen experimental assay. Potential leakage induced by LPC must also be checked. This can be done by using liposomes containing 40 m$M$ calcein or TMRD (20 mg/ml). At concentrations of LPC that do not solubilize liposomes, a slight leakage is detected with $C_{12}$ and $C_{14}$ LPCs but not with LPCs with longer chains.[19] This is consistent with the results of Fujii et al.,[36] who have shown that transbilayer movement of LPC increases with decreasing acyl chain length.

An inhibitory effect of LPC on the fusion process (as shown for rabies virus in Fig. 4A) is generally considered a strong argument in favor of the stalk hypothesis: the presence of LPC in the outer leaflet would inhibit stalk formation (Fig. 2A).[2,37] However, the mechanism of LPC inhibition of fusion has been a matter of debate. An alternative interpretation is that LPC binds to fusion peptides of viral glycoproteins and inhibits their interaction with target membranes.[38] Thus, the hydrophobic binding of the virions (via their fusion peptides) to target liposomes should be compared in the presence and absence of LPC. This can be done by using liposomes containing radioactive phospholipids, provided that conditions that allow hydrophobic binding without membrane merger have been defined. A mixture of virus and radioactive liposomes incubated under such prefusion conditions (generally in the cold at suboptimal pH) in the presence or in the absence of LPC can be layered onto a cold 30% glycerol cushion at the same pH and then centrifuged. Then, the amount of liposomes associated with the virus in the pellet can be determined by scintillation counting.

The large amount of LPC that is needed to inhibit viral fusion has also raised some questions. In particular, in the case of rabies virus, the total lauroyl LPC concentration giving 50% inhibition is 25 times higher than the lipid concentration in the target liposomes.[19] However, what must be taken into account is not the total concentration of LPC but its membrane concentration, and it is well documented that partitioning of LPCs into membranes increases with their hydrocarbon chain length. The increase in NBD fluorescence due to LPC incorporation into the liposomes can be used to estimate the amount of LPC incorporated. A calibration curve can be made from liposomes containing various dilutions of NBD-PE and Rho-PE. The fluorescence of NBD must be measured before ($F$) and after ($F_0$) addition of 0.8% Triton X-100. The plot of $F_0/F$ versus the fluorescent probe concentration can then be used to evaluate the incorporation of LPC in liposomes (Fig. 3B). From such estimates, it appears that, like in

[36] T. Fujii, A. Tamura, and T. Yamane, *J. Biochem.* **98,** 1221 (1985).
[37] L. V. Chernomordik, E. Leikina, M. M. Kozlov, V. A. Frolov, and J. Zimmerberg, *Mol. Membr. Biol.* **16,** 33 (1999).
[38] S. Günther-Ausborn and T. Stegmann, *Virology* **235,** 201 (1997).

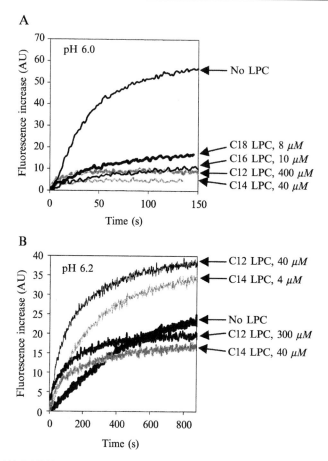

FIG. 4. (A) Inhibition of rabies virus-induced membrane fusion by LPC under optimal fusion conditions (pH 6.0 and 25°). LPC samples with saturated aliphatic chains of various lengths were added to the liposomes (10 $\mu M$ phospholipid) at the indicated concentrations before addition of the virus (about 30 $\mu$g of viral protein) at $t = 0$. (B) Effects of short-chain LPCs on rabies virus-induced membrane fusion under suboptimal fusion conditions (pH 6.2 and 25°). LPC was added to the liposomes (10 $\mu M$ phospholipid) at the indicated concentrations before addition of the virus (about 30 $\mu$g of viral proteins) at $t = 0$. Fusion was assayed with the fluorescent probes NBD-PE and Rho-PE[5] (see Technical Approaches).

a cell–cell fusion system,[2] LPCs with shorter acyl chains require lower membrane concentrations to inhibit fusion. As these LPCs have a more pronounced inverted cone shape, this result is also consistent with the stalk hypothesis.[19]

It is worthwhile to study the effect of LPCs under various pH conditions because the results can be different if the experiments are performed under optimal or suboptimal pH conditions.[2,19] In the case of rabies virus (Fig. 4), under suboptimal fusion conditions (pH 6.2 and 23°), long-chain LPCs ($C_{16}$ and $C_{18}$) inhibit fusion in a dose-dependent manner, whereas short-chain LPCs ($C_{12}$ and $C_{14}$), although inhibiting fusion at high concentrations, increase the number of fusion events at low concentrations. Furthermore, in the presence of lauroyl and myristoyl LPCs, the lag time that precedes the increase in fluorescence is no longer detected and the initial rate of fusion is much faster (Fig. 4B). Considering the fusion model proposed in Fig. 1, the only step that can be favored in the presence of LPC is the formation and/or the enlargement of a fusion pore (Fig. 2B). However, this step is favored only if LPCs have been translocated to the inner leaflet of the membrane, a hypothesis that is consistent with the slight leakage of liposomes observed in the presence of short-chain LPC. The disappearance (or its considerable shortening) of the lag time clearly indicates that it is not due to slow binding of the virions to the liposomes, but rather that the rate-limiting step under suboptimal conditions is the formation and/or the expansion of a fusion pore.[19]

Similar experiments have to be conducted with lipids with the opposite shape (i.e., cone shaped) such as cis-unsaturated fatty acid. These fatty acids can be prepared as ethanolic solutions. The lipids can be added directly to the cuvette containing the liposomes in the buffer at the required pH, before addition of the virus. It should be noted that these lipids are efficiently translocated to the inner leaflet.[39] In the case of rabies virus,[19] oleic acid (OA; up to 10 $\mu M$) is without effect under optimal fusion conditions (pH 5.85, 23°) but inhibits fusion under suboptimal conditions (pH 6.2, 23°). Because OA present in the outer leaflet of the membrane should promote stalk formation (Fig. 2A), this suggested that, at pH 5.85, in this assay, stalk formation is an efficient process that cannot be further enhanced. The inhibition of fusion observed at pH 6.2 is probably due to OA translocated to the inner leaflet, which inhibits pore formation and its enlargement by promoting negative curvature.

Finally, considering the additivity of lipid effects on the spontaneous curvature of lipid monolayers,[40] it may be interesting to see whether the inhibiting effect of LPC can, at least under some conditions, be compensated by addition of OA. This is indeed the case at pH 5.85 for rabies virus, indicating that OA is able to promote stalk formation when this process is not so efficient (due to the presence of LPC in the outer leaflet).[19]

To summarize, a typical protocol to characterize the fusion pathway, using lysophosphatidylcholine and cis-unsaturated fatty acids, is as follows.

[39] K. Broring, C. W. Haest, and B. Deuticke, *Biochim. Biophys. Acta* **986,** 321 (1989).
[40] T. D. Madden and P. R. Cullis, *Biochim. Biophys. Acta* **684,** 149 (1989).

Stock solutions of 5 mM $C_{12}$ LPC, 1 mM $C_{14}$ LPC, and 1 mM $C_{16}$ LPC (obtained from Avanti Polar Lipids, Alabaster, AL) are prepared in water and stock solutions of 1 mM OA (supplied by Sigma-Aldrich, St. Louis, MO) and 1 mM $C_{18}$ LPC (purchased from Avanti Polar Lipids) are prepared in ethanol. Ten micrograms of liposomes (containing 0.6% NBD-PE and 0.6% Rho-PE) is added to a thermostatted spectrofluorometer cuvette containing phosphate–citrate buffer at the required pH (prepared from 100 mM citric acid and 200 mM dibasic sodium phosphate solution). Exogenous lipids from stock solutions (or, in control experiments, the corresponding volume of water or ethanol) are added to the cuvette. After 2 min of incubation, virus (about 30 $\mu$g of viral proteins in 10–20 $\mu$l) is added (the final volume of the mixture in the cuvette is 1 ml) and the increase in NBD fluorescence is monitored continuously. Excitation is at 455 nm, and emission is at 535 nm. The mixture is kept under continuous stirring during the experiment. The ranges of final concentrations of exogenous lipids that must be explored are 0–400 $\mu$M for $C_{12}$ LPC, 0–40 $\mu$M for $C_{14}$ LPC, 0–15 $\mu$M for $C_{16}$ LPC, 0–10 $\mu$M for $C_{18}$ LPC, and 0–20 $\mu$M for OA.

*Trapping Fusion Intermediates*

The exact role of the structural rearrangements of the fusogenic proteins in the formation of the putative intermediates described in the Fig. 1 is unknown. Many groups have captured intermediate states that are partway toward fusion and have tried to characterize the structure of the fusion protein and the lipidic organization at these intermediate stages. Experimentally, these intermediates are captured under conditions that are suboptimal. These conditions are obtained by working at low temperature, at pH near or even slightly above the pH threshold for fusion. In cells expressing fusion proteins, modifications of the density of fusion proteins or of lipid composition have also been used to trap some intermediates.

Intermediates have been identified with influenza HA,[2,34,41,42] vesicular stomatitis virus and rabies glycoprotein G,[19,27,43] HIV gp41,[44–46] and baculovirus GP64.[47] In this quest of fusion intermediates, it could be

[41] C. Schoch, R. Blumenthal, and M. J. Clague, *FEBS Lett.* **311,** 221 (1992).

[42] G. B. Melikyan, R. M. Markosyan, M. G. Roth, and F. S. Cohen, *Mol. Biol. Cell* **11,** 3765 (2000).

[43] C. C. Pak, A. Puri, and R. Blumenthal, *Biochemistry* **36,** 8890 (1997).

[44] T. K. Hart, A. Bruneh, and P. J. Bugelski, *AIDS Res. Hum. Retroviruses* **12,** 1305 (1996).

[45] R. A. Furuta, C. T. Wild, Y. Weng, and C. D. Weiss, *Nat. Struct. Biol.* **5,** 276 (1998).

[46] G. B. Melikyan, R. M. Markosyan, H. Hemmati, M. K. Delmedico, D. M. Lambert, and F. S. Cohen, *J. Cell Biol.* **151,** 413 (2000).

[47] D. H. Kingsley, A. Behbahani, A. Rashtian, G. W. Blissard, and J. Zimmerberg, *Mol. Biol. Cell* **10,** 4191 (1999).

of interest to work with different strains of the same virus. In the case of influenza virus, at pH 5.0 and $0°$, in a first, kinetically distinct step, the viruses bind irreversibly to the target membrane via their fusion peptide. Fusion itself starts only after a lag phase of several minutes with X-31 strain viruses ($H_3N_2$ subtype) but is completely arrested with PR8/34 strain viruses ($H_1N_1$ subtype) starting only after raising the temperature.[26]

Once an intermediate has been trapped, the next step is to identify the stage at which fusion is arrested. First, the nature of the binding of the virions to the target membranes must be analyzed. In general, in virus–liposome fusion assays, this binding is due to the low pH-induced insertion of the fusion peptide into the target membrane. However, as phospholipids or gangliosides (in the case of influenza virus) may play the role of low-affinity receptors for the virus, the insertion of the fusion peptide must be checked. In many cases, this has been done by hydrophobic photolabeling.[26,27] In the case of influenza virus, binding to gangliosides containing liposomes (or to cellular membranes) becomes resistant to neuraminidase on insertion of the fusion peptide in the target membrane but remains sensitive to proteinase K treatment.[2,34] Second, the organization of the lipid in the intermediate must be characterized. In particular, it is interesting to test the effect of LPC on fusion once this intermediate state is reached. In the case of rabies virus, an intermediate stage called RVPC (for rabies virus prefusion complex) has been captured when viruses are incubated for 3 min with liposomes at pH 6.45 (a pH slightly above the pH threshold for fusion) at $0°$. When these complexes are diluted in a cuvette containing phosphate–citrate buffer at fusogenic pH and temperature, lipid mixing is then largely insensitive to the presence of $C_{16}$ and $C_{18}$ LPCs. This is not the case when the virions are preincubated with the liposomes above neutral pH and $0°$ (i.e., when the RVPC is not formed).[19] This indicates that at pH 6.45 and $0°$, fusion is blocked at a stage downstream of the LPC-sensitive stage and, thus, according to the stalk–pore model, downstream stalk formation. Finally, once this intermediate stage is reached at pH 6.45 and $0°$, incubation of the virus–liposome complexes at pH 6.2 and $23°$ results in a disappearance of the lag time preceding the increase of NBD fluorescence (Fig. 5), indicating that most of the events preceding lipid mixing have already occurred. As the rate-limiting step responsible for the lag time at pH 6.2 seems to be due to the formation of a fusion pore and/or its enlargement in a hemifusion diaphragm (see p. 205), this suggests that at pH 6.45 and $0°$, fusion is blocked at a well-advanced stage when a putative hemifusion diaphragm is already destabilized. Note that, if correct, this interpretation indicates a strong restriction of lipid diffusion from one membrane to another, in agreement with the results obtained by hydrophobic photolabeling, showing that only

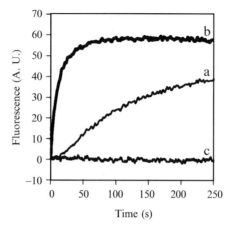

Time (s)

FIG. 5. Trapping rabies virus prefusion complex (RVPC). Curve a: Virions were added directly to the cuvette containing liposomes at pH 6.2 and 25°. Curve b: Virions were first incubated on ice with the liposomes at pH 6.45 for 3 min before dilution at pH 6.2 and 25° in the spectrofluorometer cuvette. Curve c: No fusion is detected when viruses are directly added to the cuvette containing liposomes at pH 6.45 and 0°. Fusion was assayed with the fluorescent probes NBD-PE and Rho-PE[5] (see Technical Approaches).

the fusion peptide, and not the transmembrane domain, was labeled at this stage.[27]

Although to our knowledge this has not been performed in liposome assays, it may also be possible to investigate the structure of the glycoproteins inside these captured intermediates. In particular, it seems attractive to see whether the fusion is still sensitive to a given monoclonal antibody (or a Fab) once the intermediate is reached. For some viruses, peptides inhibiting the structural transition of the fusogenic proteins have also been identified. They have been used to characterize the structure of the fusogenic glycoprotein of paramyxovirus SV5[48] and HIV[46] inside captured intermediates in cellular assays. Such approaches could be adapted to virus–liposome fusion assays provided inhibitory peptides (or any inhibitory molecule binding the fusogenic proteins) have been identified and characterized.

*Evolution and Stability of Captured Intermediates*

Once an intermediate has been captured and characterized, it is interesting to see how it proceeds on various treatments. In the case of rabies virus,[19] low pH is still required for RVPC to proceed to fusion. Furthermore, when RVPC, kept at pH 6.45 and 0°, is treated with proteinase

[48] C. J. Russell, T. S. Jardetzky, and R. A. Lamb, *EMBO J.* **20,** 4024 (2001).

K, this results in a disruption of the complex without inducing complete fusion. Both results indicate that the transition from RVPC to complete fusion is still a glycoprotein-dependent process.

Finally, even if the intermediate is blocked at a stage downstream of the LPC-sensitive stage, it is possible that its lipid organization is still sensitive to prolonged incubations in the presence of LPC. This is the case for some intermediates identified for influenza virus[49] and for RVPC,[19] revealing the dynamic character of these fusion intermediates that can reverse back to previous stages of the fusion pathway. During the incubation of the virus–liposome prefusion complexes with LPC, it is important to check for LPC-induced solubilization of the membranes and, thus, that the disruption of the fusion intermediates is due only to the modification of its lipid organization and not to detergent effects of LPC.

## Conclusions and Perspectives

The mechanism of membrane rearrangements leading to fusion seems to be similar for different viral families. However, depending on the experimental conditions, fusion can be blocked at different stages that are not the same from one system to another. The question arises about the relevance of these captured intermediates in the process of physiological fusion. It must be kept in mind that suboptimal fusion conditions used to trap these intermediates deepen or even create local energy minima that do not exist under optimal fusion conditions. It is thus possible that the captured intermediates are different from the complexes of fusion glycoproteins and lipids that occur when fusion is triggered under physiological conditions. Nevertheless, as mentioned by others,[50] the fact that these intermediates can proceed on to fusion indicates that their molecular configurations are probably not too different from those of intermediates encountered in the physiological fusion pathway. Thus, dissecting the fusion pathway of various viral systems and identifying stable intermediates that can be characterized structurally (using electron microscopy or conformational tools such as monoclonal antibodies) can lead to a detailed cinematic view of the fusion mechanism, showing how proteins induce this complex process.

## Acknowledgments

This work is supported by CNRS (UPR 9053). We thank Rob Ruigrok for careful reading of the manuscript.

[49] E. Leikina and L. V. Chernomordik, *Mol. Biol. Cell* **11**, 2359 (2000).
[50] R. M. Markosyan, G. B. Melikyan, and F. S. Cohen, *Biophys. J.* **80**, 812 (2001).

# [24]　Pore Formation in Target Liposomes by Viral Fusion Proteins

*By* Pierre Bonnafous *and* Toon Stegmann

## Introduction

Virus-induced membrane fusion may occur either between the plasma membrane of a host cell and the viral membrane [e.g., in the case of human immunodeficiency virus (HIV) and paramyxoviruses] or between the endosomal and viral membranes after endocytosis of the virus by the host cell. In the latter case fusion is often triggered by the low pH inside the endosome. The interaction of influenza virus, or its fusion protein hemagglutinin (HA), with erythrocytes at low pH was shown to lead to hemolysis.[1,2] Cells expressing HA, or the fusion proteins of Semliki Forest virus (SFV) or vesicular stomatitis virus (VSV), which also induce fusion at low pH, were shown to induce changes in cell membrane permeability at low pH.[3] We and others have developed methodologies to characterize these membrane interactions in a model system, using liposomes as target membranes.[4-6]

We have focused on HA, a trimeric integral membrane protein that protrudes from the viral membrane as a spike, with a long stem and a globular tip. The monomers of HA consist of two disulfide-linked subunits. The $HA_1$ subunit forms most of the tip. $HA_2$, the smaller subunit, is anchored in the viral membrane at the C terminus. Its N terminus, present in the stem of the protein at neutral pH, consists of a 20-amino acid amphiphilic sequence known as the fusion peptide. Low endosomal pH causes a conformational change, which moves the fusion peptide to the outside of the trimer.[7] Experiments with a photoactivatable phospholipid derivative have shown that on mixing influenza virus with liposomes at low but not at neutral pH, the fusion peptide enters the liposomal membrane.[8,9] The

---

[1] T. Maeda, K. Kawasaki, and S. I. Ohnishi, *Proc. Natl. Acad. Sci. USA* **78,** 4133 (1981).

[2] S. B. Sato, K. Kawasaki, and S. I. Ohnishi, *Proc. Natl. Acad. Sci. USA* **80,** 3153 (1983).

[3] F. Käsermann and C. Kempf, *J. Gen. Virol.* **77,** 3025 (1996).

[4] T. Shangguan, D. Alford, and J. Bentz, *Biochemistry* **35,** 4956 (1996).

[5] R. Jiricek, G. Schwarz, and T. Stegmann, *Biochim. Biophys. Acta* **1330,** 17 (1997).

[6] P. Bonnafous and T. Stegmann, *J. Biol. Chem.* **275,** 6160 (2000).

[7] P. A. Bullough, F. M. Hughson, J. J. Skehel, and D. C. Wiley, *Nature* **371,** 37 (1994).

[8] T. Stegmann, J. M. Delfino, F. M. Richards, and A. Helenius, *J. Biol. Chem.* **266,** 18404 (1991).

[9] M. Tsurudome, R. Glück, R. Graf, R. Falchetto, U. Schaller, and J. Brunner, *J. Biol. Chem.* **267,** 20225 (1992).

METHODS IN ENZYMOLOGY, VOL. 372

interaction of this peptide with the liposomal membrane is most likely the starting point for membrane permeabilization, given the kinetic coincidence of entry and leakage[5,6,8,9] and the fact that synthetic fusion peptides modeled after the N terminus of $HA_2$ readily induce leakage of substances across liposomal membranes.[10–12] Other viruses, such as HIV and paramyxoviruses, have similar N-terminal fusion peptides, but SFV and VSV are thought to have internal fusion peptides, as shown for VSV by photolabeling experiments.[13]

Using liposomes as target membranes, it was found that whole virus, as well as bromelain-released HA (BHA) ectodomains or intact purified HA, induce the complete leakage of small water-soluble fluorescent probes across the liposomal membrane,[6] and that a single BHA trimer suffices to empty the liposome completely.[5] In principle, leakage could result from target membrane rupture, or could occur through pores with a defined geometry. To distinguish between the two, the pore size should first be determined. It was found that BHA induces the leakage of small molecules such as calcein (MW 623) but not that of MW 3000 dextrans, and thus that BHA induces pores of a defined size, rather than leading to membrane rupture. However, pore sizes can evolve over time. Therefore, kinetic measurements of pore size are required. Small pores were found to be formed initially by BHA, intact HA, or virus, and the latter two preparations also induced the leakage of much larger dextrans.[6] If the pores have a defined size and prolonged lifetime, they are not simply lipid defects, given the flexibility and fast diffusion of lipids, but are most likely proteinaceous. This chapter describes the measurement of pore size and lifetime, determination of the number of proteins required to form a pore, and related methodologies.

## Determination of Pore Size

To estimate the size of the pores, liposomes containing self-quenching concentrations of suitable fluorescent molecules of different sizes, such as calcein (estimated size, 13Å[14]) or tetramethylrhodamine coupled to MW 3000 dextran (TMRD-3000, estimated size, 26Å[15]) or MW 10,000 dextran

[10] M. Rafalski, A. Ortiz, A. Rockwell, L. C. van Ginkel, J. D. Lear, W. F. DeGrado, and J. Wilschut, *Biochemistry* **30**, 10211 (1991).

[11] N. Düzgüneş and F. Gambale, *FEBS Lett.* **227**, 110 (1988).

[12] N. Düzgüneş and S. A. Shavnin, *J. Membr. Biol.* **128**, 71 (1992).

[13] P. Durrer, Y. Gaudin, R. W. H. Ruigrok, R. Graf, and J. Brunner, *J. Biol. Chem.* **270**, 17575 (1995).

[14] K. Matsuzaki, S. Yoneyama, O. Murase, and K. Miyajima, *Biochemistry* **35**, 8450 (1996).

[15] C. Nicholson and L. Tao, *Biophys. J.* **65**, 2277 (1993).

(estimated size, $46\mathring{A}^{15}$). Calculations of the size of dextrans are usually based on the assumption that these molecules are perfectly spherical, although the smaller dextrans (MW <10,000) are more likely oblate in solution. Calcein is preferred over carboxyfluorescein for viruses that produce their effect on the membrane at low pH, because the $pK_a$ of the former is lower, so it does not cross the liposomal membrane at the pH of the experiment. Since TMRD cannot be included at high concentrations, because of its solubility limit, liposomes 400 nm in diameter are preferred over 100-nm liposomes to maximize fluorescence intensity. Larger fluorescently labeled dextrans (MW >10,000) cannot be incorporated at self-quenching concentrations easily. The fluorescent probes are encapsulated into multilamellar liposomes by hydrating a lipid film in buffer containing either 75 m$M$ calcein in 85 m$M$ NaCl, 2.5 m$M$ 4-(2-hydroxyethyl)-1-piperazine ethanesulfonic acid (HEPES), 1 m$M$ EDTA, pH 7.4 or by hydrating TMRDs (20 mg/ml) in 145 m$M$ NaCl, 2.5 m$M$ HEPES, 1 m$M$ EDTA, pH 7.4 buffer, giving solutions of about 310 mOsmol/kg. To produce liposomes from these suspensions, they are then frozen and thawed five times, and large unilamellar vesicles are made from the resulting multilamellar vesicles by repeated (5–10 times) extrusion through 0.4-$\mu$m defined pore polycarbonate filters (Whatman Nuclepore, Scarsborough, ME).[16] Free dye is removed subsequently by molecular sieve filtration on a 23 × 1.5 cm Sephadex G-75 column (medium mesh) for calcein and TMRD-3000 or on a Sephadex G-200 column for TMRD-10,000, using 145 m$M$ NaCl, 2.5 m$M$ HEPES, 1 m$M$ EDTA, pH 7.4 buffer for elution (buffer A). The fluorescent fractions around the void volume are pooled. Because extrusion still will have left a small percentage of the liposomes multilamellar, these are then removed by pelleting at 30,000 $g$ for 15 min. To determine the concentration of the liposomes, phospholipid phosphate is determined.[17]

Relief of self-quenching due to dilution on leakage or fusion is measured by monitoring calcein fluorescence at 515 nm, with excitation at 495 nm, or TMRD fluorescence at 580 nm, with excitation at 530 nm (Fig. 1). EDTA (1 m$M$) is required in the buffers if working with calcein. Fluorescence data are normalized by setting the initial fluorescence intensity of TMRD- or calcein-loaded liposomes to zero and the intensity of dequenched fluorophores, obtained after lysis of the liposomes with Triton X-100 [0.5% (v/v), final concentration), to 100%. Triton X-100 will not interfere with the fluorescence of these probes. Using different preparations of the HA of influenza virus, we found striking differences in pore size. Using BHA, a preparation that was made essentially as described by

[16] L. D. Mayer, M. J. Hope, and P. R. Cullis, *Biochim. Biophys. Acta* **858,** 161 (1986).
[17] C. J. F. Böttcher, C. M. Van Gent, and C. Fries, *Anal. Chim. Acta* **24,** 203 (1961).

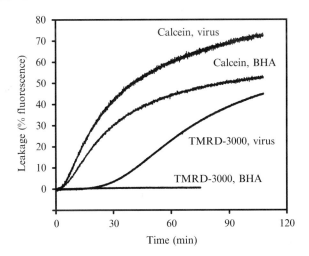

F<small>IG</small>. 1. Leakage of dextran and calcein across liposomal membranes at pH 5.1, 0°. Liposomes containing calcein or TMRD-3000 were incubated with virus or BHA, and dequenching of the fluorescent probes due to leakage was measured as described in text. A fluorescence level of 100% corresponds to fluorescence after lysis of the liposomes with Triton X-100 (0.5%, v/v). Virus and liposome concentrations were 5 $\mu M$ (lipid phosphate). These measurements were carried out in a buffer that is isotonic with the solution inside the liposomes, containing 135 m$M$ NaCl, 15 m$M$ sodium citrate, 10 m$M$ 2-($N$-morpholino) ethanesulfonic acid (MES), 5 m$M$ HEPES, and 1 m$M$ EDTA.

Brand and Skehel,[18] and consisting of individual BHA trimers that lack the membrane anchor, leakage of calcein was induced at the pH of HA-induced fusion, pH 5.1, but no leakage of the TMRDs was seen. Using intact HA or whole virus, all three fluorescent probes were found to cross the liposomal membrane (Fig. 1). Intact HA was produced essentially as described in Bonnafous and Stegmann,[6] by lysis of the virus with a detergent, followed by purification of HA, and dialysis of the detergent; the membrane anchors of the protein aggregate, and six to eight HA trimers, form a rosette. No increase in fluorescence was seen when BHA, HA, or virus was incubated with liposomes at pH 7.4, where HA is not active. Because calcein leakage was induced by BHA, whereas TMRD-3000 was retained in the liposomes, the pores formed by BHA are estimated to be between 13 and 26 Å.

[18] C. M. Brand and J. J. Skehel, *Nat. New Biol.* **238**, 145 (1972).

Determination of Number of Fusion Protein Ectodomains
Involved in Pore Formation by Soluble Ectodomain of HA

To determine the minimum number of BHA trimers per liposome required to form a pore, we assume that the binding of the protein to liposomes is noncooperative, and that, as we had found previously,[5] at low BHA concentrations, one BHA trimer induces leakage in no more than one liposome. The fraction of emptied liposomes can then be calculated as a function of the protein:liposome ratio, and because extruded liposomes have a narrow size distribution, the fraction $P_\mu(m)$ of liposomes with exactly $m$ proteins will likely be Poisson distributed according to

$$P_\mu(m) = e^{-\mu}\mu^m/m! \tag{1}$$

where $\mu$ is the average number of BHA trimers per liposome. The fraction of liposomes with one or more pores can then be calculated, provided the mode of leakage is "all or none"; that is, liposomes are either empty or full, not half-filled with probe after their interaction with the viral fusion protein.

To determine whether this is the case, the self-quenching of calcein incorporated in liposomes at a range of concentrations is first measured. Liposomes containing 75 m$M$ calcein are then prepared as described above, and incubated at low pH with various concentrations of BHA. Subsequently, they are separated from the free probe by gel filtration through a 23 × 1.5 cm Sephadex G-75 column (medium mesh) containing buffer A, and the quenching of the calcein remaining in the liposomes is determined. With BHA, it was found that, even under conditions in which more than 80% of the calcein had leaked out of the liposomes, the quenching in those that still contained the dye was still close to the original quenching, clearly indicating an "all or none" mode of leakage.[5] In this case, the fraction of emptied liposomes $1-E(\mu, m_p)$ can be simulated according to

$$1 - E(\mu, m_p) = 1 - \sum P_\mu(i), \quad i = 0 \ldots m_p - 1 \tag{2}$$

Experimental fluorescence data are then compared with simulations according to Eq. (2). For BHA, best fits were obtained with $m_p = 1$, yielding an exponential function with a mean square residue of 0.009, whereas for $m_p = 2$ the mean square residue was 0.015, and for $m_p = 3$ it was 0.021 (Jiricek et al.[5]). These data suggest that the minimum number of BHA trimers required to form a pore is one. These experiments cannot be done with fusion protein preparation such as rosettes, or with whole virus particles, because liposome binding by these proteins would likely be cooperative.

Evolution of Pores over Time

At 0°, influenza HA-induced fusion is slowed and preceded by a lag phase of several minutes following the low pH-induced conformational change,[19] allowing a precise determination of the kinetics of fusion, and the biochemical changes in the protein that are associated with it. At this temperature, the leakage of calcein induced by virus or intact HA was found to precede the leakage of TMRD-3000 (Fig. 1). An increase in fluorescence was observed with TMRD-3000 or TMRD-10,000, starting about 8 min after the onset of calcein leakage. Such data indicate that large pores develop after small pores, and suggest but do not necessarily indicate an evolution of individual pore sizes. However, because HA not bound to a liposome rapidly loses its fusion activity, HA-induced pores are probably indeed increasing in size. If fusion of virus with liposomes was measured with a lipid-mixing assay under the same circumstances, fusion started after about 10 min. Therefore, kinetically, for influenza virus, the dequenching of the fluorescence of large molecules coincided strikingly with the onset of fusion. A resonance energy transfer assay between fluorescent phospholipids can be used to test whether pore enlargement leads to lysis of the membranes; for HA, it did not.[6]

Transfer of Liposome-Entrapped Fluorescent Probes as a Result of Fusion versus Leakage

With whole virus, the increase in fluorescence observed for the dextrans could be due to transfer of dextrans from the liposomes to the viral interior during fusion, or to leakage to the buffer. To determine the amount of leakage versus fusion, virus is mixed with TMRD-10,000-containing liposomes at a virus:liposome ratio that allows the fusion of close to 100% of the liposomes. After fusion, free TMRD-10,000 is separated from dextran entrapped in liposomes and fusion products by gel filtration on a Sephadex G-200 column (medium mesh, approximately 30 × 1.5 cm). In the experiments shown in Fig. 2, 43% of the dextran was found to elute with the void volume, whereas 53% eluted some 10–15 ml later, indicating that it had leaked from the liposomes. On addition of a detergent to fractions from the column, it was found that the dextran included in the void volume was partially quenched, and was thus present in liposomes or fused membranes, retaining two thirds of the original quenching (Fig. 2). Given that parallel fusion experiments indicated that, at this virus:liposome ratio, 92% of the liposomes participated in fusion, about 40% of the void volume

[19] T. Stegmann, J. M. White, and A. Helenius, *EMBO J.* **9**, 4231 (1990).

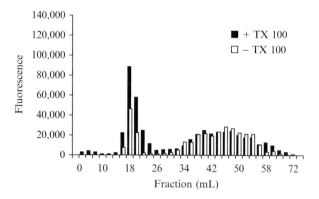

FIG. 2. Transfer of TMRD-10,000 from the liposomal to the viral interior versus leakage. Products of fusion between liposomes containing TMRD-10,000 and virus at pH 5.1, 37° (final concentrations were 16.5 $\mu M$ for liposomes and 50 $\mu M$ for virus, based on lipid phosphate) were separated on a Sephadex G-200 column, and TMRD-10,000 fluorescence was measured before (open columns) and after addition of Triton X-100 (solid columns). The void volume peak on the left contains fluorescent material that remains quenched, as evidenced by the difference in fluorescence before and after addition of Triton.

fluorescence was thus from TMRD present in fusion products, and 3% must have been in unfused liposomes. Therefore, the increase in TMRD fluorescence as shown in Fig. 1 was due to transfer of dextrans from liposomes to virus by fusion as well as by leakage, if whole virus is used. Using calcein as a probe, we do not usually find residual quenched material in fusion products, indicating that small pores permeate the membranes of fusion products. For influenza, these findings are probably strain dependent, as more extensive dextran leakage, and little transfer, was reported by Shangguan et al.[4] for a different strain of virus, in similar experiments under comparable conditions.

## Measuring Time Course of Pore Formation

To characterize the lifetime of the pores produced in liposomes, we have developed an assay that measures the amount of time for which the interior of the liposomes remains accessible to membrane-impermeant molecules.[6] This assay does not distinguish between pores that stay open all the time and those that keep opening and closing rapidly. Asymmetric liposomes are first produced from liposomes symmetrically labeled with 0.6 mol% each of the head group-labeled fluorescent phospholipid analogs $N$-(Lissamine rhodamine B sulfonyl) phosphatidylethanolamine ($N$-Rh-PE) and $N$-(7-nitro-2,1,3-benzoxadiazol-4-yl) phosphatidylethanolamine

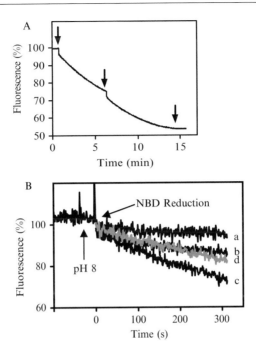

Fɪɢ. 3. Determination of pore lifetimes with dithionite. (A) Reduction of *N*-NBD-PE fluorescence by addition of 20-μl aliquots of dithionite solutions (arrowheads) as described in text, at pH 10. Liposome concentration, 1 m*M*. After complete reduction of *N*-NBD-PE in the outer leaflet of the membrane, further additions no longer affect fluorescence (rightmost arrowhead). (B) Asymmetric liposomes, made as indicated in (A), were incubated with BHA at pH 5.1, 0° for 2 (b), 60 (c), or 120 (d) min or for 15 min without BHA (a) and subsequently brought up to pH 8, after which dithionite was added for reduction.

(*N*-NBD-PE), by reduction of the *N*-NBD-PE present in the outer leaflet to a nonfluorescent product.[20] The presence of *N*-Rh-PE in these membranes allows the measurement of membrane fusion with the same liposome preparation, but it is not necessary for the experiments described below. Twenty-microliter aliquots of a freshly made solution of sodium dithionite (the stock of dithionite will keep for 6 months under argon), 1 *M* in 1 *M* Tris, pH 10, are added to liposomes in the cuvette of a fluorometer in a pH 10 buffer, until the *N*-NBD-PE fluorescence, measured at an excitation wavelength of 465 nm and an emission wavelength of 530 nm, is reduced by 50% (Fig. 3A). Dithionite, a strongly negatively charged molecule at the pH of the experiment, cannot pass membranes and thus

[20] J. C. McIntyre and R. G. Sleight, *Biochemistry* **30,** 11819 (1991).

reduces only the outer leaflet N-NBD-PE. At the point at which further additions of dithionite no longer affect the fluorescence (Fig. 3A), indicating the reduction of all of the outer leaflet N-NBD-PE, the liposomes are separated from the dithionite by column chromatography as described for the fluorescent probes above. The elution buffer in this column can be adjusted to pH 5 or 7.4. The liposomes will remain asymmetric for at least 2 h. Eventually, flip–flop of the N-NBD-PE will occur, translocating the lipid to the outer leaflet, but we have observed that this takes days rather than hours.

These asymmetric liposomes are then resuspended in the cuvette of a fluorometer, and the fusion protein or virus is added. After incubation for the desired amount of time (with BHA between 30 s and 2 h), the solution in the cuvette is brought up to pH 8 with small aliquots of 1 M Tris, pH 10 buffer, and a 20-$\mu$l aliquot of dithionite, as described above, is added. The entry of dithionite into the liposomes via the induced pores will now result in a decrease in fluorescence due to dithionite reduction of the inner leaflet N-NBD-PE (Fig. 3B). With BHA, a decrease in fluorescence is seen only after an incubation of BHA and liposomes at pH 5.1, and not at pH 7.4. No decrease is seen unless dithionite is added, or if inactivated BHA (produced by incubation of BHA at low pH in the absence of liposomes) is added to liposomes at pH 5.1, indicating that dithionite enters through BHA-induced pores. The fluorescence can be normalized between the initial fluorescence after neutralization (no pore formation), corrected for dilution by the introduction of dithionite, and zero fluorescence (complete access to the interior), as indicated in Fig. 3B. The slope of the decrease will depend predominantly on the number of liposomes with pores rather than the pore size, because the entry of dithionite through pores and the ensuing reduction in N-NBD fluorescence of an individual liposome are rapid compared with the development of large pores (Fig. 1). The pores formed by BHA are found to be remarkably stable after neutralization: if dithionite is added for up to 15 min after increasing the pH to 8, it still enters the liposomes. This may be specific for the influenza fusion protein.

### Transmembrane Movement of Fluorescent Phospholipids versus Entry of Dithionite through Pores

In principle, reduction can also result from the transmembrane movement of N-NBD-PE from the inside to the outside leaflet, but neither HA-induced membrane fusion[21] nor HA-rosette-induced pore formation[6] has been found to give rise to such transmembrane movement.

[21] K.-H. Klotz, I. Bartoldus, and T. Stegmann, *J. Biol. Chem.* **271**, 2383 (1996).

Transmembrane movement can be tested with the membrane-impermeant phospholipase D (from *Streptomyces* species), a large molecule that clearly does not pass through the 13- to 26-Å pores induced by BHA. The enzyme efficiently removes the fluorescent head group of *N*-NBD-PE from the lipid, hydrolyzing the bond between phosphate of the phospholipid and the NBD-ethanolamine. The head group does not fluoresce in water, as it is hydrophobic, aggregating in an aqueous environment, resulting in complete self-quenching.[22] If the liposomes remain asymmetric after their interaction with fusion proteins, the enzyme will not reduce NBD fluorescence. Phospholipases D from other sources than *Streptomyces* will not cleave the head group of lipids in intact liposomes efficiently. A positive control for transmembrane movement can be obtained by freezing and thawing asymmetric liposomes several times.

Conclusions

The mechanism and role of conformational changes in viral fusion proteins in fusion is now beginning to be understood, as are many of the lipid intermediates that are important for membrane fusion. We do not know, however, how the energy released in the conformational change is used for the induction of fusion. The only measurable interaction between fusion peptide and membranes occurs just before the formation of lipid intermediates, namely insertion of the fusion peptide into the target membrane.

We and others have found that this interaction induces pores in the target membrane, as fluorescent probes leak out of liposomes,[4-6] and erythrocytes hemolyse.[1,2] As hemoglobin is too small to pass through the small pores formed by BHA, but BHA is capable of inducing hemolysis, the latter is probably a secondary effect due to osmotic swelling of the erythrocytes as buffer continues to flow into these cells through the pores. The methodology in this chapter allows for a simple and efficient characterization of such pores; other techniques that could potentially be used, such as patch-clamp, require more sophisticated equipment and dexterity. These methods should now be used to determine the role of pore formation in fusion. In the case of influenza HA, the small pores that are formed shortly after the conformational change clearly stem from fusion peptide insertion into the target membrane; with virus, >50% of the calcein is found outside the liposomes before fusion even starts. Most likely, pores increase in size gradually by recruiting additional fusion proteins around the pores. Such

[22] A. Chattopadhyay, *Chem. Phys. Lipids* **53**, 1 (1990).

behavior strongly resembles the expansion of the lipidic structure joining two fusing membranes, the "stalk,"[23] suggesting that stalks and pores may be related structures. It is clear that for fusion, two membranes first need to aggregate, and the molecular contact between the membranes should be established, but these steps do not suffice to induce fusion; instead a membrane perturbation, allowing the membrane to locally and temporarily deviate from a bilayer structure, is thought to be necessary. Therefore, although some have argued that the pores formed by HA are excess stalks,[24] fusion peptide insertion could perturb the target membrane precisely in such a way, triggering membrane fusion and accompanied by leakage.

### Acknowledgments

This research was supported by the Région Midi-Pyrénées, the Agence Nationale pour la Recherche sur le SIDA (ANRS), the Fondation pour la Recherche Médicale (FRM), the Association pour la Recherche sur le Cancer (ARC), and the Comité Scientifique SIDACTION of the FRM.

[23] L. Chernomordik, V. E. Leikina, M. M. Kozlov, V. Frolov, and J. Zimmerberg, *Mol. Membr. Biol.* **16,** 33 (1999).
[24] R. Blumenthal and S. J. Morris, *Mol. Membr. Biol.* **1,** 43 (1999).

## [25] Reconstitution of Recombinant Viral Envelope Proteins

*By* CHRISTIAN OKER-BLOM and MATTI VUENTO

### Introduction

The baculovirus expression vector system (BEVS), based mainly on the use of the *Autographa californica* multiple nuclear polyhedrosis virus (*Ac*MNPV) and lepidopteran insect cells, was developed during the 1980s.[1–5] The system has since been used in a wide range of applications.

[1] R. R. Granados, K. A. Lawler, and J. P. Burand, *Intervirology* **16,** 71 (1981).
[2] G. E. Smith, M. D. Summers, and M. J. Fraser, *Mol. Cell. Biol.* **3,** 2156 (1983).
[3] M. D. Summers and G. E. Smith, Texas Agricultural Experiment Station Bulletin No. 1555 (1987).
[4] L. K. Miller, *Biotechnology* **10,** 457 (1988).
[5] L. K. Miller, *Bioessays* **11,** 91 (1989).

Numerous studies have shown that complex proteins of animal, human, and viral origin, which require folding, subunit assembly, and extensive posttranslational modification, can be expressed successfully with this system. Posttranslational modifications such as phosphorylation,[6] N- and O-glycosylation,[7,8] acylation,[9] and α-amidation[10] have been studied extensively with respect to complex recombinant eukaryotic proteins. The combination of high protein yields and authentic protein production has made baculovirus-mediated recombinant protein products an attractive choice for therapeutic use in animals as well as humans. Clearly, the system offers a potential alternative for production and subsequent purification and reconstitution of viral envelope proteins.

Liposomes comprising lipid bilayer membranes enclosing aqueous internal compartments serve as attractive models for reconstitution of a variety of membrane proteins such as viral envelope proteins. The lipid membranes consist of amphipathic lipids that are similar to those found in authentic cell membranes (see Kim[11] for review). Today, preparation of liposomes is a routine procedure and the possibility of engineering liposomes of varying size, as well as incorporating different membrane proteins on their surface,[12–15] has made them useful for many biotechnological applications including vaccine development,[11,16–18] drug delivery,[11,14] and gene delivery.[19,20]

In this chapter, production, purification, and reconstitution of the envelope glycoproteins E1 and E2 of rubella virus (RV) are described.[21–24] The corresponding viral envelope proteins are engineered to display a

[6] B. T. Lin, S. Gruenwald, A. O. Morla, W. H. Lee, and J. Y. Wang, *EMBO J.* **10**, 857 (1991).

[7] P. Kulakosky, M. L. Shuler, and H. A. Wood, *In Vitro Cell. Dev. Biol. Anim.* **34**, 101 (1998).

[8] E. Grabenhorst, B. Hofer, M. Nimtz, V. Jager, and H. S. Conradt, *Eur. J. Biochem.* **215**, 189 (1993).

[9] H. Reverey, M. Veit, E. Ponimaskin, and M. F. Schmidt, *J. Biol. Chem.* **271**, 23607 (1996).

[10] K. Suzuki, H. Shimoi, Y. Iwasaki, T. Kawahara, Y. Matsuura, and Y. Nishikawa, *EMBO J.* **9**, 4259 (1990).

[11] S. Kim, *Drugs* **46**, 618 (1993).

[12] I. Ahmad, M. Longenecker, J. Samuel, and T. M. Allen, *Cancer Res.* **53**, 1484 (1993).

[13] M. Friede, M. H. V. Van Regenmortel, and F. Schuber, *Anal. Biochem.* **211**, 117 (1993).

[14] G. Gregoriadis and A. T. Florence, *Drugs* **45**, 15 (1993).

[15] D. L. Urdal and S. Hakomori, *J. Biol. Chem.* **255**, 10509 (1980).

[16] L. Locascio-Brown, A. L. Plant, R. Chesler, M. Kroll, M. Rubbel, and R. A. Durst, *Clin. Chem.* **39**, 386 (1993).

[17] M. A. Roberts and R. A. Durst, *Anal. Chem.* **67**, 482 (1995).

[18] M. Trudel, F. Nadon, R. Comtois, M. Ravaoarinoro, and P. Payment, *Antiviral Res.* **2**, 347 (1982).

[19] K. Ramani, R. S. Bora, M. Kumar, S. K. Tyagi, and D. P. Sarkar, *FEBS Lett.* **404**, 164 (1997).

[20] C.-Y. Wang and L. Huang, *Proc. Natl. Acad. Sci. USA* **84**, 7851 (1987).

FLAG epitope[25] and a polyhistidine tag[26] at their N and C termini for specific and sensitive immunological identification, as well as for gentle and easy purification. The genes encoding the modified viral proteins are transferred into the baculovirus genome under the transcriptional regulation of the polyhedrin gene promoter and produced in infected *Sf9* insect cell suspension cultures, using the resultant baculovirus expression vectors. The recombinant proteins are purified by immobilized metal ion affinity chromatography and reconstituted into liposomes via their hydrophobic transmembrane anchors.

## Experimental Procedures

### Culture of Sf9 Insect Cells

Lepidopteran *Spodoptera frugiperda* insect cells (*Sf9*: ATCC CRL-1711) normally proliferate every 18–24 h. The cells can be maintained at 27–28° or at room temperature either as monolayer cultures in plastic tissue culture flasks or in suspension, using differently sized Erlenmeyer flasks or bioreactors. For maintenance of *Sf9* cells in suspension, they should be kept at densities of about $1.5 \times 10^6$ viable cells per milliliter of culture medium on orbital shakers (120–135 rpm). Cells can be grown successfully in medium supplemented with up to 10% fetal bovine serum (FBS) such as TNM-FH (Sigma, St. Louis, MO) or in optimized serum-free media such as HyQ SFM (HyClone, Logan, UT) or SF900 II SFM (Invitrogen Life Technologies, Gaithersburg, MD).

### Construction of Recombinant Baculovirus

Transfer of the selected gene into the baculovirus genome is quite time-consuming and laborious, and may even be the biggest obstacle for using the original transfer vectors. The vectors are based on placing the gene of choice under the control of the polyhedrin gene promoter. This is accomplished by homologous recombination by cotransfection of insect cells with wild-type *Ac*MNPV DNA and the transfer vector containing

[21] C. Oker-Blom, N. Kalkkinen, L. Kääriäinen, and R. F. Pettersson, *J. Virol.* **46,** 964 (1983).
[22] T. K. Frey, *Adv. Virus Res.* **44,** 69 (1994).
[23] J. S. Wolinsky, "Virology" (B. N. Fields, D. M. Knipe, and P. M. Howley, eds.). Lippincott-Raven, Baltimore, MD, 1996.
[24] A. Orellana, D. Mottershead, I. van der Linden, K. Keinänen, and C. Oker-Blom, *J. Biotechnol.* **75,** 209 (1999).
[25] T. P. Hopp, K. S. Prockett, V. L. Price, R. T. Libby, C. J. March, D. P. Cerretti, D. L. Urdal, and P. J. Conlon, *Bio/Technology* **6,** 1204 (1988).
[26] E. S. Hemdan, Y.-J. Zhao, E. Sulkowski, and J. Porath, *Proc. Natl. Acad. Sci. USA* **86,** 1811 (1989).

the foreign gene. The procedure relies on several plaque purification steps, in which the recombinant, occlusion-negative viral plaques are selected visually by light microscopy.[3] However, at present several commercially available methods are available that reduce the time required to produce a recombinant baculovirus from over 1 month to about 10 days. One of these is the Bac-to-Bac system (Life Technologies), which is based on site-specific transposition of the foreign gene into the baculovirus genome in *Escherichia coli*.[27] In addition, universal vector systems are available that make expression easier in different host cells, including bacterial, yeast, insect, and mammalian cells (Invitrogen, Carlsbad, CA). The choice of vector system determines strictly the method to be used for generating the recombinant baculovirus.[28]

Here, we use the Bac-to-Bac system for generation of the recombinant baculovirus. The procedure is started by amplifying the gene encoding the chosen viral membrane protein by polymerase chain reaction (PCR), using standard procedures.[29] The PCR primers are designed to encode a FLAG epitope (5' primer) and a polyhistidine tag (3' primer) including appropriate restriction enzyme sites.[30] This strategy places the corresponding tags at the N and C termini of the selected envelope protein. It is essential to omit the stop codon from the 3' primer before the polyhistidine tag. In some cases it may be necessary to place a linker peptide between the C terminus of the protein and the polyhistidine tag. If needed, a protein cleavage site for proteolytic removal of the polyhistidine tag can also be included. Briefly, the amplified gene is cloned into, for example, the pFastBac1 vector, and the corresponding gene is transposed into the baculovirus shuttle vector (bacmid) as described in detail in the user's manual for the Bac-to-Bac system. Then, the composite bacmid DNA from the bacterial cells is isolated and used to transfect *Sf*9 cells with liposomes such as Cellfectin (Life Technologies). The viral supernatant is collected from the transfected insect cells at 3–5 days posttransfection, and the recombinant is amplified.

*Amplification of Recombinant Baculovirus*

For preparation of first stock virus, 100 $\mu$l of the transfection supernatant is transferred into a 75-cm$^2$ tissue culture flask seeded with $6 \times 10^6$ cells in 12 ml of culture medium ($5 \times 10^5$ cells/ml). The cells are

[27] V. A. Luckow, S. C. Lee, G. F. Barry, and P. O. Olins, *J. Virol.* **67**, 4566 (1993).

[28] D. R. O'Reilly, V. A. Luckow, and L. K. Miller, "Baculovirus Expression Vectors: A Laboratory Manual." W. H. Freeman, New York, 1992.

[29] R. K. Saiki, S. Scharf, F. Faloona, K. B. Mullis, G. T. Horn, H. A. Erlich, and N. Arnheim, *Science* **230**, 1350 (1985).

[30] A. Kuusinen, M. Arvola, C. Oker-Blom, and K. Keinänen, *Eur. J. Biochem.* **233**, 720 (1995).

incubated for 48–72 h, and the virus titer is determined by plaque assays or end-point dilution assays.[3,28,31] Especially when initiating "scale-up," it is of the utmost importance to know the virus titers in order to be able to optimize the protein production system. An aliquot of 100 ml of Sf9 cells (1.5 × 10^6 viable cells per milliliter of growth medium) in 250-ml Erlenmeyer flasks is infected with the first viral stock at a multiplicity of infection (MOI) of 0.5 plaque-forming unit (PFU) per cell. The caps of the flasks should be kept loosened to facilitate oxygen exchange. The cells are incubated at 28° on orbital shakers set at 120–135 rpm. The budded progeny of extracellular virus (ECV) is harvested at 48 h postinfection by low-speed centrifugation. Heat-inactivated fetal bovine serum (FBS) is added to the viral supernatant to a concentration of 2%, and the virus is stored at 4°. The titer of the virus should be about 1 × 10^8 PFU/ml.

### Production of Recombinant Viral Envelope Proteins

For production of the recombinant envelope protein, 1000 ml of Sf9 cell culture in a 2.8-L Fernbach flask (Fisher Scientific, Pittsburgh, PA) at a density of 2.5–3.0 × 10^6 viable cells per milliliter is infected with the recombinant virus at an MOI of 2–3 PFU per cell. The infected cell culture is incubated on an orbital shaker (120–135 rpm) for about 72 h. If a convenient assay is available to monitor the expression level of the recombinant protein as a function of time, it could be used to select the appropriate time point for collecting the cells. However, cells should be harvested before lysis, which usually begins to take place after 3 days or more. After incubation, the cells are harvested by low-speed centrifugation (800 rpm for 10 min) and used for purification procedures. If the purification must be postponed, cells can be kept frozen at, for example, −20° and the protein may be purified at a later time point.

### Purification of Envelope Protein by IMAC

Because of the presence of the polyhistidine tag, purification of the recombinant proteins is carried out by immobilized metal ion affinity chromatography (IMAC). After centrifugation, the cell pellet is washed with ice-cold phosphate-buffered saline (PBS, pH 7.4). The cells are carefully resuspended in ice-cold lysis buffer [20 mM HEPES (pH 7.8), 0.3 M NaCl, glycerol (10%, v/v), 10 mM imidazole, 1% Triton X-100 (1%, w/v), and 0.4 mM phenylmethylsulfonyl fluoride (PMSF)] to a concentration of 1–2 × 10^7 cells/ml. After incubation (15–20 min) on ice, the cell lysate is

---

[31] L. A. King and R. D. Possee, Chapman & Hall, London (1992).

clarified by high-speed centrifugation (10,000 g) for 30 min. The supernatant fraction containing the recombinant envelope protein is incubated with 2–4 ml of $Ni^{2+}$-nitrilotriacetate resin (Ni–NTA; Qiagen, Germantown, MD) under constant mixing overnight (4°). The resin is transferred into a column, and the unbound protein is removed from the resin by extensive washing with lysis buffer. The histidine-tagged protein is eluted in a stepwise manner with buffer [20 m$M$ HEPES (pH 7.8), 0.3 $M$ NaCl, glycerol (10%, v/v), Triton X-100 (1%, w/v), glycerol (10%, v/v) pH 7.8] containing 250 and 500 m$M$ imidazole. Fractions of 1 ml are collected and analyzed for the presence of recombinant protein in each fraction. The analysis is preferably carried out by sodium dodecyl sulfate–polyacrylamide gel electrophoresis (SDS–PAGE) and immunoblot analysis, using an appropriate antibody such as M1 directed against the FLAG epitope tag (IBI-Eastman Kodak, New Haven, CT). The purity of the protein in each fraction is determined by SDS–PAGE followed by Coomassie Brilliant Blue staining. The fractions containing the recombinant protein are pooled, and the total amount of protein present in the sample is determined by standard procedures.

*Reconstitution of Purified Envelope Proteins*

Five milligrams of a yolk lipid mixture, phosphatidylcholine–phosphatidylethanolamine–cholesterol at a 10:1:5 molar ratio [Avanti Polar Lipids (Alabaster, AL); Sigma (St. Louis, MO)] is dissolved in 5 ml of 1% (w/v) $n$-octyl $\beta$-D-glucopyranoside in the presence of 50 $\mu$g (protein-to-lipid mass ratio, 1:10) of purified recombinant viral envelope protein in buffer A (150 m$M$ NaCl, 20 m$M$ HEPES, pH 7.4). For example, a Liposomat dialyzer (Dianorm, Munich, Germany) equipped with a cellulose membrane (cutoff limit, 10 kDa) can be used for the dialysis.[32] The concentration of the europium–diethylenetriaminepentanacetic acid (Eu–DTPA) chelate is kept equal (1–50 m$M$) on both sides of the membrane during the first 60 min of the dialysis. The nonencapsulated Eu–DTPA is removed by dialysis against buffer A containing 10 m$M$ DTPA for 3 h. The Eu–DTPA-loaded liposomes are concentrated by ultracentrifugation at 150,000 g for 2 h at 4°. An additional resuspension and ultracentrifugation step is recommended for complete removal of the nonencapsulated Eu–DTPA. The final preparation is resuspended in 500 $\mu$l of buffer A.

[32] A. Orellana, M. L. Laukkanen, and K. Keinänen, *Biochim. Biophys. Acta* **1284**, 29 (1996).

## Ficoll Gradient Centrifugation

One-milliliter volumes of liposome suspension are mixed with 2 ml of 30% (w/v) Ficoll to give a final concentration of 20% Ficoll in buffer A. The corresponding liposome suspensions are transferred into 12-ml ultracentrifuge tubes, and a 6-ml layer of Ficoll (10%, w/v) is added carefully on top of each. Finally, the tubes are filled with 2 ml of buffer A and the tubes are ultracentrifuged, using a Beckman Coulter (Fullerton, CA) SW 41 Ti rotor, at 100,000 g for 30 min at room temperature. About 500-$\mu$l fractions are taken from the top of the gradients, either by using a pipette or a peristaltic pump. Each fraction is analyzed with respect to the presence of the recombinant protein by immunoblot analysis.

## Binding Protocols and Detection

Monitoring of the binding of the Eu–DTPA chelate-loaded liposomes is carried out by time-resolved (TR) fluorescence assays by adopting the DELFIA system (Wallac, Turku, Finland). The E1 and E2 liposomes, as well as the control liposomes (20 $\mu$g/ml), are incubated with antibodies directed against E1 or E2 (1:500)[33,34] or against the FLAG epitope tag M1 (5.8 $\mu$g/ml) (M1; IBI-Eastman Kodak) for 1 h at room temperature in Eppendorf tubes. After incubation with the primary antibody, protein A–Sepharose (100 $\mu$l) is added and the immune complexes are separated by centrifugation (20 sec) at maximum rpm, using an Eppendorf centrifuge (Fig. 1). The pellet should be washed several times with 1 ml of buffer before the addition of 200 $\mu$l of DELFIA enhancement solution containing Triton X-100 and $\beta$-diketone. The samples are finally incubated for 10 min, and the released europium forms a highly fluorescent complex with $\beta$-diketone, which can be measured with an ARCUS 1230 time-resolved fluorometer (Wallac).

Alternatively, the liposomes are bound to streptavidin-coated microtiter plate wells via a biotinylated anti-mouse antibody. The biotinylated antibody (100 $\mu$l, 10 $\mu$g/ml) in 150 m$M$ NaCl, 20 m$M$ HEPES, pH 7.4 (HBS), is added to each well and incubated for 2 h at room temperature. The wells are then blocked with liposomes (20 $\mu$g/ml) without any label, and the antibodies specific for the proteins or the FLAG epitope are added at the concentrations mentioned above. A 100-$\mu$l volume of liposomes is added, and the plates are incubated for 2 h at room temperature. Plates

[33] C. Lindqvist, M. Schmidt, J. Heinola, R. Jaatinen, M. Österblad, A. Salmi, S. Keränen, K. Åkerman, and C. Oker-Blom, J. Clin. Microbiol. 32, 2192 (1994).
[34] J. S. Wolinsky, M. McCarthy, O. Allen-Cannady, W. T. Moore, R. Jin, S. N. Cao, A. Lovett, and D. Simmons, J. Virol. 65, 3986 (1991).

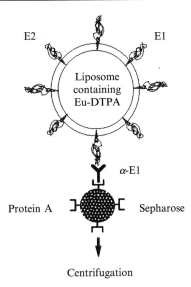

E2                 E1

Liposome
containing
Eu-DTPA

α-E1

Protein A           Sepharose

Centrifugation

FIG. 1. Schematic representation of the procedure used for specific identification and concentration of the europium–DTPA chelate-loaded liposomes containing the rubella virus envelope proteins E1 and E2.

need to be washed carefully between each step, using $1 \times$ PBS (0.9% NaCl, 50 m$M$ phosphate buffer, pH 7.4) and, if available, by using an automatic plate washer. Finally, the DELFIA enhancement solution is added and the TR fluorescence is measured.

Analysis of Results and Considerations

*Production and Purification of Viral Envelope Proteins*

At present, the generation of recombinant baculoviruses is fairly easy. Relatively high production levels of folded viral envelope proteins, such as those presented here, can be achieved with recombinant virus-infected insect cells. However, purification and subsequent reconstitution may in many cases be a restriction. The genetic constructs presented here are therefore designed to enable two consecutive protein purification steps if needed. The carboxy-terminal polyhistidine tag allows for gentle and relatively inexpensive purification procedures based on IMAC, whereas the amino-terminal FLAG epitope enables a perhaps more expensive, but also gentle, immunological purification protocol. Both tags are convenient for immunological detection during the purification procedure. At 3 days postinfection, when the infected cells are harvested, aliquots are exposed to immunoblot analysis using monoclonal antibodies specific for the corresponding tags. This confirms that the N and C termini of the proteins are intact. The recombinant rubella virus envelope proteins E1 and E2 migrate with apparent molecular masses of 58 and 42 kDa, respectively. Because the proteins are harvested from the intracellular space, heterogeneously glycosylated proteins can often be identified. This is observed normally as a smear that appears between the glycosylated and the nonglycosylated forms of the glycoprotein.

Once the protein is characterized, cytoplasmic extracts are prepared and the clarified lysates are exposed to Ni–NTA resin. For scale-up purposes the expanded bed adsorption system from Amersham Biosciences (Uppsala, Sweden) is a potential alternative. Whatever strategy is selected for immobilization of the proteins by IMAC, extensive washing with buffer containing 10–20 m$M$ imidazole is recommended before elution of the protein. Two consecutive elution steps using 250 and 500 m$M$ imidazole are usually sufficient for appropriate elution. The elution fractions are then analyzed with respect to the protein contents, and the protein-containing fractions are pooled. If highly purified protein is needed, a second purification step utilizing the FLAG epitope and M1 antibodies is recommended. However, the loss of protein in each step should be considered.

### Incorporation of E1 and E2 into Liposomes

The presence of recombinant envelope proteins in the europium–DTPA chelate-labeled liposomes can be confirmed by ultracentrifugation and Ficoll flotation gradient analysis followed by fractionation of the gradient. The samples are exposed to immunoblot analysis using either protein-specific antibodies, anti-FLAG, or antibodies directed against the polyhistidine tag (Fig. 2). In parallel, the samples should be studied with respect to the presence of europium–DTPA chelate by TR fluorescence assays (Fig. 3). The pellets of each preparation after ultracentrifugation should contain the recombinant proteins, and only trace amounts should be detectable in the supernatant fractions. Similarly, the fluorescence should be found in the protein-containing pellets when monitored by TR fluorescence. In addition, the presence and purity of the protein contents should be determined by either silver or Coomassie Brilliant Blue staining. If needed, the size of the protein/Eu–DTPA-containing liposomes can be determined by dynamic laser light scattering. When prepared as described here, the liposomes should have a diameter of 170 ± 40 nm.

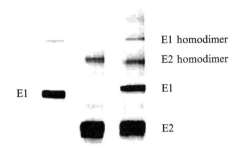

FIG. 2. Identification of the rubella virus-specific recombinant envelope proteins E1 and E2 by immunoblot analysis. The europium–DTPA chelate-loaded liposomes were collected by ultracentrifugation, and the viral proteins were probed with antibodies directed against the N-terminally located FLAG epitope tag.

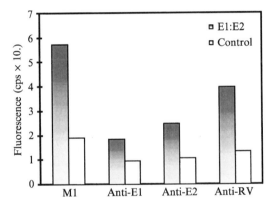

FIG. 3. The europium–DTPA chelate-loaded liposomes collected with the protein-specific antibodies were monitored by time-resolved fluorescence.

## Surface Display of E1 and E2 Proteins on Liposomes

Protein A–Sepharose is able to precipitate the E1- or E2-containing liposomes (20 $\mu$g of phospholipids and 3.6 $\mu$g of protein) in the presence of monoclonal anti-FLAG or antibodies directed against the rubella virus-specific epitopes. Further, the fluorescent label is coprecipitated with these antibodies, showing that the envelope proteins are displayed on the surface of the europium-loaded liposomes.

Also, if a biotinylated anti-mouse antibody is bound to streptavidin-coated microtier plate wells, the liposomes are bound to the wells via the specific antibody, such as anti-FLAG. This is shown by quantification of the amount of europium released from the bound liposomes after addition of the DELFIA enhancement solution containing detergent.

## Concluding Remarks

The one-step purification of proteins appears efficient and incorporation of the two proteins into liposomes seems quantitative, as the pellets, after ultracentrifugation, contain both fluorescent label and RV proteins. Coexpression of the two proteins in insect cells should enable formation of heterodimeric complexes, which would broaden the use of this type of RV-specific virosome, entrapping a lanthanide chelate as a label. In addition to the potential use of this kind of particle in diagnostics, this strategy may also enable studies aimed at solving problems related to basic biology.

## Acknowledgment

The authors acknowledge the support of the European Union (Contract BIO-2CT94-3069).


# [26] Receptor-Activated Binding of Viral Fusion Proteins to Target Membranes

*By* Laurie J. Earp, Lorraine D. Hernandez, Sue E. Delos, and Judith M. White

## Introduction

All known viral membrane fusion proteins are type I integral membrane proteins, and each contains a fusion peptide within its ectodomain. A special feature of viral fusion proteins is that they reside on the virion surface as metastable entities. For the three viral fusion proteins whose metastable structures have been probed by high-resolution X-ray crystallography or electron microscopy—the influenza hemagglutinin (HA),[1] the E glycoprotein of tick-borne encephalitis virus (TBE),[2] and the E1 glycoprotein of Semliki Forest virus (SFV)[3]—the fusion peptide is buried

[1] P. A. Bullough, F. M. Hughson, J. J. Skehel, and D. C. Wiley, *Nature* **371,** 37 (1994).
[2] F. A. Rey, F. X. Heinz, C. Mandl, C. Kunz, and S. C. Harrison, *Nature* **375,** 291 (1995).

within the ectodomain of the protein. Consequently, the ectodomains of these proteins are fully water soluble when released from their transmembrane domains.

Viral fusion proteins have been divided into two classes.[4] The influenza HA is the prototype of class I viral fusion proteins, which employ coiled-coil motifs for fusion.[1,5] The TBE E and SFV E1 glycoproteins are class II viral fusion proteins that employ alternate motifs for fusion.[4] Nonetheless, all three glycoproteins undergo dramatic conformational changes at low pH, a condition that these viruses encounter in cellular endosomes after receptor-mediated endocytosis (Table I). Although the specifics of the conformational changes differ between class I and class II viral fusion proteins, the major consequence of the low pH-induced conformational changes is the same: the formerly water-soluble protein ectodomains now bind tenaciously and hydrophobically to target membranes.[6–10] In all three cases, hydrophobic binding to target membranes has been shown to occur through the fusion peptide. Low pH-induced hydrophobic binding of the glycoproteins to target bilayers is an obligate prerequisite for membrane fusion. Hence, the key common step among viral fusion glycoproteins is conversion of their ectodomains from hydrophilic to hydrophobic entities by virtue of an exposed fusion peptide[11] (Fig. 1, step 1).

Although many enveloped viruses, including serious pathogens such as influenza virus, Ebola virus, and West Nile virus, are activated for fusion by exposure to low pH, many others, including serious pathogens such as the human immunodeficiency virus (HIV), herpesviruses, and respiratory syncytial virus (RSV), fuse with cells at neutral pH.[12,13] The current concept is that instead of being activated by low pH, these viruses are activated by specific interaction(s) between the viral glycoprotein(s) and their cognate host cell receptor(s) (Table I). This concept first emerged from studies on

[3] J. Lescar, A. Roussel, M. W. Wien, J. Navaza, S. D. Fuller, G. Wengler, and F. A. Rey, *Cell* **105**, 137 (2001).

[4] F. X. Heinz and S. L. Allison, *Curr. Opin. Microbiol.* **4**, 450 (2001).

[5] C. M. Carr and P. S. Kim, *Cell* **73**, 823 (1993).

[6] C. Harter, P. James, T. Bachi, G. Semenza, and J. Brunner, *J Biol. Chem.* **264**, 6459 (1989).

[7] M. R. Klimjack, S. Jeffrey, and M. Kielian, *J. Virol.* **68**, 6940 (1994).

[8] J. J. Skehel, P. M. Bayley, E. B. Brown, S. R. Martin, M. D. Waterfield, J. M. White, I. A. Wilson, and D. C. Wiley, *Proc. Natl. Acad. Sci. USA* **79**, 968 (1982).

[9] R. W. Doms, A. Helenius, and J. White, *J. Biol. Chem.* **260**, 2973 (1985).

[10] K. Stiasny, S. L. Allison, J. Schalich, and F. X. Heinz, *J. Virol.* **76**, 3784 (2002).

[11] J. M. White and I. A. Wilson, *J. Cell Biol.* **105**, 2887 (1987).

[12] E. Hunter, *in* "Retroviruses" (J. M. Coffin, ed.), p. 71. Cold Spring Harbor Laboratory Press, Cold Spring Harbor, NY, 1997.

[13] L. D. Hernandez, L. R. Hoffman, T. G. Wolfsberg, and J. M. White, *Annu. Rev. Cell Dev. Biol.* **12**, 627 (1996).

TABLE I

PROPERTIES OF ENVELOPED VIRUS FUSION PROTEINS

| Enveloped virus family | Fusion pH | No. of viral proteins needed for fusion | Coiled-coil?/ Type I or type II | No. of receptors | Fusion protein binds liposomes (+/−R) | Virus binds liposomes (+/−R) |
|---|---|---|---|---|---|---|
| Orthomyxovirus | Low | 1 (HA) | Yes/I | 1 | Yes (−R)[a] | Yes (−R)[b] |
| Togavirus | Low | 2 (SFV; E1/E2) | No/II | ND | Yes (−R)[c] | Yes (−R)[d] |
| Flavivirus | Low | 1 (E) | No/II | ND | ND | Yes (−R)[e] |
| Rhabdovirus | Low | 1 (VSV: G) | No/? | ND | ND | Yes (−R)[f] |
| Bunyavirus | Low | 2 (G1/G2)[g] | ND | ND | ND | ND |
| Arenavirus | Low | ND[h] | ND | 1 | ND | ND |
| Filovirus | Low | 1 (Ebola; GP) | Yes/I | ND | ND | ND |
| Retrovirus | Neutral?[i] | 1 (Env) | Yes/I | 1 (ASLV); 2 (HIV; most strains) | Yes (+R)[j] | Yes (+R)[j] |
| Paramyxovirus | Neutral | 1 or 2 (F and in some cases HN) | Yes/I | 1 | ND | Yes (+R)[k] |
| Herpesvirus | Neutral | 4 (gB, gD, gH, gL) | ND | 1 or 2 | ND | ND |
| Coronavirus | Neutral | 2 (S1/S2) | ND | 1 | ND | ND |

| | | | | | | |
|---|---|---|---|---|---|---|
| Poxvirus | Neutral | ND | ND | ND | ND | ND |
| Hepadnavirus | Neutral | ND (S)[l] | ND | ND | ND | Yes[m] |
| Iridovirus | ND | ND | ND | ND | ND | ND |

*Abbreviations:* ND, not determined; R, receptor.

[a] C. Harter, P. James, T. Bachi, G. Semenza, and J. Brunner, *J. Biol. Chem.* **264**, 6459 (1989).

[b] T. Stegmann, F. P. Booy, and J. Wilschut, *J. Biol. Chem.* **262**, 17744 (1987).

[c] M. R. Klimjack, S. Jeffrey, and M. Kielian, *J. Virol.* **68**, 6940 (1994).

[d] R. Bron, J. M. Wahlberg, H. Garoff, and J. Wilschut, *EMBO J.* **12**, 693 (1993).

[e] S. L. Allison, J. Schalich, K. Stiasny, C. W. Mandl, and F. X. Heinz, *J. Virol.* **75**, 4268 (2001).

[f] S. Yamada and S. Ohnishi, *Biochemistry* **25**, 3708 (1986).

[g] The exact role of G2 in fusion is still in debate.

[h] Arenaviruses have two surface glycoproteins with two candidate fusion peptides in GP2 [S. E. Glushakova, I. S. Lukashevich, and L. A. Baratova, *FEBS Lett.* **269**, 145 (1990)]; neither has been characterized.

[i] Previous evidence supports ASLV entry at neutral pH; however, one study (Mothes *et al.*[17]) proposes a two-step entry mechanism requiring receptor binding at neutral pH, followed by exposure to low pH. Most retroviruses (except MMTV) have been shown to fuse at neutral pH.

[j] L. J. Earp, S. E. Delos, R. C. Netter, P. Bates, and J. M. White, *J. Virol.* **77**, 3058 (2003).

[k] Y. S. Tsao and L. Huang, *Biochemistry* **25**, 3971 (1986).

[l] The S protein contains a stretch of amino acids predicted to be a fusion peptide, but has not been further characterized.

[m] E. V. Grgacic and H. Schaller, *J. Virol.* **74**, 5116 (2000).

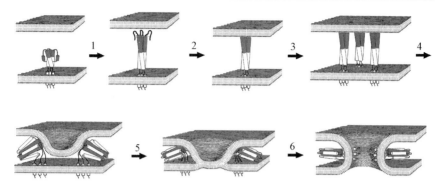

Fɪɢ. 1. Proposed model for ASLV-A fusion with target membranes. In the native membrane, EnvA exists in a trimeric, metastable form in which the fusion peptides are buried. The SU domains (not shown for clarity) are thought to act as a clamp, which maintains the TM domains in the metastable state. After engaging its receptor, Tva, at $T \geq 22°$ (step 1), EnvA undergoes a conformational change in which the fusion peptides are exposed and can penetrate the target membrane. After penetration of the target membrane by the fusion peptides (step 2), trimers may cluster (step 3). Further conformational changes occur in which EnvA begins (step 4) to form, and then forms (step 5) a six-helix bundle, which mediates lipid mixing of the outer leaflets of the target and viral membranes. Action of the fusion peptides and transmembrane domains on the hemifusion diaphragm would then open the fusion pore (step 6). This model is based on similar models for the influenza HA. In the case of HA, the trigger for step 1 is exposure to low pH.

a simple model alpharetrovirus, the avian sarcoma/leukosis virus (ASLV, formerly known as Rous sarcoma virus). The receptor for this virus is a small soluble type I integral membrane glycoprotein called Tva, an acronym denoting that it is the receptor for the avian tumor virus, subtype A.[14] In 1997, it was shown that after binding to the glycoprotein at 4°, a small (83-amino acid) soluble form of the Tva ectodomain, sTva, could induce a soluble trimeric form of the ASLV subtype A (ASLV-A) envelope glycoprotein ectodomain (EnvA-PI) to bind hydrophobically to a target membrane at neutral pH. Binding to target membranes occurred at $T > 22°$ and was dependent on the wild-type fusion peptide sequence.[15] Similar results were obtained subsequently by another laboratory.[16] As yet, and most likely because of the relative simplicity of the avian retrovirus system (one viral glycoprotein and one simple host cell receptor; Table I), ASLV EnvA is the only viral glycoprotein that has been proven

[14] P. Bates, L. Rong, H. E. Varmus, J. A. Young, and L. B. Crittenden, *J. Virol.* **72,** 2505 (1998).

[15] L. D. Hernandez, R. J. Peters, S. E. Delos, J. A. Young, D. A. Agard, and J. M. White, *J. Cell Biol.* **139,** 1455 (1997).

[16] R. L. Damico, J. Crane, and P. Bates, *Proc. Natl. Acad. Sci. USA* **95,** 2580 (1998).

to bind to target membranes after a temperature-dependent interaction with its host cell receptor.[15–16a]

Work suggested that ASLV Env is not fully activated for fusion by receptor binding at neutral pH, but rather requires receptor binding at neutral pH (and elevated temperature) followed by exposure to low pH.[17] We have found, however, that neither hydrophobic binding of the subtype A envelope glycoprotein to target membranes, binding of ASLV-A virus particles to target membranes, nor lipid mixing of the viral (ASLV-A) and cellular membranes requires exposure to low pH.[15–16a] Therefore, the critical step of activation of the ASLV-A viral fusion protein occurs at neutral pH. In the remaining sections of this chapter, we describe three assays that we have developed to monitor receptor-induced association of ASLV EnvA with target bilayers: the original assay for monitoring binding of the EnvA ectodomain (EnvA-PI) to target membranes (liposomes), a modified and miniaturized EnvA-PI-liposome binding assay, and an assay to measure binding of intact ASLV-A virus particles to target membranes.

## Methods

### Preparation of Reagents

*EnvA-PI.* EnvA-PI is prepared as described previously.[15] Briefly, a stable NIH 3T3 cell line expressing glycosylphosphatidylinositol (GPI)-anchored EnvA [which contains the GPI addition signal from decay-accelerating factor (DAF) in place of the transmembrane domain (TM) and cytoplasmic tail; Gilbert *et al.*[18]] is maintained in Dulbecco's modified Eagle's medium (DMEM)–10% bovine calf serum (supplemental) (HyClone, Logan, UT), 1× each of glutamine, penicillin–streptomycin, and sodium pyruvate, and G418 [0.33 g/500 ml (0.25 g active units); GIBCO, Grand Island, NY]. Sixteen to 18 h before harvest, GPI-anchored EnvA cells are treated with 5 m*M* sodium butyrate to induce EnvA expression. For EnvA-PI-liposome binding procedure I, cells are labeled with NHS-LC biotin. Unlabeled cells are used for EnvA-PI-liposome binding procedure II. GPI-linked proteins are released with phosphatidyl-inositol-specific phospholipase C (PI-PLC) as described previously.[18] Two cocktails of protease inhibitors are also included in the reaction: (1) phenylmethylsulfonyl fluoride (PMSF, 8.7 mg/ml) and pepstatin A

[16a] L. J. Earp, S. E. Delos, R. C. Netter, P. Bates, and J. M. White, *J. Virol.* **77**, 3058 (2003).

[17] W. Mothes, A. L. Boerger, S. Narayan, J. M. Cunningham, and J. A. Young, *Cell* **103**, 679 (2000).

[18] J. M. Gilbert, L. D. Hernandez, T. Chernov-Rogan, and J. M. White, *J. Virol.* **67**, 6889 (1993).

(0.5 mg/ml) and (2) leupeptin (1 mg/ml), aprotinin (2 mg/ml), antipain (5 mg/ml), benzamidine (25 mg/ml), soybean trypsin inhibitor (STI, 5 mg/ml), and iodoactamide (50 mg/ml).

*Virus.* ASLV-A is produced from DF-1 cells that are chronically infected with RCASBP(A)GFP.[19] RCASBP(A)GFP/DF-1 cells are cultured in DMEM–10% FBS and 1× antibiotic–antimycotic. Before harvesting virus, the medium is replaced. After 18 h, the medium is collected, clarified at 1250*g* for 15 min, and concentrated by centrifugation (2.5 h, 4°, at 82,700 *g*) through 5mL 15% sucrose in an SW 28 rotor. Viral pellets are resuspended in 200 $\mu$l of 20 m*M* morpholineethanesulforic acid (MES), 20 m*M* HEPES, pH 7.4 (MES–HEPES) overnight at 4°. A fresh preparation of virus is used for each experiment.

*Liposomes.* L-$\alpha$-Phosphatidylcholine (egg) (PC; Avanti Polar Lipids, Alabaster, AL) and cholesterol (chol; Sigma, St. Louis, MO) are stored as described previously.[15] For EnvA-PI-liposome binding procedure I,[15] liposomes are prepared as described.[20] They are composed of phosphatidylcholine and cholesterol at a 2:1 molar ratio. More recently, lipids (PC:chol molar ratio, 2:1) are mixed and dried under a stream of nitrogen in a glass test tube. After lyophilization overnight, liposomes are created by the addition of 1 ml of MES–HEPES with vortexing and sonication in a sonicator bath. Liposomes are extruded subsequently through a 0.1-$\mu$m pore size filter in an Avanti Mini-Extruder, stored at 4°, and used within 1 week of preparation.

*EnvA-PI-Liposome Binding Procedure I*

Biotinylated EnvA-PI and soluble receptor, either sTva[21] or sTva47,[22] are incubated on ice for 15 min in a final volume of 60 $\mu$l. After addition of 40 $\mu$l of liposomes, samples are maintained at 4° or incubated at 37° for 10–30 min. In a clear polycarbonate ultracentrifuge tube, samples are brought to a final concentration of 50% sucrose by the addition of 300 $\mu$l of 67% sucrose. The sample is overlaid by a sucrose step gradient composed of 300 $\mu$l of 25% sucrose and 200 $\mu$l of 10% sucrose. Samples are centrifuged at 200,000 *g* for 3 h at 4° (Fig. 2). Fractions (9 × 100 $\mu$l) are collected from the top of the gradient, and samples are immunoprecipitated with anti-DAF, which recognizes the GPI anchor addition signal from decay-accelerating factor (DAF) that is present at the C-terminal end of

[19] M. J. Federspiel and S. H. Hughes, *Methods Cell Biol.* **52,** 179 (1997).

[20] J. White and A. Helenius, *Proc. Natl. Acad. Sci. USA* **77,** 3273 (1980).

[21] J. W. Balliet, J. Berson, C. M. D'Cruz, J. Huang, J. Crane, J. M. Gilbert, and P. Bates, *J. Virol.* **73,** 3054 (1999).

[22] M. Tonelli, R. J. Peters, T. L. James, and D. A. Agard, *FEBS Lett.* **509,** 161 (2001).

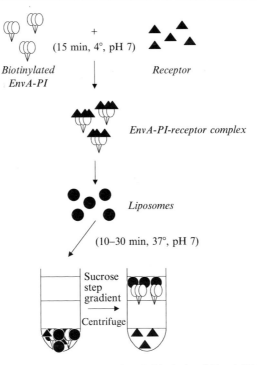

Fig. 2. EnvA-PI-liposome binding procedure I. Biotinylated EnvA-PI is incubated with soluble receptor (sTva47) on ice for 15 min. Liposomes are added and samples are shifted to 37° for 10–30 min. Samples are overlaid with a sucrose step gradient, centrifuged, and fractionated as described in Hernandez *et al.*[15]

EnvA-PI. This is done in a final volume of 400 $\mu$l (after the addition of lysis buffer). Samples are subjected to sodium dodecyl sulfate (SDS)–12.5% polyacrylamide gel electrophoresis (PAGE), transferred to nitrocellulose, and probed with streptavidin–horseradish peroxidase (HRP). Blots are developed by enhanced chemiluminescence (ECL).[15]

*Results.* When incubated with soluble receptor at 37°, EnvA-PI colocalizes with liposomes in the top three fractions of the gradient. EnvA-PI that has not been exposed to soluble receptor remains at the bottom of the gradient. See, for example, Fig. 1 in Hernandez *et al.*[15]

### EnvA-PI-Liposome Binding Procedure II

Twelve microliters of EnvA-PI and 0.29 $\mu$g of sTva47 are incubated in the bottom of a TLA100 centrifuge tube (7 × 20 mm) (343775; Beckman Coulter, Fullerton, CA) on ice for 15 min. Eight microliters of liposomes are added, and samples are either kept at 4° or incubated at 37° for 10 min. Tubes are then placed on ice, and the sample is brought to a final concentration of 50% sucrose by adding 60 $\mu$l of 67% sucrose. A step gradient is then layered on top of the sample with the addition of 60 $\mu$l of 25% sucrose and 60 $\mu$l of 10% sucrose. All sucrose solutions are made weight per weight in PBS (without $Ca^{2+}$ or $Mg^{2+}$). Samples are centrifuged at 197,000 $g$ for 1 h at 4° in a TLA100 rotor (Fig. 3A). Six 30-$\mu$l fractions are collected from the top of the gradient, boiled in 30 $\mu$l of sample buffer containing dithiothreitol (DTT), and subjected to SDS–12% PAGE. After transferring to nitrocellulose, fractions are probed for the location of EnvA-PI, using an antibody to the N terminus of the gp37 subunit (anti-Ngp37). Blots are developed by ECL.

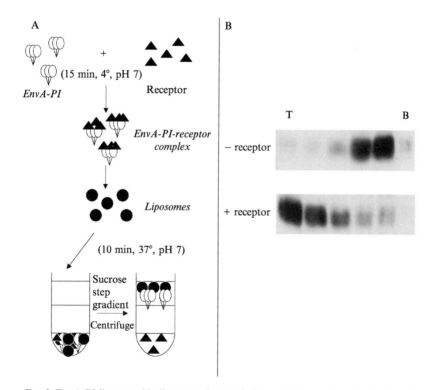

FIG. 3. EnvA-PI-liposome binding procedure II. (A) EnvA-PI is incubated with sTva47 on ice for 15 min. Liposomes are added, and samples are incubated at 37° for 10 min. Samples are overlaid with sucrose step gradients and centrifuged, fractionated, and subjected to SDS–PAGE, as described in Methods. (B) In the absence of receptor at pH 7 and 37°, EnvA-PI stays in the bottom of the sucrose gradient *(top panel)*. When incubated with receptor at pH 7 and 37°, EnvA-PI floats to the top fractions of the gradient with liposomes *(bottom panel)*.

*Results.* When exposed to sTva47 at 37°, EnvA-PI floats to the top of the gradient with liposomes, similar to the results seen with procedure I.[15] In the absence of sTva47, EnvA-PI remains in the bottom of the gradient (Fig. 3B), while the liposomes float to the top fraction (data not shown).

## Virus–Liposome Binding Assay

In an Eppendorf tube, 25 $\mu$l of RCASBP(A)GFP and 0.5 $\mu$g of sTva47 are mixed on ice and incubated for 15 min. Liposomes (25 $\mu$l) are added, and samples are kept at 4° or incubated at 37° for 30 min and then placed on ice. Sucrose step gradients are created in Beckman Coulter polycarbonate centrifuge tubes (7 × 20 mm): 25 $\mu$l of 50% sucrose is overlaid with 75 $\mu$l of 25% sucrose and 50 $\mu$l of 20% sucrose. A 50 $\mu$l sample is layered on top of the gradient, and samples are centrifuged at 197,000 $g$ for 1 h at 4° in a TLA100 rotor. Three fractions are collected from the top of each gradient such that each fraction contains an interface between sucrose concentrations (fraction 1, 75 $\mu$l; fraction 2, 50 $\mu$l; fraction 3, 65 $\mu$l) (Fig. 4A). Samples are boiled in 3× sample buffer containing DTT (fraction 1, 37.5 $\mu$l; fraction 2, 25 $\mu$l; fraction 3, 32.5 $\mu$l). Of each fraction, 25% is loaded onto an SDS–12% polyacrylamide gel and subjected to electrophoresis. The gel is then transferred to nitrocellulose, and probed with a polyclonal antibody to the matrix (MA) protein (obtained from V. Vogt, Cornell University, Ithaca, NY). Blots are developed by ECL.

*Results.* When ASLV-A particles are mixed with liposomes and incubated at 37° in the absence of sTva47, most of the ASLV-A sediments to the bottom (third) fraction of the gradient, on top of the 50% sucrose cushion (Fig. 4B, top). In contrast, if ASLV-A is preincubated with sTva47 (at 4°), before the incubation with liposomes at 37°, the virus particles are largely retained in the top two fractions of the gradient (Fig. 4B, bottom). An initial study employing this assay has been conducted.[16a]

## Conclusions

The assays described here have proven their utility for studying receptor (Tva)-induced activation of ASLV EnvA.[15,16a] They should also be useful for studying other receptor-activated viral fusion proteins. When one viral glycoprotein and one "simple" host cell receptor are involved, it should be possible to develop assays directly analogous to those described above for studying Tva-induced binding of the EnvA ectodomain (EnvA-PI) to target membranes. A general prerequisite for a fusion protein/target membrane binding assay is a soluble and correctly oligomeric form of the viral fusion protein ectodomain. The simplest host cell

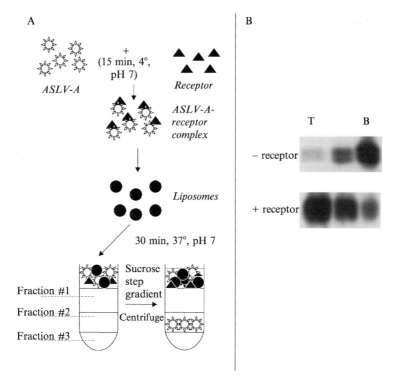

Fɪɢ. 4. Virus–liposome binding procedure. (A) ASLV-A is incubated with or without sTva47 on ice for 15 min. Liposomes are added, and samples are incubated at 37° for 30 min. Samples are placed on top of a sucrose step gradient and centrifuged, fractionated, and processed as described in Methods. (B) In the absence of receptor at pH 7 and 37°, ASLV-A pellets to the bottom of the gradient *(top panel)*. When incubated with receptor at pH 7 and 37°, ASLV-A is largely retained in the top fraction of the gradient *(bottom panel)*.

receptors that would be amenable to this type of analysis are type I or type II integral membrane proteins (i.e., proteins whose ectodomains are tethered to the host cell membrane at only one end, either N or C terminal). Soluble versions of the ectodomains of these receptors, produced by genetic engineering or proteolytic release (Fig. 5, image 1), could then be used to trigger the cognate fusion protein. The methodology could, similarly, be applicable to multimembrane-spanning host cell receptors when the functional part of the receptor is tethered at only one end (Fig. 5, image 2) or where an ectodomain loop (Fig. 5, image 3) preserves enough structure to function as a soluble analog, perhaps by generating a cyclic peptide analog of the loop. The same "receptor reagents" (Fig. 5) could be employed for intact virus particle/target membrane binding assays (Fig. 4).

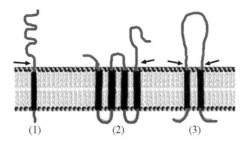

(1)                    (2)                    (3)

FIG. 5. Receptor species for use in liposome binding assays. (1) A type I or type II integral membrane protein with a functional domain that is tethered to the membrane at only one end; (2) a multimembrane-spanning protein that possesses a functional domain tethered to the membrane at one end; (3) a multimembrane-spanning protein with an extracellular loop that retains or can be engineered so as to retain enough structure to serve as a functional domain when cleaved from its transmembrane domains. Arrows indicate possible sites at which to cleave the functional receptor domain from the parent protein. Alternatively, secreted functional domains could be produced by genetic engineering or, where applicable, could be mimicked by synthetic peptides.

When a single obligate receptor does not meet the above-cited criteria (as is the case for most strains of HIV), it may be necessary to reconstitute one receptor (e.g., the chemokine receptor for HIV) into target liposomes. For HIV, a soluble ectodomain fragment of CD4, as well as the target liposomes containing the chemokine receptor, could then be used to recapitulate the target membrane binding step of the fusion reaction (Fig. 1, step 2). In addition, HIV Env presents another challenge in the development of a membrane binding assay: an envelope glycoprotein possessing two subunits that are not covalently associated. Because the gp120 and gp41 subunits of HIV Env are not associated in such a way, it has been shown that the gp120 subunit is shed after binding to its receptor (CD4).[23] Therefore, it is necessary to take this into consideration when making a soluble form of this glycoprotein.

In the case of most paramyxoviruses, two viral glycoproteins are necessary for optimal fusion, the actual fusion protein (F) and the cognate "HN" protein, which contains host cell receptor binding activity. For such systems, it may not be possible to induce binding of an isolated fusion protein ectodomain to target liposomes even in the presence of a correctly oligomeric soluble form of the HN protein; the fusion and HN proteins may need to be in a precise configuration *vis-à-vis* one another in the plane of a membrane for the HN protein to be able to activate the F protein. In

[23] M. Thali, C. Furman, E. Helseth, H. Repke, and J. Sodroski, *J. Virol.* **66,** 5516 (1992).

this case it may be easiest, at least for initial purposes, to measure receptor-induced binding of intact viral (or subviral) particles to target membranes (Fig. 4A). This would, however, require a soluble form of the host cell receptor as described above. We note that, in addition to our studies demonstrating binding of ASLV-A particles to target liposomes, several low pH-triggered viruses have been demonstrated to bind to liposomes, either as whole virus particles or as subviral particles; these cases include Sindbis virus, TBE, and influenza (Table I).

In summary, the key principle that underlies the operation of enveloped virus fusion proteins is conversion of the fusion protein ectodomain from a hydrophilic to a hydrophobic entity by virtue of an exposed fusion peptide (Fig. 1, step 2). In many cases (Table I) this event appears to be induced by binding to the host cell receptor under fusion-permissive conditions (e.g., physiologic temperature). Derivatives of the assays that we have described in this chapter should prove useful in monitoring this critical unifying event of receptor-induced fusion: receptor-induced binding of the fusion protein to target membranes. It is possible to do this with either the free (oligomeric) fusion protein ectodomain or with intact viral particles or subviral particles containing the viral fusion protein. Once a target membrane binding assay is established, it should provide a powerful tool with which to dissect further requirements for a key step in the fusion cascade: hydrophobic binding of the fusion protein to the host cell bilayer (Fig. 1, step 2).

# Author Index

Numbers in parentheses are footnote reference numbers and indicate that an author's work is referred to although the name is not cited in the text.

## A

Abe, T., 249, 250(8)
Abeywardena, M. Y., 69
Abrami, L., 102, 104
Abrams, C. K., 323
Abrams, F. S., 108
Ackerman, K. E., 216
Adachi, K., 292
Adams, S. L., 193, 203(31)
Adesnik, M., 246
Adkisson, H. D., 50
Agard, D. A., 432, 433(15), 434, 434(15), 435(15), 437(15)
Agarraberes, F. A., 105
Agarwal, S., 322
Aguilera, O., 126
Aharoni, A., 362
Ahkong, Q. F., 301
Ahl, P. L., 189, 191
Ahmad, I., 419
Akada, K. Y., 131
Åkerman, K., 424
Akuta, T., 247, 339(3), 340
Alberts, B., 350
Aleström, P., 357, 360(31)
Alfonzo, M., 70
Alford, D., 188, 261(20), 262, 274, 277, 310, 312(25), 323, 395, 398(18), 399(18), 408, 414(4), 417(4)
Ali, S., 270, 371
Allard, J., 241, 245(9), 247, 247(9)
Allen, T. M., 68, 69, 69(13), 205, 207(55), 268(51), 269, 270(51), 272, 301, 304(13), 419
Allen-Cannady, O., 424
Allietta, M. M., 38
Allison, S. L., 376, 396, 429, 431

Alonso, A., 3, 5, 6, 7(6), 8(3), 9, 10, 11(10), 12(12), 14, 14(10; 12), 18, 266
Alouf, J. E., 99, 323
Alpers, D. H., 229
Alt, T., 221(20), 222
Altenbach, C., 213
Alvarez, C., 106(45), 107, 114(48), 115(48), 117(45), 121(48)
Alvarez, L., 4(2), 5
Alving, C. R., 320
Amato, I., 143, 144(32)
Amherdt, M., 153, 157, 158(10; 12), 163(10), 249, 254(7)
Anantharamaiah, G. M., 107
Anderluh, G., 103, 108, 110
Anderson, H. A., 247
Anderson, R. A., 259
Anderson, R. G. W., 245, 246, 247(22), 248
Andrade, J., 162
Andre, J. C., 41
Andrews, N. W., 323
Angelova, M. I., 41
Angrand, M., 218, 224(9)
Annand, R., 22
Anselmi, C., 121
Antimisiaris, S. G., 247
Antonny, B., 151, 152, 153, 153(4), 157(4), 158, 158(4), 159(4), 160(25; 26), 161(25; 26), 163(4; 13), 164(4; 13)
Anzlovar, S., 115, 121(98)
Apel, C. L., 134, 135, 135(12), 136, 136(12), 137(23), 138(12), 141(23), 144(12), 148(12), 151(23)
Apell, H. J., 171, 172
Apitz-Castro, R., 23(26), 24
Apostolov, O., 128, 128(19), 129
Arai, H., 32
Arai, Y., 229
Aranda, F. J., 262

441

Szostak, J. W., 150
Szponarski, W., 168

# T

Tachibana, R., 321(15b), 322, 349, 350
Tack, B. F., 129
Tagaya, M., 261
Tait, J. F., 196, 197, 197(41)
Takagishi, T., 243, 272
Takahashi, M., 213, 214(79), 215(79), 275, 276(12)
Takamura, T., 32
Takeda, K., 247, 339(3), 340
Takei, K., 249, 250, 250(8; 10), 251, 251(12; 17), 256(10)
Takeshima, K., 132
Takeshita, K., 155
Takizawa, T., 220
Talbot, W. A., 262
Talmon, Y., 72, 221(19), 222, 222(19)
Tamm, L. K., 111
Tamura, A., 399
Tanford, C., 84
Tang, X.-W., 340
Tao, L., 409, 410(15)
Taraschi, T. F., 189, 191
Tatulian, S. A., 111
Tauber, A. I., 267, 269
Taussig, M. J., 217(5), 218
Tawfik, D. S., 150
Taylor, W., 63
Tejuca, M., 106(44; 45), 107, 114(48), 115(44; 48), 117(45), 121(48)
te Kaat, K., 121
Telford, J., 70
Tempst, P., 275, 276(6)
ter Beest, M. B., 48(11), 49
Terras, F. R. G., 101
Teulières, C., 210
Thali, M., 439
Theilen, U., 209
Theopold, K. H., 33
Thevissen, K., 101
Thiel, G., 169
Thomas, P., 168
Thum, O., 259
Tian, P., 272
Tiruppathi, C., 229

Tocanne, J. F., 210, 212(75)
Toda, M., 203, 205(52)
Todd, R. F. III, 301, 303(4)
Tokuda, H., 88
Tom-Kun, J., 204
Tonelli, M., 434
Torchilin, V. P., 339, 340, 343, 343(8), 346(8), 348(8), 350
Toribara, T. Y., 52
Touraine, B., 168
Tralka, T. S., 276
Tran, L. T., 103
Transue, A., 63, 301, 310(6), 312(6), 313(6; 7), 315(8), 316(6–8), 317(6), 318(7)
Traynor, A. E., 301, 303(1)
Tridente, G., 121
Trudel, M., 419
Tsao, Y. S., 431
Tsernoglou, D., 110
Tsiaras, W., 161
Tsien, R. Y., 203
Tsurudome, M., 396, 404(26), 405(26), 408, 409(9)
Tucker, A. D., 110
Tung, C.-H., 339(4), 340
Turk, D., 110
Turner, A. J., 220
Tweten, R. K., 102, 105(21), 108, 109, 110, 113, 323
Tyagi, S. K., 419

# U

Uchida, K., 153
Uchida, T., 322, 356
Ugalde, U., 300
Uhler, M. D., 302
Ullman, E. F., 267
Unanue, E. R., 322
Urbaneja, M. A., 18
Urbanus, M. L., 87
Urdal, D. L., 419, 420
Uster, P. S., 267, 393

# V

Vail, W. J., 188
Valent, Q. A., 87
Valentin, E., 29

# Subject Index

## A

ADP-ribosylating factor
GTP cycling, 152
liposome studies of GDP/GTP cycle
  buffers in analysis, 156–157
  fluorescence resonance energy transfer,
    158–159, 163–164
  liposome preparation
    azolectin liposomes, 154
    buffers, 156
    extrusion, 155
    liposomes of defined composition,
      154–155
    overview, 153–154
  sedimentation analysis of GTP-
    dependent translocation, 157–158
  tryptophan fluorescence activation and
    deactivation assays, 159–163
Aminonaphthalene trisulfonic acid–p-xylene
  bis(pyridinium) bromide, liposome
  fusion assays, 265–267, 273–274
Annexins
  annexin folding analysis with liposomes
    circular dichroism, 212–213
    electron paramagnetic resonance, 213
    tryptophan fluorescence, 213–215
  annexin V–liposome interactions
    buffers, 52
    electrostatic interactions, 50–51, 61
    fluorescence correlation spectroscopy
      autocorrelation function, 58–59, 61
      calcium effects, 60
      instrumentation, 59–60
      number of associated lipids per bound
        protein, 61
      pH effects, 61–62
      prospects, 64
      tetramethylrhodamine labeling of
        protein, 58
    fluorescence resonance energy transfer
      of protein binding
      calcium effects, 57–58, 63

calculations, 56
instrumentation, 55
pH effects, 57–58, 63–64
tryptophan as donor, 56–57
fusion induction, 48–49
hydrophobic interactions, 51
large unilamellar vesicle preparation, 52
microelectrophoresis of liposomes
  appararent surface charge density,
    53–54
  calcium effects, 54
  electrophoretic mobility, 53
  pH effects, 54–55
  surface potential, 53
  zeta potential, 53
multilamellar vesicle preparation, 52
protein purification, 51–52
small unilamellar vesicle preparation, 52
binding assays with liposomes
  calcium stoichiometry determination in
    complex, 199–200
  centrifugation assay with direct
    determination of protein, 192–194
  fluorescence quenching, 196–197
  fluorescence resonance energy transfer,
    195–196, 198
  isothermal titration calorimetry, 200
  kinetic analysis, 198
  light scattering, 194–195, 198
  phospholipase $A_2$ activity assay, 197–198
calcium binding, 49–50, 186
calcium influx in liposomes, FURA-2 assay
  burst analysis, 205
  continuous monitoring, 203–205
  principles, 203
functions, 50
liposome aggregation and fusion assays
  contents intermixing assay, 207–208
  phospholipid intermixing assay, 206–207
  turbidity assay, 205–206
liposome preparation for interaction
  studies
  advantages as model system, 215

469